Graduate Texts in Mathematics 247

Graduate Texts in Mathematics

(continued after index)

Christian Kassel
Vladimir Turaev

Braid Groups

With the graphical assistance of Olivier Dodane

Christian Kassel
Institut de Recherche Mathématique Avancée
CNRS et Université Louis Pasteur
7 rue René Descartes
67084 Strasbourg
France
kassel@math.u-strasbg.fr

Vladimir Turaev
Department of Mathematics
Indiana University
Bloomington, IN 47405
USA
vtouraev@indiana.edu

ISBN 978-1-4419-2220-5   e-ISBN: 978-0-387-68548-9
DOI: 10.1007/978-0-387-68548-9

Mathematics Subject Classification (2000): 20F36, 57M25, 37E30, 20C08, 06F15, 20F60, 55R80

Preface

The theory of braid groups is one of the most fascinating chapters of low-dimensional topology. Its beauty stems from the attractive geometric nature of braids and from their close relations to other remarkable geometric objects such as knots, links, homeomorphisms of surfaces, and configuration spaces. On a deeper level, the interest of mathematicians in this subject is due to the important role played by braids in diverse areas of mathematics and theoretical physics. In particular, the study of braids naturally leads to various interesting algebras and their linear representations.

Braid groups first appeared, albeit in a disguised form, in an article by Adolf Hurwitz published in 1891 and devoted to ramified coverings of surfaces. The notion of a braid was explicitly introduced by Emil Artin in the 1920s to formalize topological objects that model the intertwining of several strings in Euclidean 3-space. Artin pointed out that braids with a fixed number n of strings form a group, called the nth braid group and denoted by B_n. Since then, the braids and the braid groups have been extensively studied by topologists and algebraists. This has led to a rich theory with numerous ramifications.

In 1983, Vaughan Jones, while working on operator algebras, discovered new representations of the braid groups, from which he derived his celebrated polynomial of knots and links. Jones's discovery resulted in a strong increase of interest in the braid groups. Among more recent important results in this field are the orderability of the braid group B_n, proved by Patrick Dehornoy in 1991, and the linearity of B_n, established by Daan Krammer and Stephen Bigelow in 2001–2002.

The principal objective of this book is to give a comprehensive introduction to the theory of braid groups and to exhibit the diversity of their facets. The book is intended for graduate and postdoctoral students, as well as for all mathematicians and physicists interested in braids. Assuming only a basic knowledge of topology and algebra, we provide a detailed exposition of the more advanced topics. This includes background material in topology and algebra that often goes beyond traditional presentations of the theory of braids.

In particular, we present the basic properties of the symmetric groups, the theory of semisimple algebras, and the language of partitions and Young tableaux.

We now detail the contents of the book. Chapter 1 is concerned with the foundations of the theory of braids and braid groups. In particular, we describe the connections with configuration spaces, with automorphisms of free groups, and with mapping class groups of punctured disks.

In Chapter 2 we study the relation between braids and links in Euclidean 3-space. The central result of this chapter is the Alexander–Markov description of oriented links in terms of Markov equivalence classes of braids.

Chapter 3 is devoted to two remarkable representations of the braid group B_n: the Burau representation, introduced by Werner Burau in 1936, and the Lawrence–Krammer–Bigelow representation, introduced by Ruth Lawrence in 1990. We use the technique of Dehn twists to show that the Burau representation is nonfaithful for large n, as was first established by John Moody in 1991. We employ the theory of noodles on punctured disks introduced by Stephen Bigelow to prove the Bigelow–Krammer theorem on the faithfulness of the Lawrence–Krammer–Bigelow representation. In this chapter we also construct the one-variable Alexander–Conway polynomial of links.

Chapter 4 is concerned with the symmetric groups and the Iwahori–Hecke algebras, both closely related to the braid groups. As an application, we construct the two-variable Jones–Conway polynomial of links, also known as the HOMFLY or HOMFLY-PT polynomial, which extends two fundamental one-variable link polynomials, namely the aforementioned Alexander–Conway polynomial and the Jones polynomial.

Chapter 5 is devoted to a classification of the finite-dimensional representations of the generic Iwahori–Hecke algebras in terms of Young diagrams. As an application, we show that the (reduced) Burau representation of B_n is irreducible. We also discuss the Temperley–Lieb algebras and classify their finite-dimensional representations.

Chapter 6 presents the Garside solution of the conjugacy problem in the braid groups. Following Patrick Dehornoy and Luis Paris, we introduce the concept of a Garside monoid, which is a monoid with appropriate divisibility properties. We show that the braid group B_n is the group of fractions of a Garside monoid of positive braids on n strings. We also describe similar results for the generalized braid groups associated with Coxeter matrices.

Chapter 7 is devoted to the orderability of the braid groups. Following Dehornoy, we prove that the braid group B_n is orderable for every n.

The book ends with four short appendices: Appendix A on the modular group $\mathrm{PSL}_2(\mathbf{Z})$, Appendix B on fibrations, Appendix C on the Birman–Murakami–Wenzl algebras, and Appendix D on self-distributive sets.

The chapters of the book are to a great degree independent. The reader may start with the first section of Chapter 1 and then freely explore the rest of the book.

The theory of braids is certainly too vast to be covered in a single volume. One important area entirely skipped in this book concerns the connections with mathematical physics, quantum groups, Hopf algebras, and braided monoidal categories. On these subjects we refer the reader to the monographs [Lus93], [CP94], [Tur94], [Kas95], [Maj95], [KRT97], [ES98].

Other areas not presented here include the homology and cohomology of the braid groups [Arn70], [Vai78], [Sal94], [CS96], automatic structures on the braid groups [ECHLPT92], [Mos95], and applications to cryptography [SCY93], [AAG99], [KLCHKP00].

For further aspects of the theory of braids, we refer the reader to the following monographs and survey articles: [Bir74], [BZ85], [Han89], [Kaw96], [Mur96], [MK99], [Ver99], [Iva02], [BB05].

This book grew out of the lectures [Kas02], [Tur02] given by the authors at the Bourbaki Seminar in 1999 and 2000 and from graduate courses given by the first-named author at Université Louis Pasteur, Strasbourg, in 2002–2003 and by the second-named author at Indiana University, Bloomington, in 2006.

Acknowledgments. It is a pleasure to thank Patrick Dehornoy, Nikolai Ivanov, and Hans Wenzl for valuable discussions and useful comments. We owe special thanks to Olivier Dodane, who drew the figures and guided us through the labyrinth of LaTeX formats and commands.

Strasbourg   *Christian Kassel*
March 3, 2008   *Vladimir Turaev*

Contents

1

Braids and Braid Groups

In this chapter we discuss the basics of the theory of braids and braid groups.

1.1 The Artin braid groups

We introduce the braid groups and discuss some of their simple properties.

1.1.1 Basic definition

We give an algebraic definition of the braid group B_n for any positive integer n. The definition is formulated in terms of a group presentation by generators and relations.

Definition 1.1. *The Artin braid group B_n is the group generated by $n - 1$ generators $\sigma_1, \sigma_2, \ldots, \sigma_{n-1}$ and the "braid relations"*

$$\sigma_i \sigma_j = \sigma_j \sigma_i$$

for all $i, j = 1, 2, \ldots, n - 1$ with $|i - j| \geq 2$, and

$$\sigma_i \sigma_{i+1} \sigma_i = \sigma_{i+1} \sigma_i \sigma_{i+1}$$

for $i = 1, 2, \ldots, n - 2$.

By definition, $B_1 = \{1\}$ is a trivial group. The group B_2 is generated by a single generator σ_1 and an empty set of relations. This is an infinite cyclic group. As we shall see shortly, the groups B_n with $n \geq 3$ are nonabelian.

Given a group homomorphism f from B_n to a group G, the elements $\{s_i = f(\sigma_i)\}_{i=1,\ldots,n-1}$ of G satisfy the braid relations

$$s_i s_j = s_j s_i$$

for all $i, j = 1, 2, \ldots, n - 1$ with $|i - j| \geq 2$, and

C. Kassel, V. Turaev, *Braid Groups*, DOI: 10.1007/978-0-387-68548-9_1,
© Springer Science+Business Media, LLC 2008

$$s_i s_{i+1} s_i = s_{i+1} s_i s_{i+1}$$

for $i = 1, 2, \ldots, n - 2$. There is a converse, which we record in the following lemma.

Lemma 1.2. *If $s_1, \ldots, s_{n-1}$ are elements of a group G satisfying the braid relations, then there is a unique group homomorphism $f : B_n \to G$ such that $s_i = f(\sigma_i)$ for all $i = 1, 2, \ldots, n - 1$.*

Proof. Let F_n be the free group generated by the set $\{\sigma_1, \ldots, \sigma_{n-1}\}$. There is a unique group homomorphism $\bar{f} : F_n \to G$ such that $\bar{f}(\sigma_i) = s_i$ for all $i = 1, 2, \ldots, n - 1$. This homomorphism induces a group homomorphism $f : B_n \to G$ provided $\bar{f}(r^{-1} r') = 1$ or, equivalently, provided $\bar{f}(r) = \bar{f}(r')$ for all braid relations $r = r'$. For the first braid relation, we have

$$\bar{f}(\sigma_i \sigma_j) = \bar{f}(\sigma_i) \bar{f}(\sigma_j) = s_i s_j = s_j s_i = \bar{f}(\sigma_j) \bar{f}(\sigma_i) = \bar{f}(\sigma_j \sigma_i).$$

For the second braid relation, we similarly have

$$\bar{f}(\sigma_i \sigma_{i+1} \sigma_i) = s_i s_{i+1} s_i = s_{i+1} s_i s_{i+1} = \bar{f}(\sigma_{i+1} \sigma_i \sigma_{i+1}). \qquad \square$$

1.1.2 Projection to the symmetric group

We apply the previous lemma to the symmetric group $G = \mathfrak{S}_n$. An element of $\mathfrak{S}_n$ is a permutation of the set $\{1, 2, \ldots, n\}$. Consider the *simple transpositions* $s_1, \ldots, s_{n-1} \in \mathfrak{S}_n$, where s_i permutes i and $i + 1$ and leaves all the other elements of $\{1, 2, \ldots, n\}$ fixed. It is an easy exercise to verify that the simple transpositions satisfy the braid relations. By Lemma 1.2, there is a unique group homomorphism $\pi : B_n \to \mathfrak{S}_n$ such that $s_i = \pi(\sigma_i)$ for all $i = 1, 2, \ldots, n-1$. This homomorphism is surjective because, as is well known, the simple transpositions generate $\mathfrak{S}_n$. (For more on the structure of $\mathfrak{S}_n$, see Section 4.1.)

Lemma 1.3. *The group B_n with $n \geq 3$ is nonabelian.*

Proof. The group $\mathfrak{S}_n$ with $n \geq 3$ is nonabelian because $s_1 s_2 \neq s_2 s_1$. Since the projection $B_n \to \mathfrak{S}_n$ is surjective, B_n is nonabelian for $n \geq 3$. $\qquad \square$

1.1.3 Natural inclusions

From the defining relations of Definition 1.1 it is clear that the formula $\iota(\sigma_i) = \sigma_i$ with $i = 1, 2, \ldots, n - 1$ defines a group homomorphism

$$\iota : B_n \to B_{n+1}.$$

As will be proven in Corollary 1.14, the homomorphism ι is injective. It is called the *natural inclusion*.

It is sometimes convenient to view B_n as a subgroup of B_{n+1} via ι. In this way we obtain an increasing chain of groups $B_1 \subset B_2 \subset B_3 \subset \cdots$.

Composing ι with the projection $\pi : B_{n+1} \to \mathfrak{S}_{n+1}$, we obtain the composition of $\pi : B_n \to \mathfrak{S}_n$ with the canonical inclusion $\mathfrak{S}_n \hookrightarrow \mathfrak{S}_{n+1}$. (The latter inclusion extends each permutation of $\{1, 2, \ldots, n\}$ to a permutation of $\{1, 2, \ldots, n+1\}$ fixing $n+1$.) This gives a commutative diagram

$$
\begin{array}{ccc}
B_n & \longrightarrow & \mathfrak{S}_n \\
\iota \downarrow & & \downarrow \\
B_{n+1} & \longrightarrow & \mathfrak{S}_{n+1}
\end{array}
\tag{1.1}
$$

1.1.4 The group B_3

Already the simplest noncommutative braid group B_3 presents considerable interest. This group is generated by two generators σ_1, σ_2, and the unique relation $\sigma_1\sigma_2\sigma_1 = \sigma_2\sigma_1\sigma_2$. Setting $x = \sigma_1\sigma_2\sigma_1$ and $y = \sigma_1\sigma_2$, we obtain generators x, y of B_3 subject to the unique relation $x^2 = y^3$ (verify). This relation implies in particular that $x^2 = (\sigma_1\sigma_2\sigma_1)^2$ lies in the center of B_3. (We shall compute the center of B_n for all n in Section 1.3.3.)

The group B_3 admits a homomorphism to $\mathrm{SL}(2, \mathbf{Z})$ sending σ_1, σ_2 to the matrices

$$
\begin{pmatrix} 1 & 1 \\ 0 & 1 \end{pmatrix} \quad \text{and} \quad \begin{pmatrix} 1 & 0 \\ -1 & 1 \end{pmatrix},
$$

respectively. This homomorphism is surjective and its kernel is the infinite cyclic group generated by $(\sigma_1\sigma_2\sigma_1)^4$. For a proof, see [Mil71, Th. 10.5] or Appendix A.

The group B_3 appears in knot theory as the fundamental group of the complement of the trefoil knot $K \subset S^3$. The trefoil K can be defined as the subset of $S^3 = \{(z_1, z_2) \in \mathbf{C}^2 \,|\, |z_1|^2 + |z_2|^2 = 1\}$ consisting of (z_1, z_2) such that $z_1^2 + z_2^3 = 0$; see Figure 2.1 for a picture of K. The isomorphism

$$
\pi_1(S^3 - K) \cong \langle x, y \,|\, x^3 = y^2 \rangle = B_3
$$

is well known in knot theory. From the algebraic viewpoint, the key phenomenon underlying this isomorphism is the homeomorphism

$$
S^3 - K \approx \mathrm{SL}(2, \mathbf{R})/\mathrm{SL}(2, \mathbf{Z}) \,;
$$

see [Mil71, Sect. 10].

Exercise 1.1.1. Show that there is a group homomorphism $f : B_n \to \mathbf{Z}$ such that $f(\sigma_i) = 1$ for all $i = 1, \ldots, n-1$. Prove that f induces an isomorphism $B_n/[B_n, B_n] \cong \mathbf{Z}$, where $[B_n, B_n]$ is the commutator subgroup of B_n.

Exercise 1.1.2. Verify that the formula $\sigma_i \mapsto \sigma_i^{-1}$ for $i = 1, 2, \ldots, n-1$ defines an involutive automorphism of B_n. Prove that this automorphism is not a conjugation by an element of B_n.

Exercise 1.1.3. Verify the following relations in B_3:

$$\sigma_1 \sigma_2 \sigma_1^{-1} = \sigma_2^{-1} \sigma_1 \sigma_2\,, \quad \sigma_1^{-1} \sigma_2^{-1} \sigma_1 = \sigma_2 \sigma_1^{-1} \sigma_2^{-1}\,, \quad \sigma_1^{-1} \sigma_2^{-1} \sigma_1^{-1} = \sigma_2^{-1} \sigma_1^{-1} \sigma_2^{-1}\,,$$

$$\sigma_1 \sigma_2^{-1} \sigma_1^{-1} = \sigma_2^{-1} \sigma_1^{-1} \sigma_2\,, \quad \sigma_1^{-1} \sigma_2 \sigma_1 = \sigma_2 \sigma_1 \sigma_2^{-1}\,.$$

Exercise 1.1.4. Prove that for any $n \geq 1$, the group B_n is generated by two elements σ_1 and $\alpha = \sigma_1 \sigma_2 \cdots \sigma_{n-1}$. (Hint: $\sigma_i = \alpha^{i-1} \sigma_1 \alpha^{1-i}$ for all i.)

Exercise 1.1.5. Let f be a homomorphism from B_n to a certain group. If $f(\sigma_i)$ commutes with $f(\sigma_{i+1})$ for some i, then $f(B_n)$ is a cyclic group. If $f(\sigma_i) = f(\sigma_j)$ for some $i < j$ such that either $j \neq i + 2$ or $n \neq 4$, then $f(B_n)$ is a cyclic group.

Exercise 1.1.6. Prove that each element $\sigma_i \sigma_j^{-1}$ with $1 \leq i < j \leq n-1$ belongs to $[B_n, B_n]$ and generates $[B_n, B_n]$ as a normal subgroup of B_n provided either $j \neq i+2$ or $n \neq 4$. (Hint: Consider first the case $j = i+1$.)

Exercise 1.1.7. Verify the identity

$$\sigma_{i+2} \sigma_i^{-1} = (\sigma_i \sigma_{i+1})^{-1} [\sigma_{i+2} \sigma_i^{-1}, \sigma_i \sigma_{i+1}^{-1}] \sigma_i \sigma_{i+1}\,,$$

where $1 \leq i \leq n-3$ and $[a, b] = a^{-1} b^{-1} ab$.

Exercise 1.1.8. Prove that for $n \neq 3, 4$ the commutator subgroup of $[B_n, B_n]$ coincides with $[B_n, B_n]$.

Exercise 1.1.9. Prove that $[B_3, B_3]$ is a free group of rank two. (A topological proof: use that the trefoil is a fibered knot of genus one.)

Exercise 1.1.10. (a) Define automorphisms σ_1', σ_2', σ_3' of the free group F_2 on two generators a and b by

$$\sigma_1'(a) = a\,, \quad \sigma_1'(b) = ab\,, \quad \sigma_2'(a) = b^{-1}a\,, \quad \sigma_2'(b) = b\,, \quad \sigma_3'(a) = a\,, \quad \sigma_3'(b) = ba\,.$$

Prove that there is a group homomorphism $\psi : B_4 \to \mathrm{Aut}(F_2)$ such that $\psi(\sigma_i) = \sigma_i'$ for $i = 1, 2, 3$. Check that $\psi(\sigma_1 \sigma_3^{-1})$ is the conjugation by a and $\psi(\sigma_2 \sigma_1 \sigma_3^{-1} \sigma_2^{-1})$ is the conjugation by $b^{-1}a$ in F_2.

(b) Consider the group homomorphism $B_4 \to B_3$ sending σ_1, σ_3 to σ_1 and σ_2 to σ_2. Prove that its kernel is generated by $\sigma_1 \sigma_3^{-1}$ and $\sigma_2 \sigma_1 \sigma_3^{-1} \sigma_2^{-1}$. Deduce that this kernel is a free group of rank 2.

1.2 Braids and braid diagrams

In this section we interpret the braid groups in geometric terms. From now on, we denote by I the closed interval $[0, 1]$ in the set of real numbers $\mathbf{R}$. By a *topological interval*, we mean a topological space homeomorphic to $I = [0, 1]$.

1.2.1 Geometric braids

Definition 1.4. *A geometric braid on $n \geq 1$ strings is a set $b \subset \mathbf{R}^2 \times I$ formed by n disjoint topological intervals called the strings of b such that the projection $\mathbf{R}^2 \times I \to I$ maps each string homeomorphically onto I and*

$$b \cap (\mathbf{R}^2 \times \{0\}) = \{(1,0,0), (2,0,0), \ldots, (n,0,0)\},$$
$$b \cap (\mathbf{R}^2 \times \{1\}) = \{(1,0,1), (2,0,1), \ldots, (n,0,1)\}.$$

It is clear that every string of b meets each plane $\mathbf{R}^2 \times \{t\}$ with $t \in I$ in exactly one point and connects a point $(i,0,0)$ to a point $(s(i),0,1)$, where $i, s(i) \in \{1, 2, \ldots, n\}$. The sequence $(s(1), s(2), \ldots, s(n))$ is a permutation of the set $\{1, 2, \ldots, n\}$ called the *underlying permutation* of b.

An example of a geometric braid is given in Figure 1.1. Here x, y are the coordinates in $\mathbf{R}^2$, the x-axis is directed to the right, the y-axis is directed away from the reader, and the t-axis is directed downward. The underlying permutation of this braid is $(1, 3, 2, 4)$.

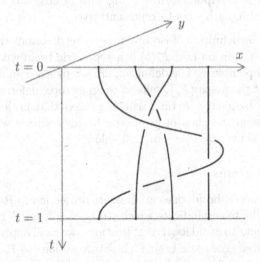

Fig. 1.1. A geometric braid on four strings

Two geometric braids b and b' on n strings are *isotopic* if b can be continuously deformed into b' in the class of braids. More formally, b and b' are isotopic if there is a continuous map $F : b \times I \to \mathbf{R}^2 \times I$ such that for each $s \in I$, the map $F_s : b \to \mathbf{R}^2 \times I$ sending $x \in b$ to $F(x, s)$ is an embedding whose image is a geometric braid on n strings, $F_0 = \mathrm{id}_b : b \to b$, and $F_1(b) = b'$. Each F_s automatically maps every endpoint of b to itself. Both the map F and the family of geometric braids $\{F_s(b)\}_{s \in I}$ are called an *isotopy* of $b = F_0(b)$ into $b' = F_1(b)$.

It is obvious that the relation of isotopy is an equivalence relation on the class of geometric braids on n strings. The corresponding equivalence classes are called *braids on n strings*.

Given two n-string geometric braids $b_1, b_2 \subset \mathbf{R}^2 \times I$, we define their product $b_1 b_2$ to be the set of points $(x, y, t) \in \mathbf{R}^2 \times I$ such that $(x, y, 2t) \in b_1$ if $0 \leq t \leq 1/2$ and $(x, y, 2t - 1) \in b_2$ if $1/2 \leq t \leq 1$. It is obvious that $b_1 b_2$ is a geometric braid on n strings. It is clear that if b_1, b_2 are isotopic to geometric braids b_1', b_2', respectively, then $b_1 b_2$ is isotopic to $b_1' b_2'$. Therefore the formula $(b_1, b_2) \mapsto b_1 b_2$ defines a multiplication on the set of braids on n strings. This multiplication is associative and has a neutral element, which is the *trivial braid* 1_n represented by the geometric braid

$$\{1, 2, \ldots, n\} \times \{0\} \times I \subset \mathbf{R}^2 \times I.$$

We shall see below that the set of braids on n strings with this multiplication is a group canonically isomorphic to the braid group B_n.

Any geometric braid is isotopic to a geometric braid $b \subset \mathbf{R}^2 \times I$ such that b is a smooth one-dimensional submanifold of $\mathbf{R}^2 \times I$ orthogonal to $\mathbf{R}^2 \times 0$ and $\mathbf{R}^2 \times 1$ near the endpoints. In working with braids, it is often convenient to restrict oneself to such smooth representatives.

Remark 1.5. The definition of isotopy for geometric braids can be weakened by replacing the condition that $F_s(b)$ is a geometric braid with the condition that F_s keeps ∂b pointwise. The definition of isotopy also can be strengthened by requiring that the maps $\{F_s\}_s$ extend to an isotopic deformation of $\mathbf{R}^2 \times I$ constant on the boundary. Artin [Art47a] proved that both resulting equivalence relations on the class of geometric braids coincide with the isotopy relation defined above; cf. Theorem 1.40 below.

1.2.2 Braid diagrams

To specify a geometric braid, one can draw its projection to $\mathbf{R} \times \{0\} \times I$ along the second coordinate and indicate which string goes "under" the other one at each crossing point. To avoid local complications, we shall apply this procedure exclusively to those geometric braids whose projections to $\mathbf{R} \times \{0\} \times I$ have only double transversal crossings. These considerations lead to a notion of a braid diagram.

A *braid diagram* on n strands is a set $\mathcal{D} \subset \mathbf{R} \times I$ split as a union of n topological intervals called the *strands* of $\mathcal{D}$ such that the following three conditions are met:

(i) The projection $\mathbf{R} \times I \to I$ maps each strand homeomorphically onto I.
(ii) Every point of $\{1, 2, \ldots, n\} \times \{0, 1\}$ is the endpoint of a unique strand.
(iii) Every point of $\mathbf{R} \times I$ belongs to at most two strands. At each intersection point of two strands, these strands meet transversely, and one of them is distinguished and said to be *undergoing*, the other strand being *overgoing*.

Note that three strands of a braid diagram $\mathcal{D}$ never meet in one point. An intersection point of two strands of $\mathcal{D}$ is called a *double point* or a *crossing* of $\mathcal{D}$. The transversality condition in (iii) means that in a neighborhood of a crossing, $\mathcal{D}$ looks, up to homeomorphism, like the set $\{(x, y) \,|\, xy = 0\}$ in $\mathbf{R}^2$. Condition (iii) and the compactness of the strands easily imply that the number of crossings of $\mathcal{D}$ is finite.

In the figures, the strand going under a crossing is graphically represented by a line broken near the crossing; the strand going over a crossing is represented by a continued line. An example of a braid diagram is given in Figure 1.2. Here the top horizontal line represents $\mathbf{R} \times \{0\}$, and the bottom horizontal line represents $\mathbf{R} \times \{1\}$. In the sequel we shall sometimes draw and sometimes omit these lines in the figures.

Fig. 1.2. A braid diagram on four strands

We now describe the relationship between braids and braid diagrams. Each braid diagram $\mathcal{D}$ presents an isotopy class of geometric braids as follows. Using the obvious identification $\mathbf{R} \times I = \mathbf{R} \times \{0\} \times I$, we can assume that $\mathcal{D}$ lies on $\mathbf{R} \times \{0\} \times I \subset \mathbf{R}^2 \times I$. In a small neighborhood of every crossing of $\mathcal{D}$ we slightly push the undergoing strand into $\mathbf{R} \times (0, +\infty) \times I$ by increasing the second coordinate while keeping the first and third coordinates. This transforms $\mathcal{D}$ into a geometric braid on n strings. Its isotopy class is a well-defined braid *presented by* $\mathcal{D}$. This braid is denoted by $\beta(\mathcal{D})$. For instance, the braid diagram in Figure 1.2 presents the braid drawn in Figure 1.1.

It is easy to see that any braid β can be presented by a braid diagram. To obtain a diagram of β, pick a geometric braid b that represents β and is generic with respect to the projection along the second coordinate. This means that the projection of b to $\mathbf{R} \times \{0\} \times I$ may have only double transversal crossings. At each crossing point of this projection choose the undergoing strand to be the one that comes from a subarc of b with larger second coordinate. The projection of b to $\mathbf{R} \times \{0\} \times I = \mathbf{R} \times I$ thus yields a braid diagram, $\mathcal{D}$, and it is clear that $\beta(\mathcal{D}) = \beta$.

Two braid diagrams $\mathcal{D}$ and $\mathcal{D}'$ on n strands are said to be *isotopic* if there is a continuous map $F : \mathcal{D} \times I \to \mathbf{R} \times I$ such that for each $s \in I$ the set $\mathcal{D}_s = F(\mathcal{D} \times s) \subset \mathbf{R} \times I$ is a braid diagram on n strands, $\mathcal{D}_0 = \mathcal{D}$, and $\mathcal{D}_1 = \mathcal{D}'$. It is understood that F maps the crossings of $\mathcal{D}$ to the crossings of $\mathcal{D}_s$ for all $s \in I$ preserving the under/overgoing data. The family of braid diagrams $\{\mathcal{D}_s\}_{s \in I}$ is called an *isotopy* of $\mathcal{D}_0 = \mathcal{D}$ into $\mathcal{D}_1 = \mathcal{D}'$. An example of an isotopy is given in Figure 1.3. It is obvious that if $\mathcal{D}$ is isotopic to $\mathcal{D}'$, then $\beta(\mathcal{D}) = \beta(\mathcal{D}')$.

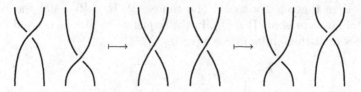

Fig. 1.3. An isotopy of braid diagrams

Given two braid diagrams $\mathcal{D}_1, \mathcal{D}_2$ on n strands, their product $\mathcal{D}_1\mathcal{D}_2$ is obtained by placing $\mathcal{D}_1$ on the top of $\mathcal{D}_2$ and squeezing the resulting diagram into $\mathbf{R} \times I$; see Figure 1.4. It is clear that if $\mathcal{D}_1$ presents a braid β_1 and $\mathcal{D}_2$ presents a braid β_2, then $\mathcal{D}_1\mathcal{D}_2$ presents the product $\beta_1\beta_2$.

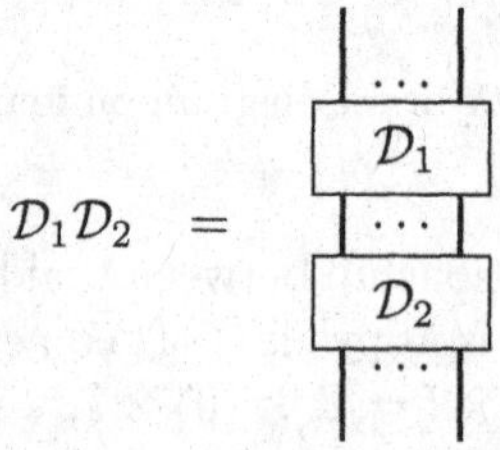

$$\mathcal{D}_1\mathcal{D}_2 \quad = \quad \boxed{\begin{array}{c} \cdots \\ \mathcal{D}_1 \\ \cdots \\ \mathcal{D}_2 \\ \cdots \end{array}}$$

Fig. 1.4. Product of braid diagrams

1.2.3 Reidemeister moves on braid diagrams

The transformations of braid diagrams Ω_2, Ω_3 shown in Figures 1.5a and 1.5b, as well as the inverse transformations $\Omega_2^{-1}, \Omega_3^{-1}$ (obtained by reversing the arrows in Figures 1.5a and 1.5b), are called *Reidemeister moves*. These moves come from the theory of knots and knot diagrams, where they were introduced by Kurt Reidemeister; see [Rei83] and Section 2.1. The moves affect only the position of a diagram in a disk inside $\mathbf{R} \times I$ and leave the remaining part of the diagram unchanged. The move Ω_2 involves two strands and creates two additional crossings (there are two types of Ω_2-moves, as shown in Figure 1.5a). The move Ω_3 involves three strands and preserves the number of crossings.

All these transformations of braid diagrams preserve the corresponding braids up to isotopy.

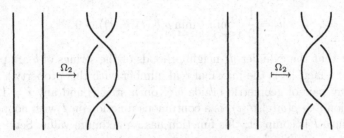

Fig. 1.5a. The Reidemeister move Ω_2

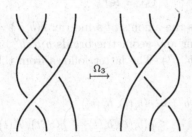

Fig. 1.5b. The Reidemeister move Ω_3

We say that two braid diagrams $\mathcal{D}, \mathcal{D}'$ are *R-equivalent* if $\mathcal{D}$ can be transformed into $\mathcal{D}'$ by a finite sequence of isotopies and Reidemeister moves $\Omega_2^{\pm 1}, \Omega_3^{\pm 1}$. It is obvious that if $\mathcal{D}, \mathcal{D}'$ are R-equivalent, then $\beta(\mathcal{D}) = \beta(\mathcal{D}')$. The following theorem asserts the converse.

Theorem 1.6. *Two braid diagrams present isotopic geometric braids if and only if these diagrams are R-equivalent.*

Proof. This theorem is an analogue for braids of the classical result of Reidemeister on knot diagrams; see [BZ85], [Mur96], and Chapter 2. The key point of Theorem 1.6 is that the diagrams of isotopic geometric braids are R-equivalent. The proof of the theorem goes in four steps.

Step 1. We introduce some notation used in the next steps. Consider a geometric braid $b \subset \mathbf{R}^2 \times I$ on n strings. For $i = 1, \ldots, n$, denote the ith string of b, that is, the string adjacent to the point $(i, 0, 0)$, by b_i. Each plane $\mathbf{R}^2 \times \{t\}$ with $t \in I$ meets b_i in one point, denoted by $b_i(t)$. In particular, we have $b_i(0) = (i, 0, 0)$.

Let ρ be the Euclidean metric on $\mathbf{R}^3$. Given a real number $\varepsilon > 0$, the *cylinder ε-neighborhood* of b_i consists of all points $(x, t) \in \mathbf{R}^2 \times I$ such that $\rho((x, t), b_i(t)) < \varepsilon$. This neighborhood meets each plane $\mathbf{R}^2 \times \{t\} \subset \mathbf{R}^2 \times I$ along the open disk of radius ε centered at $b_i(t)$.

For distinct $i, j \in \{1, \ldots, n\}$, the function $t \mapsto \rho(b_i(t), b_j(t))$ is a continuous function on I with positive values. Since I is compact, this function has a minimum value. Set

$$|b| = \frac{1}{2} \min_{1 \leq i < j \leq n} \min_{t \in I} \rho(b_i(t), b_j(t)) > 0 .$$

It is clear that the cylinder $|b|$-neighborhoods of the strings of b are pairwise disjoint. (In fact, $|b|$ is the maximal real number with this property.)

For any pair of geometric braids b, b' on n strings and any $i = 1, \ldots, n$, the function $t \mapsto \rho(b_i(t), b_i'(t))$ is a continuous function on I with nonnegative values. Since I is compact, this function has a maximum value. Set

$$\widetilde{\rho}(b, b') = \max_{1 \leq i \leq n} \max_{t \in I} \rho(b_i(t), b_i'(t)) \geq 0 .$$

The function $\widetilde{\rho}$ satisfies the axioms of a metric: $\widetilde{\rho}(b, b') = \widetilde{\rho}(b', b)$; $\widetilde{\rho}(b, b') = 0$ if and only if $b = b'$; for any geometric braids b, b', b'' on n strings, we have $\widetilde{\rho}(b, b'') \leq \widetilde{\rho}(b, b') + \widetilde{\rho}(b', b'')$. The latter follows from the fact that for some $i = 1, \ldots, n$ and $t \in I$,

$$\begin{aligned}
\widetilde{\rho}(b, b'') &= \rho(b_i(t), b_i''(t)) \\
&\leq \rho(b_i(t), b_i'(t)) + \rho(b_i'(t), b_i''(t)) \\
&\leq \widetilde{\rho}(b, b') + \widetilde{\rho}(b', b'') .
\end{aligned}$$

Note also that

$$|b| \leq |b'| + \widetilde{\rho}(b, b') . \tag{1.2}$$

Indeed, for some $t \in I$ and certain distinct $i, j = 1, \ldots, n$,

$$\begin{aligned}
|b| &= \frac{1}{2} \rho(b_i(t), b_j(t)) \\
&\leq \frac{1}{2} \Big(\rho(b_i(t), b_i'(t)) + \rho(b_i'(t), b_j'(t)) + \rho(b_j'(t), b_j(t)) \Big) \\
&\leq \frac{1}{2} \Big(\widetilde{\rho}(b, b') + 2|b'| + \widetilde{\rho}(b', b) \Big) \\
&= |b'| + \widetilde{\rho}(b, b') .
\end{aligned}$$

Step 2. A geometric braid is *polygonal* if all its strings are formed by consecutive (linear) segments; see Figure 1.6. Any geometric braid b on n strings can be approximated by polygonal braids as follows. Pick an integer $N \geq 2$ and an index $i = 1, \ldots, n$. For $k = 1, \ldots, N$, consider the segment in $\mathbf{R}^2 \times I$ with endpoints $b_i(\frac{k-1}{N})$ and $b_i(\frac{k}{N})$. The union of these N segments is a broken line, b_i^N, with endpoints $b_i^N(0) = b_i(0) = (i, 0, 0)$ and $b_i^N(1) = b_i(1)$. For sufficiently large N, this broken line lies in the cylinder $|b|$-neighborhood of b_i. Therefore for sufficiently large N, the broken lines $b_1^N, \ldots, b_n^N$ are disjoint and form a polygonal braid, b^N, *approximating* b. Moreover, for any real number $\varepsilon > 0$ and all sufficiently large N, we have $\widetilde{\rho}(b, b^N) < \varepsilon$. For instance, Figure 1.6 shows a polygonal approximation of the braid in Figure 1.1.

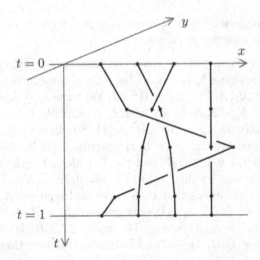

Fig. 1.6. A polygonal braid on four strands

We now reformulate the notion of isotopy of braids in the polygonal setting. To this end, we introduce so-called Δ-moves on polygonal braids. Let A, B, C be three points in $\mathbf{R}^2 \times I$ such that the third coordinate of A is strictly smaller than the third coordinate of B and the latter is strictly smaller than the third coordinate of C. The move $\Delta(ABC)$ applies to a polygonal braid $b \subset \mathbf{R}^2 \times I$ whenever this braid meets the triangle ABC precisely along the segment AC. (By the triangle ABC, we mean the linear 2-simplex with vertices A, B, C.) Under this assumption, the move $\Delta(ABC)$ on b replaces $AC \subset b$ by $AB \cup BC$, keeping the rest of b intact; see Figure 1.7, where the triangle ABC is shaded. The inverse move $(\Delta(ABC))^{-1}$ applies to a polygonal braid meeting the triangle ABC precisely along $AB \cup BC$. This move replaces $AB \cup BC$ by AC. The moves $\Delta(ABC)$ and $(\Delta(ABC))^{-1}$ are called Δ-*moves*.

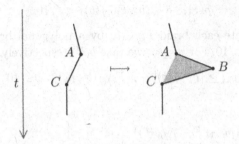

Fig. 1.7. A Δ-move

It is obvious that polygonal braids related by a Δ-move are isotopic. We establish a converse assertion.

Claim 1.7. *If polygonal braids b, b' are isotopic, then b can be transformed into b' by a finite sequence of Δ-moves.*

Proof. We first verify this claim under the assumption $\tilde{\rho}(b, b') < |b|/10$. Assume that the ith string b_i is formed by $K \geq 1$ consecutive segments with vertices $A_0 = (i, 0, 0), A_1, \ldots, A_K \in \mathbf{R}^2 \times I$. We write $b_i = A_0 A_1 \cdots A_K$. Similarly, assume that $b_i' = B_0 B_1 \cdots B_L$ with $L \geq 1$ and $B_0, B_1, \ldots, B_L \in \mathbf{R}^2 \times I$. Note that $A_0 = B_0$ and $A_K = B_L \in \mathbf{R}^2 \times \{1\}$. Subdividing b_i, b_i' into smaller segments, we can ensure that $K = L$, the points A_j, B_j have the same third coordinate for all $j = 0, 1, \ldots, K$, and the Euclidean length of the segments $A_j A_{j+1}, B_j B_{j+1}$ is smaller than $|b|/10$ for $j = 0, 1, \ldots, K - 1$. The assumption $\tilde{\rho}(b, b') < |b|/10$ implies that each horizontal segment $A_j B_j$ has length $< |b|/10$. The move $(\Delta(A_0 A_1 A_2))^{-1}$ transforms $b_i = A_0 A_1 \cdots A_K$ into the string $A_0 A_2 \cdots A_K = B_0 A_2 \cdots A_K$. The move $\Delta(B_0 B_1 A_2)$ transforms the latter in the string $B_0 B_1 A_2 \cdots A_K$. Continuing by induction and applying the moves $(\Delta(B_j A_{j+1} A_{j+2}))^{-1}$, $\Delta(B_j B_{j+1} A_{j+2})$ for $j = 0, \ldots, K - 2$, we transform b_i into b_i'. The conditions on the lengths imply that all the intermediate strings as well as the triangles $B_j A_{j+1} A_{j+2}, B_j B_{j+1} A_{j+2}$ determining these moves lie in the cylinder $|b|$-neighborhood of b_i; they are therefore disjoint from the cylinder $|b|$-neighborhoods of the other strings of b. We apply these transformations for $i = 1, \ldots, n$ and obtain thus a sequence of Δ-moves transforming b into b'.

Consider now an arbitrary pair of isotopic polygonal braids b, b'. Let $F : b \times I \to \mathbf{R}^2 \times I$ be an isotopy transforming $b = F_0(b)$ into $b' = F_1(b)$ (the braids $F_s(b)$ with $0 < s < 1$ may be nonpolygonal). The continuity of F implies that the function $I \times I \to \mathbf{R}, (s, s') \mapsto \tilde{\rho}(F_s(b), F_{s'}(b))$ is continuous. This function is equal to 0 on the diagonal $s = s'$ of $I \times I$. These facts and the inequality (1.2) imply that the function $I \to \mathbf{R}, s \mapsto |F_s(b)|$ is continuous. Since $|F_s(b)| > 0$ for all s, there is a real number $\varepsilon > 0$ such that $|F_s(b)| > \varepsilon$ for all $s \in I$. The continuity of the function $(s, s') \mapsto \tilde{\rho}(F_s(b), F_{s'}(b))$ now implies that for a sufficiently large integer N and all $k = 1, 2, \ldots, N$,

$$\tilde{\rho}(F_{(k-1)/N}(b), F_{k/N}(b)) < \varepsilon/10.$$

Let us approximate each braid $F_{k/N}(b)$ by a polygonal braid p_k such that $\tilde{\rho}(F_{k/N}(b), p_k) < \varepsilon/10$. For p_0, p_N, we take b, b', respectively. By (1.2),

$$|p_k| \geq |F_{k/N}(b)| - \tilde{\rho}(F_{k/N}(b), p_k) > 9\varepsilon/10.$$

At the same time,

$$\tilde{\rho}(p_{k-1}, p_k) \leq \tilde{\rho}(p_{k-1}, F_{(k-1)/N}(b))$$
$$+ \tilde{\rho}(F_{(k-1)/N}(b), F_{k/N}(b)) + \tilde{\rho}(F_{k/N}(b), p_k) < 3\varepsilon/10.$$

Therefore $\tilde{\rho}(p_{k-1}, p_k) < |p_k|/2$ for $k = 1, \ldots, N$. By the previous paragraph, p_{k-1} can be transformed into p_k by a sequence of Δ-moves. Composing these transformations $b = p_0 \mapsto p_1 \mapsto \cdots \mapsto p_N = b'$, we obtain a required transformation $b \mapsto b'$. This completes the proof of Claim 1.7. $\qquad\square$

Step 3. A polygonal braid is *generic* if its projection to $\mathbf{R} \times I = \mathbf{R} \times \{0\} \times I$ along the second coordinate has only double transversal crossings. Slightly deforming the vertices of a polygonal braid b (keeping ∂b), we can approximate this braid by a generic polygonal braid. Moreover, if b, b' are generic polygonal braids related by a sequence of Δ-moves, then slightly deforming the vertices of the intermediate polygonal braids, we can ensure that these polygonal braids are also generic. Note the following corollary of this argument and Claim 1.7.

Claim 1.8. *If generic polygonal braids b, b' are isotopic, then b can be transformed into b' by a finite sequence of Δ-moves such that all the intermediate polygonal braids are generic.*

To present generic polygonal braids, we can apply the technique of braid diagrams. The diagrams of generic polygonal braids are the braid diagrams, whose strands are formed by consecutive straight segments. Without loss of generality, we can always assume that the vertices of these segments do not coincide with the crossing points of the diagrams.

Claim 1.9. *The diagrams of two generic polygonal braids related by a Δ-move are R-equivalent.*

Proof. Consider a Δ-move $\Delta(ABC)$ on a generic polygonal braid b producing a generic polygonal braid b'. Pick points A', C' inside the segments AB, BC, respectively. Pick a point D inside the segment AC such that the third coordinate of D lies strictly between the third coordinates of A' and C'. Applying to b the moves $\Delta(AA'D)$, $\Delta(DC'C)$, we transform the segment AC into the broken line $AA'DC'C$. Further applying the moves $(\Delta(A'DC'))^{-1}$ and $\Delta(A'BC')$, we obtain b'. This shows that the move $\Delta(ABC)$ can be replaced by a sequence of four Δ-moves along smaller triangles (one should choose the points A', C', D so that the intermediate polygonal braids are generic). This expansion of the move $\Delta(ABC)$ can be iterated. In this way, subdividing the triangle ABC into smaller triangles and expanding Δ-moves as compositions of Δ-moves along the smaller triangles, we can reduce ourselves to the case in which the projection of ABC to $\mathbf{R} \times I$ meets the rest of the diagram of b either along a segment or along two segments with one crossing point.

Consider the first case. If both endpoints of the segment in question lie on $AB \cup BC$, then the diagram of b is transformed under $\Delta(ABC)$ by Ω_2. If one endpoint of the segment lies on AC and the other one lies on $AB \cup BC$, then the diagram is transformed by an isotopy.

If the projection of ABC to $\mathbf{R} \times I$ meets the rest of the diagram along two segments having one crossing, then we can similarly distinguish several subcases. Subdividing if necessary the triangle ABC into smaller triangles and expanding our Δ-move as a composition of Δ-moves along the smaller triangles, we can reduce ourselves to the case in which the move preserves the part of the diagram lying outside a small disk in $\mathbf{R} \times I$ and changes the diagram inside this disk via one of the following six formulas:

$$d_1^+ d_2^+ d_1^+ \leftrightarrow d_2^+ d_1^+ d_2^+, \quad d_1^+ d_2^+ d_1^- \leftrightarrow d_2^- d_1^+ d_2^+, \quad d_1^- d_2^+ d_1^+ \leftrightarrow d_2^+ d_1^- d_2^-,$$

$$d_1^- d_2^- d_1^- \leftrightarrow d_2^- d_1^- d_2^-, \quad d_1^+ d_2^- d_1^- \leftrightarrow d_2^- d_1^- d_2^+, \quad d_1^- d_2^+ d_1^+ \leftrightarrow d_2^+ d_1^+ d_2^-.$$

Here $d_1^\pm$ and $d_2^\pm$ are the braid diagrams on three strands shown in Figure 1.8; for the definition of the product of braid diagrams, see Figure 1.4. The reader is encouraged to draw the pictures of these transformations. It remains to prove that for each of them, the diagrams on the left-hand and right-hand sides are R-equivalent. The transformation $d_1^+ d_2^+ d_1^+ \mapsto d_2^+ d_1^+ d_2^+$ is just Ω_3. For the other five transformations, the R-equivalence is established by the following sequences of moves:

$$\omega = (d_1^+ d_2^+ d_1^- \xrightarrow{\Omega_2} d_2^- d_2^+ d_1^+ d_2^+ d_1^- \xrightarrow{\Omega_3^{-1}} d_2^- d_1^+ d_2^+ d_1^+ d_1^- \xrightarrow{\Omega_2^{-1}} d_2^- d_1^+ d_2^+),$$

$$\gamma = (d_1^- d_2^- d_1^+ \xrightarrow{\Omega_2} d_1^- d_2^- d_1^+ d_2^+ d_2^- \xrightarrow{\omega^{-1}} d_1^- d_1^+ d_2^+ d_1^- d_2^- \xrightarrow{\Omega_2^{-1}} d_2^+ d_1^- d_2^-),$$

$$\mu = (d_1^- d_2^- d_1^- \xrightarrow{\Omega_2} d_2^- d_2^+ d_1^- d_2^- d_1^- \xrightarrow{\gamma^{-1}} d_2^- d_1^- d_2^- d_1^+ d_1^- \xrightarrow{\Omega_2^{-1}} d_2^- d_1^- d_2^-),$$

$$d_1^+ d_2^- d_1^- \xrightarrow{\Omega_2} d_1^+ d_2^- d_1^- d_2^- d_2^+ \xrightarrow{\mu^{-1}} d_1^+ d_1^- d_2^- d_1^- d_2^+ \xrightarrow{\Omega_2^{-1}} d_2^- d_1^- d_2^+,$$

$$d_1^- d_2^+ d_1^+ \xrightarrow{\Omega_2} d_1^- d_2^+ d_1^+ d_2^+ d_2^- \xrightarrow{\Omega_3^{-1}} d_1^- d_1^+ d_2^+ d_1^+ d_2^- \xrightarrow{\Omega_2^{-1}} d_2^+ d_1^+ d_2^-.$$

This completes the proof of Claim 1.9. □

Fig. 1.8. The diagrams d_1^+, d_1^-, d_2^+, d_2^-

Step 4. We can now complete the proof of Theorem 1.6. It is obvious that R-equivalent braid diagrams present isotopic braids. To prove the converse, consider two braid diagrams $\mathcal{D}_1, \mathcal{D}_2$ presenting isotopic braids. For $i = 1, 2$, straightening $\mathcal{D}_i$ near its crossing points and approximating the rest of $\mathcal{D}_i$ by broken lines as at Step 2, we obtain a diagram, $\mathcal{D}_i'$, of a generic polygonal braid, b^i. If the approximation is close enough, then $\mathcal{D}_i'$ is isotopic to $\mathcal{D}_i$ (cf. Exercise 1.2.1 below). Then the braids b^1, b^2 are isotopic. Claim 1.8 implies that b^1 can be transformed into b^2 by a finite sequence of Δ-moves in the class of generic polygonal braids. Claim 1.9 implies that the diagrams $\mathcal{D}_1', \mathcal{D}_2'$ are R-equivalent. Therefore the diagrams $\mathcal{D}_1, \mathcal{D}_2$ are R-equivalent. □

In the next exercise we use the notation introduced at Step 1 of the proof.

Exercise 1.2.1. Any geometric braids b, b' with the same number of strings and such that $\widetilde{\rho}(b, b') < |b|$ are isotopic to each other.

Solution. The required isotopy $F : b \times I \to \mathbf{R}^2 \times I$ can be obtained by pushing each point $b_i(t)$ into $b'_i(t)$ along the line connecting these points. Thus,

$$F(b_i(t), s) = s\, b_i(t) + (1 - s)\, b'_i(t),$$

for $t, s \in I$ and $i = 1, \ldots, n$, where n is the number of strings of b. To see that F is an isotopy of b into b', it is enough to check that for all $s \in I$, the map $F_s : b \to \mathbf{R}^2 \times I$ sending $b_i(t)$ to $s\, b_i(t) + (1 - s)\, b'_i(t)$ is an embedding. Since the points $b_i(t), b'_i(t)$ have the third coordinate t, so does the point $s\, b_i(t) + (1 - s)\, b'_i(t)$. Therefore the restriction of F_s to any string b_i of b is an embedding. Moreover,

$$\rho(b_i(t), F_s(b_i(t))) \le \rho(b_i(t), b'_i(t)) \le \widetilde{\rho}(b, b') < |b|.$$

Therefore the image of b_i under F_s lies in the cylinder $|b|$-neighborhood of b_i. This implies that the images of distinct strings of b under F_s are disjoint.

1.2.4 The group of braids

Denote by $\mathcal{B}_n$ the set of braids on n strings with multiplication defined above. The next lemma implies that $\mathcal{B}_n$ is a group.

Lemma 1.10. *Each $\beta \in \mathcal{B}_n$ has a two-sided inverse β^{-1} in $\mathcal{B}_n$.*

Proof. For $i = 1, 2, \ldots, n - 1$, we define two elementary braids σ_i^+ and σ_i^- represented by diagrams with only one crossing shown in Figure 1.9. We claim that the braids $\sigma_1^+, \ldots, \sigma_{n-1}^+, \sigma_1^-, \ldots, \sigma_{n-1}^- \in \mathcal{B}_n$ generate $\mathcal{B}_n$ as a monoid. To see this, consider a braid β on n strings represented by a braid diagram $\mathcal{D}$. By a slight deformation of $\mathcal{D} \subset \mathbf{R} \times I$ in a neighborhood of its crossing points, we may arrange that distinct crossings of $\mathcal{D}$ have distinct second coordinates. Then there are real numbers

$$0 = t_0 < t_1 < \cdots < t_{k-1} < t_k = 1$$

such that the intersection of $\mathcal{D}$ with each strip $\mathbf{R} \times [t_j, t_{j+1}]$ has exactly one crossing lying inside this strip. This intersection is then a diagram of σ_i^+ or σ_i^- for some $i = 1, 2, \ldots, n-1$. The resulting splitting of $\mathcal{D}$ as a product of k braid diagrams shows that

$$\beta = \beta(\mathcal{D}) = \sigma_{i_1}^{\varepsilon_1} \sigma_{i_2}^{\varepsilon_2} \cdots \sigma_{i_k}^{\varepsilon_k}, \tag{1.3}$$

where each ε_j is either $+$ or $-$ and $i_1, \ldots, i_k \in \{1, 2, \ldots, n - 1\}$.

Clearly, $\sigma_i^+ \sigma_i^- = \sigma_i^- \sigma_i^+ = 1$ for all i. (The corresponding braid diagrams are related by Ω_2.) Therefore $\beta^{-1} = \sigma_{i_k}^{-\varepsilon_k} \cdots \sigma_{i_2}^{-\varepsilon_2} \sigma_{i_1}^{-\varepsilon_1}$ is a two-sided inverse of β in $\mathcal{B}_n$ (here we use the convention $-+ = -$ and $-- = +$). $\qquad\square$

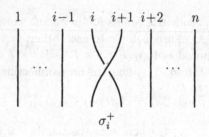

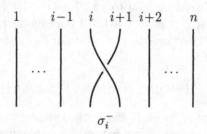

Fig. 1.9. The elementary braids σ_i^+ and σ_i^-

Lemma 1.11. *The elements* $\sigma_1^+, \ldots, \sigma_{n-1}^+ \in \mathcal{B}_n$ *satisfy the braid relations, that is,* $\sigma_i^+ \sigma_j^+ = \sigma_j^+ \sigma_i^+$ *for all* $i, j = 1, 2, \ldots, n-1$ *with* $|i - j| \geq 2$, *and* $\sigma_i^+ \sigma_{i+1}^+ \sigma_i^+ = \sigma_{i+1}^+ \sigma_i^+ \sigma_{i+1}^+$ *for* $i = 1, 2, \ldots, n-2$.

Proof. The first relation follows from the fact that its sides are represented by isotopic diagrams. The diagrams representing the sides of the second relation differ by the Reidemeister move Ω_3. $\qquad\square$

Theorem 1.12. *For* $\varepsilon = \pm$, *there is a unique homomorphism* $\varphi_\varepsilon : \mathcal{B}_n \to \mathcal{B}_n$ *such that* $\varphi_\varepsilon(\sigma_i) = \sigma_i^\varepsilon$ *for all* $i = 1, 2, \ldots, n-1$. *The homomorphism* φ_ε *is an isomorphism.*

Proof. For concreteness, we take $\varepsilon = +$ (the case $\varepsilon = -$ can be treated similarly or reduced to the case $\varepsilon = +$ using Exercise 1.1.2). The existence and uniqueness of φ_+ follow directly from Lemmas 1.2 and 1.11. The proof of Lemma 1.10 shows that $\sigma_1^+, \ldots, \sigma_{n-1}^+$ generate $\mathcal{B}_n$ as a group. These generators belong to the image of φ_+. Therefore, φ_+ is surjective.

We now construct a set-theoretic map $\psi : \mathcal{B}_n \to \mathcal{B}_n$ such that $\psi \circ \varphi_+ = \mathrm{id}$. This will imply that φ_+ is injective. As in the proof of Lemma 1.10, we represent any $\beta \in \mathcal{B}_n$ by a braid diagram $\mathcal{D}$ whose crossings have distinct second coordinates. This leads to an expansion of the form (1.3). Set

$$\psi(\mathcal{D}) = (\sigma_{i_1})^{\varepsilon_1} (\sigma_{i_2})^{\varepsilon_2} \cdots (\sigma_{i_k})^{\varepsilon_k} \in \mathcal{B}_n \,,$$

where

$$(\sigma_i)^+ = \sigma_i \quad \text{and} \quad (\sigma_i)^- = \sigma_i^{-1}.$$

We claim that $\psi(\mathcal{D})$ depends only on β. By Theorem 1.6 we need only verify that $\psi(\mathcal{D})$ does not change under isotopies of $\mathcal{D}$ and the Reidemeister moves on $\mathcal{D}$. Isotopies of $\mathcal{D}$ keeping the order of the double points of $\mathcal{D}$ with respect to the second coordinate keep the expansion (1.3) and therefore preserve $\psi(\mathcal{D})$. An isotopy exchanging the order of two double points of $\mathcal{D}$ (as in Figure 1.3) replaces the term $\sigma_i^{\varepsilon_i}\sigma_j^{\varepsilon_j}$ in (1.3) by $\sigma_j^{\varepsilon_j}\sigma_i^{\varepsilon_i}$ for some $i,j \in \{1,2,\ldots,n-1\}$ with $|i-j| \geq 2$. Under ψ, these expressions are sent to the same element of B_n because of the first braid relation of Definition 1.1.

The move Ω_2 (resp. Ω_2^{-1}) on $\mathcal{D}$ inserts (resp. removes) in the expansion (1.3) a term $\sigma_i^+\sigma_i^-$ or $\sigma_i^-\sigma_i^+$. Clearly, this preserves $\psi(\mathcal{D})$.

The move Ω_3 on $\mathcal{D}$ replaces a sequence $\sigma_i^+\sigma_{i+1}^+\sigma_i^+$ in (1.3) by $\sigma_{i+1}^+\sigma_i^+\sigma_{i+1}^+$. Under ψ, these expressions are sent to the same element of B_n because of the second braid relation of Definition 1.1. The move Ω_3^{-1} is considered similarly.

This shows that ψ is a well-defined map from $\mathcal{B}_n$ to B_n. By construction, $\psi \circ \varphi_+ = \mathrm{id}$. Hence φ_+ is both surjective and injective.   □

Conventions 1.13. From now on, we shall identify the groups B_n and $\mathcal{B}_n$ via φ_+. The elements of B_n henceforth will be called *braids on n strings*. We shall write σ_i for the braid σ_i^+. In this notation, $\sigma_i^- = (\sigma_i^+)^{-1} = \sigma_i^{-1}$.

The projection to the symmetric group $\pi : B_n \to \mathfrak{S}_n$ can be easily described in geometric terms. For a geometric braid b on n strings, the permutation $\pi(b) \in \mathfrak{S}_n$ sends each $i \in \{1,2,\ldots,n\}$ to the only $j \in \{1,2,\ldots,n\}$ such that the string of b attached to $(i,0,0)$ has the second endpoint at $(j,0,1)$.

Corollary 1.14. *The natural inclusion $\iota : B_n \to B_{n+1}$ is injective for all n.*

Proof. In geometric language, $\iota : B_n \to B_{n+1}$ adds to a geometric braid b on n strings a vertical string on its right completely unlinked from b. Denote the resulting braid on $n+1$ strings by $\iota(b)$. If b_1, b_2 are two geometric braids on n strings such that $\iota(b_1)$ is isotopic to $\iota(b_2)$, then restricting the isotopy to the leftmost n strings, we obtain an isotopy of b_1 into b_2. Therefore ι is injective.   □

Remarks 1.15. (a) Some authors, including Artin [Art25], use φ_- to identify B_n and $\mathcal{B}_n$. We follow [Art47a], where these groups are identified via φ_+.

(b) In the definition of geometric braids on n strings we chose the set of endpoints to be $\{1,2,\ldots,n\} \times \{0\} \times \{0,1\}$. Instead of $\{1,2,\ldots,n\}$ we can use an arbitrary set of n distinct real numbers. This gives the same group of braids, since such a set can be continuously deformed into $\{1,2,\ldots,n\}$ in $\mathbf{R}$.

Exercise 1.2.2. Prove that for an arbitrary geometric braid $b \subset \mathbf{R}^2 \times I$, there is a Euclidean disk $U \subset \mathbf{R}^2$ such that $b \subset U \times I$. (Hint: The projection of b to the plane $\mathbf{R}^2$ is a compact set.)

Exercise 1.2.3. Prove that for an arbitrary isotopy of braids $\{b_s\}_{s\in I}$, there is a Euclidean disk $U \subset \mathbf{R}^2$ such that $b_s \subset U \times I$ for all $s \in I$.

Exercise 1.2.4. Let U be an open Euclidean disk in $\mathbf{R}^2$ containing the points $(1,0),\ldots,(n,0)$. Prove that any geometric braid on n strings $b \subset \mathbf{R}^2 \times I$ is isotopic to a geometric braid lying in $U \times I$.

Solution. By Exercise 1.2.2, there is a disk $U_1 \subset \mathbf{R}^2$ such that $b \subset U_1 \times I$. Enlarging U_1, we may assume that $U_1 \supset U$. There is a small $\varepsilon > 0$ such that

$$b \cap (\mathbf{R}^2 \times [0,\varepsilon]) \subset U \times [0,\varepsilon] \quad \text{and} \quad b \cap (\mathbf{R}^2 \times [1-\varepsilon,1]) \subset U \times [1-\varepsilon,1].$$

Keeping fixed the part of b lying in

$$(\mathbf{R}^2 \times [0,\varepsilon/2]) \cup (\mathbf{R}^2 \times [1-\varepsilon/2,1])$$

and squeezing $U_1 \times [\varepsilon, 1-\varepsilon]$ into $U \times [\varepsilon, 1-\varepsilon]$, we obtain a geometric braid in $U \times [0,1]$ isotopic to b.

Exercise 1.2.5. For U as in Exercise 1.2.4, prove that any two geometric braids lying in $U \times I$ and isotopic in $\mathbf{R}^2 \times I$ are isotopic already in $U \times I$.

Exercise 1.2.6. For a geometric braid $b \subset \mathbf{R}^2 \times I$ on n strings, denote by $\bar{b}$ the image of b under the involution of $\mathbf{R}^2 \times I$ mapping (x,y,t) to $(x,y,1-t)$, where $x,y \in \mathbf{R}, t \in I$. Verify that $\bar{b}$ is a geometric braid. Show that if b represents $\beta \in \mathcal{B}_n$, then $\bar{b}$ represents β^{-1}. Deduce that if β is represented by a braid diagram $\mathcal{D}$, then β^{-1} is represented by the image of $\mathcal{D}$ under the reflection in the line $\mathbf{R} \times \{1/2\}$.

1.3 Pure braid groups

In this section we introduce so-called pure braids and use them to establish important algebraic properties of the braid groups.

1.3.1 Pure braids

The kernel of the natural projection $\pi : B_n \to \mathfrak{S}_n$ is called the *pure braid group* and is denoted by P_n:

$$P_n = \mathrm{Ker}\,(\pi : B_n \to \mathfrak{S}_n).$$

The elements of P_n are called *pure braids on n strings*. A geometric braid on n strings represents an element of P_n if and only if for all $i = 1,\ldots,n$, the string of this braid attached to $(i,0,0)$ has the second endpoint at $(i,0,1)$. Such geometric braids are said to be *pure*.

An important role in the sequel will be played by the pure n-string braid $A_{i,j}$ shown in Figure 1.10, where $1 \leq i < j \leq n$. This braid can be expressed via the generators $\sigma_1,\ldots,\sigma_{n-1}$ by

$$A_{i,j} = \sigma_{j-1}\sigma_{j-2}\cdots\sigma_{i+1}\sigma_i^2\sigma_{i+1}^{-1}\cdots\sigma_{j-2}^{-1}\sigma_{j-1}^{-1}.$$

The braids $\{A_{i,j}\}_{i,j}$ are conjugate to each other in B_n. Indeed, set

$$\alpha_{i,j} = \sigma_{j-1}\sigma_{j-2}\cdots\sigma_i$$

for any $1 \le i < j \le n$. It is a simple pictorial exercise to check that for any $1 \le i < j < k \le n$,

$$\alpha_{j,k}\, A_{i,j}\, \alpha_{j,k}^{-1} = A_{i,k} \quad \text{and} \quad \alpha_{i,k}\, A_{i,j}\, \alpha_{i,k}^{-1} = A_{j,k} \,. \tag{1.4}$$

We shall see shortly that the braids $\{A_{i,j}\}_{i,j}$ are not mutually conjugate in P_n.

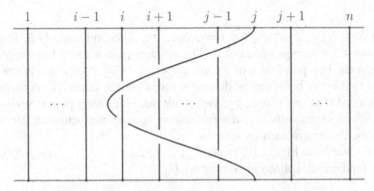

Fig. 1.10. The n-string braid $A_{i,j}$ with $1 \le i < j \le n$

The commutativity of the diagram (1.1) implies that the inclusion homomorphism $\iota : B_n \to B_{n+1}$ maps P_n to P_{n+1}. The homomorphism $P_n \to P_{n+1}$ induced by ι will be denoted by the same symbol ι. In geometric language, $\iota : P_n \to P_{n+1}$ adds to a pure geometric braid b on n strings a vertical string on its right completely unlinked from b. By Corollary 1.14, $\iota : P_n \to P_{n+1}$ is injective. It is sometimes convenient to view P_n as a subgroup of P_{n+1} via ι. In this way we obtain an increasing chain of groups $P_1 \subset P_2 \subset P_3 \subset \cdots$. It is clear that $P_1 = \{1\}$ and P_2 is an infinite cyclic group generated by $A_{1,2} = \sigma_1^2$.

1.3.2 Forgetting homomorphisms

We define a *forgetting homomorphism* $f_n : P_n \to P_{n-1}$ as follows. Represent an element of P_n by a geometric braid b. The ith string of b connects $(i, 0, 0)$ to $(i, 0, 1)$ for $i = 1, 2, \ldots, n$. Removing the nth string from b, we obtain a braid $f_n(b)$ on $n-1$ strings. It is obvious that if b is isotopic to b', then $f_n(b)$ is isotopic to $f_n(b')$. Passing to isotopy classes, we obtain a well-defined map $f_n : P_n \to P_{n-1}$. From the definition of multiplication for geometric braids, it is clear that f_n is a group homomorphism. From the geometric description of the natural inclusion $\iota : P_{n-1} \to P_n$, it is clear that $f_n \circ \iota = \mathrm{id}_{P_{n-1}}$. This yields another proof of the injectivity of ι and of Corollary 1.14, and also implies that the homomorphism f_n is surjective.

For $n \geq 2$, set

$$U_n = \operatorname{Ker}\left(f_n : P_n \to P_{n-1}\right).$$

Note that since f_n has a section, P_n is isomorphic to the semidirect product of P_{n-1} by U_n. Any pure braid $\beta \in P_n$ can be expanded uniquely in the form

$$\beta = \iota(\beta')\beta_n \qquad (1.5)$$

with $\beta' \in P_{n-1}$ and $\beta_n \in U_n$. Here $\beta' = f_n(\beta)$ and $\beta_n = \iota(\beta')^{-1}\beta$. Applying this expansion inductively, we conclude that β can be written uniquely as

$$\beta = \beta_2\,\beta_3\,\cdots\,\beta_n\,, \qquad (1.6)$$

where $\beta_j \in U_j \subset P_j \subset P_n$ for $j = 2, 3, \ldots, n$. The expansion (1.6) is called the *combed* (or *normal*) form of β. The authors cannot resist the temptation to quote the last paragraph of Artin's paper [Art47a]: "Although it has been proved that every braid can be deformed into a similar normal form the writer is convinced that any attempt to carry this out on a living person would only lead to violent protests and discrimination against mathematics. He would therefore discourage such an experiment."

It is clear from Figure 1.10 that $A_{i,n} \in U_n$ for $i = 1, 2, \ldots, n-1$. We state now a fundamental theorem computing U_n.

Theorem 1.16. *For all $n \geq 2$, the group U_n is free on the $n-1$ generators $\{A_{i,n}\}_{i=1,2,\ldots,n-1}$.*

A proof of Theorem 1.16 will be given in Section 1.4. The rest of this section will be devoted to corollaries of Theorem 1.16.

Corollary 1.17. *The group P_n admits a normal filtration*

$$1 = U_n^{(0)} \subset U_n^{(1)} \subset \cdots \subset U_n^{(n-1)} = P_n$$

such that $U_n^{(i)}/U_n^{(i-1)}$ is a free group of rank $n-i$ for all i.

Proof. Set $U_n^{(0)} = \{1\}$ and for $i = 1, 2, \ldots, n-1$ set

$$U_n^{(i)} = \operatorname{Ker}\left(f_{n-i+1}\cdots f_{n-1}f_n : P_n \to P_{n-i}\right).$$

Then

$$U_n^{(i)}/U_n^{(i-1)} \cong \operatorname{Ker}\left(f_{n-i+1} : P_{n-i+1} \to P_{n-i}\right) = U_{n-i+1}. \qquad \square$$

Corollary 1.18. *The group P_n is torsion free, i.e., it has no nontrivial elements of finite order.*

This follows directly from Corollary 1.17, since free groups are torsion free. The braid group B_n is also torsion free; this will be proven in Section 1.4.3 using a different method.

Corollary 1.19. P_n *is generated by the* $n(n-1)/2$ *elements* $\{A_{i,j}\}_{1\leq i<j\leq n}$.

This directly follows from formula (1.6) and Theorem 1.16.

Here is a list of defining relations for the generators $\{A_{i,j}\}_{1\leq i<j\leq n}$ of P_n:

$$A_{r,s}^{-1} A_{i,j} A_{r,s} = \qquad\qquad\qquad\qquad\qquad\qquad\qquad (1.7)$$

$$\begin{cases} A_{i,j} & \text{if } s < i \text{ or } i < r < s < j, \\ A_{r,j} A_{i,j} A_{r,j}^{-1} & \text{if } s = i, \\ A_{r,j} A_{s,j} A_{i,j} A_{s,j}^{-1} A_{r,j}^{-1} & \text{if } i = r < s < j, \\ A_{r,j} A_{s,j} A_{r,j}^{-1} A_{s,j}^{-1} A_{i,j} A_{s,j} A_{r,j} A_{s,j}^{-1} A_{r,j}^{-1} & \text{if } r < i < s < j. \end{cases}$$

That these relations hold in P_n can be verified directly by drawing the corresponding pictures. That all relations between $\{A_{i,j}\}_{1\leq i<j\leq n}$ follow from the relations in this list can be verified using the Reidemeister–Schreier rewriting process; see Appendix 1 to [Han89], written by Lars Gæde. In this book we use relations (1.7) only once, in Section 7.2.3.

Corollary 1.20. *We have* $P_n/[P_n, P_n] \cong \mathbf{Z}^{n(n-1)/2}$.

Proof. By Corollary 1.19, the abelian group $P_n/[P_n, P_n]$ is generated by the elements represented by $A_{i,j}$, where $1 \leq i < j \leq n$. To prove that these elements are linearly independent, it suffices to construct for each pair $1 \leq i < j \leq n$ a group homomorphism $l_{i,j} : P_n \to \mathbf{Z}$ such that $l_{i,j}(A_{i,j}) = 1$ and $l_{i,j}(A_{r,s}) = 0$ for all pairs (r, s) distinct from (i, j).

Pick $\beta \in P_n$ and represent it by a braid diagram $\mathcal{D}$. Orient all strands of $\mathcal{D}$ from the top (the level $t = 0$) to the bottom (the level $t = 1$). Let $l_{i,j}^+(\mathcal{D})$ be the number of crossings of $\mathcal{D}$, where the ith strand goes over the jth strand from left to right. Let $l_{i,j}^-(\mathcal{D})$ be the number of crossings of $\mathcal{D}$, where the ith strand goes over the jth strand from right to left. Set

$$l_{i,j}(\beta) = l_{i,j}^+(\mathcal{D}) - l_{i,j}^-(\mathcal{D}).$$

It is straightforward to check that $l_{i,j}(\beta)$ is invariant under isotopies and Reidemeister moves on $\mathcal{D}$. By Theorem 1.6, $l_{i,j}(\beta)$ is a well-defined invariant of β. (This invariant can be also defined as the linking number of the ith and jth components of the link in $\mathbf{R}^3$ obtained by closing β; cf. Chapter 2.) The map $l_{i,j} : P_n \to \mathbf{Z}$ is a group homomorphism taking the value $+1$ on $A_{i,j}$ and the value 0 on all $A_{r,s}$ with $(r, s) \neq (i, j)$. $\qquad\square$

Corollary 1.21. *The group* B_n *and all its subgroups are residually finite.*

Proof. Recall that a group G is *residually finite* if for each $\beta \in G - \{1\}$, there is a homomorphism f from G to a finite group such that $f(\beta) \neq 1$. It is known that free groups are residually finite (see [LS77, Chap. IV, Sect. 4], [MKS66, Sect. 6.5]) and a semidirect product of two finitely generated residually finite groups is residually finite (the latter fact is due to Maltsev [Mal40]). Therefore Theorem 1.16 implies by induction on n that P_n is residually finite.

Note that any extension (not necessarily semidirect) of a residually finite group P by a finite group is residually finite. This can be easily deduced from the fact that the intersection of a finite family of subgroups of P of finite index is a subgroup of P of finite index. Since B_n is an extension of P_n by $\mathfrak{S}_n$ and P_n is residually finite, so is B_n. It remains to observe that all subgroups of a residually finite group are residually finite. □

A group is *Hopfian* if all its surjective endomorphisms are injective.

Corollary 1.22. *The group B_n and all its finitely generated subgroups are Hopfian.*

Proof. A finitely generated residually finite group is Hopfian (see [LS77, Chap. IV, Th. 4.10], [Neu67]). □

Corollary 1.23. *For $i = 1, 2, \ldots, n$, forgetting the ith string defines a group homomorphism $f_n^i : P_n \to P_{n-1}$. The kernel of f_n^i is a free group of rank $n-1$ with free generators $A_{1,i}, \ldots, A_{i-1,i}, A_{i,i+1}, \ldots, A_{i,n}$.*

Proof. Set $\alpha_{i,n} = \sigma_{n-1}\sigma_{n-2}\cdots\sigma_i$ and observe that for any $\beta \in P_n$, forgetting the nth string of $\alpha_{i,n}\beta\alpha_{i,n}^{-1}$ yields the braid

$$1_{n-1}\, f_n^i(\beta)\, 1_{n-1} = f_n^i(\beta)\,.$$

Hence, $f_n^i(\beta) = f_n(\alpha_{i,n}\beta\alpha_{i,n}^{-1})$, where $f_n = f_n^n$. Therefore,

$$\operatorname{Ker} f_n^i = \alpha_{i,n}^{-1}\left(\operatorname{Ker} f_n\right)\alpha_{i,n} = \alpha_{i,n}^{-1} U_n\, \alpha_{i,n}\,.$$

It remains to use Theorem 1.16 and to observe that conjugation by $\alpha_{i,n}^{-1}$ transforms the set $\{A_{j,n}\}_{j=1,2,\ldots,n-1}$ into the set

$$\{A_{1,i}, \ldots, A_{i-1,i}, A_{i,i+1}, \ldots, A_{i,n}\}\,,$$

as is clear from (1.4). □

1.3.3 The center of B_n

The *center* of a group G is the subgroup of G consisting of all $g \in G$ such that $gx = xg$ for every $x \in G$. The center of a group G is denoted by $Z(G)$.

Theorem 1.24. *If $n \geq 3$, then $Z(B_n) = Z(P_n)$ is an infinite cyclic group generated by $\theta_n = \Delta_n^2$, where*

$$\Delta_n = (\sigma_1\sigma_2\cdots\sigma_{n-1})(\sigma_1\sigma_2\cdots\sigma_{n-2})\cdots(\sigma_1\sigma_2)\,\sigma_1 \in B_n\,.$$

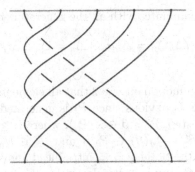

Fig. 1.11. The braid Δ_5

Proof. The braid Δ_n can be obtained from the trivial braid 1_n by a half-twist achieved by keeping the top of the braid fixed and turning over the row of the lower ends by an angle of π. See Figure 1.11 for a diagram of Δ_5. The braid $\theta_n = \Delta_n^2$ can be obtained from the trivial braid 1_n by a full twist achieved by keeping the top of the braid fixed and turning over the row of the lower ends by an angle of 2π. We have

$$\pi(\Delta_n) = (n, n-1, \ldots, 1) \in \mathfrak{S}_n.$$

Hence $\theta_n \in P_n$. It is a simple exercise to compute θ_n inductively from $\iota(\theta_{n-1})$, where $\iota : P_{n-1} \to P_n$ is the natural inclusion. Namely, $\theta_n = \iota(\theta_{n-1})\gamma$, where

$$\gamma = \gamma_n = A_{1,n} A_{2,n} \cdots A_{n-1,n} \in P_n;$$

see Figure 1.12 for a diagram of γ_5.

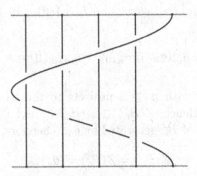

Fig. 1.12. The braid γ_5

Sliding a crossing along the strands of the diagram of Δ_n from top to bottom, one easily obtains for all $i = 1, 2, \ldots, n-1$,

$$\sigma_i \Delta_n = \Delta_n \sigma_{n-i}. \tag{1.8}$$

This implies that θ_n commutes with all the generators of B_n:

$$\sigma_i\,\theta_n = \sigma_i\,\Delta_n\Delta_n = \Delta_n\sigma_{n-i}\,\Delta_n = \Delta_n\Delta_n\,\sigma_i = \theta_n\,\sigma_i\,.$$

Hence, $\theta_n \in Z(B_n)$.

We now prove by induction on $n \geq 2$ that all elements of $Z(P_n)$ are powers of θ_n. For $n = 2$, this is obvious since P_2 is generated by $A_{1,2} = \theta_2 = \sigma_1^2$. Here is the inductive step. Pick $\beta \in Z(P_n)$, where $n \geq 3$. By formula (1.5), $\beta = \iota(\beta')\beta_n$, where $\beta' = f_n(\beta) \in P_{n-1}$ and $\beta_n \in U_n$. An easy geometric argument shows that the braid $\gamma = \gamma_n$ introduced above commutes with any element of $\iota(P_{n-1}) \subset P_n$ and in particular with $\iota(\beta')$. Since β lies in the center of P_n, it commutes with γ. Hence, γ commutes with $\beta_n = \iota(\beta')^{-1}\beta$. The group $G \subset U_n$ generated by β_n and γ is therefore abelian. By Theorem 1.16, the group U_n is free and therefore all its subgroups are free. This implies that G is an infinite cyclic group. Recall now the homomorphism $l_{i,j} : P_n \to \mathbf{Z}$ defined in the proof of Corollary 1.20 for all $1 \leq i < j \leq n$. Clearly $l_{1,n}(\gamma) = 1$, so that γ has to be a generator of G. Thus, $\beta_n = \gamma^k$ for some integer k. Since the forgetting homomorphism $f_n : P_n \to P_{n-1}$ is onto, $\beta' = f_n(\beta) \in Z(P_{n-1})$. By the induction assumption, $\beta' = (\theta_{n-1})^m$ for some integer m. We prove below that $m = k$. Since γ commutes with $\iota(\theta_{n-1})$, this will give

$$\beta = \iota(\beta')\beta_n = \iota((\theta_{n-1})^m)\,\gamma^k = \iota((\theta_{n-1})^k)\,\gamma^k = (\iota(\theta_{n-1})\,\gamma)^k = (\theta_n)^k\,.$$

It follows from the definitions and the expansion $\beta = \iota((\theta_{n-1})^m)\,\gamma^k$ that $l_{i,n}(\beta) = k$ for all $i = 1, 2, \ldots, n-1$. In particular, $l_{i,n}(\beta)$ does not depend on i. Since β lies in $Z(P_n)$, so does $\sigma_{n-1}\beta\sigma_{n-1}^{-1}$. By the result above, the integer $l_{i,n}(\sigma_{n-1}\beta\sigma_{n-1}^{-1})$ does not depend on $i = 1, 2, \ldots, n-1$. Computing from the definitions and using the expansion $\beta = \iota((\theta_{n-1})^m)\,\gamma^k$, we obtain

$$l_{1,n}(\sigma_{n-1}\beta\sigma_{n-1}^{-1}) = l_{1,n-1}(\beta) = m$$

and

$$l_{n-1,n}(\sigma_{n-1}\beta\sigma_{n-1}^{-1}) = l_{n-1,n}(\beta) = k\,.$$

Thus, $m = k$.

The center of B_n with $n \geq 3$ projects to the trivial subgroup of $\mathfrak{S}_n$ since $Z(\mathfrak{S}_n) = \{1\}$. Hence, $Z(B_n) \subset Z(P_n) \subset (\theta_n) \subset Z(B_n)$, where (θ_n) is the cyclic subgroup of B_n generated by θ_n. Therefore,

$$Z(B_n) = Z(P_n) = (\theta_n)\,.$$

By Corollary 1.18, (θ_n) is an infinite cyclic group. $\square$

Corollary 1.25. *For $m \neq n$, the groups B_m and B_n are not isomorphic.*

Proof. Theorem 1.24 implies that the image of $Z(B_n)$ in $B_n/[B_n, B_n] \cong \mathbf{Z}$ is a subgroup of $\mathbf{Z}$ of index $n(n-1)$. If B_m is isomorphic to B_n, then we must have $m(m-1) = n(n-1)$, and hence $m = n$. $\square$

Exercise 1.3.1. Deduce Corollary 1.20 from the presentation of P_n given by the generators $\{A_{i,j}\}_{1\leq i<j\leq n}$ and the relations (1.7).

Exercise 1.3.2. Verify that $\Delta_n^2 = (\sigma_1\sigma_2\cdots\sigma_{n-1})^n$.

Exercise 1.3.3. Verify that P_n is the minimal normal subgroup of B_n containing $\sigma_1^2 = A_{1,2}$.

Exercise 1.3.4. Verify (1.4) using only the expression of $A_{i,j}$ via $\sigma_1,\ldots,\sigma_{n-1}$ and the braid relations between these generators.

Exercise 1.3.5. Show that any nontrivial subgroup of P_n has a nontrivial homomorphism onto $\mathbf{Z}$. (Hint: Any free group has a normal filtration with free abelian consecutive quotients.)

1.4 Configuration spaces

We discuss here an approach to braids based on configuration spaces. As an application, we prove Theorem 1.16.

1.4.1 Configuration spaces of ordered sets of points

Let M be a topological space and let

$$M^n = M \times M \times \cdots \times M$$

be the product of $n \geq 1$ copies of M with the product topology. Set

$$\mathcal{F}_n(M) = \{(u_1, u_2, \ldots, u_n) \in M^n \mid u_i \neq u_j \text{ for all } i \neq j\}.$$

This subspace of M^n is called the *configuration space* of ordered n-tuples of (distinct) points in M.

If M is a topological manifold (possibly with boundary ∂M), then the configuration space $\mathcal{F}_n(M)$ is a topological manifold of dimension $n\dim(M)$. Clearly, any ordered n-tuple of points in M can be deformed into an ordered n-tuple of points in the interior $M^\circ = M - \partial M$ of M. If $\dim(M) \geq 2$ and M is connected, then any ordered n-tuple of points in M° can be deformed into any other such tuple. Therefore for such M, the manifold $\mathcal{F}_n(M)$ is connected. Its fundamental group is called the *pure braid group* of M on n strings.

For $M = \mathbf{R}^2$, we recover the same pure braid group P_n as above. To see this, assign to a pure geometric braid $b \subset \mathbf{R}^2 \times I$ the path $I \to \mathcal{F}_n(\mathbf{R}^2)$ sending $t \in I$ into the tuple $(u_1(t), u_2(t), \ldots, u_n(t))$ defined by the condition that the ith string of b meets $\mathbf{R}^2 \times \{t\}$ at the point $(u_i(t), t)$ for all $i = 1, 2, \ldots, n$. This path begins and ends at the n-tuple

$$q_n = ((1,0), (2,0), \ldots, (n,0)) \in \mathcal{F}_n(\mathbf{R}^2).$$

Conversely, any path $(\alpha_1, \alpha_2, \ldots, \alpha_n) : I \to \mathcal{F}_n(\mathbf{R}^2)$ beginning and ending at q_n gives rise to the pure geometric braid

$$\bigcup_{i=1}^{n} \bigcup_{t \in I} (\alpha_i(t), t).$$

These constructions are mutually inverse and yield a bijective correspondence between pure geometric braids and loops in $(\mathcal{F}_n(\mathbf{R}^2), q_n)$. Under this correspondence the isotopy of braids corresponds to the homotopy of loops. Thus, $P_n = \pi_1(\mathcal{F}_n(\mathbf{R}^2), q_n)$. The braid group B_n admits a similar interpretation, which will be discussed in Section 1.4.3.

Coming back to an arbitrary connected topological manifold M of dimension ≥ 2, it is useful to generalize the definition of $\mathcal{F}_n(M)$ by prohibiting several points in $M^\circ = M - \partial M$. More precisely, fix a finite set $Q_m \subset M^\circ$ of $m \geq 0$ points and set

$$\mathcal{F}_{m,n}(M) = \mathcal{F}_n(M - Q_m).$$

The topological type of this space depends on M, m, and n, but not on the choice of Q_m. Clearly, $\mathcal{F}_{0,n}(M) = \mathcal{F}_n(M)$ and $\mathcal{F}_{m,1}(M) = M - Q_m$.

To describe the relationship between various configuration spaces, we need the notion of a locally trivial fibration. For the convenience of the reader, we recall this notion in Appendix B.

Lemma 1.26. *Let M be a connected topological manifold of dimension ≥ 2 with $\partial M = \emptyset$. For $n > r \geq 1$, the forgetting map $p : \mathcal{F}_n(M) \to \mathcal{F}_r(M)$ defined by $p(u_1, \ldots, u_n) = (u_1, \ldots, u_r)$ is a locally trivial fibration with fiber $\mathcal{F}_{r,n-r}(M)$.*

Proof. Pick a point $u^0 = (u_1^0, \ldots, u_r^0) \in \mathcal{F}_r(M)$. The fiber $p^{-1}(u^0)$ consists of the tuples $(u_1^0, \ldots, u_r^0, v_1, \ldots, v_{n-r}) \in M^r$, where all $u_1^0, \ldots, u_r^0, v_1, \ldots, v_{n-r}$ are distinct. Setting $Q_r = \{u_1^0, \ldots, u_r^0\}$, we obtain

$$\mathcal{F}_{r,n-r}(M) = \{(v_1, \ldots, v_{n-r}) \in (M - Q_r)^{n-r} \mid v_i \neq v_j \text{ for } i \neq j\}.$$

It is obvious that the formula $(u_1^0, \ldots, u_r^0, v_1, \ldots, v_{n-r}) \mapsto (v_1, \ldots, v_{n-r})$ defines a homeomorphism $p^{-1}(u^0) \approx \mathcal{F}_{r,n-r}(M)$.

We shall prove the local triviality of p in a neighborhood of u^0. For each $i = 1, 2, \ldots, r$, pick an open neighborhood $U_i \subset M$ of u_i^0 such that its closure $\overline{U}_i$ is a closed ball with interior U_i. Since the points $u_1^0, \ldots, u_r^0$ are distinct, we may assume that $U_1, \ldots, U_r$ are mutually disjoint. Then

$$U = U_1 \times U_2 \times \cdots \times U_r$$

is a neighborhood of u^0 in $\mathcal{F}_r(M)$. We shall see that the restriction of p to U is a trivial bundle, i.e., that there is a homeomorphism $p^{-1}(U) \to U \times \mathcal{F}_{r,n-r}(M)$ commuting with the projections to U.

We construct below for each $i = 1, 2, \ldots, r$ a continuous map

$$\theta_i : U_i \times \overline{U}_i \to \overline{U}_i$$

such that for every $u \in U_i$, the map $\theta_i^u : \overline{U}_i \to \overline{U}_i$ sending $v \in \overline{U}_i$ to $\theta_i(u, v)$ is a homeomorphism sending u_i^0 to u and fixing the boundary sphere $\partial \overline{U}_i$ pointwise. For $u = (u_1, \ldots, u_r) \in U$, define a map $\theta^u : M \to M$ by

$$\theta^u(v) = \begin{cases} \theta_i(u_i, v) & \text{if } v \in U_i \text{ for some } i = 1, 2, \ldots, r, \\ v & \text{if } v \in M - \bigcup_i U_i. \end{cases}$$

It is clear that $\theta^u : M \to M$ is a homeomorphism continuously depending on u and sending the points $u_1^0, \ldots, u_r^0$ to $u_1, \ldots, u_r$, respectively. The formula

$$(u, v_1, \ldots, v_{n-r}) \mapsto (u, \theta^u(v_1), \ldots, \theta^u(v_{n-r}))$$

defines a homeomorphism $U \times \mathcal{F}_{r,n-r}(M) \to p^{-1}(U)$ commuting with the projections to U. The inverse homeomorphism is defined by

$$(u, v_1, \ldots, v_{n-r}) \mapsto (u, (\theta^u)^{-1}(v_1), \ldots, (\theta^u)^{-1}(v_{n-r})).$$

Thus, $p|_U : p^{-1}(U) \to U$ is a trivial fibration.

To construct θ_i, we may assume that $U_i = U$ is the open unit ball in Euclidean space $\mathbf{R}^{\dim M}$ with center at the origin $u_i = 0$. Fix a smooth function of two variables $\lambda : [0, 1[\times [0, 1] \to \mathbf{R}$ such that $\lambda(x, y) = 1$ if $x \geq y$ and $\lambda(x, y) = 0$ if $(x + 1)/2 \leq y$, where $x \in [0, 1[$ and $y \in [0, 1]$. For $u \in U$, define a vector field f^u on the closed unit ball $\overline{U} = \{v \in \mathbf{R}^{\dim M} \,|\, \|v\| \leq 1\}$ by

$$f^u(v) = \lambda(\|u\|, \|v\|) \, u.$$

The choice of λ ensures that $f^u = u$ on the ball of radius $\|u\|$ with center at the origin and $f^u = 0$ outside the ball of radius $(\|u\| + 1)/2$ with center at the origin. Let $\{\theta^{u,t} : \overline{U} \to \overline{U}\}_{t \in \mathbf{R}}$ be the flow determined by f^u, that is, the (unique) family of self-diffeomorphisms of $\overline{U}$ such that $\theta^{u,0} = \mathrm{id}$ and $d\theta^{u,t}(v)/dt = f^u(v)$ for all $v \in \overline{U}$, $t \in \mathbf{R}$. The diffeomorphism $\theta^{u,t}$ smoothly depends on u, t, fixes the sphere $\partial \overline{U}$ pointwise, and sends the origin to tu. Therefore the map $\theta_i : U \times \overline{U} \to \overline{U}$ defined by $\theta_i(u, v) = \theta^{u,1}(v)$ for $u \in U$, $v \in \overline{U}$ satisfies all the required conditions. $\qquad\square$

Lemma 1.27. *Let M be a connected topological manifold of dimension ≥ 2 with $\partial M = \emptyset$. For any $m \geq 0$, $n > r \geq 1$, the forgetting map*

$$p : \mathcal{F}_{m,n}(M) \to \mathcal{F}_{m,r}(M)$$

defined by $p(u_1, \ldots, u_n) = (u_1, \ldots, u_r)$ is a locally trivial fibration with fiber $\mathcal{F}_{m+r,n-r}(M)$.

Proof. This lemma is obtained by applying Lemma 1.26 to $M - Q_m$. $\qquad\square$

Recall that a connected manifold M is *aspherical* if its universal covering is contractible or, equivalently, if its homotopy groups $\pi_i(M)$ vanish for all $i \geq 2$.

Lemma 1.28. *For any $m \geq 0, n \geq 1$, the manifold $\mathcal{F}_{m,n}(\mathbf{R}^2)$ is aspherical.*

Proof. Consider the fibration $\mathcal{F}_{m,n}(\mathbf{R}^2) \to \mathcal{F}_{m,1}(\mathbf{R}^2) = \mathbf{R}^2 - Q_m$ with fiber $\mathcal{F}_{m+1,n-1}(\mathbf{R}^2)$ defined above. The homotopy sequence of this fibration gives an exact sequence

$$\cdots \longrightarrow \pi_{i+1}(\mathbf{R}^2 - Q_m) \longrightarrow \pi_i(\mathcal{F}_{m+1,n-1}(\mathbf{R}^2))$$
$$\longrightarrow \pi_i(\mathcal{F}_{m,n}(\mathbf{R}^2)) \longrightarrow \pi_i(\mathbf{R}^2 - Q_m) \longrightarrow \cdots.$$

Observe that $\mathbf{R}^2 - Q_m$ contains a wedge of m circles as a deformation retract. A wedge of circles is aspherical since its universal covering is a tree and hence is contractible. Therefore $\mathbf{R}^2 - Q_m$ is aspherical, so that $\pi_i(\mathbf{R}^2 - Q_m) = 0$ for $i \geq 2$. We conclude that for all $i \geq 2$,

$$\pi_i(\mathcal{F}_{m,n}(\mathbf{R}^2)) \cong \pi_i(\mathcal{F}_{m+1,n-1}(\mathbf{R}^2)).$$

An inductive argument shows for all $i \geq 2$,

$$\pi_i(\mathcal{F}_{m,n}(\mathbf{R}^2)) \cong \pi_i(\mathcal{F}_{m+n-1,1}(\mathbf{R}^2)) \cong \pi_i(\mathbf{R}^2 - Q_{m+n-1}) = 0. \qquad \square$$

1.4.2 Proof of Theorem 1.16

Applying Lemma 1.26 to $M = \mathbf{R}^2$, we obtain a locally trivial fibration $p : \mathcal{F}_n(\mathbf{R}^2) \to \mathcal{F}_{n-1}(\mathbf{R}^2)$ with fiber $\mathcal{F}_{n-1,1}(\mathbf{R}^2)$. This gives a short exact sequence

$$1 \longrightarrow \pi_1(\mathcal{F}_{n-1,1}(\mathbf{R}^2)) \longrightarrow \pi_1(\mathcal{F}_n(\mathbf{R}^2)) \overset{p_\#}{\longrightarrow} \pi_1(\mathcal{F}_{n-1}(\mathbf{R}^2)) \longrightarrow 1, \qquad (1.9)$$

where we use the triviality of $\pi_2(\mathcal{F}_{n-1}(\mathbf{R}^2))$ (by Lemma 1.28) and the triviality of $\pi_0(\mathcal{F}_{n-1,1}(\mathbf{R}^2))$ (since $\mathcal{F}_{n-1,1}(\mathbf{R}^2)$ is connected).

Under the isomorphisms $\pi_1(\mathcal{F}_n(\mathbf{R}^2)) \cong P_n$ and $\pi_1(\mathcal{F}_{n-1}(\mathbf{R}^2)) \cong P_{n-1}$, the homomorphism $p_\#$ in (1.9) is identified with the forgetting homomorphism $f_n : P_n \to P_{n-1}$ of Section 1.3.2. We can rewrite (1.9) as

$$1 \longrightarrow \pi_1(\mathcal{F}_{n-1,1}(\mathbf{R}^2)) \longrightarrow P_n \overset{f_n}{\longrightarrow} P_{n-1} \longrightarrow 1. \qquad (1.10)$$

To compute $\pi_1(\mathcal{F}_{n-1,1}(\mathbf{R}^2)) = \pi_1(\mathbf{R}^2 - Q_{n-1})$, we take as $Q_{n-1} \subset \mathbf{R}^2$ the set $\{(1,0),(2,0),\ldots,(n-1,0)\}$ and take $a_0 = (n,0)$ as the base point of $\mathbf{R}^2 - Q_{n-1}$. Clearly, the group $\pi_1(\mathbf{R}^2 - Q_{n-1}, a_0)$ is a free group of rank $n-1$ with the free generators $x_1, \ldots, x_{n-1}$, shown in Figure 1.13.

The homomorphism $\pi_1(\mathcal{F}_{n-1,1}(\mathbf{R}^2)) \to P_n = \pi_1(\mathcal{F}_n(\mathbf{R}^2))$ in (1.10) is induced by the inclusion $\mathbf{R}^2 - Q_{n-1} = \mathcal{F}_{n-1,1}(\mathbf{R}^2) \hookrightarrow \mathcal{F}_n(\mathbf{R}^2)$ assigning to a point $a \in \mathbf{R}^2 - Q_{n-1}$ the tuple of n points $((1,0),(2,0),\ldots,(n-1,0),a)$. Comparing Figures 1.10 and 1.13, we observe that this homomorphism sends x_i to $A_{i,n}$ for all i. Now the exact sequence (1.10) directly implies the claim of Theorem 1.16. $\qquad \square$

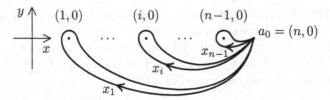

Fig. 1.13. The generators $x_1, \ldots, x_{n-1}$ of $\pi_1(\mathbf{R}^2 - Q_{n-1}, a_0)$

1.4.3 Configuration spaces of nonordered sets of points

Consider again the configuration space $\mathcal{F}_{m,n}(M)$ associated with integers $m \geq 0$, $n \geq 1$ and a connected topological manifold M of dimension ≥ 2. The symmetric group $\mathfrak{S}_n$ acts on $\mathcal{F}_{m,n}(M) = \mathcal{F}_n(M - Q_m)$ by permutation of the coordinates. Consider the quotient space

$$\mathcal{C}_{m,n}(M) = \mathcal{F}_{m,n}(M)/\mathfrak{S}_n.$$

Since the action of $\mathfrak{S}_n$ on $\mathcal{F}_{m,n}(M)$ is fixed-point free, the natural projection $\mathcal{F}_{m,n}(M) \to \mathcal{C}_{m,n}(M)$ is a covering. Hence $\pi_i(\mathcal{F}_{m,n}(M)) \cong \pi_i(\mathcal{C}_{m,n}(M))$ for all $i \geq 2$, and $\mathcal{C}_{m,n}(M)$ is a connected topological manifold of dimension $n \dim(M)$. The points of $\mathcal{C}_{m,n}$ are nonordered sets of n distinct points in $M - Q_m$. The group $\pi_1(\mathcal{C}_{m,n}(M))$ is called the *braid group* of $M - Q_m$ on n strings. We shall write $\mathcal{C}_n(M)$ for $\mathcal{C}_{0,n}(M)$.

For $M = \mathbf{R}^2$, we recover in this way the Artin braid group B_n. Indeed, B_n is canonically isomorphic to $\pi_1(\mathcal{C}_n(\mathbf{R}^2), q)$, where q is the point of $\mathcal{C}_n(\mathbf{R}^2) = \mathcal{C}_{0,n}(\mathbf{R}^2)$ represented by the unordered set

$$\{(1,0), (2,0), \ldots, (n,0)\} \subset \mathbf{R}^2.$$

The isomorphism is obtained by assigning to a geometric braid $b \subset \mathbf{R}^2 \times I$ the loop $I \to \mathcal{C}_n(\mathbf{R}^2)$ sending $t \in I$ into the unique n-point set $b_t \subset \mathbf{R}^2$ such that $b \cap (\mathbf{R}^2 \times \{t\}) = b_t \times \{t\}$.

Corollary 1.29. *For any $n \geq 1$, the braid group B_n is torsion free.*

Proof. Lemma 1.28 with $m = 0$ implies that $\mathcal{F}_n(\mathbf{R}^2)$ is aspherical. Therefore $\pi_i(\mathcal{C}_n(\mathbf{R}^2)) = \pi_i(\mathcal{F}_n(\mathbf{R}^2)) = 0$ for all $i \geq 2$. The following classical argument uses the integral homology of spaces and groups to deduce that $B_n \cong \pi_1(\mathcal{C}_n(\mathbf{R}^2), q)$ is torsion free. If B_n contains a nontrivial finite cyclic subgroup A, then there is a covering $\widetilde{C} \to \mathcal{C}_n(\mathbf{R}^2)$ with $\pi_1(\widetilde{C}) = A$. We have $\pi_i(\widetilde{C}) = \pi_i(\mathcal{C}_n(\mathbf{R}^2)) = 0$ for all $i \geq 2$, so that $\widetilde{C}$ is an Eilenberg–MacLane space $K(A, 1)$. The integral homology groups of $\widetilde{C}$ satisfy $H_i(\widetilde{C}) = H_i(A) = A$ for all odd $i \geq 1$. This contradicts the fact that $\widetilde{C}$ is a manifold of dimension $2n$. $\qquad\Box$

Remark 1.30. Corollary 1.29 can be reformulated by saying that if $\alpha \in B_n$ is an mth root of the trivial braid (i.e., $\alpha^m = 1$) with $m \geq 1$, then $\alpha = 1$. In general, the roots of nontrivial braids are not unique. For example, we have $(\sigma_1\sigma_2)^3 = (\sigma_2\sigma_1)^3$ although $\sigma_1\sigma_2 \neq \sigma_2\sigma_1$. It is known that the mth root of a braid is unique up to conjugacy; see [Gon03].

1.4.4 The space $\mathcal{C}_n(\mathbf{R}^2)$ as a space of polynomials

There is a beautiful description of the configuration space $\mathcal{C}_n(\mathbf{R}^2)$ in terms of polynomials. Identifying $\mathbf{R}^2 = \mathbf{C}$, we obtain

$$\mathcal{F}_n(\mathbf{R}^2) = \mathcal{F}_n(\mathbf{C}) = \{(u_1, u_2, \ldots, u_n) \in \mathbf{C}^n \mid u_i \neq u_j \text{ for } i \neq j\}.$$

Recall the elementary symmetric polynomial of n complex variables

$$p_k(u) = (-1)^k \sum_{1 \leq i_1 < i_2 < \cdots < i_k \leq n} u_{i_1} u_{i_2} \cdots u_{i_k},$$

where $k = 1, 2, \ldots, n$. We consider $p_1, p_2, \ldots, p_n$ as functions on $\mathcal{F}_n(\mathbf{C})$. These functions are invariant under the action of $\mathfrak{S}_n$ on $\mathcal{F}_n(\mathbf{C})$ by permutation of coordinates and thus induce a map $\mathcal{C}_n(\mathbf{R}^2) = \mathcal{C}_n(\mathbf{C}) \rightarrow \mathbf{C}^n$. This map is a homeomorphism onto the set of all $(z_1, z_2, \ldots, z_n) \in \mathbf{C}^n$ such that the polynomial $t^n + z_1 t^{n-1} + z_2 t^{n-2} + \cdots + z_n$ has no multiple roots. The inverse map assigns to each such tuple $(z_1, z_2, \ldots, z_n)$ the nonordered set of roots of the polynomial $t^n + z_1 t^{n-1} + z_2 t^{n-2} + \cdots + z_n$.

Exercise 1.4.1. Prove the following generalization of Lemma 1.28. Let M be a connected surface with $\partial M = \emptyset$ and $m \geq 0$ an integer (if M is homeomorphic to S^2 or to the real projective plane RP^2, then we assume that $m > 0$). Then $\mathcal{F}_{m,n}(M)$ and $\mathcal{C}_{m,n}(M)$ are aspherical for all $n \geq 1$. (Hint: The universal covering of any connected surface $\neq S^2, RP^2$ is homeomorphic to $\mathbf{R}^2$ and therefore is contractible.) Deduce that the groups $\pi_1(\mathcal{F}_{m,n}(M))$ and $\pi_1(\mathcal{C}_{m,n}(M))$ are torsion free.

Exercise 1.4.2. Verify that $\pi_1(\mathcal{F}_2(S^2)) = \{1\}$. (Hint: Use the forgetting fibration $\mathcal{F}_2(S^2) \rightarrow \mathcal{F}_1(S^2) = S^2$.) Deduce that $\pi_1(\mathcal{C}_2(S^2)) \cong \mathbf{Z}/2\mathbf{Z}$.

Exercise 1.4.3. Verify that the map $SO(3) \rightarrow \mathcal{F}_3(S^2)$ sending an element g of the special orthogonal group $SO(3)$ to the triple of vectors

$$(g(1,0,0), g(0,1,0), g(0,0,1)) \in S^2$$

is a homotopy equivalence. Deduce that

$$\pi_1(\mathcal{F}_3(S^2)) \cong \mathbf{Z}/2\mathbf{Z} \quad \text{and} \quad \operatorname{card} \pi_1(\mathcal{C}_3(S^2)) = 12$$

(for a computation of $\pi_1(\mathcal{C}_n(S^2))$ for all n, see [FV62]).

Exercise 1.4.4. Let $U \subset \mathbf{R}^2$ be an open disk. Prove that the inclusion homomorphism $\pi_1(\mathcal{C}_n(U), q) \to \pi_1(\mathcal{C}_n(\mathbf{R}^2), q)$ is an isomorphism for any $q \in \mathcal{C}_n(U)$.

Exercise 1.4.5. Let b be a pure geometric braid in $\mathbf{R}^2 \times I$ and let b' be a "subbraid" formed by several strings of b. Prove that any isotopy of b' in the class of geometric braids extends to an isotopy of b in the class of geometric braids. (Hint: Use Lemma 1.27.)

1.5 Braid automorphisms of free groups

In this section we realize the braid group B_n as a group of automorphisms of the free group F_n on n generators $x_1, x_2, \ldots, x_n$.

1.5.1 Braid automorphisms of F_n

We say that an automorphism φ of F_n is a *braid automorphism* if it satisfies the following two conditions:

(i) there is a permutation μ of the set $\{1, 2, \ldots, n\}$ such that $\varphi(x_k)$ is conjugate in F_n to $x_{\mu(k)}$ for all $k \in \{1, 2, \ldots, n\}$;

(ii) $\varphi(x_1 x_2 \cdots x_n) = x_1 x_2 \cdots x_n$.

To give examples of braid automorphisms of F_n, observe that an endomorphism of F_n is entirely determined by its action on the generators $x_1, \ldots, x_n$. It is straightforward to check that the following formulas define two mutually inverse braid automorphisms $\widetilde{\sigma}_i$ and $\widetilde{\sigma}_i^{-1}$ of F_n for $i = 1, 2, \ldots, n-1$:

$$\widetilde{\sigma}_i(x_k) = \begin{cases} x_{k+1} & \text{if } k = i, \\ x_k^{-1} x_{k-1} x_k & \text{if } k = i+1, \\ x_k & \text{otherwise,} \end{cases}$$

$$\widetilde{\sigma}_i^{-1}(x_k) = \begin{cases} x_k x_{k+1} x_k^{-1} & \text{if } k = i, \\ x_{k-1} & \text{if } k = i+1, \\ x_k & \text{otherwise.} \end{cases}$$

Denote the set of braid automorphisms of F_n by $\widetilde{B}_n$. It follows from the definitions that the inverse of a braid automorphism and the composition of two braid automorphisms are again braid automorphisms. Therefore $\widetilde{B}_n$ is a group with respect to composition $\varphi\psi = \varphi \circ \psi$ for any $\varphi, \psi \in \widetilde{B}_n$.

We now state the main theorem relating braids to braid automorphisms.

Theorem 1.31. *The formula* $\sigma_i \mapsto \widetilde{\sigma}_i$ *with* $i = 1, 2, \ldots, n-1$ *defines a group isomorphism* $B_n \to \widetilde{B}_n$.

The image of $\beta \in B_n$ under the isomorphism $B_n \to \widetilde{B}_n$ will be denoted by $\widetilde{\beta}$. In the proof of Theorem 1.31 we shall give a direct definition of $\widetilde{\beta}$. Yet another interpretation of $\widetilde{\beta}$ will be given in Section 1.6.

Theorem 1.31 gives a solution to the word problem in B_n. For a group defined by generators and relations, the *word problem* consists in finding an algorithm that allows one to decide whether a given word in the generators represents the neutral element of the group. By Theorem 1.31, a braid $\beta \in B_n$ is equal to 1 if and only if $\widetilde{\beta} = \mathrm{id}$. The latter condition can be easily verified; it suffices to check that $\widetilde{\beta}(x_k) = x_k$ for all $k = 1, 2, \ldots, n$.

Abelianizing the action of $B_n \cong \widetilde{B}_n$ on F_n, we obtain an action of B_n on the lattice $F_n/[F_n, F_n] = \mathbf{Z}^n$ with basis $\dot{x}_1, \ldots, \dot{x}_n$ determined by $x_1, \ldots, x_n$. The latter action is determined by the canonical projection $\pi : B_n \to \mathfrak{S}_n$. Indeed, the automorphism of $\mathbf{Z}^n$ induced by $\widetilde{\sigma}_i$ is the transposition of the vectors $\dot{x}_i, \dot{x}_{i+1}$. Therefore for any $\beta \in B_n$, the automorphism of $\mathbf{Z}^n$ induced by $\widetilde{\beta}$ acts as the permutation $\pi(\beta)$ on the vectors $\dot{x}_1, \ldots, \dot{x}_n$.

1.5.2 Proof of Theorem 1.31

The braid relations for $\widetilde{\sigma}_1, \ldots, \widetilde{\sigma}_{n-1} \in \widetilde{B}_n$ can be verified by a direct computation (they follow also from further arguments in this paragraph). Therefore the formula $\sigma_i \mapsto \widetilde{\sigma}_i$ defines a group homomorphism $B_n \to \widetilde{B}_n$. We give now another definition of this homomorphism. Recall the natural inclusion $\iota : B_n \to B_{n+1}$, the group of pure braids $P_{n+1} \subset B_{n+1}$, and the forgetting homomorphism $f_{n+1} : P_{n+1} \to P_n$. If $\beta \in B_n$ and $u \in U_{n+1} = \mathrm{Ker}\, f_{n+1}$, then $\iota(\beta)\, u\, \iota(\beta)^{-1} \in P_{n+1}$ since P_{n+1} is a normal subgroup of B_{n+1}. Moreover, it follows from the definition of f_{n+1} that

$$\iota(\beta)\, u\, \iota(\beta)^{-1} \in U_{n+1}.$$

Therefore the formula $u \mapsto \iota(\beta)\, u\, \iota(\beta)^{-1}$ defines an automorphism of U_{n+1}. We obtain thus a group homomorphism, ξ, from B_n to the group $\mathrm{Aut}\, U_{n+1}$ of automorphisms of U_{n+1}. By Theorem 1.16, we can identify U_{n+1} with F_n by setting $x_k = A_{k,n+1} \in U_{n+1}$ for $k = 1, 2, \ldots, n$. Under this identification, $\xi(\beta) = \widetilde{\beta}$ for all $\beta \in B_n$. Indeed, it suffices to verify this equality for the generators $\sigma_1, \ldots, \sigma_{n-1}$ of B_n. This amounts to checking the equalities

$$\iota(\sigma_i)\, A_{k,n+1}\, \iota(\sigma_i)^{-1} = \begin{cases} A_{k+1,n+1} & \text{if } k = i, \\ A_{k,n+1}^{-1}\, A_{k-1,n+1}\, A_{k,n+1} & \text{if } k = i+1, \\ A_{k,n+1} & \text{otherwise.} \end{cases}$$

These equalities are verified by drawing the corresponding braid diagrams and checking that the diagrams on both sides present isotopic braids.

Let us prove the injectivity of the homomorphism $\beta \mapsto \widetilde{\beta} : B_n \to \widetilde{B}_n$. Consider a braid $\beta \in B_n$ such that $\widetilde{\beta} = 1$. Abelianizing $\widetilde{\beta}$, we obtain the identity automorphism of $U_{n+1}/[U_{n+1}, U_{n+1}]$. Hence, $\pi(\beta) = 1$, so that

$\beta \in P_n \subset B_n$. By formula (1.6), $\beta = \beta_2 \beta_3 \cdots \beta_n$, where $\beta_j \in U_j \subset P_j \subset P_n$ for $j = 2, 3, \ldots, n$. If $\beta \neq 1$, then take the largest $i \leq n$ such that $\beta_i \neq 1$. Then $\beta = \beta_2 \beta_3 \cdots \beta_i$. Since $\widetilde{\beta} = 1$, we must have $\xi(\beta) = 1$, so that $\iota(\beta) \in P_{n+1}$ commutes with all elements of U_{n+1} and in particular with $A_{i,n+1}$. Note that $\beta_2, \beta_3, \ldots, \beta_{i-1}$ are braids on the leftmost $i - 1$ strings. Therefore they commute with $A_{i,n+1}$. Hence β_i commutes with $A_{i,n+1}$. By Corollary 1.23, the braids $A_{1,i}, \ldots, A_{i-1,i}, A_{i,i+1}, \ldots, A_{i,n+1}$ are free generators of a free subgroup of P_{n+1}. We know that β_i commutes with $A_{i,n+1}$ and lies in the group $U_i \subset P_i \subset P_{n+1}$ generated by $A_{1,i}, \ldots A_{i-1,i}$. This is possible only if $\beta_i = 1$, which contradicts the choice of i. Hence, $\beta = 1$.

Let us prove the surjectivity of the homomorphism $\beta \mapsto \widetilde{\beta} : B_n \to \widetilde{B}_n$. Let φ be a nontrivial braid automorphism of F_n. Suppose that

$$\varphi(x_k) = A_k \, x_{\mu(k)} \, A_k^{-1},$$

where $k = 1, 2, \ldots, n$ and A_k is a word in the alphabet $x_1^{\pm 1}, \ldots, x_n^{\pm 1}$. We can always choose A_k so that the product $A_k \, x_{\mu(k)} \, A_k^{-1}$ is reduced, that is, does not contain consecutive entries $x_r x_r^{-1}$ or $x_r^{-1} x_r$. By the definition of a braid automorphism,

$$A_1 \, x_{\mu(1)} \, A_1^{-1} A_2 \, x_{\mu(2)} \, A_2^{-1} \cdots A_n \, x_{\mu(n)} \, A_n^{-1} = x_1 x_2 \cdots x_n. \qquad (1.11)$$

We claim that there exist $j \in \{1, 2, \ldots, n-1\}$ and a word A (possibly empty) in $x_1^{\pm 1}, \ldots, x_n^{\pm 1}$ satisfying one of the following two conditions:
 (a) we have an equality of words $A_j = A_{j+1} \, x_{\mu(j+1)} \, A$,
 (b) we have an equality of words $A_{j+1} = A_j \, x_{\mu(j)}^{-1} \, A$.

This claim will imply that φ lies in the image of the homomorphism $\beta \mapsto \widetilde{\beta}$. To see this, define the *length* of φ to be the sum of the letter lengths of the words $A_k \, x_{\mu(k)} \, A_k^{-1}$ over $k = 1, 2, \ldots, n$. If (a) holds, then the homomorphism

$$\varphi \widetilde{\sigma}_j = \varphi \circ \widetilde{\sigma}_j : F_n \to F_n$$

can be computed as follows. Both φ and $\varphi \widetilde{\sigma}_j$ have the same effect on x_k for $k \neq j, j+1$ and

$$\varphi \widetilde{\sigma}_j(x_j) = \varphi(x_{j+1}) = A_{j+1} \, x_{\mu(j+1)} \, A_{j+1}^{-1},$$

$$\begin{aligned}
\varphi \widetilde{\sigma}_j(x_{j+1}) &= \varphi(x_{j+1}^{-1} x_j x_{j+1}) \\
&= A_{j+1} \, x_{\mu(j+1)}^{-1} \, A_{j+1}^{-1} A_j \, x_{\mu(j)} A_j^{-1} \, A_{j+1} \, x_{\mu(j+1)} \, A_{j+1}^{-1} \\
&= A_{j+1} \, x_{\mu(j+1)}^{-1} \, A_{j+1}^{-1} \\
&\quad \times A_{j+1} \, x_{\mu(j+1)} \, A \, x_{\mu(j)} \, A^{-1} \, x_{\mu(j+1)}^{-1} \, A_{j+1}^{-1} A_{j+1} \, x_{\mu(j+1)} \, A_{j+1}^{-1} \\
&= A_{j+1} A \, x_{\mu(j)} \, A^{-1} A_{j+1}^{-1}.
\end{aligned}$$

The word $A_{j+1} A$ is shorter than $A_j = A_{j+1} x_{\mu(j+1)} A$. Therefore $\varphi \widetilde{\sigma}_j$ has shorter length than φ. Similarly, if (b) holds, then $\varphi \widetilde{\sigma}_j^{-1}$ has shorter length than φ. This implies that φ can be reduced to the identity by repeated composition with appropriate $\widetilde{\sigma}_j$ or $\widetilde{\sigma}_j^{-1}$. Thus, φ is a power product of the $\widetilde{\sigma}_j$. Hence φ lies in the image of the homomorphism $\beta \mapsto \widetilde{\beta}$.

It remains to prove the claim stated above. Let us call the term $x_{\mu(k)}$ appearing in the middle of $A_k\, x_{\mu(k)}\, A_k^{-1}$ *special*. Each letter $x_1, \ldots, x_n$ appears as a special term on the left-hand side of (1.11) exactly once. Equality (1.11) implies that the left-hand side of (1.11) reduces to the right-hand side after all possible free reductions (i.e., cancellations $x_r x_r^{-1} = x_r^{-1} x_r = 1$). Suppose that a special term $x_{\mu(k)}$ is canceled with a letter $x_{\mu(k)}^{-1}$ during these reductions. This $x_{\mu(k)}^{-1}$ cannot come from the word $A_k\, x_{\mu(k)}\, A_k^{-1}$, which was assumed to be reduced. If this $x_{\mu(k)}^{-1}$ comes from A_{k-1}^{-1}, then we must have an equality of words $A_{k-1}^{-1} = B\, x_{\mu(k)}^{-1}\, A_k^{-1}$ for some word B. Then (a) holds for $j = k-1$ and $A = B^{-1}$. If the letter $x_{\mu(k)}^{-1}$ canceling the special term $x_{\mu(k)}$ comes from the right of the special term $x_{\mu(k+1)}$, then we must have (a) for $j = k$. Similarly, if $x_{\mu(k)}^{-1}$ comes from A_{k+1} or from the left of the special term $x_{\mu(k-1)}$, then (b) holds. If the special terms on the left-hand side of (1.11) do not cancel with other letters, then we must have $\mu(k) = k$ for all k, A_1 and A_n are empty words, and each pair $A_k^{-1} A_{k+1}$ cancels out, so that $A_k = A_{k+1}$ for all k. Then $\varphi = \mathrm{id}$, which contradicts our choice of φ. $\square$

Remark 1.32. Theorem 1.31 yields another proof of the residual finiteness of B_n (Corollary 1.21). Indeed, by the Baumslag–Smirnov theorem [Bau63], [Smi63], the group of automorphisms of an arbitrary residually finite group is residually finite. Since F_n is residually finite, its group of automorphisms and all subgroups of this group are residually finite.

Exercise 1.5.1. For any integer r, let $\widetilde{\sigma}_i^{(r)}$ be the automorphism of F_n defined by the same formulas as $\widetilde{\sigma}_i$ with the only difference that

$$\widetilde{\sigma}_i^{(r)}(x_{i+1}) = x_{i+1}^{-r} x_i x_{i+1}^r\,.$$

Verify that $\widetilde{\sigma}_1^{(r)}, \widetilde{\sigma}_2^{(r)}, \ldots, \widetilde{\sigma}_{n-1}^{(r)}$ satisfy the braid relations. (The resulting representation $B_n \to \mathrm{Aut}(F_n)$ is faithful for all $r \neq 0$; see [Shp01].)

Exercise 1.5.2. Let F_{2n} be the free group on $2n$ generators $a_1, \ldots, a_n$, $b_1, \ldots, b_n$. For $j = 1, \ldots, 2n+1$, define an automorphism σ_j' of F_{2n} as follows. For even $j = 2i$, set $\sigma_j'(a_i) = b_i^{-1} a_i$, $\sigma_j'(a_k) = a_k$ for $k \neq i$, and $\sigma_j'(b_k) = b_k$ for all k. If j is odd, then $\sigma_j'(a_k) = a_k$ for all k. Also, $\sigma_1'(b_1) = a_1 b_1$ and $\sigma_1'(b_k) = b_k$ for $k > 1$; $\sigma_{2n+1}'(b_n) = b_n a_n$ and $\sigma_{2n+1}'(b_k) = b_k$ for $k < n$. For other odd $j = 2i + 1$, set $\sigma_j'(b_i) = b_i a_i a_{i+1}^{-1}$, $\sigma_j'(b_{i+1}) = a_{i+1} a_i^{-1} b_{i+1}$, and $\sigma_j'(b_k) = b_k$ for $k \neq i, i+1$. Verify that $\sigma_1', \ldots, \sigma_{2n+1}'$ satisfy the braid relations. Check that the corresponding group homomorphism $B_{2n+2} \to \mathrm{Aut}(F_{2n})$ sends the center of B_{2n+2} to the identity. (For $n = 1$, we recover the formulas of Exercise 1.1.10.)

1.6 Braids and homeomorphisms

We discuss an approach to braids based on their interpretation as isotopy classes of homeomorphisms of a 2-dimensional disk.

1.6.1 Mapping class groups

Let M be an oriented topological manifold (possibly with boundary ∂M). Let Q be a finite (possibly empty) subset of the interior $M^\circ = M - \partial M$ of M. By a *self-homeomorphism* of the pair (M, Q) we mean a homeomorphism $f : M \to M$ that fixes ∂M pointwise, fixes Q setwise, and preserves the orientation of M. The first two conditions mean that $f(x) = x$ for all $x \in \partial M$ and $f(Q) = Q$. Any self-homeomorphism of (M, Q) induces a permutation on Q, which may be trivial or not. Note that if M is connected and has a nonempty boundary, then any homeomorphism $M \to M$ fixing ∂M pointwise automatically preserves the orientation of M.

Two self-homeomorphisms of (M, Q) are *isotopic* if they can be included in a continuous one-parameter family of self-homeomorphisms of (M, Q). More precisely, two self-homeomorphisms f_0, f_1 of (M, Q) are isotopic if they can be included in a family $\{f_t\}_{t \in I}$ of self-homeomorphisms of (M, Q) such that the map $M \times I \to M$ sending (x, t) with $x \in M, t \in I$ into $f_t(x)$ is continuous. Such a family is called an *isotopy* of f_0 into f_1. It is clear that the isotopy of self-homeomorphisms of (M, Q) is an equivalence relation and that isotopic self-homeomorphisms induce the same permutation on Q.

The *mapping class group* $\mathfrak{M}(M, Q)$ of (M, Q) is the group of isotopy classes of self-homeomorphisms of (M, Q) with multiplication determined by composition: $fg = f \circ g$ for $f, g \in \mathfrak{M}(M, Q)$. Set $\mathfrak{M}(M) = \mathfrak{M}(M, \emptyset)$.

An important example, in which the group $\mathfrak{M}(M)$ can be easily computed, is that of a ball. For a closed ball $D = D^n$ of dimension $n \geq 0$, we have $\mathfrak{M}(D) = \{1\}$. This follows from the classical Alexander–Tietze theorem: any self-homeomorphism of D is isotopic to the identity in the class of self-homeomorphisms of D. Here is a proof of this theorem. We can assume D to be the unit ball in $\mathbf{R}^n$ centered at the origin 0. Denote the Euclidean norm of a vector $z \in \mathbf{R}^n$ by $|z|$. For any self-homeomorphism h of D, the formula

$$h_t(z) = \begin{cases} z & \text{if } t \leq |z| \leq 1, \\ t\, h(z/t) & \text{if } |z| < t \end{cases}$$

defines an isotopy $\{h_t : D \to D\}_{t \in I}$ of $h_0 = \mathrm{id}$ to $h_1 = h$. Note that if $h(0) = 0$, then $h_t(0) = 0$ for all $t \in I$. Therefore we also have $\mathfrak{M}(D, \{0\}) = \{1\}$.

The study of the mapping class groups leads to a vast and ramified theory; see [Iva02] for a recent survey of the mapping class groups of surfaces. We shall focus on one series of mapping class groups arising when M is a 2-disk and Q is an n-point subset of M°, where $n = 1, 2, \ldots$. It turns out that the resulting group $\mathfrak{M}(M, Q)$ is nothing but the braid group B_n. The rest of this section is devoted to an exact formulation of this claim.

1.6.2 Half-twists

Let M be an oriented surface (possibly with boundary) and let Q be a finite subset of M°. By a *spanning arc* on (M, Q), we mean a subset of M homeomorphic to $I = [0, 1]$ and disjoint from $Q \cup \partial M$ except at its two endpoints, which should lie in Q. Let us stress that all arcs considered here are simple, i.e., have no self-intersections.

Let $\alpha \subset M$ be a spanning arc on (M, Q). The *half-twist*

$$\tau_\alpha : (M, Q) \to (M, Q)$$

is obtained as the result of the isotopy of the identity map id $: M \to M$ rotating α in M about its midpoint by the angle π in the direction provided by the orientation of M. The half-twist τ_α is the identity outside a small neighborhood of α in M. Clearly, $\tau_\alpha(\alpha) = \alpha$, $\tau_\alpha(Q) = Q$, and τ_α induces a transposition on Q permuting the endpoints of α. Note that rotating α as above but in the opposite direction, we obtain τ_α^{-1}.

For completeness, we give a more formal definition of τ_α. Let us identify a small neighborhood U of α with the open unit disk $\{z \in \mathbf{C} \,|\, |z| < 1\}$ so that $\alpha = [-1/2, 1/2]$ and the orientation in M corresponds to the counterclockwise orientation in $\mathbf{C}$. The homeomorphism $\tau_\alpha : M \to M$ is the identity outside U, sends any $z \in \mathbf{C}$ with $|z| \leq 1/2$ to $-z$, and sends any $z \in \mathbf{C}$ with $1/2 \leq |z| < 1$ to $\exp(-2\pi i |z|)z$. Clearly, $\tau_\alpha \in \mathfrak{M}(M, Q)$ does not depend on the choice of U. The action of τ_α on a curve on M transversely meeting α in one point is shown in Figure 1.14.

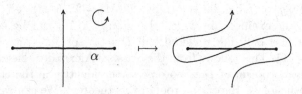

Fig. 1.14. The action of τ_α on a transversal curve

We state a few properties of half-twists.

(i) If $f : (M, Q) \to (M', Q')$ is an orientation-preserving homeomorphism of two pairs as above and α is a spanning arc on (M, Q), then $f(\alpha)$ is a spanning arc on (M', Q') and $\tau_{f(\alpha)} = f \tau_\alpha f^{-1} \in \mathfrak{M}(M', Q')$.

This property is obvious. Informally speaking, it says that applying the construction of a half-twist on two copies of the same surface, we obtain two copies of the same homeomorphism.

(ii) If two spanning arcs α, α' on (M, Q) are isotopic in the class of spanning arcs on (M, Q) (in particular they must have the same endpoints), then $\tau_\alpha = \tau_{\alpha'}$ in $\mathfrak{M}(M, Q)$.

Indeed, if α, α' are isotopic, then there is a self-homeomorphism f of (M, Q) that is the identity on Q, is isotopic to the identity, and sends α onto α'. By (i),

$$\tau_{\alpha'} = \tau_{f(\alpha)} = f\tau_\alpha f^{-1}.$$

Since f is isotopic to the identity, $f\tau_\alpha f^{-1} = \tau_\alpha$.

(iii) A self-homeomorphism of (M, Q) induces a self-homeomorphism of M by forgetting Q. The resulting group homomorphism $\mathfrak{M}(M, Q) \to \mathfrak{M}(M)$ sends τ_α to 1. This is clear from the definitions.

(iv) If α, β are disjoint spanning arcs on (M, Q), then

$$\tau_\alpha \tau_\beta = \tau_\beta \tau_\alpha \in \mathfrak{M}(M, Q). \tag{1.12}$$

This is obtained by using disjoint neighborhoods of α, β in the construction of τ_α, τ_β.

(v) For any two spanning arcs α, β on (M, Q) that share one common endpoint and are disjoint otherwise,

$$\tau_\alpha \tau_\beta \tau_\alpha = \tau_\beta \tau_\alpha \tau_\beta \in \mathfrak{M}(M, Q). \tag{1.13}$$

To prove this fundamental formula, we begin with the equality

$$\tau_\alpha(\beta) = \tau_\beta^{-1}(\alpha),$$

which can be verified by drawing the arcs $\tau_\alpha(\beta)$ and $\tau_\beta^{-1}(\alpha)$. The equality here is understood as isotopy in the class of spanning arcs on (M, Q). By (ii),

$$\tau_{\tau_\alpha(\beta)} = \tau_{\tau_\beta^{-1}(\alpha)}.$$

By (i), this implies $\tau_\alpha \tau_\beta \tau_\alpha^{-1} = \tau_\beta^{-1} \tau_\alpha \tau_\beta$. This is equivalent to (1.13).

1.6.3 The isomorphism $B_n \cong \mathfrak{M}(D, Q_n)$

For $n \geq 1$, let $Q_n \subset \mathbf{R}^2$ be the n-point set $\{(1,0), (2,0), \ldots, (n,0)\}$. Let D be a closed Euclidean disk in $\mathbf{R}^2$ containing the set Q_n in its interior. We orient D counterclockwise. For every $i = 1, 2, \ldots, n-1$, consider the arc

$$\alpha_i = [i, i+1] \times \{0\} \subset D.$$

This arc meets Q_n only at its endpoints and hence gives rise to a half-twist

$$\tau_{\alpha_i} \in \mathfrak{M}(D, Q_n).$$

Formulas (1.12) and (1.13) imply that $\tau_{\alpha_1}, \ldots, \tau_{\alpha_{n-1}}$ satisfy the braid relations of Section 1.1. By Lemma 1.2, there is a group homomorphism

$$\eta : B_n \to \mathfrak{M}(D, Q_n)$$

such that $\eta(\sigma_i) = \tau_{\alpha_i}$ for all $i = 1, \ldots, n-1$.

Recall the group of braid automorphisms $\widetilde{B}_n$ defined in Section 1.5.1. We now define a certain group homomorphism $\rho : \mathfrak{M}(D, Q_n) \to \widetilde{B}_n$. Pick a base point $d \in \partial D$ as in Figure 1.15. It is clear that the fundamental group $\pi_1(D - Q_n, d)$ is a free group F_n of rank n with generators $x_1, x_2, \ldots, x_n$ represented by the loops $X_1, X_2, \ldots, X_n$ shown in Figure 1.15. Every self-homeomorphism f of (D, Q_n) can be restricted to $D - Q_n$ and yields in this way a self-homeomorphism of $D - Q_n$. The latter sends $d \in \partial D$ to itself and induces a group automorphism $\rho(f)$ of $F_n = \pi_1(D - Q_n, d)$. This automorphism depends only on the isotopy class of f: if two self-homeomorphisms of (D, Q_n) are isotopic, then their restrictions to $D - Q_n$ are isotopic relative to ∂D, and therefore they induce the same automorphism of F_n.

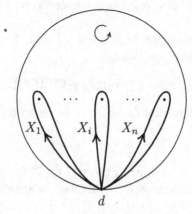

Fig. 1.15. The loops $X_1, \ldots, X_n$ on $D - Q_n$

Let us verify that $\rho(f)$ is a braid automorphism of F_n. The loop X_k in Figure 1.15 can be deformed in $D - Q_n$ into a small loop encircling clockwise the point $(k, 0)$. The homeomorphism f maps the latter loop onto a small loop encircling clockwise the point $(\mu(k), 0)$ for some $\mu(k) \in \{1, 2, \ldots, n\}$. This small loop can be deformed into the loop $X_{\mu(k)}$ in $D - Q_n$. Hence, the loop $f(X_k)$ can be deformed into $X_{\mu(k)}$ in $D - Q_n$. (Under the deformation, the base point $f(d) = d$ may move in $D - Q_n$.) This implies that the homotopy classes of these two loops $\rho(f)(x_k)$ and $x_{\mu(k)}$ are conjugate in $\pi_1(D - Q_n, d)$. This verifies Condition (i) in the definition of a braid automorphism. Condition (ii) follows from the fact that the product $x_1 x_2 \cdots x_n$ is represented by the loop ∂D based at d. This loop is preserved by f pointwise, and therefore its homotopy class in $\pi_1(D - Q_n, d)$ is invariant under $\rho(f)$.

We conclude that the formula $f \mapsto \rho(f)$ defines a map ρ from $\mathfrak{M}(D, Q_n)$ to $\widetilde{B}_n$. This map is a group homomorphism, since

$$\rho(fg) = \rho(f \circ g) = \rho(f) \circ \rho(g) = \rho(f)\,\rho(g),$$

for any $f, g \in \mathfrak{M}(D, Q_n)$.

We can now state the main theorem relating braids to homeomorphisms.

Theorem 1.33. *For any $n \geq 1$, the homomorphisms η and ρ are isomorphisms. The following diagram is commutative:*

$$\begin{array}{ccc} B_n & & \\ \eta \downarrow & \searrow^{\beta \mapsto \widetilde{\beta}} & \\ \mathfrak{M}(D, Q_n) & \xrightarrow{\rho} & \widetilde{B}_n \end{array} \qquad (1.14)$$

where $\beta \mapsto \widetilde{\beta} : B_n \to \widetilde{B}_n$ is the isomorphism defined in Section 1.5.

This fundamental theorem allows us to identify B_n with the mapping class group $\mathfrak{M}(D, Q_n)$. We have by now three different geometric interpretations of B_n: via geometric braids on n strings, via the configuration space of n points in the plane, and via the group of homeomorphisms of a 2-disk with n distinguished points. It is this variety of geometric facets of B_n that makes this group so appealing.

The commutativity of the diagram (1.14) means that $\widetilde{\beta} = \rho(\eta(\beta))$ for any $\beta \in B_n$. This can be verified at once. Since ρ, η, and $\beta \mapsto \widetilde{\beta}$ are group homomorphisms, it suffices to verify this equality for the generators $\sigma_1, \ldots, \sigma_{n-1}$. We need to check that $\rho(\tau_{\alpha_i}) = \widetilde{\sigma}_i$ for $i = 1, 2, \ldots, n-1$. The formulas $\rho(\tau_{\alpha_i})(x_k) = x_k$ for $k \neq i, i+1$ and $\rho(\tau_{\alpha_i})(x_i) = x_{i+1}$ follow directly from the definition of τ_{α_i}. The equality $\rho(\tau_{\alpha_i})(x_{i+1}) = x_{i+1}^{-1} x_i x_{i+1}$ can be verified directly or deduced from the formula $\rho(\tau_{\alpha_i})(x_1 \cdots x_n) = x_1 \cdots x_n$. Hence, we have $\rho(\tau_{\alpha_i}) = \widetilde{\sigma}_i$. In view of the commutativity of the diagram (1.14) and Theorem 1.31, to prove Theorem 1.33 we need only show that η is an isomorphism. This will be done in Section 1.7.

It is clear that for all $i = 1, \ldots, n-1$, the half-twist $\tau_{\alpha_i} : D \to D$ is a diffeomorphism with respect to the standard smooth structure on D induced by that on $\mathbf{R}^2$. Integral powers of diffeomorphisms and their products are also diffeomorphisms. Therefore the surjectivity of η implies the following assertion.

Corollary 1.34. *An arbitrary self-homeomorphism of the pair (D, Q_n) is isotopic in the class of self-homeomorphisms of this pair to a diffeomorphism $(D, Q_n) \to (D, Q_n)$.*

Exercise 1.6.1. Let M, Q be as in Section 1.6.2.

(a) Consider an embedded r-gon $P \subset M$ (with $r \geq 3$) meeting Q precisely in its vertices. Moving along ∂P in the direction provided by the orientation of M, we meet consecutively r edges, say $\alpha_1, \alpha_2, \ldots, \alpha_r$, of P. Each α_i is a spanning arc on (M, Q). Prove that

$$\tau_{\alpha_1} \tau_{\alpha_2} \cdots \tau_{\alpha_{r-1}} = \tau_{\alpha_2} \tau_{\alpha_3} \cdots \tau_{\alpha_r} .$$

(Hint: For $r = 3$ rewrite as $\tau_{\alpha_2}^{-1} \tau_{\alpha_1} \tau_{\alpha_2} = \tau_{\alpha_3}$; for $r \geq 4$ use induction.)

(b) Consider $r \geq 2$ spanning arcs on (M, Q) with one common endpoint $a \in Q$ and disjoint otherwise. Moving around a in the direction given by the orientation of M, denote these arcs by $\alpha_1, \alpha_2, \ldots, \alpha_r$. Prove that

$$\tau_{\alpha_1}^{-1} \tau_{\alpha_2} \tau_{\alpha_1} = \tau_{\alpha_2} \tau_{\alpha_1} \tau_{\alpha_2}^{-1}$$

commutes with τ_{α_i} for $3 \leq i \leq r$. Deduce that

$$\tau_{\alpha_1} \beta \, \tau_{\alpha_2} \tau_{\alpha_1} = \tau_{\alpha_2} \tau_{\alpha_1} \beta \, \tau_{\alpha_2}$$

for any element β of the group generated by $\tau_{\alpha_3}, \tau_{\alpha_4}, \ldots, \tau_{\alpha_r}$.

Exercise 1.6.2. Prove that $\mathfrak{M}(S^1) = \{1\}$. (Hint: Composing an arbitrary self-homeomorphism f of S^1 with a rotation of S^1 into itself, we can assume that f has a fixed point. Cutting out S^1 at a fixed point of f, we obtain a self-homeomorphism of a closed interval, which, as we know, is isotopic to the identity.)

1.7 Groups of homeomorphisms vs. configuration spaces

We discuss groups of homeomorphisms of manifolds, their relations to configuration spaces, and applications to braids.

1.7.1 Groups of homeomorphisms

Let M be a compact connected oriented topological manifold (possibly with boundary) and let Q be a finite subset of $M^\circ = M - \partial M$. Denote by $\mathrm{Top}(M, Q)$ the group of all self-homeomorphisms of (M, Q), i.e., the group of all orientation-preserving self-homeomorphisms of M that fix ∂M pointwise and fix Q setwise. The multiplication in $\mathrm{Top}(M, Q)$ is given by composition: $fg = f \circ g$ for $f, g \in \mathrm{Top}(M, Q)$. We provide $\mathrm{Top}(M, Q)$ with the compact-open topology. For completeness, we recall the definition and basic properties of this topology, referring for proofs to [FR84, Sect. 2.7 of Chap. 1 and Sect. 2 of Chap. 4] or [Kel55]. For a compact set $K \subset M$ and an open set $U \subset M$, put

$$N(K, U) = \{f \in \mathrm{Top}(M, Q) \mid f(K) \subset U\}.$$

Such sets $N(K, U)$ as well as the intersections of a finite number of such sets and arbitrary unions of such intersections are declared *open subsets* of $\mathrm{Top}(M, Q)$. This defines the *compact-open* topology on $\mathrm{Top}(M, Q)$ and makes $\mathrm{Top}(M, Q)$ into a topological group. Here, the inversion $f \mapsto f^{-1}$ in $\mathrm{Top}(M, Q)$ is continuous because of the obvious equality

$$\{f^{-1} \mid f \in N(K, U)\} = N(M - U, M - K).$$

It is here that we need the compactness of M.

It is known that a map f from a topological space X to $\mathrm{Top}(M,Q)$ is continuous if and only if the map $X \times M \to M$ sending $(x,y) \in X \times M$ to $f(x)(y)$ is continuous. Applying this to $X = I$, we conclude that two self-homeomorphisms of (M,Q) are isotopic if and only if they can be connected by a path in $\mathrm{Top}(M,Q)$, i.e., if and only if they lie in the same connected component of $\mathrm{Top}(M,Q)$. Therefore,

$$\mathfrak{M}(M,Q) = \pi_0(\mathrm{Top}(M,Q)). \tag{1.15}$$

Set $\mathrm{Top}(M) = \mathrm{Top}(M,\emptyset)$. The obvious embedding $\mathrm{Top}(M,Q) \hookrightarrow \mathrm{Top}(M)$ makes $\mathrm{Top}(M,Q)$ into a closed subgroup of the topological group $\mathrm{Top}(M)$.

The group $\mathrm{Top}(M)$ is intimately related to the configuration spaces of nonordered points of M° introduced in Section 1.4.3. For $n \geq 1$, set

$$\mathcal{C}_n = \mathcal{C}_n(M^\circ) = \mathcal{F}_n(M^\circ)/\mathfrak{S}_n.$$

To describe the relation between $\mathrm{Top}(M)$ and $\mathcal{C}_n$, pick a set $Q \subset M^\circ$ consisting of n points. We define an *evaluation map* $e = e_Q : \mathrm{Top}(M) \to \mathcal{C}_n$ by $e(f) = f(Q)$, where $f \in \mathrm{Top}(M)$. It is easy to deduce from the definitions that e is a surjective continuous map.

Lemma 1.35. *The map* $e : \mathrm{Top}(M) \to \mathcal{C}_n$ *is a locally trivial fibration with fiber* $\mathrm{Top}(M,Q)$.

Proof. Let $\mathcal{F}_n = \mathcal{F}_n(M^\circ)$ be the configuration space of n ordered points in M°. We can factor e as the composition of a map $c : \mathrm{Top}(M) \to \mathcal{F}_n$ with the covering $\mathcal{F}_n \to \mathcal{C}_n$. To construct c, fix an order in the set Q and define c by $c(f) = f(Q)$, where $f \in \mathrm{Top}(M)$ and the order in $f(Q)$ is induced by the one in Q. To prove the lemma, it suffices to prove that c is a locally trivial fibration. The proof of the latter is very similar to the proof of Lemma 1.26. Let us prove the local triviality of c in a neighborhood of a point $u^0 = (u_1^0, \ldots, u_n^0) \in \mathcal{F}_n$. For each $i = 1, 2, \ldots, n$, pick an open neighborhood $U_i \subset M^\circ$ of u_i^0 such that its closure $\overline{U}_i$ is a closed ball with interior U_i. Since the points $u_1^0, \ldots, u_n^0$ are distinct, we can assume that $U_1, \ldots, U_n$ are mutually disjoint. Then $U = U_1 \times U_2 \times \cdots \times U_n$ is a neighborhood of u^0 in $\mathcal{F}_n$. We shall prove that the restriction of c to U is a trivial bundle. For every $i = 1, 2, \ldots, n$, there is a continuous map $\theta_i : U_i \times \overline{U}_i \to \overline{U}_i$ such that setting $\theta_i^u(v) = \theta_i(u,v)$, we obtain a homeomorphism $\theta_i^u : \overline{U}_i \to \overline{U}_i$ that sends u_i^0 to u and fixes the sphere $\partial \overline{U}_i$ pointwise (see the proof of Lemma 1.26). For $u = (u_1, \ldots, u_n) \in U$, we define a homeomorphism $\theta^u : M \to M$ by $\theta^u(v) = \theta_i^{u_i}(v)$ if $v \in U_i$ with $i = 1, 2, \ldots, n$ and $\theta^u(v) = v$ if $v \in M - \bigcup_i U_i$. It is clear that $\theta^u : M \to M$ sends $u_1^0, \ldots, u_n^0$ to $u_1, \ldots, u_n$, respectively. Observe that $c^{-1}(u^0)$ is the closed subgroup of $\mathrm{Top}(M)$ consisting of all $f \in \mathrm{Top}(M)$ such that $f(u_i^0) = u_i^0$ for $i = 1, 2, \ldots, n$. The formula $(u,f) \mapsto \theta^u f$ defines a homeomorphism $U \times c^{-1}(u^0) \to c^{-1}(U)$ commuting with the projections to U. The inverse homeomorphism sends any $g \in c^{-1}(U)$ to the pair $(c(g), (\theta^{c(g)})^{-1} g) \in U \times c^{-1}(u^0)$. $\square$

Remark 1.36. Two elements of $\text{Top}(M)$ have the same image under the evaluation map e if and only if they lie in the same left coset of $\text{Top}(M, Q)$ in $\text{Top}(M)$. Although we shall not need it, note that e induces a homeomorphism from the quotient homogeneous space $\text{Top}(M)/\text{Top}(M, Q)$ onto $\mathcal{C}_n$.

1.7.2 Parametrizing isotopies

We show here that geometric braids naturally give rise to one-parameter families of homeomorphisms of the 2-disk. This construction will be instrumental in the sequel.

Let $n \geq 1$ and $D \subset \mathbf{R}^2$ be a closed Euclidean disk containing the set

$$Q = Q_n = \{(1,0), (2,0), \ldots, (n,0)\} \tag{1.16}$$

in its interior. An isotopy $\{f_t : D \to D\}_{t \in I}$ in the class of self-homeomorphisms of D is *normal* if $f_0(Q) = Q$ and $f_1 = \text{id}_D$. In other words, a normal isotopy is a path in $\text{Top}(D)$ leading from a point of $\text{Top}(D, Q)$ to the identity homeomorphism $\text{id}_D \in \text{Top}(D)$. For any normal isotopy $\{f_t : D \to D\}_{t \in I}$, the set

$$\bigcup_{t \in I} (f_t(Q), t) \subset \mathbf{R}^2 \times I$$

is a geometric braid on n strings. We say that the isotopy $\{f_t\}_{t \in I}$ *parametrizes* this geometric braid.

Lemma 1.37. *For any geometric braid $b \subset D^\circ \times I$ on n strings, there is a normal isotopy parametrizing b.*

Proof. Consider the evaluation map $e = e_Q : \text{Top}(D) \to \mathcal{C}_n = \mathcal{C}_n(D^\circ)$ sending $f \in \text{Top}(D)$ to $f(Q)$. As already observed in Section 1.4.3, the braid b gives rise to a loop $f^b : I \to \mathcal{C}_n$ sending any $t \in I$ into the unique n-point subset b_t of $\mathbf{R}^2$ such that $b \cap (\mathbf{R}^2 \times \{t\}) = b_t \times \{t\}$. This loop begins and ends at the point $q = e(\text{id}_D) \in \mathcal{C}_n$ represented by Q. By Lemma 1.35 and the homotopy lifting property of locally trivial fibrations (see Appendix B), the loop f^b lifts to a path $\widehat{f^b} : I \to \text{Top}(D)$ beginning at a point of $e^{-1}(q) = \text{Top}(D, Q)$ and ending at id_D. The path $\widehat{f^b}$ is a normal isotopy and the equality $e\widehat{f^b} = f^b$ means that this isotopy parametrizes b. $\square$

1.7.3 Proof of Theorem 1.33

Let D, $Q = Q_n$, $\mathcal{C}_n = \mathcal{C}_n(D^\circ)$, $e = e_Q : \text{Top}(D) \to \mathcal{C}_n$, and $q = e(\text{id}_D) \in \mathcal{C}_n$ be the same objects as in Section 1.7.2. By Lemma 1.35, e is a locally trivial fibration with $e^{-1}(q) = \text{Top}(D, Q)$. This fibration induces a mapping

$$\partial : \pi_1(\mathcal{C}_n, q) \to \pi_0(\text{Top}(D, Q)) = \mathfrak{M}(D, Q).$$

Recall the definition of ∂ following Appendix B. Let $\beta \in \pi_1(\mathcal{C}_n, q)$ be represented by a loop $f : I \to \mathcal{C}_n$ beginning and ending at q. By the homotopy lifting property of e, this loop lifts to a path $\widehat{f} : I \to \mathrm{Top}(D)$ beginning at a point of $e^{-1}(q) = \mathrm{Top}(D, Q)$ and ending at id_D. Then $\partial(\beta) = [\widehat{f}(0)] \in \pi_0(\mathrm{Top}(D, Q))$ is the homotopy class of $\widehat{f}(0)$. That $\partial(\beta)$ depends only on β can be seen directly: if f' is another loop representing β, then a homotopy between f, f' lifts to a homotopy between arbitrary lifts $\widehat{f}, \widehat{f'}$ in $\mathrm{Top}(D)$. This homotopy yields a path in $\mathrm{Top}(D, Q)$ connecting $\widehat{f}(0)$ to $\widehat{f'}(0)$. Hence $[\widehat{f}(0)] = [\widehat{f'}(0)]$.

The mapping $\partial : \pi_1(\mathcal{C}_n, q) \to \mathfrak{M}(D, Q)$ is a group homomorphism. Indeed, consider two loops f, g in $\mathcal{C}_n$ beginning and ending at q and representing $\beta, \gamma \in \pi_1(\mathcal{C}_n, q)$, respectively. Let $\widehat{f}, \widehat{g} : I \to \mathrm{Top}(D)$ be lifts of f, g ending at id_D. Observe that for any $t \in I$,

$$e\left(\widehat{f}(t)\,\widehat{g}(0)\right) = \widehat{f}(t)\,\widehat{g}(0)(Q) = \widehat{f}(t)(Q) = f(t).$$

Therefore the path $t \mapsto \widehat{f}(t)\,\widehat{g}(0) : I \to \mathrm{Top}(D)$ is a lift of f ending at $\widehat{f}(1)\,\widehat{g}(0) = \widehat{g}(0)$. The product of this path with $\widehat{g}$ is a lift of $fg : I \to \mathcal{C}_n$ ending at id_D and beginning at $\widehat{f}(0)\,\widehat{g}(0)$. Thus,

$$\partial(\beta\gamma) = [\widehat{f}(0)\,\widehat{g}(0)] = [\widehat{f}(0)]\,[\widehat{g}(0)] = \partial(\beta)\,\partial(\gamma).$$

Recall that $B_n = \pi_1(\mathcal{C}_n, q)$; see Exercise 1.4.4. The homomorphism $\partial : B_n = \pi_1(\mathcal{C}_n, q) \to \mathfrak{M}(D, Q)$ can be described in terms of parametrizing isotopies as follows. If b is a geometric braid representing $\beta \in B_n$, then for any normal isotopy $\{f_t : D \to D\}_{t \in I}$ parametrizing b as in Section 1.7.2, $\partial(\beta) \in \mathfrak{M}(D, Q)$ is the isotopy class of $f_0 : (D, Q) \to (D, Q)$.

We claim that $\partial = \eta$, where $\eta : B_n \to \mathfrak{M}(D, Q)$ is the homomorphism introduced in Section 1.6.3. It suffices to verify that ∂ and η coincide on the generators σ_i, where $i = 1, 2, \ldots, n-1$. Since $\eta(\sigma_i) = \tau_{\alpha_i}$, we need only check that $\partial(\sigma_i) = \tau_{\alpha_i}$. Let $\{g_t : D \to D\}_{t \in I}$ be the isotopy of the identity map $g_0 = \mathrm{id} : D \to D$ into $g_1 = \tau_{\alpha_i}$ obtained by rotating α_i in D about its midpoint counterclockwise. Then

$$\{f_t = g_{1-t} : D \to D\}_{t \in I}$$

is an isotopy of $f_0 = \tau_\alpha$ into $f_1 = \mathrm{id}$. It is easy to see that the geometric braid

$$\bigcup_{t \in I} (f_t(Q), t) \subset \mathbf{R}^2 \times I$$

represents $\sigma_i \in B_n$. Thus, $\partial(\sigma_i) = [f_0] = \tau_{\alpha_i}$.

By the Alexander–Tietze theorem (Section 1.6.1), any point of the set $\mathrm{Top}(D, Q) \subset \mathrm{Top}(D)$ can be connected to $\mathrm{id}_D \in \mathrm{Top}(D)$ by a path in $\mathrm{Top}(D)$. This implies that the homomorphism

$$\eta = \partial : \pi_1(\mathcal{C}_n, q) \to \pi_0(\mathrm{Top}(D, Q)) = \mathfrak{M}(D, Q)$$

is surjective. The commutativity of the diagram (1.14) and Theorem 1.31 imply that η is injective. Therefore η is an isomorphism. $\qquad\square$

Remark 1.38. The proof of the Alexander–Tietze theorem in Section 1.6.1 actually shows that the point $\{\mathrm{id}_D\}$ is a deformation retract of $\mathrm{Top}(D)$. Therefore, $\pi_i(\mathrm{Top}(D)) = 0$ for all $i \geq 0$ and the homotopy sequence of the fibration $e : \mathrm{Top}(D) \to \mathcal{C}_n(D^\circ)$ directly implies that the homomorphism $\partial : \pi_1(\mathcal{C}_n, q) \to \pi_0(\mathrm{Top}(D, Q))$ is an isomorphism.

1.7.4 Applications

We state two further applications of the techniques introduced above.

Theorem 1.39. *For any geometric braid b on n strings, the topological type of the pair $(\mathbf{R}^2 \times I, b)$ depends only on n.*

Proof. Pick a disk $D \subset \mathbf{R}^2$ such that $b \subset D^\circ \times I$. Then the set $Q = Q_n$ defined by (1.16) lies in D°. By Lemma 1.37, there is a normal isotopy $\{f_t : D \to D\}_{t \in I}$ parametrizing b. The formula $(x, t) \mapsto (f_t(x), t)$ defines a homeomorphism $F : D \times I \to D \times I$ mapping $Q \times I$ onto b and keeping $\partial D \times I$ pointwise. Extending F by the identity on $(\mathbf{R}^2 - D) \times I$, we obtain a homeomorphism $\mathbf{R}^2 \times I \to \mathbf{R}^2 \times I$ mapping $Q \times I$ onto b. Note that this homeomorphism is level-preserving in the sense that it commutes with the projection to I. □

Theorem 1.40. *Every isotopy of a geometric braid in $\mathbf{R}^2 \times I$ extends to an isotopy of $\mathbf{R}^2 \times I$ in itself constant on the boundary.*

Proof. Set $T = \mathbf{R}^2 \times I$. Let $b \subset T$ be a geometric braid on n strings and let $F : b \times I \to T$ be an isotopy of b. Thus, for each $s \in I$, the map $F_s : b \to T$ sending $x \in b$ to $F(x, s)$ is an embedding whose image is a geometric braid and $F_0 = \mathrm{id}_b$. We shall construct a (continuous) map $G : T \times I \to T$ such that for each $s \in I$, the map $G_s : T \to T$ sending $x \in T$ to $G(x, s)$ is a homeomorphism fixing ∂T pointwise and extending F_s, and $G_0 = \mathrm{id}_T$.

Let $Q \subset \mathbf{R}^2$ be the set $\{(1, 0), (2, 0), \ldots, (n, 0)\}$ and let D be a closed Euclidean disk in $\mathbf{R}^2$ such that $Q \subset D^\circ$ and $F(b \times I) \subset D^\circ \times I$. For any $s, t \in I$, denote by $f(s, t)$ the unique n-point subset of D° such that

$$F_s(b) \cap (D \times \{t\}) = f(s, t) \times \{t\}.$$

The formula $(s, t) \mapsto f(s, t)$ defines a continuous map $f : I^2 \to \mathcal{C}_n(D^\circ)$. Clearly, $f(s, 0) = f(s, 1) = Q$ for all $s \in I$ and $b = \bigcup_{t \in I} f(0, t) \times \{t\}$.

Consider the evaluation fibration $e = e_Q : \mathrm{Top}(D) \to \mathcal{C}_n(D^\circ)$. By the homotopy lifting property of e, the loop $t \mapsto f(0, t)$ lifts to a path $t \mapsto \widehat{f}(0, t)$ in $\mathrm{Top}(D)$ ending at id_D and beginning at a point of $\mathrm{Top}(D, Q)$. By the homotopy lifting property of e with respect to the pair $(I, \partial I)$, the latter path extends to a lift $\widehat{f} : I^2 \to \mathrm{Top}(D)$ of f such that $\widehat{f}(s, 1) = \mathrm{id}_D$ and $\widehat{f}(s, 0) = \widehat{f}(0, 0)$ for all $s \in I$.

We define a homeomorphism $g(s,t) : \mathbf{R}^2 \to \mathbf{R}^2$ to be the identity on $\mathbf{R}^2 - D$ and to be $\widehat{f}(s,t) \circ (\widehat{f}(0,t))^{-1}$ on D. It is clear that $g(s,t)$ is a continuous function of $s, t \in I$ and

$$g(0,t) = g(s,0) = g(s,1) = \mathrm{id}$$

for all $s, t \in I$. We have

$$g(s,t)(f(0,t)) = g(s,t)(\widehat{f}(0,t)(Q)) = \widehat{f}(s,t)(Q) = f(s,t).$$

It is now straightforward to check that the map $G : T \times I \to T$ sending (a, t, s) to $(g(s,t)(a), t)$ for $a \in \mathbf{R}^2$, $s, t \in I$ has all the required properties. $\square$

Exercise 1.7.1. Let f be a self-homeomorphism of the 2-sphere S^2 fixing a point $a \in S^2$ and isotopic to the identity id : $S^2 \to S^2$. Prove that f is isotopic to the identity in the class of self-homeomorphisms of S^2 fixing a.

Solution. Applying Lemma 1.35 to $M = S^2$, $Q = \{a\}$, $n = 1$, we obtain a locally trivial fibration $\mathrm{Top}(S^2) \to S^2$ with fiber $\mathrm{Top}(S^2, \{a\})$. Since $\pi_0(\mathrm{Top}(S^2, \{a\})) = \mathfrak{M}(S^2, \{a\})$ and $\pi_0(\mathrm{Top}(S^2)) = \mathfrak{M}(S^2)$, this fibration yields an exact sequence

$$\pi_1(S^2) \to \mathfrak{M}(S^2, \{a\}) \to \mathfrak{M}(S^2).$$

Since $\pi_1(S^2) = 0$, the kernel of the homomorphism $\mathfrak{M}(S^2, \{a\}) \to \mathfrak{M}(S^2)$ is trivial. This implies the required property of self-homeomorphisms of S^2.

Notes

The definition of braids and braid groups as well as a considerable part of the results of this chapter are due to Emil Artin [Art25], [Art47a], [Art47b]. This includes, among other things, the standard presentation of braid groups by generators and relations and the theory of braid automorphisms of Section 1.5. It should be noted that the braid automorphisms of free groups were studied by Hurwitz [Hur91] in 1891; see also [Mag72], [Bri88].

The generators $A_{i,j}$ of P_n and the defining relations for them were introduced by Burau [Bur32]; see also [Mar45], [Art47a], [Cho48]. Theorem 1.16 is due to Fröhlich [Frö36], Markov [Mar45], Artin [Art47a]. The combed form of braids was discovered by Markov [Mar45] and Artin [Art47a]. Theorem 1.24 was obtained by Artin [Art47a] and Chow [Cho48]. Corollary 1.25 is due to Artin [Art47a].

The theory of braids from the viewpoint of configuration spaces was first studied by Fox and Neuwirth [FoN62] and Fadell and Neuwirth [FaN62]. Definitions and results of Section 1.4 are taken from [FaN62]. The interpretation of $C_n(\mathbf{R}^2)$ in terms of polynomials was pointed out by Arnold [Arn70].

The Alexander–Tietze theorem used in Section 1.7.3 was established by Tietze [Tie14] and Alexander [Ale23b]. Lemma 1.35 is due to Birman [Bir69a]. Theorem 1.40 is due to Artin [Art47a].

Exercises 1.1.4 and 1.1.5 are due to Artin [Art25], [Art47b]. Exercises 1.1.6 and 1.1.8 are due to Gorin and Lin [GL69]; see also [Lin96]. Exercise 1.1.7 is due to Gorin. Exercise 1.1.10 is due to Kassel and Reutenauer [KR07] (see also [Gas62] and [GL69] for a proof of the freeness of the kernel of $B_4 \to B_3$). Exercise 1.4.3 is due to Fadell and Van Buskirk [FV62]. Exercise 1.5.1 is due to Wada [Wad92]. Exercise 1.6.1 is due to Sergiescu [Ser93].

2

Braids, Knots, and Links

In this chapter we study the relationship between braids, knots, and links. Throughout the chapter, we denote by I the closed interval $[0,1]$ in $\mathbf{R}$.

2.1 Knots and links in three-dimensional manifolds

We briefly discuss the notions from knot theory needed for the sequel. For detailed expositions of knot theory, the reader is referred to the monographs [BZ85], [Kaw96], [Mur96], [Rol76].

2.1.1 Basic definitions

Let M be a 3-dimensional topological manifold, possibly with boundary ∂M. A *geometric link* in M is a locally flat closed 1-dimensional submanifold of M. Recall that a manifold is *closed* if it is compact and has an empty boundary. A closed 1-dimensional submanifold $L \subset M$ is *locally flat* if every point of L has a neighborhood $U \subset M$ such that the pair $(U, U \cap L)$ is homeomorphic to the pair $(\mathbf{R}^3, \mathbf{R} \times \{0\} \times \{0\})$. This condition implies that $L \subset M^\circ = M - \partial M$ and excludes all kinds of locally wild behavior of L inside M°.

Being a closed 1-dimensional manifold, a geometric link in M must consist of a finite number of components homeomorphic to the standard unit circle

$$S^1 = \{z \in \mathbf{C} \mid |z| = 1\}.$$

A space homeomorphic to S^1 is called a (topological) circle. A geometric link consisting of $n \geq 1$ circles is called an *n-component link*. For example, the boundary of n disjoint embedded 2-disks in M° is a *trivial n-component link* in M.

One-component geometric links are called *geometric knots*. Examples of nontrivial knots and links in $\mathbf{R}^3$ are shown in Figure 2.1, which presents the trefoil knot, the figure-eight knot, and the Hopf link.

C. Kassel, V. Turaev, *Braid Groups*, DOI: 10.1007/978-0-387-68548-9_2,
© Springer Science+Business Media, LLC 2008

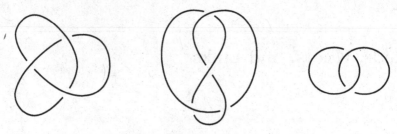

Fig. 2.1. Knots and links in $\mathbf{R}^3$

Two geometric links L and L' in M are *isotopic* if L can be deformed into L' by an isotopy of M into itself. Here by an *isotopy* of M (into itself), we mean a continuous family of homeomorphisms $\{F_s : M \to M\}_{s \in I}$ such that $F_0 = \mathrm{id}_M : M \to M$. The continuity of this family means that the mapping $I \to \mathrm{Top}(M)$, $s \mapsto F_s$ is continuous or, equivalently, the mapping

$$M \times I \to M, \quad (x, s) \mapsto F_s(x),$$

where $x \in M, s \in I$, is continuous; see Section 1.7.1. An isotopy $\{F_s\}_{s \in I}$ of M is said to be an *isotopy of L into L'* if $F_1(L) = L'$. The links L and L' are isotopic if there is an isotopy of L into L'. Isotopic geometric links have the same number of components. In other words, the number of components is an isotopy invariant of geometric links.

The relation of isotopy is obviously an equivalence relation in the class of geometric links in M. The corresponding equivalence classes are called *links* in M. The links having only one component are called *knots*. The ultimate goal of knot theory is a classification of knots and links.

If M has a smooth structure, then any geometric link in M is isotopic to a geometric link whose underlying 1-dimensional manifold is a smooth submanifold of M. Therefore working with links in smooth 3-dimensional manifolds, we can always restrict ourselves to smooth representatives.

2.1.2 Link diagrams

The technique of braid diagrams discussed in Chapter 1 can be extended to links. We shall restrict ourselves to the case in which the ambient 3-manifold is the product of a surface Σ (possibly with boundary $\partial\Sigma$) with I. For $n \geq 1$, a *link diagram* on Σ with n components is a set $\mathcal{D} \subset \Sigma - \partial\Sigma$ obtained as a union of n circles with a finite number of intersections and self-intersections such that each (self-)intersection is a meeting point of exactly two branches of $\mathcal{D}$, one of these branches being distinguished and called *undergoing*, the other one being *overgoing*. In a neighborhood of a point, $\mathcal{D}$ looks like a straight line in $\mathbf{R}^2$ or like the set $\{(x, y) \mid xy = 0\} \subset \mathbf{R}^2$, where one of the branches $x = 0, y = 0$ is distinguished. The circles forming $\mathcal{D}$ are called the *components* of $\mathcal{D}$. The (self-)intersections of these circles are called *crossings* or *double points* of $\mathcal{D}$. Note that three components of $\mathcal{D}$ never meet in a point.

The branch of a link diagram going under a crossing is graphically represented by a broken line. The pictures in Figure 2.1 can be considered as link diagrams in the plane.

Each link diagram $\mathcal{D}$ on a surface Σ presents a link

$$L(\mathcal{D}) \subset \Sigma \times I.$$

It is obtained from $\mathcal{D} \subset \Sigma = \Sigma \times \{1/2\}$ by pushing the undergoing branches into $\Sigma \times [1/2, 1)$. The link $L(\mathcal{D})$ is well defined up to isotopy.

Observe that any link in $\Sigma \times I$ can be presented by a link diagram on Σ. To see this, represent the given link by a geometric link $L \subset \Sigma \times I$ whose projection to Σ has only double transversal crossings. At each of the crossings choose the undergoing branch to be the one that comes from the subarc of L with bigger I-coordinate. This gives a link diagram on Σ representing the isotopy class of L.

Two link diagrams $\mathcal{D}$ and $\mathcal{D}'$ on Σ are *isotopic* if there is an isotopy of Σ into itself transforming $\mathcal{D}$ into $\mathcal{D}'$. More precisely, $\mathcal{D}$ and $\mathcal{D}'$ are *isotopic* if there is a continuous family of homeomorphisms $\{F_s : \Sigma \to \Sigma\}_{s \in I}$ such that $F_0 = \mathrm{id}_\Sigma$ and $F_1(\mathcal{D}) = \mathcal{D}'$. It is understood that F_1 maps the crossings of $\mathcal{D}$ to the crossings of $\mathcal{D}'$, preserving the under/overgoing data. Clearly, if $\mathcal{D}$ and $\mathcal{D}'$ are isotopic, then $L(\mathcal{D})$ and $L(\mathcal{D}')$ are isotopic in $\Sigma \times I$.

The transformations of link diagrams $\Omega_1, \Omega_2, \Omega_3$ shown in Figures 1.5a, 1.5b, and 2.2 below (as well as the inverse transformations) are called *Reidemeister moves*. These moves affect only a part of the diagram lying in a disk in Σ and preserve the rest of the diagram. Note that to apply these moves, we identify the disk in Σ with a disk in the plane of the pictures. If Σ is oriented, then we use only identifications transforming the orientation of Σ into the counterclockwise orientation in the plane of the pictures. For nonoriented Σ, we use arbitrary identifications.

In comparison to the theory of braid diagrams, we need here two additional moves Ω_1 shown in Figure 2.2. These moves add a "curl" or "kink" to the diagram. The inverse moves Ω_1^{-1} remove such kinks from link diagrams. On the other hand, in the setting of link diagrams, one Ω_2-move is sufficient: the two Ω_2-moves in Figure 1.5a can be obtained from each other by an isotopy in Σ rotating a small 2-disk in Σ affected by the move to an angle of $180°$.

The classical Reidemeister theorem for link diagrams on $\mathbf{R}^2$ generalizes to diagrams on Σ: two link diagrams on Σ represent isotopic links in $\Sigma \times I$ if and only if these diagrams are related by a finite sequence of isotopies and Reidemeister moves $\Omega_1^{\pm 1}, \Omega_2^{\pm 1}, \Omega_3^{\pm 1}$. Indeed, any isotopy of a geometric link in $\Sigma \times I$ may be split as a composition of a finite number of "local" isotopies changing the link only inside a cylinder of type $U \times I$, where U is a small open disk in Σ. Since the pair $(U \times I, U \times \{1/2\})$ is homeomorphic to the pair $(\mathbf{R}^2 \times I, \mathbf{R}^2 \times \{1/2\})$, we can apply the standard Reidemeister theory to the part of the link diagram lying in U. This implies that under such a local isotopy the diagram is changed via a sequence of moves $\Omega_1^{\pm 1}, \Omega_2^{\pm 1}, \Omega_3^{\pm 1}$.

Fig. 2.2. The moves Ω_1

2.1.3 Ordered and oriented links

Links admit a number of natural additional structures. Here we describe two such structures: order and orientation. An n-component geometric link is *ordered* if its components are numbered by $1, 2, \ldots, n$. By *isotopies* of ordered links, we mean order-preserving isotopies. The order is easily exhibited on link diagrams: it suffices to attach the numbers $1, 2, \ldots, n$ to the components of the diagram and to keep these numbers unchanged under isotopy.

An *orientation* of a geometric link L in a 3-dimensional manifold M is an orientation of the underlying 1-dimensional manifold L. In the figures, the orientation is indicated by arrows on the link components. By *isotopies* of oriented links, one means orientation-preserving isotopies. Each oriented link $L \subset M$ is a 1-cycle and represents a homology class

$$[L] \in H_1(M) = H_1(M; \mathbf{Z}).$$

This class is an isotopy invariant of L. Indeed, the components of two isotopic oriented links are pairwise homotopic and consequently pairwise homologous.

To exhibit the orientation of the link presented by a link diagram on a surface it suffices to orient all components of the diagram. Each Reidemeister move gives rise to several *oriented Reidemeister moves* on oriented link diagrams keeping the orientations of the strands. Specifically, orienting all the strands in Figure 2.2 in the same direction (up or down), we obtain four oriented Ω_1-moves. Similarly, the two moves Ω_2 in Figure 1.5a give rise to eight oriented Ω_2-moves. In two of them, both strands are directed down (before and after the move). These two oriented Ω_2-moves are said to be *braidlike* and are denoted by Ω_2^{br}. The two oriented Ω_2-moves in which the strands are directed up can be expressed as compositions of Ω_2^{br} and isotopies rotating a 2-disk by the angle $180°$. The remaining oriented Ω_2-moves, in which the strands are directed in opposite directions, are said to be *nonbraidlike*. In a similar way, the move Ω_3 in Figure 1.5b gives rise to eight oriented Ω_3-moves. Any seven of them can be expressed as compositions of the eighth move and oriented Ω_2-moves (see [Tur88] or [Tra98]). Therefore it is enough to consider only the oriented Ω_3-move in which all three strands are directed down. This move is said to be *braidlike* and is denoted by Ω_3^{br}.

The Reidemeister theorem mentioned at the end of Section 2.1.2 implies that two oriented link diagrams on a surface Σ present isotopic oriented links in $\Sigma \times I$ if and only if these diagrams are related by a finite sequence of orientation-preserving isotopies and oriented Reidemeister moves.

2.1.4 The linking number

As an application of link diagrams, we define the integral linking number of knots in $\Sigma \times I$, where Σ is an arbitrary oriented surface (for nonoriented Σ, the linking number is defined only mod 2). Let L_1, L_2 be disjoint oriented knots in $\Sigma \times I$. Let us present the ordered oriented 2-component link $L_1 \cup L_2$ by a diagram on Σ. Let l^+ (resp. l^-) be the number of crossings of this diagram where a strand representing L_1 goes over a strand representing L_2 from left to right (resp. from right to left). Here the left and right sides of an oriented strand s are defined by the condition that the pair (a positively oriented vector tangent to s, a vector directed from the right of s to the left, of s) determines the orientation of Σ. It is straightforward to check that the *linking number*

$$\mathrm{lk}(L_1, L_2) = l^+ - l^- \in \mathbf{Z}$$

is invariant under isotopies and oriented Reidemeister moves in the diagram. Hence $\mathrm{lk}(L_1, L_2)$ is a well-defined isotopy invariant of the link $L_1 \cup L_2$.

Exercise 2.1.1. Prove that an arbitrary geometric knot L in an orientable 3-dimensional manifold has an open neighborhood $U \supset L$ such that the pair (U, L) is homeomorphic to $(\mathbf{R}^2 \times S^1, \{x\} \times S^1)$, where $x \in \mathbf{R}^2$.

Exercise 2.1.2. Prove that two oriented link diagrams on $\mathbf{R}^2$ isotopic in the 2-sphere $S^2 = \mathbf{R}^2 \cup \{\infty\}$ represent isotopic oriented links in $\mathbf{R}^3$. (Hint: It suffices to verify this for an isotopy pushing a branch of the diagram across the point $\infty \in S^2$.)

Exercise 2.1.3. For any oriented surface Σ and any two disjoint oriented knots $L_1, L_2 \subset \Sigma \times I$,

$$\mathrm{lk}(L_1, L_2) - \mathrm{lk}(L_2, L_1) = [L_1] \cdot [L_2],$$

where $[L_1] \cdot [L_2] \in \mathbf{Z}$ is the intersection number of $[L_1], [L_2] \in H_1(\Sigma)$. (Hint: This equality is obvious if

$$L_1 \subset \Sigma \times [0, 1/2], \quad L_2 \subset \Sigma \times [1/2, 1]$$

and is preserved when a branch of L_1 is pushed across a branch of L_2.) Deduce that if Σ embeds in S^2, then $\mathrm{lk}(L_1, L_2) = \mathrm{lk}(L_2, L_1)$.

2.2 Closed braids in the solid torus

We introduce certain links in the solid torus, called closed braids, and classify them in terms of braids.

2.2.1 Solid tori

By a *solid torus*, we mean the product $V = D \times S^1$, where D is a closed 2-disk and $S^1 = \{z \in \mathbf{C} \,|\, |z| = 1\}$. The solid torus V is a compact connected orientable 3-dimensional manifold with boundary

$$\partial V = \partial D \times S^1 \approx S^1 \times S^1 .$$

Clearly,

$$V^\circ = V - \partial V = D^\circ \times S^1 ,$$

where $D^\circ = D - \partial D$. The solid torus naturally arises in knot theory as a closed regular neighborhood of any knot in an orientable 3-dimensional manifold. Using a homeomorphism $D \approx I \times I$, we obtain

$$V \approx I \times I \times S^1 \approx S^1 \times I \times I .$$

The technique of link diagrams of Section 2.1 allows us to present links in V by diagrams on the annulus $S^1 \times I$.

2.2.2 Closed braids

A geometric link L in the solid torus $V = D \times S^1$ is called a *closed n-braid* with $n \geq 1$ if L meets each 2-disk $D \times \{z\}$ with $z \in S^1$ transversely in n points. It is clear that the projection on the second factor $V \to S^1$ restricted to L yields an (unramified) n-fold covering $L \to S^1$. We shall always provide L with the *canonical orientation* obtained as the lift of the counterclockwise orientation on S^1. Thus, a point moving along a component of L in the positive direction projects to a point moving along S^1 counterclockwise without ever stopping or going backward. The homology class $[L] \in H_1(V) = \mathbf{Z}$ of the oriented link $L \subset V$ is computed by $[L] = n\,[\{x\} \times S^1]$ for any $x \in D$.

For example, if Q is a finite subset of D°, then the link $Q \times S^1 \subset V$ is a closed n-braid, where $n = \operatorname{card}(Q)$. A closed 3-braid is drawn in Figure 2.3. Our interest in closed braids is due to their connection with braids. This connection will be discussed in the next subsections.

Two closed braids in V are *isotopic* if they are isotopic as oriented links. Note that the intermediate links appearing during an isotopy are not required to be closed braids. By abuse of language, isotopy classes of closed braids in V will be also called closed braids in V.

In general, a link in V is not isotopic to a closed braid in V. For instance, a link lying inside a small 3-ball in V is never isotopic to a closed braid. More generally, an oriented link in V homological to $m\,[\{x\} \times S^1]$ with $m \leq 0$, $x \in D$ is not isotopic to a closed braid in V. Another obstruction will be discussed in Exercise 2.2.4.

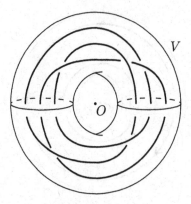

Fig. 2.3. A closed 3-braid in V

2.2.3 Closure of braids

Every braid β on n strings gives rise to a closed n-braid in the solid torus as follows. Fix a closed Euclidean disk $D \subset \mathbf{R}^2$ containing the set $Q = \{(1,0),(2,0),\ldots,(n,0)\}$ in its interior. Observe that the solid torus $V = D \times S^1$ can be obtained from the cylinder $D \times I$ by the identification $(x,0) = (x,1)$ for all $x \in D$. (Here we identify $I/\partial I$ with S^1 via the standard homeomorphism $t \mapsto \exp(2\pi i t) : I/\partial I \to S^1$.) Pick a geometric braid $b \subset D^\circ \times I$ representing β (for the existence of such b, see Exercise 1.2.4). Let $\widehat{b} \subset V$ be the image of b under the projection $D \times I \to V$. It is obvious that $\widehat{b}$ is a closed n-braid in V. The canonical orientation of $\widehat{b}$ is determined by the direction on b leading from $Q \times \{0\}$ to $Q \times \{1\}$. If $b' \subset D \times I$ is another geometric braid representing β, then b is isotopic to b' in $D \times I$ (cf. Exercise 1.2.5). By (the proof of) Theorem 1.40, there is an isotopy of $D \times I$ constant on the boundary and transforming b into b'. This isotopy induces an isotopy between $\widehat{b}$ and $\widehat{b'}$ in V. Therefore the isotopy class of $\widehat{b}$ depends only on β. This class is called the *closure* of β and denoted by $\widehat{\beta}$.

Note that any closed n-braid $L \subset V$ is isotopic to $\widehat{\beta}$ for a certain $\beta \in B_n$. Indeed, we can deform L in the class of closed braids so that

$$L \cap (D \times \{1\}) = Q \times \{1\}.$$

Cutting V open along $D \times \{1\}$, we obtain a braid in $D \times I$ with closure L.

The description of $\widehat{\beta}$ given above is somewhat awkward from the point of view of drawing pictures. The following equivalent description is often more convenient. Observe that gluing two copies of $D \times I$ along $D \times \partial I = D \times \{0,1\}$, we again obtain V. Gluing a geometric braid representing β in the first $D \times I$ with the trivial braid $Q \times I$ in the second $D \times I$, we obtain $\widehat{\beta}$; see Figure 2.4. A link diagram in $S^1 \times I$ presenting $\widehat{\beta}$ is obtained by closing a diagram of β as in Figure 2.5.

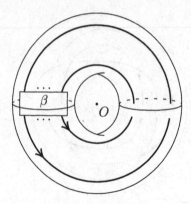

Fig. 2.4. Closing a braid β

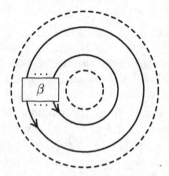

Fig. 2.5. A diagram of $\widehat{\beta}$

Theorem 2.1. *For any $n \geq 1$ and any β, $\beta' \in B_n$, the closed braids $\widehat{\beta}$, $\widehat{\beta'}$ are isotopic in the solid torus if and only if β and β' are conjugate in B_n.*

Theorem 2.1 gives an isotopy classification of closed n-braids in the solid torus: the isotopy classes of closed n-braids correspond bijectively to the conjugacy classes in B_n. In particular, any conjugacy invariant of elements of B_n determines an isotopy invariant of closed n-braids. For instance, the characteristic polynomial of a finite-dimensional linear representation of B_n yields an invariant of closed n-braids. Theorem 2.1 raises the problem of finding an algorithm to decide whether two given elements of B_n are conjugate. We shall address this problem in Chapter 6.

2.2.4 Proof of Theorem 2.1

Observe first that conjugate elements of B_n give rise to isotopic closed braids. In other words, $\widehat{\alpha\beta\alpha^{-1}} = \widehat{\beta}$ for any $\alpha, \beta \in B_n$. This is obtained by forming a diagram of $\alpha\beta\alpha^{-1}$ from three diagrams representing the three factors, pushing the upper diagram representing α along the n parallel strings on the right so that eventually it comes to the diagram of α^{-1} from below. This gives $\widehat{\alpha\beta\alpha^{-1}} = \widehat{\beta\alpha\alpha^{-1}} = \widehat{\beta}$.

We prove the converse: any braids with isotopic closures in $V = D \times S^1$ are conjugate. To this end, we need to study closed braids in V in more detail. Set $\overline{V} = D \times \mathbf{R}$. Multiplying D by the universal covering

$$t \mapsto \exp(2\pi it) : \mathbf{R} \to S^1,$$

we obtain a universal covering $\overline{V} \to V$. Denote by T the covering transformation $\overline{V} \to \overline{V}$ sending (x, t) to $(x, t+1)$ for all $x \in D$, $t \in \mathbf{R}$.

If L is a closed n-braid in V, then its preimage $\overline{L} \subset \overline{V}$ is a 1-dimensional manifold meeting each disk $D \times \{t\}$ with $t \in \mathbf{R}$ transversely in n points. This implies that $\overline{L}$ consists of n components homeomorphic to $\mathbf{R}$. More information about $\overline{L}$ can be obtained by presenting L as the closure of a geometric braid $b \subset D \times I$, where we identify $I/\partial I = S^1$. Then

$$\overline{L} = \bigcup_{m \in \mathbf{Z}} T^m(b).$$

By Section 1.7.2, $b = \bigcup_{t \in I} (f_t(Q), t)$ for a continuous family of homeomorphisms $\{f_t : D \to D\}_{t \in I}$ such that $f_0(Q) = Q$, $f_1 = \mathrm{id}_D$, and all f_t fix ∂D pointwise. We define a level-preserving self-homeomorphism of $\overline{V} = D \times \mathbf{R}$ by

$$(x, t) \mapsto (f_{t-[t]} f_0^{-[t]}(x), t),$$

where $x \in D$, $t \in \mathbf{R}$, and $[t]$ is the greatest integer less than or equal to t. This homeomorphism fixes $\partial \overline{V} = \partial D \times \mathbf{R}$ pointwise and sends $Q \times \mathbf{R}$ onto $\overline{L}$; see Figure 2.6. The induced homeomorphism $(D - Q) \times \mathbf{R} \approx \overline{V} - \overline{L}$ shows that $D - Q = (D - Q) \times \{0\} \subset \overline{V} - \overline{L}$ is a deformation retract of $\overline{V} - \overline{L}$.

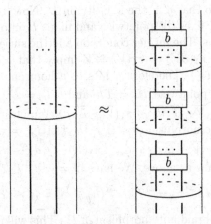

Fig. 2.6. A homeomorphism $(D \times \mathbf{R}, Q \times \mathbf{R}) \approx (D \times \mathbf{R}, \overline{L})$

Pick a point $d \in \partial D$ as in Figure 1.15 and set $\overline{d} = (d, 0) \in \overline{V}$. It is clear that the inclusion homomorphism $i : \pi_1(D - Q, d) \to \pi_1(\overline{V} - \overline{L}, \overline{d})$ is an isomorphism. By definition, $T(\overline{d}) = (d, 1)$. The covering transformation T restricted to $\overline{V} - \overline{L}$ induces an isomorphism $\pi_1(\overline{V} - \overline{L}, \overline{d}) \to \pi_1(\overline{V} - \overline{L}, T(\overline{d}))$. Consider the isomorphism $\pi_1(\overline{V} - \overline{L}, T(\overline{d})) \to \pi_1(\overline{V} - \overline{L}, \overline{d})$ obtained by conjugating the loops by the path $d \times [0, 1] \subset \partial D \times \mathbf{R} \subset \overline{V} - \overline{L}$. The composition $T_\#$ of these two isomorphisms is an automorphism of $\pi_1(\overline{V} - \overline{L}, \overline{d})$. It follows from the description of $\overline{L}$ given above that the following diagram is commutative:

$$
\begin{array}{ccc}
\pi_1(D - Q, d) & \xrightarrow{\;\;i\;\;} & \pi_1(\overline{V} - \overline{L}, \overline{d}) \\
{\scriptstyle \rho(f_0)} \downarrow & & \downarrow {\scriptstyle T_\#} \\
\pi_1(D - Q, d) & \xrightarrow{\;\;i\;\;} & \pi_1(\overline{V} - \overline{L}, \overline{d})
\end{array}
$$

where $\rho(f_0)$ is the automorphism of $\pi_1(D - Q, d)$ induced by the restriction of f_0 to $D - Q$; cf. Section 1.6.3. Therefore $i^{-1}T_\# i = \rho(f_0)$. Indeed, the proof of Theorem 1.33 shows that the group homomorphism $\eta : B_n \to \mathfrak{M}(D, Q)$ introduced in Section 1.6.3 sends the braid $\beta \in B_n$ represented by b to the isotopy class of $f_0 : (D, Q) \to (D, Q)$. Identifying $\pi_1(D - Q, d)$ with the free group F_n on n generators $x_1, x_2, \ldots, x_n$ as in Section 1.6.3 and applying Theorem 1.33, we conclude that $\rho(f_0) = \rho\eta(\beta) = \widetilde{\beta}$ is the braid automorphism of F_n corresponding to β. Thus, $i^{-1}T_\# i = \widetilde{\beta}$.

Suppose now that $\beta, \beta' \in B_n$ are two braids with isotopic closures in V. Present them by geometric braids $b, b' \subset D \times I$ and let $L, L' \subset V$ be their respective closures. By assumption, there is a homeomorphism $g : V \to V$ such that g maps L onto L' preserving their canonical orientations and g is isotopic to the identity $\mathrm{id}_V : V \to V$. The latter condition implies that the restriction of g to ∂V is isotopic to the identity $\mathrm{id} : \partial V \to \partial V$. Therefore $g|_{\partial V}$ extends to a homeomorphism $g' : V \to V$ equal to the identity outside a narrow tubular neighborhood of ∂V in V. We can assume that this neighborhood is disjoint from L', so that g' is the identity on L'. Now, the homeomorphism $h = (g')^{-1} g : V \to V$ fixes ∂V pointwise and maps L onto L' preserving their canonical orientations. The former condition and the surjectivity of the inclusion homomorphism $\pi_1(\partial V) \to \pi_1(V) \cong \mathbf{Z}$ imply that h induces an identity automorphism of $\pi_1(V)$. Therefore h lifts to a homeomorphism $\overline{h} : \overline{V} \to \overline{V}$ such that $\overline{h}$ fixes $\partial \overline{V}$ pointwise, $\overline{h}T = T\overline{h}$, and $\overline{h}(\overline{L}) = \overline{L}'$. Hence $\overline{h}$ induces an isomorphism $\overline{h}_\# : \pi_1(\overline{V} - \overline{L}, \overline{d}) \to \pi_1(\overline{V} - \overline{L}', \overline{d})$ commuting with $T_\#$.

Consider the automorphism $\varphi = (i')^{-1}\overline{h}_\# i$ of $F_n = \pi_1(D - Q, d)$, where $i : \pi_1(D - Q, d) \to \pi_1(\overline{V} - \overline{L}, \overline{d})$ and $i' : \pi_1(D - Q, d) \to \pi_1(\overline{V} - \overline{L}', \overline{d})$ are the inclusion isomorphisms as above. We have $\widetilde{\beta}' = (i')^{-1}T_\# i'$ and

$$
\varphi \widetilde{\beta} \varphi^{-1} = (i')^{-1}\overline{h}_\# i i^{-1} T_\# i i^{-1} (\overline{h}_\#)^{-1} i' = (i')^{-1} T_\# i' = \widetilde{\beta}' .
$$

We claim that φ is a braid automorphism of F_n. This will imply that $\widetilde{\beta}$ and $\widetilde{\beta}'$ are conjugate in the group of braid automorphisms of F_n. By Theorem 1.31, this will imply that β and β' are conjugate in B_n.

By definition, the conjugacy classes of the generators

$$x_1, x_2, \ldots, x_n \in F_n = \pi_1(D - Q, d)$$

are represented by small loops encircling the points of Q in D. The inclusion $D - Q = (D - Q) \times \{0\} \subset \overline{V} - \overline{L}$ maps these loops to small loops in $\overline{V} - \overline{L}$ encircling the components of $\overline{L}$. The homeomorphism $\overline{h} : \overline{V} \to \overline{V}$ transforms these loops into small loops in $\overline{V} - \overline{L}'$ encircling the components of $\overline{L}'$. The latter represent the conjugacy classes of the images of $x_1, x_2, \ldots, x_n$ under the inclusion $D - Q = (D - Q) \times \{0\} \subset \overline{V} - \overline{L}'$. Hence φ transforms the conjugacy classes of $x_1, \ldots, x_n$ into themselves, up to permutation. This verifies the first condition in the definition of a braid automorphism. The second condition says that $\varphi(x) = x$, where

$$x = x_1 x_2 \cdots x_n \in F_n = \pi_1(D - Q, d).$$

Observe that x is represented by the loop ∂D based at d. The inclusion of $D - Q = (D - Q) \times \{0\}$ into $\overline{V} - \overline{L}$ maps this loop to $\partial D \times \{0\}$. Since $\overline{h}$ fixes $\partial \overline{V}$ pointwise, $\overline{h}_\# i(x) = i'(x)$ and therefore $\varphi(x) = x$. $\qquad\square$

2.2.5 Closed braid diagrams

A *closed braid diagram* in the annulus $S^1 \times I$ is an oriented link diagram $\mathcal{D}$ in $S^1 \times I$ such that whenever a point moves along $\mathcal{D}$ in the positive direction, its projection to S^1 moves along S^1 counterclockwise without ever stopping or going backward. In other words, the projection $S^1 \times I \to S^1$ restricted to $\mathcal{D}$ is an orientation-preserving covering of S^1 (ramified at the crossings of $\mathcal{D}$). The number of points of $\mathcal{D}$ projecting to a given point on S^1 does not depend on the choice of that point, provided the crossings of $\mathcal{D}$ are counted with multiplicity 2. This number is called the *number of strands* of $\mathcal{D}$. Examples of closed braid diagrams on n strands in $S^1 \times I$ can be obtained by closing usual braid diagrams on n strands as in Figure 2.5.

Every closed braid diagram in $S^1 \times I$ presents a closed braid in the solid torus $S^1 \times I \times I$ in the obvious way; cf. Section 2.1.2. Clearly, every closed braid in $S^1 \times I \times I$ can be presented by a closed braid diagram in $S^1 \times I$.

We can apply to a closed braid diagram the moves $\Omega_2^{br}, \Omega_3^{br}$ and their inverses. These moves act as in Figures 1.5a and 1.5b, where the projections on the horizontal and vertical axes in the plane of the picture correspond to the projections to I and S^1, respectively. These moves keep the diagram in the class of closed braid diagrams and preserve the isotopy class of the closed braid represented by the diagram.

Lemma 2.2. *Two closed braid diagrams $\mathcal{D}, \mathcal{D}'$ in $S^1 \times I$ represent isotopic closed braids in the solid torus $S^1 \times I \times I$ if and only if $\mathcal{D}$ can be transformed into $\mathcal{D}'$ by a finite sequence of isotopies (in the class of closed braid diagrams) and moves $(\Omega_2^{br})^{\pm 1}, (\Omega_3^{br})^{\pm 1}$.*

Proof. We need only prove that if $\mathcal{D}, \mathcal{D}'$ represent isotopic closed braids in the solid torus, then $\mathcal{D}$ can be transformed into $\mathcal{D}'$ by a finite sequence of isotopies and moves $(\Omega_2^{\mathrm{br}})^{\pm 1}, (\Omega_3^{\mathrm{br}})^{\pm 1}$. Pick a point $z \in S^1$ such that the interval $\{z\} \times I$ does not meet the crossings of $\mathcal{D}$ or $\mathcal{D}'$. Cutting open $\mathcal{D}, \mathcal{D}'$ along this interval, we obtain two braid diagrams b, b', respectively. By Theorem 2.1, they represent conjugate braids. Applying a Ω_2^{br}-move to $\mathcal{D}$ in a neighborhood of $\{z\} \times I$, we can transform b into $\sigma_i b \sigma_i^{-1}$ and $\sigma_i^{-1} b \sigma_i$ for any $i = 1, 2, \ldots, n - 1$. Applying such moves recursively, we can transform b into an arbitrary conjugate diagram. Thus we can assume that b and b' represent isotopic braids. Then, by Theorem 1.6, these diagrams can be related by a finite sequence of isotopies and braidlike moves. This induces a sequence of isotopies and braidlike moves transforming $\mathcal{D}$ into $\mathcal{D}'$. $\qquad\square$

Exercise 2.2.1. Verify that for any $\beta \in B_n$, the number of components of the closed braid $\widehat{\beta}$ is equal to the number of cycles in the decomposition of the permutation $\pi(\beta) \in \mathfrak{S}_n$ as a product of commuting cycles.

Exercise 2.2.2. The closure of a pure braid $\beta \in P_n$ is an ordered n-component link: its ith component is the closure of the ith string of β for $i = 1, 2, \ldots, n$. Prove that for any $\beta, \beta' \in P_n$, the links $\widehat{\beta}, \widehat{\beta}'$ are isotopic in the solid torus in the class of ordered oriented links if and only if β and β' are conjugate in P_n.

Exercise 2.2.3. Prove that if two closed braids $L, L' \subset V = D \times S^1$ are isotopic, then they are isotopic in the class of closed braids in V, that is, there is an isotopy $\{F_s : V \to V\}_{s \in I}$ of L into L' such that $F_s(L)$ is a closed braid for all $s \in I$. (Hint: Use Theorem 2.1.)

Exercise 2.2.4. Let $L \subset V$ be a closed braid. Prove that the kernel of the inclusion homomorphism $\pi_1(V - L) \to \pi_1(V) = \mathbf{Z}$ is a free group. (Hint: In the notation of Section 2.2.4, this kernel is isomorphic to $\pi_1(\overline{V} - \overline{L}, \overline{d})$.)

2.3 Alexander's theorem

We establish here a fundamental theorem, due to J. W. Alexander, asserting that all links in $\mathbf{R}^3$ are isotopic to closed braids.

2.3.1 Closed braids in $\mathbf{R}^3$

Pick a Euclidean circle in the plane $\mathbf{R}^2 \times \{0\} \subset \mathbf{R}^3$ with center at the origin $O = (0, 0, 0)$. We identify a closed cylindrical neighborhood of this circle in $\mathbf{R}^3$ with the solid torus $V = D \times S^1$. By a *closed n-braid in $\mathbf{R}^3$*, we shall mean an oriented geometric link in $\mathbf{R}^3$ lying in $V \subset \mathbf{R}^3$ as a closed n-braid with its canonical counterclockwise orientation (cf. Figure 2.3, where the plane $\mathbf{R}^2 \times \{0\}$ is the plane of the picture).

In particular, for any $\beta \in B_n$, the closed braid $\widehat{\beta} \subset V$ yields a closed braid in $\mathbf{R}^3$ via the inclusion $V \subset \mathbf{R}^3$; cf. Figure 2.4. The latter closed braid is also denoted by $\widehat{\beta}$ and is called the *closure* of β. A diagram of $\widehat{\beta}$ is obtained from a diagram of β by connecting the bottom endpoints with the top endpoints by n standard arcs; cf. Figure 2.5, where the dotted circles should be disregarded. We stress that closed braids in $\mathbf{R}^3$ are *oriented* geometric links.

For example, the closure of the trivial braid on n strings is a trivial n-component link. The closure of $\sigma_1^{\pm 1} \in B_2$ is a trivial knot. The closure of $\sigma_1^{\pm 2} \in B_2$ is an oriented Hopf link. More generally, the closure of $\sigma_1^m \in B_2$ with $m \in \mathbf{Z}$ is a so-called *torus* $(2,m)$-*link*. It has two components for even m and one component for odd m.

We can give an equivalent but more direct definition of closed braids in $\mathbf{R}^3$. Consider the coordinate axis $\ell = \{(0,0)\} \times \mathbf{R} \subset \mathbf{R}^3$ meeting the plane $\mathbf{R}^2 \times \{0\}$ at the origin $O = (0,0,0)$. The counterclockwise rotation about O in the plane $\mathbf{R}^2 \times \{0\}$ determines a positive direction of rotation about ℓ. An oriented geometric link $L \subset \mathbf{R}^3 - \ell$ is a *closed n-braid* if the vector from O to any point $X \in L$ rotates in the positive direction about ℓ when X moves along L in the direction determined by the orientation of L. The equivalence of this definition with the previous one can be seen as follows. Pick a Euclidean disk D lying in an open half-plane bounded by ℓ in $\mathbf{R}^3$ and having its center in $\mathbf{R}^2 \times \{0\}$. Rotating D around ℓ, we sweep a solid torus $V = D \times S^1$ as above. Taking D big enough, we can assume that a given link $L \subset \mathbf{R}^3 - \ell$ lies in V. It is clear that L is a closed braid in the sense of the first definition if and only if L is a closed braid in the sense of the second definition.

Theorem 2.3 (J. W. Alexander). *Any oriented link in $\mathbf{R}^3$ is isotopic to a closed braid.*

Proof. By a *polygonal link*, we shall mean a geometric link in $\mathbf{R}^3$ whose components are closed broken lines. By *vertices* and *edges* of a polygonal link, we mean the vertices and the edges of its components. It is well known that any geometric link in $\mathbf{R}^3$ is isotopic to a polygonal link (cf. the proof of Theorem 1.6). We need only to prove that any oriented polygonal link $L \subset \mathbf{R}^3$ is isotopic to a closed braid. Moving slightly the vertices of L in $\mathbf{R}^3$, we obtain a polygonal link isotopic to L. We use such small deformations to ensure that $L \subset \mathbf{R}^3 - \ell$ and that the edges of L do not lie in planes containing the axis $\ell = \{(0,0)\} \times \mathbf{R}$. Let

$$AC \subset L \subset \mathbf{R}^3 - \ell$$

be an edge of L, where L is oriented from A to C. The edge AC is said to be *positive* (resp. *negative*) if the vector from the origin $O \in \ell$ to a point $X \in AC$ rotates in the positive (resp. negative) direction about ℓ when X moves from A to C. The assumption that AC does not lie in a plane containing ℓ implies that AC is necessarily positive or negative. The edge AC of L is said to be *accessible* if there is a point $B \in \ell$ such that the triangle ABC meets L only along AC.

If all edges of L are positive, then L is a closed braid and there is nothing to prove. Consider a negative edge AC of L. We replace AC with a sequence of positive edges as follows. If AC is accessible, then we pick $B \in \ell$ such that the triangle ABC meets L only along AC. In the plane ABC we take a slightly bigger triangle $AB'C$ containing B in its interior, meeting ℓ only at B, and meeting L only along AC; see Figure 2.7. We apply to L the Δ-move $\Delta(AB'C)$ replacing AC with two positive edges AB' and $B'C$ (see Section 1.2.3 for similar moves on geometric braids; in contrast to the setting of braids, we impose here no conditions on the third coordinates of the vertices). The resulting polygonal link is isotopic to L and has one negative edge fewer than L.

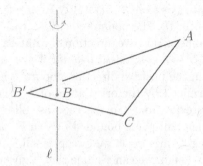

Fig. 2.7. The triangle $AB'C$

Suppose that the edge AC is not accessible. Note that every point P of AC is contained in an accessible subsegment of AC. (To see this, pick $B \in \ell$ such that the segment PB meets L only at P and then slightly "thicken" this segment inside the triangle ABC to obtain a triangle P^-BP^+ meeting L along its side $P^-P^+ \subset AC$ containing P. Then P^-P^+ is an accessible subsegment of AC.) Since AC is compact, we can split it into a finite number of consecutive accessible subsegments. We apply to each of them the Δ-move as above choosing the corresponding points $B \in \ell$ distinct and choosing B' close enough to B to stay away from other edges of L. Since AC does not lie in a plane containing ℓ, the triangles determining these Δ-moves meet only at the common vertices of the consecutive subsegments of AC (to see this, consider the projections of these segments and triangles to the plane $\{0\} \times \mathbf{R}^2$ orthogonal to ℓ). Therefore these Δ-moves do not hinder each other and may be performed in an arbitrary order. They replace $AC \subset L$ with a finite sequence of positive edges, beginning at A and ending at C. The resulting polygonal link is isotopic to L in $\mathbf{R}^3$. Applying this procedure inductively to all negative edges of L, we obtain a closed braid isotopic to L. $\square$

Exercise 2.3.1. Verify that the oriented 2-component links obtained by closing $\sigma_1^2 \in B_2$ and $\sigma_1^{-2} \in B_2$ are not isotopic, while the underlying unoriented
links are isotopic. (Hint: Consider the linking number of the components.)

Exercise 2.3.2. Observe that the closure of σ_1^3 is the trefoil knot shown on the
left of Figure 2.1 and endowed with an orientation. Observe that the closure
of $\sigma_1^{-1}\sigma_2\sigma_1^{-1}\sigma_2$ is the figure-eight knot shown in Figure 2.1 and endowed with
an orientation.

Exercise 2.3.3. Verify that the oriented link in $\mathbf{R}^3$ obtained by inverting the
orientation of all components of the closure of a braid $\sigma_{i_1}^{r_1}\sigma_{i_2}^{r_2}\cdots\sigma_{i_m}^{r_m}$, where
$r_1, r_2, \ldots, r_m \in \mathbf{Z}$, is isotopic to the closure of $\sigma_{i_m}^{r_m}\cdots\sigma_{i_2}^{r_2}\sigma_{i_1}^{r_1}$.

2.4 Links as closures of braids: an algorithm

By Alexander's theorem, every oriented link $L \subset \mathbf{R}^3$ is isotopic to a closed
braid. It is useful to be able to find such a braid starting from a diagram of L.
The proof of Alexander's theorem given above is not of much help: in the
course of the proof, the diagram is modified by global transformations over
which we have little control. In this section we give a simple algorithm deriving
from any diagram of L a braid whose closure is isotopic to L. Incidentally,
this will give another proof of Alexander's theorem.

2.4.1 Preliminaries

We observe first that any two disjoint oriented (topological) circles on the
sphere S^2 bound an annulus in S^2. These circles are said to be *incompatible* if
their orientation is induced by an orientation of this annulus. Otherwise, these
circles are *compatible*. For instance, two oriented concentric cirles in $\mathbf{R}^2 \subset S^2$
are compatible if they both are oriented clockwise or both counterclockwise.

Consider an oriented link diagram $\mathcal{D}$ in $\mathbf{R}^2$. Near each crossing point x
of $\mathcal{D}$ the diagram looks either like the 2-braid σ_1 or like the 2-braid σ_1^{-1}. A
smoothing of $\mathcal{D}$ at x replaces this 2-braid with a trivial 2-braid and keeps the
rest of $\mathcal{D}$ untouched; see Figure 2.8. Smoothing $\mathcal{D}$ at all crossings, we obtain a
closed oriented 1-dimensional submanifold of $\mathbf{R}^2$. It consists of a finite number
of disjoint oriented (topological) circles called the *Seifert circles* of $\mathcal{D}$. The
number of Seifert circles of $\mathcal{D}$ is denoted by $n(\mathcal{D})$. Two Seifert circles of $\mathcal{D}$ are
compatible (resp. *incompatible*) if they are compatible (resp. incompatible) in
$S^2 = \mathbf{R}^2 \cup \{\infty\}$. The number of pairs of incompatible Seifert circles of $\mathcal{D}$ is
denoted by $h(\mathcal{D})$ and is called the *height* of $\mathcal{D}$. Clearly, $0 \leq h(\mathcal{D}) \leq n(n-1)/2$,
where $n = n(\mathcal{D})$. Both numbers $n(\mathcal{D})$ and $h(\mathcal{D})$ are isotopy invariants of $\mathcal{D}$.

An oriented link diagram $\mathcal{D}$ in $\mathbf{R}^2$ is a *closed braid diagram on n strands*
if it lies in an annulus $S^1 \times I \in \mathbf{R}^2$ and is a closed braid diagram in this
annulus in the sense of Section 2.2.5. It is understood that all strands of $\mathcal{D}$
are oriented counterclockwise.

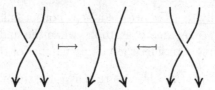

Fig. 2.8. Smoothing of a crossing

Examples of such $\mathcal{D}$ are obtained from braid diagrams on n strands by connecting the bottom and top endpoints by n disjoint arcs in $\mathbf{R}^2$ as in Figure 2.5, with orientation of the strands induced by the orientation on the braid from the top to the bottom. Smoothing a closed braid diagram $\mathcal{D}$ on n strands at all crossings, we obtain a closed braid diagram on n strands without crossings. Such a diagram consists of n disjoint concentric circles in $\mathbf{R}^2$ with counterclockwise orientation. Thus, $n(\mathcal{D}) = n$ and $h(\mathcal{D}) = 0$.

2.4.2 Bending and tightening of link diagrams

Consider an oriented link diagram $\mathcal{D}$ in $\mathbf{R}^2$. Let

$$|\mathcal{D}| \subset \mathbf{R}^2$$

be the union of the components of $\mathcal{D}$ with the over/undercrossing data forgotten. This is a 4-valent graph in $\mathbf{R}^2$ whose vertices are the crossings of $\mathcal{D}$. By an *edge* of $\mathcal{D}$, we mean a connected component of the complement of the set of crossings in $|\mathcal{D}|$. Edges of $\mathcal{D}$ are embedded arcs or circles in $\mathbf{R}^2$ (the circles arise from the components of $\mathcal{D}$ having no crossings). By a *face* of $\mathcal{D}$, we mean a connected component of $\mathbf{R}^2 - |\mathcal{D}|$. We say that a face f of $\mathcal{D}$ is *adjacent* to an edge a of $\mathcal{D}$ if a is contained in the closure of f. We say that f is *adjacent* to a Seifert circle S of $\mathcal{D}$ if f is adjacent to at least one edge of $\mathcal{D}$ contained in S. A face f of $\mathcal{D}$ is a *defect face* if f is adjacent to distinct edges a_1, a_2 of $\mathcal{D}$ such that the Seifert circles S_1, S_2 of $\mathcal{D}$ going along a_1, a_2 are distinct and incompatible. An oriented embedded arc $c \subset \mathbf{R}^2$ leading from a point of a_1 to a point of a_2 and lying (except the endpoints) in f is called a *reduction arc* of $\mathcal{D}$ in f. The incompatibility of S_1, S_2 may be reformulated by saying that one of the edges a_1, a_2 crosses c from right to left and the other one crosses c from left to right. Given such a_1, a_2, c, we can apply to $\mathcal{D}$ the second Reidemeister move pushing a subarc of a_1 along c and then sliding it over a_2; see Figure 2.9. We call this move a *bending* of $\mathcal{D}$ along c *involving the (incompatible) Seifert circles* S_1, S_2. This move produces a diagram of an isotopic link. The inverse move is called a *tightening*.

For example, consider the diagram $\mathcal{D}$ of a trivial knot in $\mathbf{R}^3$ shown on the left of Figure 2.10. The underlying graph $|\mathcal{D}|$ has two vertices and four edges. Smoothing $\mathcal{D}$ at both crossings, we obtain three Seifert circles. All three are oriented counterclockwise and one of them encloses the other two.

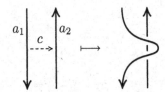

Fig. 2.9. Bending along an arc c

The two smaller circles are incompatible with each other and compatible with the bigger circle. Thus, $n(\mathcal{D}) = 3$ and $h(\mathcal{D}) = 1$. The diagram $\mathcal{D}$ has one defect face. A reduction arc in this face is represented by the dotted arrow on the left-hand side of Figure 2.10. Bending $\mathcal{D}$ along this arc, we obtain the diagram on the right-hand side of Figure 2.10. This diagram is a closed braid diagram in the annulus bounded by the dotted circles. (This diagram is isotopic to the closure of $\sigma_1\sigma_2\sigma_1\sigma_2^{-1}$.) As we shall see below, this example is typical in the sense that any oriented link diagram can be transformed by a sequence of bendings and isotopies into a closed braid diagram.

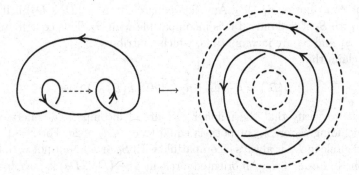

Fig. 2.10. Example of a bending

The following three lemmas give a key to the transformation of link diagrams into closed braid diagrams.

Lemma 2.4. *If $\mathcal{D}'$ is obtained from an oriented link diagram $\mathcal{D}$ in $\mathbf{R}^2$ by a bending, then $n(\mathcal{D}') = n(\mathcal{D})$ and $h(\mathcal{D}') = h(\mathcal{D}) - 1$.*

Proof. Let S_1, S_2 be the incompatible (distinct) Seifert circles of $\mathcal{D}$ involved in the bending; see Figure 2.11. The small biangle created by the bending gives rise to a Seifert circle of $\mathcal{D}'$, denoted by S_0. The remaining parts of S_1, S_2 give rise to a Seifert circle of $\mathcal{D}'$, denoted by S_∞. All other Seifert circles of $\mathcal{D}$ survive in $\mathcal{D}'$ without changes. Therefore $n(\mathcal{D}') = n(\mathcal{D})$. Note that the Seifert circles of $\mathcal{D}$ and $\mathcal{D}'$ do not pass through the shaded areas in Figure 2.11.

We now compare the heights $h(\mathcal{D})$ and $h(\mathcal{D}')$. Observe first that the Seifert circles S_1, S_2 bound respective disjoint disks D_1, D_2 in $S^2 = \mathbf{R}^2 \cup \{\infty\}$.

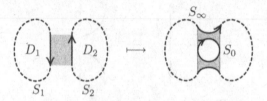

Fig. 2.11. Seifert circles before and after bending

For $i = 1, 2$, let d_i denote the number of Seifert circles of $\mathcal{D}$ lying in the open disk $D_i^\circ = D_i - \partial D_i$. Let d be the number of Seifert circles of $\mathcal{D}$ lying in the annulus $S^2 - (D_1 \cup D_2)$ and incompatible with S_1. Finally, let h be the number of pairs of incompatible Seifert circles of $\mathcal{D}$ both distinct from S_1, S_2. We claim that

$$h(\mathcal{D}) = h + d_1 + d_2 + 2d + 1.$$

It suffices to verify that the number of pairs of incompatible Seifert circles of $\mathcal{D}$ including S_1 or S_2 or both is equal to $d_1 + d_2 + 2d + 1$. For $i = 1, 2$, an oriented circle in D_i° is incompatible with S_1 or S_2, but not with both. This gives the contribution $d_1 + d_2$. An oriented circle in $S^2 - (D_1 \cup D_2)$ is incompatible with S_1 if and only if it is incompatible with S_2. This contributes $2d$. Finally, S_1 and S_2 are incompatible, which contributes 1.

We claim that

$$h(\mathcal{D}') = h + d_1 + d_2 + 2d = h(\mathcal{D}) - 1.$$

It suffices to verify that the number of pairs of incompatible Seifert circles of $\mathcal{D}'$ including S_0 or S_∞ or both is equal to $d_1 + d_2 + 2d$. For $i = 1, 2$, an oriented circle in D_i° is always incompatible with S_0 or S_∞, but not with both. This contributes $d_1 + d_2$. An oriented circle in $S^2 - (D_1 \cup D_2)$ is incompatible with S_0 if and only if it is incompatible with S_∞ and if and only if it is incompatible with S_1. This contributes $2d$. Finally, S_0 and S_∞ are compatible. Hence $h(\mathcal{D}') = h(\mathcal{D}) - 1$. ☐

Lemma 2.5. *An oriented link diagram $\mathcal{D}$ in $\mathbf{R}^2$ has a defect face if and only if $h(\mathcal{D}) \neq 0$.*

Proof. Cutting S^2 open along the Seifert circles of $\mathcal{D}$, we obtain a compact surface Σ with boundary. For a crossing x of $\mathcal{D}$, denote by γ_x a line segment near x joining the Seifert circles as in Figure 2.12. These segments are all disjoint and each of them lies in a component of Σ.

If $\mathcal{D}$ has a defect face, then clearly $h(\mathcal{D}) > 0$. We prove the converse: if $h(\mathcal{D}) > 0$, then $\mathcal{D}$ has a defect face. We first prove that there are a component F of Σ and two Seifert circles in ∂F whose orientation is induced by an orientation on F. Pick two incompatible Seifert circles S_1, S_2 of $\mathcal{D}$ and consider an oriented embedded arc $c \subset \mathbf{R}^2$ leading from a point of S_1 to a point of S_2. We can assume that c meets each Seifert circle of $\mathcal{D}$ transversely in at

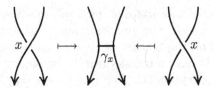

Fig. 2.12. The segment γ_x

most one point. The crossings of c with these circles form a finite subset of c including the endpoints. At each of the crossings, the corresponding Seifert circle is directed to the left or to the right of c. The incompatibility of S_1, S_2 means that their directions at the endpoints of c are opposite: one of these circles is directed to the left of c and the other one is directed to the right of c. Therefore, among the crossings of c with the Seifert circles of $\mathcal{D}$, there are two that lie *consecutively* on c and at which the directions of the corresponding Seifert circles are opposite. The component F of Σ containing the subarc of c between two such crossings satisfies the requirements above. Warning: this subarc may meet certain segments γ_x; then it does not lie in a face of $\mathcal{D}$.

Consider in more detail a component F of Σ such that there are at least two Seifert circles in ∂F whose orientation is induced by an orientation on F. Fix such an orientation on F. Let us call a Seifert circle in ∂F *positive* if its orientation is induced by the one on F and *negative* otherwise. By assumption, there are at least two positive Seifert circles in ∂F. If F contains no segments γ_x, then $F^\circ = F - \partial F$ is a face of $\mathcal{D}$ adjacent to ≥ 2 positive Seifert circles in ∂F. Hence this face is a defect face. Suppose that F contains certain segments γ_x. Removing them all from F, we obtain a subsurface $F' \subset F$. It is clear that any component f of F' is adjacent to at least one segment γ_x and the interior of f is a face of $\mathcal{D}$. Each $\gamma_x \subset F$ connects a positive Seifert circle in ∂F with a negative one. Therefore f is adjacent to at least one positive and at least one negative Seifert circle. If f is adjacent to at least two positive or to at least two negative Seifert circles, then f is a defect face. Suppose that each component f of F' is adjacent to exactly one positive and exactly one negative Seifert circle. Note that moving from f to a neighboring component of F' across some $\gamma_x \subset F$, we meet the same Seifert circles. Since F is connected, we can move in this way from any component of F' to any other component. Therefore ∂F contains exactly one positive and one negative Seifert circle. This contradicts our assumptions. Hence $\mathcal{D}$ has a defect face. $\square$

Lemma 2.6. *An oriented link diagram $\mathcal{D}$ in $\mathbf{R}^2$ with $h(\mathcal{D}) = 0$ is isotopic in the sphere $S^2 = \mathbf{R}^2 \cup \{\infty\}$ to a closed braid diagram in $\mathbf{R}^2$.*

Proof. Let Σ and $\{\gamma_x\}_x$ be the same objects as in the proof of the previous lemma. Suppose that $h(\mathcal{D}) = 0$. We must prove that $\mathcal{D}$ is isotopic in S^2 to a closed braid diagram in the plane $\mathbf{R}^2 = S^2 - \{\infty\}$. If a certain component of the surface Σ has three or more boundary components, then two of them must be incompatible in S^2, which contradicts our assumption $h(\mathcal{D}) = 0$.

A compact connected subsurface of the 2-sphere whose boundary has one or two components is a disk or an annulus. Thus, Σ consists only of disks and annuli. An induction on the number of annuli components of Σ shows that the Seifert circles of $\mathcal{D}$ can be transformed by an isotopy of S^2 into a union of disjoint concentric circles in $\mathbf{R}^2$. Applying this isotopy of S^2 to $\mathcal{D}$, we can assume from the very beginning that the Seifert circles of $\mathcal{D}$ are concentric circles in $\mathbf{R}^2$. The equality $h(\mathcal{D}) = 0$ implies that all these circles are oriented in the same direction, either clockwise or counterclockwise. In the first case we apply to $\mathcal{D}$ an additional isotopy pushing all its Seifert circles across $\infty \in S^2$ so that in the final position the Seifert circles of $\mathcal{D}$ become concentric circles in $\mathbf{R}^2$ with counterclockwise orientation. With a further isotopy of $\mathcal{D}$, we can additionally ensure that these circles are concentric Euclidean circles and the segments γ_x are radial, i.e., are contained in some radii. The resulting link diagram is transversal to all radii and therefore is a closed braid diagram. $\square$

2.4.3 The algorithm

Now we can describe an algorithm transforming any diagram $\mathcal{D}$ of an oriented link L in $\mathbf{R}^3$ into a closed braid diagram of L. It suffices to perform a bending on the diagram each time there is a defect face. By Lemmas 2.4 and 2.5, this process stops after $h(\mathcal{D})$ steps and yields a diagram $\mathcal{D}'$ of L with $n(\mathcal{D}') = n(\mathcal{D})$ and $h(\mathcal{D}') = 0$. By Lemma 2.6, $\mathcal{D}'$ is isotopic in S^2 to a closed braid diagram, $\mathcal{D}_0$, in $\mathbf{R}^2$. The latter diagram also represents L; cf. Exercise 2.1.2. Since the number of Seifert circles is an isotopy invariant, $n(\mathcal{D}_0) = n(\mathcal{D}') = n(\mathcal{D})$. Thus $\mathcal{D}_0$ is a closed braid diagram on $n = n(\mathcal{D})$ strands. If $\mathcal{D}$ has k crossings, then $\mathcal{D}_0$ has $k + 2h(\mathcal{D})$ crossings. The corresponding braid is represented by a word of length $k + 2h(\mathcal{D}) \leq k + n(n-1)$ in the generators $\sigma_1^{\pm 1}, \ldots, \sigma_{n-1}^{\pm 1} \in B_n$.

Note the following corollary of this algorithm.

Corollary 2.7. *If an oriented link in $\mathbf{R}^3$ is presented by a diagram with n Seifert circles, then it is isotopic to a closed n-braid.*

The converse to this corollary is also true, since as we know, a closed braid diagram on n strands has n Seifert circles.

Exercise 2.4.1. Show that smoothing of a crossing (or of any number of crossings) on an oriented link diagram does not increase the number of defect faces.

Solution. Let $\mathcal{D}$ be an oriented link diagram and let $\mathcal{D}_x$ be the oriented link diagram obtained from $\mathcal{D}$ by smoothing at a crossing x. Observe that $\mathcal{D}$ and $\mathcal{D}_x$ have the same Seifert circles. Denote by γ_x the line segment near x joining Seifert circles as in Figure 2.12. Let f be the face of $\mathcal{D}_x$ containing γ_x. If $f - \gamma_x$ is connected, then $\mathcal{D}$ and $\mathcal{D}_x$ have the same faces. Then they have an equal number of defect faces. Suppose that γ_x splits f into two connected pieces f_1, f_2, which are then faces of $\mathcal{D}$. It suffices to prove that if f_1, f_2 are not defect faces of $\mathcal{D}$, then f is not a defect face of $\mathcal{D}_x$. Since f_1 (resp. f_2) is not a defect face, it is adjacent to at most two Seifert circles. Since $\gamma_x \subset f$ joins Seifert circles of different signs (with respect to any orientation of f), these circles are distinct and compatible. Therefore f_1 and f_2 are adjacent to the same pair of distinct compatible Seifert circles. The face f is adjacent to the same circles. Therefore f is not a defect face.

2.5 Markov's theorem

We state a fundamental theorem that allows us to describe all braids with isotopic closures in $\mathbf{R}^3$. This theorem, due to A. Markov, is based on so-called Markov moves on braids.

2.5.1 Markov moves

The presentation of an oriented link in $\mathbf{R}^3$ as a closed braid is far from being unique. As we know, if two braids $\beta, \beta' \in B_n$ are conjugate (we record it as $\beta \sim_c \beta'$), then their closures $\widehat{\beta}, \widehat{\beta'}$ are isotopic in the solid torus and a fortiori in $\mathbf{R}^3$. In general, the converse is not true. For instance, the closures of the 2-string braids σ_1, σ_1^{-1} are trivial knots although these braids are not conjugate in $B_2 \cong \mathbf{Z}$. There is another simple construction of braids with isotopic closures. For $\beta \in B_n$, consider the braids $\sigma_n \, \iota(\beta)$ and $\sigma_n^{-1}\iota(\beta)$, where ι is the natural embedding $B_n \hookrightarrow B_{n+1}$. Drawing pictures, one easily observes that the closures of $\sigma_n \, \iota(\beta)$ and $\sigma_n^{-1}\iota(\beta)$ are isotopic to $\widehat{\beta}$ in $\mathbf{R}^3$.

For $\beta, \gamma \in B_n$, the transformation $\beta \mapsto \gamma\beta\gamma^{-1}$ is called the *first Markov move* and is denoted by M_1. The transformation $\beta \mapsto \sigma_n^\varepsilon \, \iota(\beta)$ with $\varepsilon = \pm 1$ is called the *second Markov move* and is denoted by M_2. Note that the inverse to an M_1-move is again an M_1-move. We shall say that two braids β, β' (possibly with different numbers of strings) are M-*equivalent* if they can be related by a finite sequence of moves $\mathrm{M}_1, \mathrm{M}_2, \mathrm{M}_2^{-1}$, where M_2^{-1} is the inverse of an M_2-move. We record it as $\beta \sim \beta'$. It is clear that the M-equivalence $\sim$ is an equivalence relation on the disjoint union $\amalg_{n \geq 1} B_n$ of all braid groups. For example, the braids $\sigma_1, \sigma_1^{-1} \in B_2$ are M-equivalent. Indeed, using the equalities $\sigma_2^{-1}\sigma_1^{-1}\sigma_2^{-1} = \sigma_1^{-1}\sigma_2^{-1}\sigma_1^{-1}$ and $\sigma_1^{-1}\sigma_2^{-1}\sigma_1 = \sigma_2\sigma_1^{-1}\sigma_2^{-1}$, we obtain

$$\sigma_1 \sim \sigma_2^{-1}\sigma_1 \sim_c (\sigma_1\sigma_2)^{-1}(\sigma_2^{-1}\sigma_1)(\sigma_1\sigma_2)$$
$$= \sigma_2^{-1}\sigma_1^{-1}\sigma_2^{-1}\sigma_1^2\sigma_2 = \sigma_1^{-1}\sigma_2^{-1}\sigma_1^{-1}\sigma_1^2\sigma_2$$
$$= \sigma_1^{-1}\sigma_2^{-1}\sigma_1\sigma_2 = \sigma_2\sigma_1^{-1}\sigma_2^{-1}\sigma_2$$
$$= \sigma_2\sigma_1^{-1} \sim \sigma_1^{-1} \, .$$

As we saw, the closures of M-equivalent braids are isotopic as oriented links in $\mathbf{R}^3$. The following deep theorem asserts that conversely, any two braids with isotopic closures are M-equivalent.

Theorem 2.8 (A. Markov). *Two braids (possibly with different numbers of strings) have isotopic closures in Euclidean space $\mathbf{R}^3$ if and only if these braids are M-equivalent.*

The following fundamental corollary yields a description of the set of isotopy classes of oriented links in $\mathbf{R}^3$ in terms of braids.

Corollary 2.9. *Let $\mathcal{L}$ be the set of all isotopy classes of nonempty oriented links in $\mathbf{R}^3$. The mapping $\amalg_{n\geq 1} B_n \to \mathcal{L}$ assigning to a braid the isotopy class of its closure induces a bijection from the quotient set $(\amalg_{n\geq 1} B_n)/\sim$ onto $\mathcal{L}$.*

Here the surjectivity follows from Alexander's theorem, while the injectivity follows from Markov's theorem.

The proof of Theorem 2.8 starts in Section 2.5.3 and occupies the rest of the chapter.

2.5.2 Markov functions

Corollary 2.9 allows one to identify isotopy invariants of oriented links in $\mathbf{R}^3$ with functions on $\amalg_{n\geq 1} B_n$ constant on the M-equivalence classes. This leads us to the following definition.

Definition 2.10. *A Markov function with values in a set E is a sequence of set-theoretic maps $\{f_n : B_n \to E\}_{n\geq 1}$, satisfying the following conditions:*

(i) for all $n \geq 1$ and all α, $\beta \in B_n$,

$$f_n(\alpha\beta) = f_n(\beta\alpha) ; \tag{2.1}$$

(ii) for all $n \geq 1$ and all $\beta \in B_n$,

$$f_n(\beta) = f_{n+1}(\sigma_n\beta) \quad and \quad f_n(\beta) = f_{n+1}(\sigma_n^{-1}\beta). \tag{2.2}$$

For example, for any $e \in E$, the constant maps $B_n \to E$ sending B_n to e for all n form a Markov function. More interesting examples of Markov functions will be given in Chapters 3 and 4.

Any Markov function $\{f_n : B_n \to E\}_{n\geq 1}$ determines an E-valued isotopy invariant $\widehat{f}$ of oriented links in $\mathbf{R}^3$ as follows. Let L be an oriented link in $\mathbf{R}^3$. Pick a braid $\beta \in B_n$ whose closure is isotopic to L and set $\widehat{f}(L) = f_n(\beta) \in E$. Note that $\widehat{f}(L)$ does not depend on the choice of β. Indeed, if $\beta' \in B_{n'}$ is another braid whose closure is isotopic to L, then β and β' are M-equivalent (Theorem 2.8). It follows directly from the definition of M-equivalence and the definition of a Markov function that $f_n(\beta) = f_{n'}(\beta')$. The function $\widehat{f}$ is an isotopy invariant of oriented links: if L, L' are isotopic oriented links in $\mathbf{R}^3$ and $\beta \in B_n$ is a braid whose closure is isotopic to L, then the closure of β is also isotopic to L' and $\widehat{f}(L) = f_n(\beta) = \widehat{f}(L')$.

2.5.3 A pivotal lemma

We formulate an important lemma needed in the proof of Theorem 2.8. We begin with some notation. Given two braids $\alpha \in B_m$ and $\beta \in B_n$, we form their *tensor product* $\alpha \otimes \beta \in B_{m+n}$ by placing β to the right of α without any mutual intersection or linking; see Figure 2.13. Here the vertical lines represent bunches of parallel strands with the number of strands indicated near the line.

A diagram of $\alpha \otimes \beta$ is obtained by placing a diagram of β to the right of a diagram of α without mutual crossings. For example, $1_m \otimes 1_n = 1_{m+n}$, where 1_m is the trivial braid on m strands. Clearly,

$$\alpha \otimes \beta = (\alpha \otimes 1_n)(1_m \otimes \beta) = (1_m \otimes \beta)(\alpha \otimes 1_n).$$

Note also that

$$(\alpha \otimes \beta) \otimes \gamma = \alpha \otimes (\beta \otimes \gamma)$$

for any braids α, β, γ. This allows us to suppress the parentheses and to write simply $\alpha \otimes \beta \otimes \gamma$.

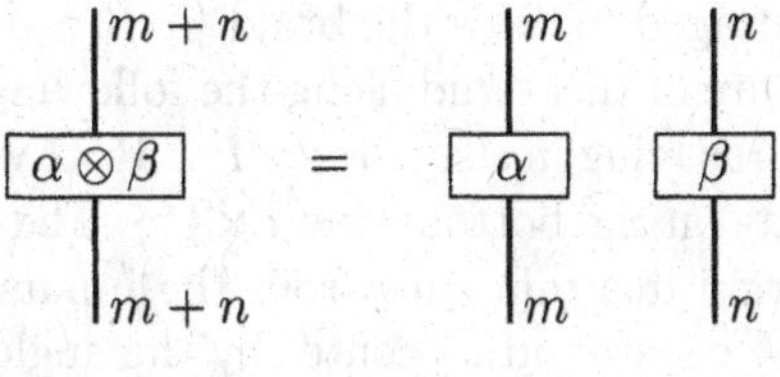

Fig. 2.13. The tensor product of braids

For a sign $\varepsilon = \pm$ and any integers $m, n \geq 0$ with $m + n \geq 1$, we define a braid $\sigma_{m,n}^{\varepsilon} \in B_{m+n}$ as follows. Consider the standard diagram of $\sigma_1 \in B_2$ consisting of two strands with one crossing. Replacing the overcrossing strand with m parallel strands running very closely to each other and similarly replacing the undercrossing strand with n parallel strands, we obtain a braid diagram with $m + n$ strands and mn crossings. This diagram represents $\sigma_{m,n}^{+} \in B_{m+n}$. Transforming all overcrossings in the latter diagram into undercrossings, we obtain a diagram of $\sigma_{m,n}^{-} \in B_{m+n}$. The braids $\sigma_{m,n}^{+}$ and $\sigma_{m,n}^{-}$ are schematically shown in Figure 2.14. In particular,

$$\sigma_{m,0}^{+} = \sigma_{m,0}^{-} = \sigma_{0,m}^{+} = \sigma_{0,m}^{-} = 1_m$$

for all $m \geq 1$. It is clear that $(\sigma_{m,n}^{\varepsilon})^{-1} = \sigma_{n,m}^{-\varepsilon}$ for all m, n, and ε.

It is convenient to introduce the symbols $\sigma_{0,0}^{+}, \sigma_{0,0}^{-}$, and 1_0; they all represent an "empty braid on zero strings" $\emptyset$, which satisfies the identities $\emptyset \otimes \alpha = \alpha \otimes \emptyset = \alpha$ for any genuine braid α.

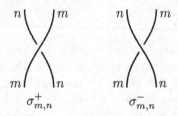

Fig. 2.14. The braids $\sigma_{m,n}^+,\ \sigma_{m,n}^- \in B_{m+n}$

Lemma 2.11. *For any integers $m, n \geq 0$, $r, t \geq 1$, signs $\varepsilon, \nu = \pm$, and braids $\alpha \in B_{n+r}$, $\beta \in B_{n+t}$, $\gamma \in B_{m+t}$, $\delta \in B_{m+r}$, consider the braid*

$$\langle \alpha, \beta, \gamma, \delta \,|\, \varepsilon, \nu \rangle = (1_m \otimes \alpha \otimes 1_t)(1_{m+n} \otimes \sigma_{t,r}^\nu)(1_m \otimes \beta \otimes 1_r)(\sigma_{n,m}^{-\varepsilon} \otimes 1_{t+r})$$
$$\times (1_n \otimes \gamma \otimes 1_r)(1_{n+m} \otimes \sigma_{r,t}^{-\nu})(1_n \otimes \delta \otimes 1_t)(\sigma_{m,n}^{\varepsilon} \otimes 1_{r+t}) \in B_{m+n+r+t}.$$

Then the M-equivalence class of $\langle \alpha, \beta, \gamma, \delta \,|\, \varepsilon, \nu \rangle$ does not depend on ε, ν, and

$$\langle \alpha, \beta, \gamma, \delta \,|\, \varepsilon, \nu \rangle \sim \langle \delta, \gamma, \beta, \alpha \,|\, \varepsilon, \nu \rangle. \tag{2.3}$$

The reader is encouraged to draw the braid $\langle \alpha, \beta, \gamma, \delta \,|\, \varepsilon, \nu \rangle$ for $\varepsilon = \nu = +$. We shall draw the closure of this braid using the following conventions. Let us think of braid diagrams as lying in a square $I \times I \subset \mathbf{R} \times I$ with inputs on the top side $I \times \{0\}$ and outputs on the bottom side $I \times \{1\}$. The standard orientation on the strands of a braid diagram runs from the inputs to the outputs. We can rotate the square $I \times I$ around its center by the angle $\pi/2$. Rotating $I \times I$ by the angle $\pi/2$ counterclockwise (resp. clockwise), we transform any picture a in $I \times I$ into a picture in $I \times I$ denoted by a_+ (resp. a_-). If a is a braid diagram, then the inputs and outputs of a_+, a_- lie on the vertical sides of the square. Note also that $a_{++} = a_{--}$, where $a_{++} = (a_+)_+$ and $a_{--} = (a_-)_-$.

Pick certain diagrams of the braids $\alpha, \beta, \gamma, \delta$, which we denote by the same letters $\alpha, \beta, \gamma, \delta$, respectively. A little contemplation should persuade the reader that Figure 2.15 represents the closure of the braid $\langle \alpha, \beta, \gamma, \delta \,|\, +, + \rangle$.

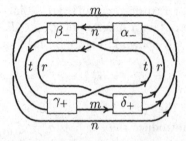

Fig. 2.15. The closure of $\langle \alpha, \beta, \gamma, \delta \,|\, +, + \rangle$

The rest of the proof of Theorem 2.8 goes as follows. In Section 2.6 we deduce this theorem from Lemma 2.11. In Section 2.7 we prove Lemma 2.11. These two sections use different techniques and can be read in any order.

Exercise 2.5.1. Verify that the braids $\langle \alpha, \beta, \gamma, \delta \mid \varepsilon, \nu \rangle$ and $\langle \alpha, \beta, \gamma, \delta \mid -\varepsilon, -\nu \rangle$ have isotopic closures. Verify that

$$\langle \alpha, \beta, \gamma, \delta \mid \varepsilon, \nu \rangle \sim_c \langle \gamma, \delta, \alpha, \beta \mid -\varepsilon, -\nu \rangle. \tag{2.4}$$

(Hint: Rotate the closed braid in Figure 2.15 through $180°$.)

Exercise 2.5.2. Verify (2.3) for $m = n = 0$.

2.6 Deduction of Markov's theorem from Lemma 2.11

We begin by introducing an additional Markov move.

2.6.1 The move M_3

By definition, the second Markov move M_2 transforms a braid $\beta \in B_n$ into $\sigma_n^\varepsilon(\beta \otimes 1_1)$ with $\varepsilon = \pm 1$. We define another move M_3 on braids transforming $\beta \in B_n$ into $\sigma_1^\varepsilon(1_1 \otimes \beta) \in B_{n+1}$. One can check directly that M_3 preserves the isotopy class of the closure.

Lemma 2.12. *The move M_3 expands as a composition of the moves M_1, M_2.*

Proof. Recall the braid $\Delta_n \in B_n$ defined in Section 1.3.3. By formula (1.8),

$$\Delta_n \sigma_i \Delta_n^{-1} = \sigma_{n-i} \in B_n \tag{2.5}$$

for all $n \geq 1$ and all $i = 1, \ldots, n-1$. In particular, $\Delta_{n+1} \sigma_1 \Delta_{n+1}^{-1} = \sigma_n \in B_{n+1}$. Taking the inverses in B_{n+1}, we obtain

$$\Delta_{n+1} \sigma_1^\varepsilon \Delta_{n+1}^{-1} = \sigma_n^\varepsilon \tag{2.6}$$

for $\varepsilon = \pm 1$. We check now that for any $\beta \in B_n$,

$$\Delta_{n+1}(1_1 \otimes \beta)\Delta_{n+1}^{-1} = \Delta_n \beta \Delta_n^{-1} \otimes 1_1. \tag{2.7}$$

Both sides of (2.7) are multiplicative with respect to β, so that it suffices to verify (2.7) for $\beta = \sigma_i \in B_n$, where $i = 1, \ldots, n-1$. We have

$$1_1 \otimes \sigma_i = \sigma_{i+1} \in B_{n+1}$$

and

$$\Delta_{n+1}(1_1 \otimes \sigma_i)\Delta_{n+1}^{-1} = \Delta_{n+1}\sigma_{i+1}\Delta_{n+1}^{-1} = \sigma_{(n+1)-(i+1)} = \sigma_{n-i} \in B_{n+1}.$$

At the same time, $\Delta_n \sigma_i \Delta_n^{-1} = \sigma_{n-i} \in B_n$ and

$$\Delta_n \sigma_i \Delta_n^{-1} \otimes 1_1 = \sigma_{n-i} \in B_{n+1}.$$

This proves (2.7). Multiplying (2.6) and (2.7), we obtain

$$\Delta_{n+1} \sigma_1^\varepsilon (1_1 \otimes \beta) \Delta_{n+1}^{-1} = \sigma_n^\varepsilon (\Delta_n \beta \Delta_n^{-1} \otimes 1_1)$$

or, equivalently,

$$\sigma_1^\varepsilon (1_1 \otimes \beta) = \Delta_{n+1}^{-1} \sigma_n^\varepsilon (\Delta_n \beta \Delta_n^{-1} \otimes 1_1) \Delta_{n+1}.$$

Hence M_3 is a composition of the conjugation by Δ_n with M_2 and with the conjugation by Δ_{n+1}^{-1}. $\square$

This lemma implies that the moves M_1, M_2, M_3 generate the same equivalence relation $\sim$ on the set $\amalg_{n\geq 1} B_n$ as M_1, M_2.

2.6.2 Reduction of Theorem 2.8 to Claim 2.15

We now reformulate Theorem 2.8 in terms of closed braids in the solid torus $V \subset \mathbf{R}^3$. Let $\widehat{M}_2$ be the transformation of closed braids in V replacing the closure of a braid β on n strings with the closure of $\sigma_n^\varepsilon (\beta \otimes 1_1)$, where $\varepsilon = \pm 1$. Let $\widehat{M}_3$ be the transformation of closed braids in V replacing the closure of a braid β on n strings with the closure of $\sigma_1^\varepsilon (1_1 \otimes \beta)$, where $\varepsilon = \pm 1$. The moves inverse to $\widehat{M}_2$, $\widehat{M}_3$ are denoted by $\widehat{M}_2^{-1}$, $\widehat{M}_3^{-1}$, respectively. By Theorem 2.1, to prove Theorem 2.8 it suffices to prove the following assertion.

Claim 2.13. *Two closed braids in V representing isotopic oriented links in $\mathbf{R}^3$ can be related by a sequence of moves $\widehat{M}_2^{\pm 1}, \widehat{M}_3^{\pm 1}$ and isotopies in V.*

Here and below all sequences of moves are finite. In Claim 2.13, by isotopy in V we mean a move replacing a closed braid in V with a closed braid in V isotopic to the first one in the class of oriented links in V.

We can reformulate Claim 2.13 in terms of closed braid diagrams in the annulus, as defined in Section 2.2.5. Let $\widetilde{M}_2$ (resp. $\widetilde{M}_3$) be the transformation of closed braid diagrams replacing the closure of a braid diagram β on n strands with the closure of $\sigma_n^\varepsilon (\beta \otimes 1_1)$ (resp. of $\sigma_1^\varepsilon (1_1 \otimes \beta)$), where $\varepsilon = \pm 1$. The moves $\widetilde{M}_2, \widetilde{M}_3$ are just the moves $\widehat{M}_2, \widehat{M}_3$ restated in terms of diagrams. The moves on closed braid diagrams inverse to $\widetilde{M}_2, \widetilde{M}_3$ are denoted by $\widetilde{M}_2^{-1}, \widetilde{M}_3^{-1}$, respectively. Recall the braidlike Reidemeister moves Ω_2^{br}, Ω_3^{br}; see Sections 2.1.3 and 2.2.5. To prove Claim 2.13 it suffices to prove the following.

Claim 2.14. *Two closed braid diagrams in an annulus $A \subset \mathbf{R}^2$ representing isotopic oriented links in $\mathbf{R}^3$ can be related by a sequence of moves $(\Omega_2^{br})^{\pm 1}$, $(\Omega_3^{br})^{\pm 1}$, $\widetilde{M}_2^{\pm 1}, \widetilde{M}_3^{\pm 1}$ and isotopies in the class of oriented link diagrams in A.*

The isotopies here should begin and end with closed braid diagrams in A (with their canonical orientation), but the intermediate oriented link diagrams in A are not required to be closed braid diagrams.

We shall now reduce Claim 2.14 to another claim formulated in terms of so-called 0-diagrams. We use the notation and the terminology introduced in Section 2.4. A 0-*diagram* is an oriented link diagram $\mathcal{D}$ in $\mathbf{R}^2$ such that $h(\mathcal{D}) = 0$ and all the Seifert circles of $\mathcal{D}$ are oriented counterclockwise. These conditions imply that the Seifert circles of $\mathcal{D}$ form a system of concentric circles in $\mathbf{R}^2$. These circles can be numbered by $1, 2, \ldots, n(\mathcal{D})$ counting from the smallest (innermost) circle toward the biggest (outermost) one. Note that the braidlike moves $\Omega_2^{\mathrm{br}}, \Omega_3^{\mathrm{br}}$ transform 0-diagrams into 0-diagrams. The move Ω_1 adding a kink on the left or on the right of a 0-diagram, generally speaking, does not yield a 0-diagram. (Here the left side and the right side of a diagram are determined by its orientation and the counterclockwise orientation in $\mathbf{R}^2$.) However, for any 0-diagram $\mathcal{D}$, the Ω_1-move adding a left kink at a point of $\mathcal{D}$ lying on the innermost Seifert circle yields a 0-diagram $\mathcal{D}'$. The kink becomes the innermost Seifert circle of $\mathcal{D}'$. Such a transformation $\mathcal{D} \mapsto \mathcal{D}'$ is denoted by Ω_1^{int}. Similarly, adding a right kink at a point of $\mathcal{D}$ lying on its outermost Seifert circle and then pushing the kink across the point $\infty \in S^2$ so that it encircles this point, we obtain again a 0-diagram $\mathcal{D}''$ in $\mathbf{R}^2$. The kink becomes the outermost Seifert circle of this diagram. Such a transformation $\mathcal{D} \mapsto \mathcal{D}''$ is denoted by Ω_1^{ext}. In the sequel, by Ω-*moves* on 0-diagrams we mean the transformations $\Omega_2^{\mathrm{br}}, \Omega_3^{\mathrm{br}}, \Omega_1^{\mathrm{int}}, \Omega_1^{\mathrm{ext}}$, the inverse transformations, and isotopies in $\mathbf{R}^2$.

Claim 2.15. *Two 0-diagrams in $\mathbf{R}^2$ representing isotopic oriented links in $\mathbf{R}^3$ can be related by a sequence of Ω-moves.*

This claim implies Claim 2.14. To see this, note first that closed braid diagrams in an annulus $A \subset \mathbf{R}^2$ are 0-diagrams and for them, $\Omega_1^{\mathrm{int}} = \widetilde{\mathrm{M}}_2$ and $\Omega_1^{\mathrm{ext}} = \widetilde{\mathrm{M}}_3$. Consider now two closed braid diagrams $\mathcal{C}, \mathcal{D}$ in A representing isotopic oriented links in $\mathbf{R}^3$. By Claim 2.15, there is a sequence of 0-diagrams $\mathcal{C} = \mathcal{C}_1, \mathcal{C}_2, \ldots, \mathcal{C}_m = \mathcal{D}$ in $\mathbf{R}^2$ such that each $\mathcal{C}_{i+1}$ is obtained from $\mathcal{C}_i$ by an Ω-move. The construction in the proof of Lemma 2.6 shows that each $\mathcal{C}_i$ is isotopic to a closed braid diagram $\mathcal{B}_i$ in A. It is clear that if $\mathcal{C}_{i+1}$ is obtained from $\mathcal{C}_i$ by $(\Omega_2^{\mathrm{br}})^{\pm 1}, (\Omega_3^{\mathrm{br}})^{\pm 1}, (\Omega_1^{\mathrm{int}})^{\pm 1}, (\Omega_1^{\mathrm{ext}})^{\pm 1}$, then $\mathcal{B}_{i+1}$ is obtained from $\mathcal{B}_i$ by $(\Omega_2^{\mathrm{br}})^{\pm 1}, (\Omega_3^{\mathrm{br}})^{\pm 1}, \widetilde{\mathrm{M}}_3^{\pm 1}, \widetilde{\mathrm{M}}_2^{\pm 1}$, respectively. A little thinking shows that if $\mathcal{C}_{i+1}$ is obtained from $\mathcal{C}_i$ by an isotopy in $\mathbf{R}^2$, then $\mathcal{B}_{i+1}$ is obtained from $\mathcal{B}_i$ by an isotopy in A. This yields Claim 2.14.

2.6.3 Reduction to Lemma 2.17

Recall the isotopies, bendings, and tightenings of link diagrams as defined in Sections 2.1.2 and 2.4.2. The proof of Claim 2.15 begins with the following lemma.

Lemma 2.16. *Let $\mathcal{E}$, $\mathcal{E}'$ be 0-diagrams in $\mathbf{R}^2$ representing isotopic oriented links in $\mathbf{R}^3$. Then there is a sequence of 0-diagrams $\mathcal{E} = \mathcal{E}_1, \mathcal{E}_2, \ldots, \mathcal{E}_m = \mathcal{E}'$ such that for all $i = 1, 2, \ldots, m - 1$, the diagram $\mathcal{E}_{i+1}$ is obtained from $\mathcal{E}_i$ by an Ω-move or by a sequence of bendings, tightenings, and isotopies in the sphere $S^2 = \mathbf{R}^2 \cup \{\infty\}$.*

Proof. Since $\mathcal{E}, \mathcal{E}'$ represent isotopic links, they can be related by a sequence of the following oriented Reidemeister moves: (a) $\Omega_1^{\pm 1}$, (b) $(\Omega_2^{\mathrm{br}})^{\pm 1}$, $(\Omega_3^{\mathrm{br}})^{\pm 1}$, isotopies in $\mathbf{R}^2$, (c) nonbraidlike moves $\Omega_2^{\pm 1}$. Note that the intermediate diagrams created by these moves may have positive height. We will transform this sequence of moves into another one consisting only of bendings, tightenings, isotopies in S^2, and Ω-moves on 0-diagrams.

Recall from Section 2.4.2 that a nonbraidlike move Ω_2 involving two distinct Seifert circles is a bending. A nonbraidlike move Ω_2 involving only one Seifert circle can be obtained as a composition of two Ω_1, a bending, and a tightening; see Figure 2.16. Therefore we can assume that in our sequence of moves, all moves of type (c) are bendings and tightenings.

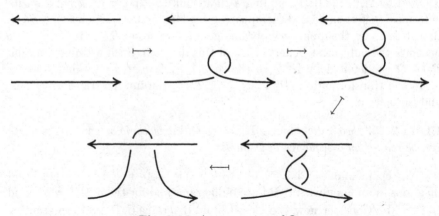

Fig. 2.16. An expansion of Ω_2

Let g be a transformation of type (b) in our sequence applied to a link diagram $\mathcal{D}$ with $h(\mathcal{D}) > 0$. Note that g preserves the set of Seifert circles of the diagram and therefore preserves its height. Since $h(\mathcal{D}) > 0$, the diagram $\mathcal{D}$ has a defect face. We can choose a reduction arc in this face disjoint from the disk where g changes $\mathcal{D}$. Let r be the corresponding bending of $\mathcal{D}$. Clearly, the transformations r and g on $\mathcal{D}$ commute. We replace the transformation $\mathcal{D} \mapsto g(\mathcal{D})$ in our sequence with the sequence

$$\mathcal{D} \xrightarrow{r} r(\mathcal{D}) \xrightarrow{g} gr(\mathcal{D}) \xrightarrow{r^{-1}} r^{-1}gr(\mathcal{D}) = g(\mathcal{D}).$$

The operation g is now performed at a lower height. Gradually we can go all the way down to height zero. Thus we can replace g with a sequence of bendings, tightenings, and a single move of type (b), say g', on a diagram, $\mathcal{D}'$, of height 0. If the Seifert circles of $\mathcal{D}'$ are oriented counterclockwise, then $\mathcal{D}'$ is a 0-diagram and g' is an Ω-move. If the Seifert circles of $\mathcal{D}'$ are oriented clockwise, then we expand g' as a composition of an isotopy of S^2 transforming $\mathcal{D}'$ into a 0-diagram (cf. the proof of Lemma 2.6), an Ω-move on the latter diagram, and the inverse isotopy.

Let $g = \Omega_1$ be an operation of type (a) in our sequence applied to a link diagram $\mathcal{D}$ in $\mathbf{R}^2$. Inserting bendings and tightenings as above, we can assume that $h(\mathcal{D}) = 0$. Conjugating if necessary g by an isotopy of S^2, we can assume that the Seifert circles of $\mathcal{D}$ are oriented counterclockwise, i.e., that $\mathcal{D}$ is a 0-diagram in $\mathbf{R}^2$. Suppose that the kink added by g to a branch a of $\mathcal{D}$ lies to its left. If a lies on the first (innermost) Seifert circle of $\mathcal{D}$, then $g = \Omega_1^{\mathrm{int}}$. If a lies on the mth Seifert circle of $\mathcal{D}$ with $m \geq 2$, then we apply $m - 1$ moves Ω_2^{br} to push a under $m - 1$ smaller Seifert circles of $\mathcal{D}$ inside the disk bounded by the innermost Seifert circle. Then we apply Ω_1^{int} on a and push the resulting kink back under the first $m - 1$ Seifert circles to the place where the original move $g = \Omega_1$ must have been applied. This pushing should be performed carefully: one first pushes all the $m - 1$ Seifert circles in question over the crossing created by Ω_1^{int}. This amounts to $m - 1$ moves

$$d_1^+ d_2^\pm d_1^- \;\mapsto\; d_2^- d_1^\pm d_2^+$$

analyzed in the proof of Theorem 1.6. (This analysis shows that these moves are compositions of $(\Omega_2^{\mathrm{br}})^{\pm 1}, (\Omega_3^{\mathrm{br}})^{\pm 1}$.) After that, one pushes these $m - 1$ Seifert circles over the remaining part of the kink, which amounts to $m - 1$ tightenings. The resulting chain of moves, schematically shown in Figure 2.17, transforms $\mathcal{D}$ into the same diagram $g(\mathcal{D})$ as g itself. Thus, we can replace the move $\mathcal{D} \mapsto g(\mathcal{D})$ with a finite sequence of moves $(\Omega_2^{\mathrm{br}})^{\pm 1}, (\Omega_3^{\mathrm{br}})^{\pm 1}, \Omega_1^{\mathrm{int}}$ on 0-diagrams followed by $m-1$ tightenings. If the kink added by g lies to the right of a, then we proceed as above but push a toward the external (infinite) face of $\mathcal{D}$ in $\mathbf{R}^2$ and then apply Ω_1^{ext}.   □

Lemma 2.17. *Two 0-diagrams in $\mathbf{R}^2$ related by a sequence of bendings, tightenings, and isotopies in S^2 can be related by a sequence of Ω-moves.*

This lemma together with the previous one implies Claim 2.15 and Theorem 2.8. The rest of the section is devoted to the proof of Lemma 2.17.

2.6.4 Proof of Lemma 2.17, part I

We consider here the simplest case of Lemma 2.17, namely the one in which the sequence relating two 0-diagrams consists solely of isotopies.

Lemma 2.18. *If two 0-diagrams are isotopic in $S^2 = \mathbf{R}^2 \cup \{\infty\}$, then they are isotopic in $\mathbf{R}^2$.*

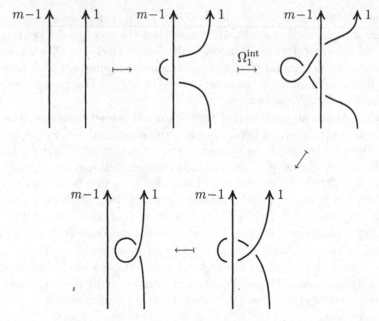

Fig. 2.17. An expansion of Ω_1

Proof. Let $\mathcal{D}, \mathcal{D}'$ be 0-diagrams in $\mathbf{R}^2$ isotopic in S^2. They have then the same number of Seifert circles $N \geq 1$. If $N = 1$, then $\mathcal{D}, \mathcal{D}'$ are embedded circles in $\mathbf{R}^2$ oriented counterclockwise. By the Jordan curve theorem, any embedded circle in $\mathbf{R}^2$ bounds a disk. This implies that such a circle is isotopic to a small metric circle in $\mathbf{R}^2$. Since any two metric circles in $\mathbf{R}^2$, endowed with counterclockwise orientation, are isotopic in $\mathbf{R}^2$, the same holds for $\mathcal{D}, \mathcal{D}'$.

Suppose that $N \geq 2$. Since $\mathcal{D}, \mathcal{D}'$ are isotopic in S^2, there is a continuous family of homeomorphisms $\{F_t : S^2 \to S^2\}_{t \in I}$ such that $F_0 = \mathrm{id}$ and F_1 transforms $\mathcal{D}$ into $\mathcal{D}'$. By continuity, all the homeomorphisms F_t are orientation preserving. The Seifert circles of $\mathcal{D}$ split S^2 into $N - 1$ annuli and two disks $D_i = D_i(\mathcal{D})$ and $D_o = D_o(\mathcal{D})$ bounded by the innermost and the outermost Seifert circles of $\mathcal{D}$, respectively. Recall that $S^2 = \mathbf{R}^2 \cup \{\infty\}$ is oriented counterclockwise and so are all Seifert circles of $\mathcal{D}$. It is clear that the orientation of the innermost Seifert circle ∂D_i is compatible with the orientation of D_i induced from the one on S^2. On the other hand, the orientation of the outermost Seifert circle ∂D_o is incompatible with the orientation of D_o induced from the one on S^2. This implies that $F_1 : S^2 \to S^2$ necessarily transforms $D_i(\mathcal{D})$ into $D_i(\mathcal{D}')$ and $D_o(\mathcal{D})$ into $D_o(\mathcal{D}')$ (and not the other way round).

We have $\infty \in D_o(\mathcal{D})$ and therefore $F_1(\infty) \in D_o(\mathcal{D}')$. Hence, there is a closed 2-disk B in the complement of $\mathcal{D}'$ in S^2 containing the points ∞ and $F_1(\infty)$. Pushing $F_1(\infty)$ toward ∞ inside B, we obtain a continuous family of homeomorphisms $\{g_t : S^2 \to S^2\}_{t \in I}$ such that $g_0 = \mathrm{id}$, $g_1(F_1(\infty)) = \infty$, and all g_t are equal to the identity outside B (cf. the proof of Lemma 1.26).

Then $g_1 F_1(\mathcal{D}) = g_1(\mathcal{D}') = \mathcal{D}'$ and the one-parameter family of homeomorphisms $\{g_t F_t : S^2 \to S^2\}_{t \in I}$ relates $g_0 F_0 = \mathrm{id}$ with $g_1 F_1$. Thus, $g_1 F_1$ is isotopic to the identity in the class of self-homeomorphisms of S^2. By Exercise 1.7.1, $g_1 F_1$ is isotopic to the identity in the class of self-homeomorphisms of S^2 keeping fixed the point ∞. Restricting all homeomorphisms in such an isotopy to $\mathbf{R}^2 = S^2 - \{\infty\}$, we obtain an isotopy of $\mathcal{D}$ into $\mathcal{D}'$ in $\mathbf{R}^2$. $\square$

2.6.5 Proof of Lemma 2.17, part II

Consider a sequence of moves as in Lemma 2.17. By a general position argument, we can assume that the intermediate diagrams created by these moves lie in $\mathbf{R}^2 = S^2 - \{\infty\}$. We will denote bendings and tightenings by arrows pointing in the direction of a lower height. Thus, the notation $C \xleftarrow{s} \mathcal{D} \xrightarrow{s'} C'$ means that the link digram C is transformed into $\mathcal{D}$ by a tightening, inverse to a bending s of $\mathcal{D}$, and $\mathcal{D}$ is transformed into C' by a bending s'. Note that $h(C) = h(C') = h(\mathcal{D}) - 1$, so that the height function h has a local maximum at $\mathcal{D}$. We call such a sequence $C \xleftarrow{s} \mathcal{D} \xrightarrow{s'} C'$ a *local maximum*. Our strategy will be to replace local maxima by (longer) sequences at a lower height.

For a local maximum $C \xleftarrow{s} \mathcal{D} \xrightarrow{s'} C'$, consider the reduction arcs of s and s'. By a general position argument, we can assume that for all local maxima in our sequence of moves, these two arcs have distinct endpoints and meet transversely in a finite number of points. This number is denoted by $s \cdot s'$.

Lemma 2.19. *For any local maximum $C \xleftarrow{s} \mathcal{D} \xrightarrow{s'} C'$ with $s \cdot s' \neq 0$, there is a sequence of bendings and tightenings*

$$C = C_1 \xleftarrow{s_1} \mathcal{D}_1 \xrightarrow{s_1'} C_2 \xleftarrow{s_2} \cdots \xrightarrow{s_{m-1}'} C_m \xleftarrow{s_m} \mathcal{D}_m \xrightarrow{s_m'} C_{m+1} = C'$$

such that $s_i \cdot s_i' = 0$ for all i.

Proof. Since the reduction arcs of link diagrams are oriented, we can speak of their left and right sides (with respect to the counterclockwise orientation in $\mathbf{R}^2$). Each reduction arc c of $\mathcal{D}$ can be pushed slightly to the left or to the right, keeping the endpoints on $\mathcal{D}$. This gives disjoint reduction arcs giving rise to the same bending (at least up to isotopy). These arcs are denoted by c_l, c_r, respectively.

Let c, c' be the reduction arcs of s, s', respectively. Let us suppose first that $s \cdot s' \geq 2$. We prove below that there is a reduction arc c'' of $\mathcal{D}$ disjoint from c' and meeting c at fewer than $s \cdot s'$ points. Consider the sequence

$$C \xleftarrow{s} \mathcal{D} \xrightarrow{s''} C'' \xleftarrow{s''} \mathcal{D} \xrightarrow{s'} C',$$

where s'' is the bending along c''. We have

$$s \cdot s'' = |c \cap c''| < s \cdot s' \quad \text{and} \quad s' \cdot s'' = |c' \cap c''| = 0.$$

Continuing in this way we can reduce the lemma to the case $s \cdot s' = 1$. We now construct c''. Let A, B be distinct points of $c \cap c'$ such that the subarc $AB \subset c$ does not meet c'. Inverting if necessary the orientations of c, c', we can assume that both c and c' are directed from A to B. Assume first that c' crosses c at A from left to right. If c' crosses c at B from right to left, then c'' is obtained by going along c_l' to its intersection point with c_l close to A, then along c_l to its intersection point with c_l' close to B, and then along c_l'. If c' crosses c at B from left to right, then c'' is obtained by going along c_l' to its intersection point with c_r close to A, then along c_r to its intersection point with c_r' close to B, and then along c_r'. It is easy to check that in both cases the arc c'' has the required properties; see Figure 2.18. The case in which c' crosses c at A from right to left is similar.

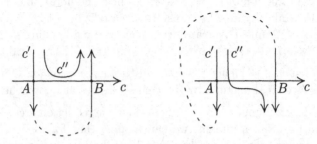

Fig. 2.18. The arc c''

It remains to consider the case $s \cdot s' = 1$. We claim that there is a reduction arc c'' of $\mathcal{D}$ disjoint from $c \cup c'$. Inserting $\mathcal{D} \xrightarrow{s''} C'' \xleftarrow{s''} \mathcal{D}$ as above, we will obtain the claim of the lemma. Let O be the unique point of $c \cap c'$ and let f be the face of $\mathcal{D}$ containing c and c' (except their endpoints). Denote the endpoints of c on $\mathcal{D}$ by A_1, A_2. Denote the endpoints of c' on $\mathcal{D}$ by A_3, A_4. Denote by S_i the Seifert circle of $\mathcal{D}$ passing through A_i. By the definition of a reduction arc, $S_1 \neq S_2$ and $S_3 \neq S_4$. Note that the arc $A_1 O \cup O A_3$ can be slightly deformed into an arc $c_{1,3}$ in $f - (c \cup c')$ leading from a point on S_1 to a point on S_3. The same arc with opposite orientation is denoted by $c_{3,1}$. We similarly define arcs $c_{1,4}$ and $c_{4,1}$; see Figure 2.19.

If two of the circles S_1, S_2, S_3, S_4 coincide, say $S_1 = S_4$, then the circles $S_1 = S_4$ and $S_3 \neq S_4$ are distinct. Since c' is a reduction arc, they are incompatible. Hence $c'' = c_{1,3}$ is a reduction arc satisfying our requirements.

Thus, we can assume that the circles S_1, S_2, S_3, S_4 are all distinct. Their topological position in $S^2 = \mathbf{R}^2 \cup \{\infty\}$ is uniquely determined: they are boundaries of four disjoint disks in S^2 meeting the crosslike graph $c \cup c'$ at its four endpoints. If S_1 and S_3 are incompatible, then $c_{1,3}$ is a reduction arc in $f - (c \cup c')$ and we are done. Assume that S_1 is compatible with S_3. Since S_4 is incompatible with S_3, the circle S_1 is compatible with S_4 as well. Note that the arcs $c_{1,3}$ and $c_{1,4}$ are not reduction arcs.

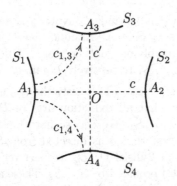

Fig. 2.19. The arcs c, c' and the circles S_1, S_2, S_3, S_4

Recall the disjoint segments γ_x connecting the Seifert circles of $\mathcal{D}$, where x runs over the crossings of $\mathcal{D}$ (see Figure 2.12). The orientation arguments show that the endpoints of each such γ_x necessarily lie on different but compatible Seifert circles. We now distinguish three cases.

Case (i): there are no segments γ_x attached to S_1. A reduction arc of $\mathcal{D}$ connecting S_3 to S_4 is obtained by first following $c_{3,1}$ to a point close to S_1, then encircling S_1 and finally moving along $c_{1,4}$. Since there are no γ_x attached to S_1, this arc lies in $f - (c \cup c')$.

Case (ii): the segments γ_x attached to S_1 connect it to one and the same Seifert circle S. Suppose first that $S \neq S_3$. A reduction arc c'' connecting S_3 to S in $f - (c \cup c')$ is obtained by first following $c_{3,1}$ to a point close to S_1, then encircling S_1 until hitting for the first time a segment γ_x attached to S_1, and then going close to this γ_x until meeting S. If $S = S_3$, then $S \neq S_4$ and we can apply the same construction with S_3 replaced by S_4.

Case (iii): the segments γ_x attached to S_1 connect it to at least two different Seifert circles. We can find two of these segments γ_1, γ_2 with endpoints e_1, e_2 on S_1 such that their second endpoints lie on different Seifert circles and the arc $d \subset S_1 - \{A_1\}$ connecting e_1 to e_2 is disjoint from all the other γ_x attached to S_1. Then a small deformation of the arc $\gamma_1 \cup d \cup \gamma_2$ gives a reduction arc c'' of $\mathcal{D}$ disjoint from $c \cup c'$. □

Lemma 2.20. *For a local maximum $C \xleftarrow{s} D \xrightarrow{s'} C'$ with $s \cdot s' = 0$, there are sequences of isotopies in S^2 and bendings $C \to \cdots \to C_*$, $C' \to \cdots \to C'_*$ such that $C_* = C'_*$ or C_*, C'_* are 0-diagrams in $\mathbf{R}^2$ related by Ω-moves.*

Proof. Let c, c' be the reduction arcs of the bendings s, s' on $\mathcal{D}$. The assumption $s \cdot s' = 0$ implies that c and c' are disjoint. Hence the bendings s and s' are performed in disjoint areas of the plane and commute with each other. Suppose that they involve different pairs of Seifert circles of $\mathcal{D}$ (these pairs may have one common circle). Then c' is a reduction arc for $C = s(\mathcal{D})$ and c is a reduction arc for $C' = s'(\mathcal{D})$. Let $\mathcal{D}'$ be the link diagram obtained by bending C along c' or, equivalently, by bending C' along c. The sequences $C \to \mathcal{D}'$ and $C' \to \mathcal{D}'$ satisfy the conditions of the lemma.

Suppose from now on that s and s' involve the same (distinct and incompatible) Seifert circles S_1, S_2 of $\mathcal{D}$. Assume that $\mathcal{D}$ has a reduction arc c_1 disjoint from $c \cup c'$ and involving another pair of Seifert circles. Then the bendings s, s', s_1 along c, c', c_1, respectively, commute with each other. The sequences $\mathcal{C} \xrightarrow{s_1} \mathcal{C}_1 \xrightarrow{s'} \mathcal{D}', \mathcal{C}' \xrightarrow{s_1} \mathcal{C}'_1 \xrightarrow{s} \mathcal{D}'$ satisfy the conditions of the lemma.

Suppose from now on that all reduction arcs of $\mathcal{D}$ disjoint from $c \cup c'$ involve the Seifert circles S_1, S_2. We choose notation so that c is directed from S_1 to S_2. Assume first that c' is directed from S_1 to S_2. The circles S_1, S_2 bound in S^2 disjoint 2-disks D_1, D_2, respectively. The arcs c, c' lie in the annulus $S^2 - (D_1^\circ \cup D_2^\circ)$ bounded by $S_1 \cup S_2$. These arcs split this annulus into two topological 2-disks D_3, D_4 where $D_3 \cap D_4 = c \cup c'$.

Observe that the Seifert circles of $\mathcal{D}$ distinct from S_1, S_2 are disjoint from $S_1 \cup S_2 \cup c \cup c'$. Therefore the Seifert circles of $\mathcal{D}$ can be partitioned into four disjoint families: the circles lying in D_1, those in D_2, those in the interior of D_3, and those in the interior of D_4. The first two families include $S_1 = \partial D_1$ and $S_2 = \partial D_2$, while the other two families may be empty. To analyze the position of Seifert circles in D_1, note that a reduction arc of $\mathcal{D}$ lying in D_1 is disjoint from $c \cup c'$ or can be made disjoint from $c \cup c'$ by a small deformation near its endpoints. Since such an arc cannot meet S_2, our assumptions imply that $\mathcal{D}$ has no reduction arcs in D_1. The same argument as in the proof of Lemma 2.6 shows that the Seifert circles of $\mathcal{D}$ lying in D_1 form a system of $t \geq 1$ concentric compatible circles with the external circle being S_1. This system of t concentric circles with the same orientation is schematically represented in Figure 2.20 by the left oval. Similar arguments show that the Seifert circles of $\mathcal{D}$ lying in D_2 (resp. in D_3, D_4) form a system of $r \geq 1$ (resp. $n \geq 0, m \geq 0$) concentric circles with the same orientation, represented in Figure 2.20 by the right (resp. upper, lower) oval. The diagram $\mathcal{D}$ is recovered from these four systems of concentric circles by inserting certain braids $\alpha \in B_{n+r}, \beta \in B_{n+t}, \gamma \in B_{m+t}, \delta \in B_{m+r}$ as in Figure 2.20, where we use the notation $\alpha_-, \beta_-, \gamma_+, \delta_+$ introduced after the statement of Lemma 2.11.

Since S_1, S_2 are incompatible they must have the same orientation (clockwise or counterclockwise). For concreteness, we assume that they are oriented counterclockwise. (The case of the clockwise orientation can be reduced to this one by reversing the orientations on $\mathcal{C}, \mathcal{D}, \mathcal{C}'$.) The circles of the other two families are then oriented clockwise: otherwise we can easily find a reduction arc connecting S_1 to one of these circles and disjoint from $c \cup c'$.

Recall that the diagram $\mathcal{C}$ is obtained from $\mathcal{D}$ by a bending s that pushes (a subarc of) S_1 toward S_2 along c and then above S_2. Consider a "superbending" along c pushing the whole band of t circles on the left along c and then over the r right circles. This superbending is a composition of rt bendings, the first of them being s. Moreover, to the resulting link diagram we can apply one more superbending along the arc in S^2 going from the bottom point of the diagram $\mathcal{D}$ down to ∞ and then from ∞ down to the top point of $\mathcal{D}$. (It is of

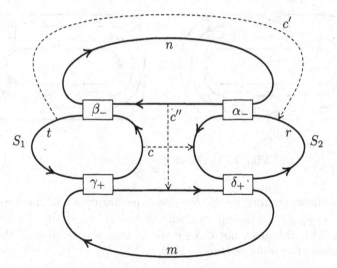

Fig. 2.20. The diagram $\mathcal{D}$

course important that we consider diagrams in S^2 so that reduction arcs and isotopies in S^2 are allowed.)

Performing these two superbendings on $\mathcal{D}$, we obtain the link diagram $\mathcal{C}_*$ drawn in Figure 2.15. (Actually, it is easier to observe the converse, i.e., that $\mathcal{C}_*$ produces $\mathcal{D}$ via two supertightenings inverse to the superbendings described above.) A remarkable although obvious fact is that $\mathcal{C}_*$ is a closed braid diagram and in particular a 0-diagram. In the notation of Lemma 2.11, $\mathcal{C}_*$ represents the closure of the braid $\langle \alpha, \beta, \gamma, \delta \mid +, + \rangle$. As we saw, there is a sequence of $rt + mn$ bendings $\mathcal{D} \xrightarrow{s} \mathcal{C} \to \cdots \to \mathcal{C}_*$ in S^2.

Similarly, we can apply a superbending to $\mathcal{D}$ along the arc c', oriented from S_1 to S_2, and then another superbending along the short vertical segment c'' leading from the bottom point of the upper oval toward the top point of the lower oval in Figure 2.20. This gives a link diagram isotopic to the link diagram $\mathcal{C}'_{**}$ drawn on the left of Figure 2.21. (Again, it is easier to check that the inverse moves transform $\mathcal{C}'_{**}$ into $\mathcal{D}$.) As above, there is a sequence of $rt + mn$ bendings $\mathcal{D} \xrightarrow{s'} \mathcal{C}' \to \cdots \to \mathcal{C}'_{**}$ in S^2.

The diagram $\mathcal{C}'_{**}$ looks like a closed braid diagram, but not quite because its Seifert circles are oriented clockwise. Pushing the lower part of $\mathcal{C}'_{**}$ across $\infty \in S^2$, we obtain that $\mathcal{C}'_{**}$ is isotopic in S^2 to a closed braid diagram $\mathcal{C}'_*$ drawn on the right of Figure 2.21. This diagram represents the closure of $\langle \delta, \gamma, \beta, \alpha \mid +, + \rangle$. By (2.3), the braids $\langle \alpha, \beta, \gamma, \delta \mid +, + \rangle$ and $\langle \delta, \gamma, \beta, \alpha \mid +, + \rangle$ are M-equivalent. Therefore the diagrams $\mathcal{C}_*$ and $\mathcal{C}'_*$, representing the closures of these braids, are related by Ω-moves. This gives the sequences of bendings and isotopies $\mathcal{C} \to \cdots \to \mathcal{C}_*$ and $\mathcal{C}' \to \cdots \to \mathcal{C}'_{**} \to \mathcal{C}'_*$ satisfying the requirements of the lemma.

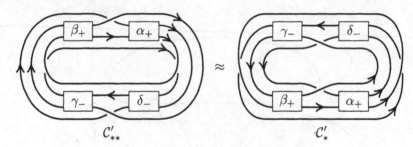

Fig. 2.21. The diagrams C'_{**} and C'_{*}

If c' is directed from S_2 to S_1, then the argument is similar, though $\langle \delta, \gamma, \beta, \alpha \,|\, +, + \rangle$ should be replaced with $\langle \delta, \gamma, \beta, \alpha \,|\, +, - \rangle$. By the first claim of Lemma 2.11, this does not change the M-equivalence class of the braid. This completes the proof of Lemma 2.20. □

2.6.6 Proof of Lemma 2.17, part III

By the *height* of a sequence of bendings, tightenings, and isotopies on link diagrams in S^2, we mean the maximal height of the diagrams appearing in this sequence. We prove the lemma by induction on the height m of the sequence relating two 0-diagrams in $\mathbf{R}^2$.

If $m = 0$, then the sequence consists solely of isotopies in S^2. In this case Lemma 2.17 follows directly from Lemma 2.18.

Assume that $m > 0$. It is clear that a transformation of a link diagram in S^2 obtained as an isotopy followed by a bending (resp. a tightening) can be also obtained as a bending (resp. a tightening) followed by an isotopy. Therefore all isotopies in our sequence of bendings, tightenings, and isotopies in S^2 can be accumulated at the end of the sequence. In particular, all diagrams of height m in this sequence appear as local maxima, i.e., are obtained by tightening from the previous diagram and yield the next diagram by bending. Lemma 2.19 shows that we can replace our sequence with another one that connects the same initial and terminal 0-diagrams, has the same height m, and additionally satisfies the condition that $s \cdot s' = 0$ in all its local maxima $C \overset{s}{\leftarrow} D \overset{s'}{\rightarrow} C'$. By Lemma 2.20, for each such local maximum, there is a sequence of isotopies, bendings, and tightenings

$$ C \to \cdots \to C_* \sim C'_* \leftarrow \cdots \leftarrow C', $$

where $\sim$ stands for the coincidence $C_* = C'_*$ or for Ω-moves transforming C_* into C'_* (which are then 0-diagrams). The height of all link diagrams in this sequence is less than or equal to $h(C) = h(C') < h(D) \le m$. Replacing every local maximum $C \overset{s}{\leftarrow} D \overset{s'}{\rightarrow} C'$ by such a sequence, we obtain a concatenation of sequences of height $\le m - 1$ with sequences of Ω-moves on 0-diagrams. By the induction assumption, this implies the claim of the lemma. □

2.7 Proof of Lemma 2.11

We begin by introducing a useful involution $\beta \mapsto \overline{\beta}$ on the set of braids.

2.7.1 The involution $\beta \mapsto \overline{\beta}$

For a braid $\beta \in B_n$, set $\overline{\beta} = \Delta_n \beta \Delta_n^{-1} \in B_n$, where $\Delta_n \in B_n$ is the braid defined in Section 1.3.3. Since Δ_n^2 lies in the center of B_n, the automorphism $\beta \mapsto \overline{\beta}$ of B_n is an involution. Formula (2.5) implies that if

$$\beta = \sigma_{i_1}^{r_1} \sigma_{i_2}^{r_2} \cdots \sigma_{i_m}^{r_m}$$

with $1 \le i_1, i_2, \ldots, i_m \le n-1$ and $r_1, r_2, \ldots, r_m \in \mathbf{Z}$, then

$$\overline{\beta} = \sigma_{n-i_1}^{r_1} \sigma_{n-i_2}^{r_2} \cdots \sigma_{n-i_m}^{r_m} .$$

This formula implies that a diagram of $\overline{\beta}$ can be obtained from a diagram of β in $\mathbf{R} \times I = \mathbf{R} \times I \times \{0\}$ by rotating about the line $\{(n+1)/2\} \times \mathbf{R} \times \{0\}$ in $\mathbf{R}^3$ by the angle π. This geometric description of the involution $\beta \mapsto \overline{\beta}$ shows that $\overline{\alpha \otimes \beta} = \overline{\beta} \otimes \overline{\alpha}$ for any braids $\alpha \in B_m$ and $\beta \in B_n$. Note for the record that $\overline{\alpha\beta} = \overline{\alpha}\overline{\beta}$ for any $\alpha, \beta \in B_n$ and $\overline{1_n} = 1_n$. It is easy to deduce from the definitions that $\overline{\sigma_{m,n}^\varepsilon} = \sigma_{n,m}^\varepsilon$ for any $m, n \ge 0$ and $\varepsilon = \pm$.

Lemma 2.21. *If two braids β, β' are M-equivalent, then the braids $\overline{\beta}, \overline{\beta'}$ are M-equivalent.*

Proof. We have $\overline{\beta} \sim_c \beta \sim \beta' \sim_c \overline{\beta'}$.   □

2.7.2 Ghost braids

We introduce a class of ghost braids. Let $\mu \in B_{n+k}$ with $n \ge 1, k \ge 0$. We say that μ is *n-right-ghost* and write $\mu \equiv 1_n$ if for any $m \ge 0$ and any $\beta \in B_{m+n}$, we have $(\beta \otimes 1_k)(1_m \otimes \mu) \sim \beta$; see Figure 2.22. Examples of right-ghost braids will be given below. Taking $m = 0$, $\beta = 1_n$, we conclude that $\mu \equiv 1_n \Rightarrow \mu \sim 1_n$. (The converse is in general not true.)

Given an n-right-ghost braid $\mu \in B_{n+k}$, we define a move (a transformation) on braids, denoted by $M(\mu)$. For any $m \ge 0$, $\alpha, \beta \in B_{m+n}$, $\rho \in B_m$, the move $M(\mu)$ transforms $\beta(\rho \otimes 1_n)\alpha$ into $(\beta \otimes 1_k)(\rho \otimes \mu)(\alpha \otimes 1_k)$; see Figure 2.23. The inverse transformation replaces the factor $\rho \otimes \mu$ with $\rho \otimes 1_n$ and deletes 1_k on the right of the other factors. The move $M(\mu)$ and its inverse preserve the M-equivalence class of the braid. Indeed,

$$\beta(\rho \otimes 1_n)\alpha \sim_c \alpha\beta(\rho \otimes 1_n)$$
$$\sim (\alpha\beta(\rho \otimes 1_n) \otimes 1_k)(1_m \otimes \mu)$$
$$= (\alpha \otimes 1_k)(\beta \otimes 1_k)(\rho \otimes 1_{n+k})(1_m \otimes \mu)$$
$$= (\alpha \otimes 1_k)(\beta \otimes 1_k)(\rho \otimes \mu)$$
$$\sim_c (\beta \otimes 1_k)(\rho \otimes \mu)(\alpha \otimes 1_k) .$$

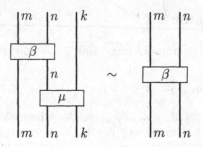

Fig. 2.22. The formula $(\beta \otimes 1_k)(1_m \otimes \mu) \sim \beta$

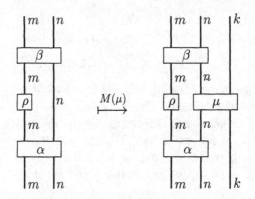

Fig. 2.23. The transformation $M(\mu)$

Given a braid $\mu \in B_{n+k}$ with $n \geq 1, k \geq 0$, we say that μ is *n-left-ghost* and write $\mu \equiv' 1_n$ if $(1_k \otimes \beta)(\mu \otimes 1_m) \sim \beta$ for any $m \geq 0$, $\beta \in B_{n+m}$. For any such μ and any $\alpha, \beta \in B_{n+m}$, $\rho \in B_m$, we denote by $M'(\mu)$ the move transforming $\beta(1_n \otimes \rho)\alpha$ into $(1_k \otimes \beta)(\mu \otimes \rho)(1_k \otimes \alpha)$. An argument similar to the one above shows that this move and its inverse preserve the M-equivalence class of the braid.

Lemma 2.22. *Let $\mu \in B_{n+k}$ with $n \geq 1, k \geq 0$. If $\mu \equiv 1_n$, then $\overline{\mu} \equiv' 1_n$.*

Proof. Pick $\beta \in B_{n+m}$ with $m \geq 0$ and set $\gamma = (1_k \otimes \beta)(\overline{\mu} \otimes 1_m)$. We must verify that $\gamma \sim \beta$. Obviously, $\gamma \sim_c \overline{\gamma} = (\overline{\beta} \otimes 1_k)(1_m \otimes \mu)$. Since $\mu \equiv 1_n$, we have $(\overline{\beta} \otimes 1_k)(1_m \otimes \mu) \sim \overline{\beta} \sim_c \beta$. Therefore $\gamma \sim \beta$. $\square$

For $n \geq 1$, set $\theta_n^+ = \Delta_n^2 \in B_n$ and $\theta_n^- = \Delta_n^{-2} \in B_n$. Clearly, for any $\varepsilon = \pm$,

$$\overline{\theta_n^\varepsilon} = \Delta_n \theta_n^\varepsilon \Delta_n^{-1} = \theta_n^\varepsilon .$$

As an exercise, the reader may check that

$$\theta_n^\varepsilon = (\theta_{n-1}^\varepsilon \otimes 1_1) \sigma_{1,n-1}^\varepsilon \sigma_{n-1,1}^\varepsilon = \sigma_{1,n-1}^\varepsilon (1_1 \otimes \theta_{n-1}^\varepsilon) \sigma_{n-1,1}^\varepsilon . \qquad (2.8)$$

The following lemma provides key examples of ghost braids. The proof of this lemma is given in an algebraic form. Here and below, the reader is strongly encouraged to draw the pictures corresponding to our formulas.

Lemma 2.23. *For any $n \geq 1$ and $\varepsilon = \pm$, set*

$$\mu_{n,\varepsilon} = (1_n \otimes \theta_n^{-\varepsilon})\,\sigma_{n,n}^{\varepsilon} = \sigma_{n,n}^{\varepsilon}(\theta_n^{-\varepsilon} \otimes 1_n) \in B_{2n}\,.$$

Then

$$\overline{\mu}_{n,\varepsilon} = (\theta_n^{-\varepsilon} \otimes 1_n)\,\sigma_{n,n}^{\varepsilon} = \sigma_{n,n}^{\varepsilon}(1_n \otimes \theta_n^{-\varepsilon}) \in B_{2n}$$

and $\mu_{n,\varepsilon} \equiv 1_n$, $\overline{\mu}_{n,\varepsilon} \equiv 1_n$, $\mu_{n,\varepsilon} \equiv' 1_n$, $\overline{\mu}_{n,\varepsilon} \equiv' 1_n$.

Proof. We shall represent θ_n^{ε} graphically by a box with ε inside. Two pictorial representations of $\mu_{n,-}$ are given in Figure 2.24. Pictures of $\mu_{n,+}$ are obtained by exchanging the over/undercrossings and replacing $+$ by $-$ in the box.

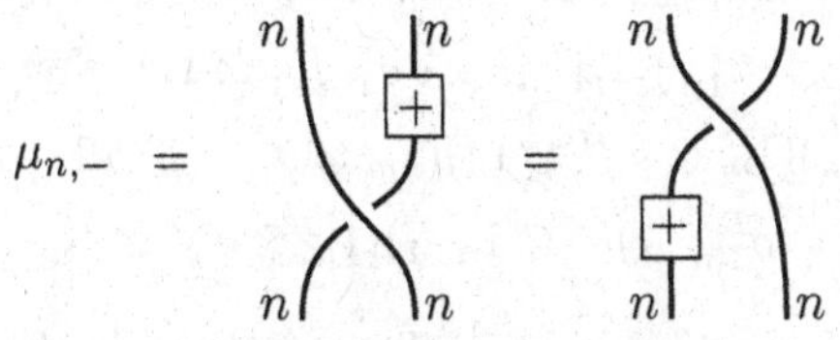

Fig. 2.24. The braid $\mu_{n,-}$

The expansions for $\overline{\mu}_{n,\varepsilon}$ in the statement of the lemma are obtained from the expansions for $\mu_{n,\varepsilon}$ and the geometric interpretation of the involution $\mu \mapsto \overline{\mu}$. By Lemma 2.22, the formulas $\mu_{n,\varepsilon} \equiv 1_n$, $\overline{\mu}_{n,\varepsilon} \equiv 1_n$ will imply that $\mu_{n,\varepsilon} \equiv' 1_n$, $\overline{\mu}_{n,\varepsilon} \equiv' 1_n$. To prove that $\mu_{n,\varepsilon}$ is n-right-ghost, we must verify that

$$(\beta \otimes 1_n)(1_m \otimes \mu_{n,\varepsilon}) \sim \beta$$

for any $\beta \in B_{m+n}$ with $m \geq 0$. Clearly,

$$\begin{aligned}(\beta \otimes 1_n)(1_m \otimes \mu_{n,\varepsilon}) &= (\beta \otimes 1_n)(1_m \otimes 1_n \otimes \theta_n^{-\varepsilon})(1_m \otimes \sigma_{n,n}^{\varepsilon}) \\ &= (\beta \otimes \theta_n^{-\varepsilon})(1_m \otimes \sigma_{n,n}^{\varepsilon}) \\ &\sim_c (1_m \otimes \sigma_{n,n}^{\varepsilon})(\beta \otimes \theta_n^{-\varepsilon})\,.\end{aligned}$$

It remains to prove that

$$(1_m \otimes \sigma_{n,n}^{\varepsilon})(\beta \otimes \theta_n^{-\varepsilon}) \sim \beta\,. \tag{2.9}$$

The formula $\overline{\mu}_{n,\varepsilon} \equiv 1_n$ also follows from (2.9), since

$$\begin{aligned}(\beta \otimes 1_n)(1_m \otimes \overline{\mu}_{n,\varepsilon}) &\sim_c (1_m \otimes \overline{\mu}_{n,\varepsilon})(\beta \otimes 1_n) \\ &= (1_m \otimes \sigma_{n,n}^{\varepsilon})(1_{m+n} \otimes \theta_n^{-\varepsilon})(\beta \otimes 1_n) \\ &= (1_m \otimes \sigma_{n,n}^{\varepsilon})(\beta \otimes \theta_n^{-\varepsilon})\,.\end{aligned}$$

The proof of the equality (2.9) goes by induction on n. For $n = 1$, we have $\theta_n^{-\varepsilon} = 1_1$ and $1_m \otimes \sigma_{n,n}^{\varepsilon} = \sigma_{m+1}^{\varepsilon}$, where $\sigma_{m+1}^{+} = \sigma_{m+1}$ and $\sigma_{m+1}^{-} = \sigma_{m+1}^{-1}$.

The transformation $\sigma^\varepsilon_{m+1}(\beta \otimes 1_1) \mapsto \beta$ is an inverse Markov move. Therefore, formula (2.9) holds for $n = 1$. In the inductive step we shall use the identity

$$\sigma^\varepsilon_{n,n} = (\sigma^\varepsilon_{n-1,n} \otimes 1_1)(1_{n-1} \otimes \sigma^\varepsilon_{1,n}).$$

For $n > 1$,

$$(1_m \otimes \sigma^\varepsilon_{n,n})(\beta \otimes \theta^{-\varepsilon}_n)$$
$$= (1_m \otimes \sigma^\varepsilon_{n,n})(1_m \otimes 1_n \otimes \theta^{-\varepsilon}_n)(\beta \otimes 1_n)$$
$$= (1_m \otimes \theta^{-\varepsilon}_n \otimes 1_n)(1_m \otimes \sigma^\varepsilon_{n,n})(\beta \otimes 1_n)$$
$$= (1_m \otimes \theta^{-\varepsilon}_n \otimes 1_n)(1_m \otimes \sigma^\varepsilon_{n-1,n} \otimes 1_1)(1_{m+n-1} \otimes \sigma^\varepsilon_{1,n})(\beta \otimes 1_n)$$
$$\sim_c (1_{m+n-1} \otimes \sigma^\varepsilon_{1,n})(\beta \otimes 1_n)(1_m \otimes \theta^{-\varepsilon}_n \otimes 1_n)(1_m \otimes \sigma^\varepsilon_{n-1,n} \otimes 1_1)$$
$$= (1_{m+2n-2} \otimes \sigma^\varepsilon_{1,1})(1_{m+n-1} \otimes \sigma^\varepsilon_{1,n-1} \otimes 1_1)$$
$$\quad \times (\beta \otimes 1_n)(1_m \otimes \theta^{-\varepsilon}_n \otimes 1_n)(1_m \otimes \sigma^\varepsilon_{n-1,n} \otimes 1_1)$$
$$\sim (1_{m+n-1} \otimes \sigma^\varepsilon_{1,n-1})(\beta \otimes 1_{n-1})(1_m \otimes \theta^{-\varepsilon}_n \otimes 1_{n-1})(1_m \otimes \sigma^\varepsilon_{n-1,n}),$$

where the last transformation is M_2^{-1}. The resulting braid is a conjugate of

$$(1_m \otimes \theta^{-\varepsilon}_n \otimes 1_{n-1})(1_m \otimes \sigma^\varepsilon_{n-1,n})(1_{m+n-1} \otimes \sigma^\varepsilon_{1,n-1})(\beta \otimes 1_{n-1})$$
$$= (1_m \otimes \theta^{-\varepsilon}_n \otimes 1_{n-1})(1_m \otimes \sigma^\varepsilon_{1,n-1} \otimes 1_{n-1})(1_m \otimes \sigma^\varepsilon_{n-1,n})(\beta \otimes 1_{n-1})$$
$$= (1_m \otimes \theta^{-\varepsilon}_n \sigma^\varepsilon_{1,n-1} \otimes 1_{n-1})(1_m \otimes \sigma^\varepsilon_{n-1,n})(\beta \otimes 1_{n-1}).$$

Substituting in the latter braid the expansion

$$\theta^{-\varepsilon}_n \sigma^\varepsilon_{1,n-1} = \theta^{-\varepsilon}_n (\sigma^{-\varepsilon}_{n-1,1})^{-1} = \sigma^{-\varepsilon}_{1,n-1}(1_1 \otimes \theta^{-\varepsilon}_{n-1}),$$

which follows from (2.8), we obtain

$$(1_m \otimes \sigma^{-\varepsilon}_{1,n-1} \otimes 1_{n-1})(1_{m+1} \otimes \theta^{-\varepsilon}_{n-1} \otimes 1_{n-1})(1_m \otimes \sigma^\varepsilon_{n-1,n})(\beta \otimes 1_{n-1})$$
$$= (1_m \otimes \sigma^{-\varepsilon}_{1,n-1} \otimes 1_{n-1})(1_m \otimes \sigma^\varepsilon_{n-1,n})(\beta \otimes \theta^{-\varepsilon}_{n-1})$$
$$\sim_c (1_m \otimes \sigma^\varepsilon_{n-1,n})(\beta \otimes \theta^{-\varepsilon}_{n-1})(1_m \otimes \sigma^{-\varepsilon}_{1,n-1} \otimes 1_{n-1})$$
$$= (1_{m+1} \otimes \sigma^\varepsilon_{n-1,n-1})(1_m \otimes \sigma^\varepsilon_{n-1,1} \otimes 1_{n-1})$$
$$\quad \times (\beta \otimes \theta^{-\varepsilon}_{n-1})(1_m \otimes \sigma^{-\varepsilon}_{1,n-1} \otimes 1_{n-1})$$
$$= (1_{m+1} \otimes \sigma^\varepsilon_{n-1,n-1})(\beta' \otimes \theta^{-\varepsilon}_{n-1}),$$

where

$$\beta' = (1_m \otimes \sigma^\varepsilon_{n-1,1})\beta(1_m \otimes \sigma^{-\varepsilon}_{1,n-1}).$$

By the induction assumption,

$$(1_{m+1} \otimes \sigma^\varepsilon_{n-1,n-1})(\beta' \otimes \theta^{-\varepsilon}_{n-1}) \sim \beta'$$
$$= (1_m \otimes \sigma^\varepsilon_{n-1,1})\beta(1_m \otimes \sigma^\varepsilon_{n-1,1})^{-1} \sim_c \beta.$$

This completes the proof of (2.9) and of the lemma.   □

Lemma 2.24. *For any integers $m, n \geq 0$, $r \geq 1$ and braids $\beta \in B_{m+r}$, $\gamma \in B_{m+n}$, the M-equivalence class of the braid*

$$\alpha_\varepsilon = (\beta \otimes 1_n)\,(1_m \otimes \sigma_{n,r}^\varepsilon)\,(\gamma \otimes 1_r)\,(1_m \otimes \sigma_{r,n}^{-\varepsilon})$$

does not depend on $\varepsilon = \pm$. (Here, if $m = n = 0$, then $\gamma = 1_0$.)

Proof. If $n = 0$, then $\sigma_{n,r}^+ = \sigma_{n,r}^-$ and hence $\alpha_+ = \alpha_-$. If $m = 0$, then $\alpha_+ = \beta \otimes \gamma = \alpha_-$. Suppose that $m \geq 1$ and $n \geq 1$. We shall prove that $\alpha_+ \sim \alpha_-$; see Figure 2.25.

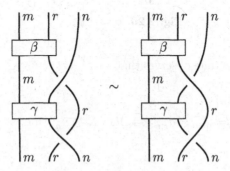

Fig. 2.25. $\alpha_+ \sim \alpha_-$

We first rewrite the factor $1_m \otimes \sigma_{n,r}^+$ of α_+ using the obvious expansion

$$1_m \otimes \sigma_{n,r}^+ = (1_m \otimes \sigma_{n,r}^+ \sigma_{r,n}^+)\,(1_{m+r} \otimes 1_n)\,(1_m \otimes \sigma_{n,r}^-)\,. \qquad (2.10)$$

By Lemma 2.23, the M-equivalence class of α_+ is preserved under the transformation replacing the term 1_n in the factor $1_{m+r} \otimes 1_n$ by the braid

$$\bar\mu_{n,-} = (\theta_n^+ \otimes 1_n)\,\sigma_{n,n}^- = \sigma_{n,n}^-(1_n \otimes \theta_n^+)$$

and tensoring on the right all the other factors in the expression for α_+ by 1_n. This transforms the right-hand side of (2.10) into the braid

$$\psi = (1_m \otimes \sigma_{n,r}^+ \sigma_{r,n}^+ \otimes 1_n)\,(1_{m+r} \otimes \theta_n^+ \otimes 1_n)\,(1_{m+r} \otimes \sigma_{n,n}^-)\,(1_m \otimes \sigma_{n,r}^- \otimes 1_n)\,.$$

Figure 2.26 shows that $\psi = \psi_1 \psi_2 \psi_3$, where

$$\psi_1 = 1_{m+r} \otimes \bar\mu_{n,-}\,, \quad \psi_2 = 1_m \otimes \sigma_{n,r}^- \otimes 1_n\,, \quad \psi_3 = 1_{m+n} \otimes \sigma_{n,r}^+ \sigma_{r,n}^+\,.$$

Therefore,

$$\begin{aligned}
\alpha_+ &\sim (\beta \otimes 1_{2n})\,\psi_1 \psi_2 \psi_3\,(\gamma \otimes 1_{r+n})\,(1_m \otimes \sigma_{r,n}^- \otimes 1_n) \\
&= \psi_1\,(\beta \otimes 1_{2n})\,\psi_2\,(\gamma \otimes 1_{r+n})\,\psi_3\,(1_m \otimes \sigma_{r,n}^- \otimes 1_n) \\
&\sim_c (\beta \otimes 1_{2n})\,\psi_2\,(\gamma \otimes 1_{r+n})\,\psi_3\,(1_m \otimes \sigma_{r,n}^- \otimes 1_n)\,\psi_1\,.
\end{aligned}$$

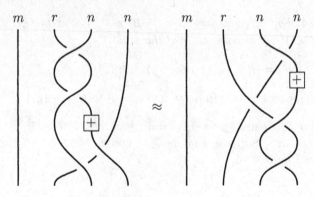

$$m \quad r \quad n \quad n \qquad m \quad r \quad n \quad n$$

Fig. 2.26. $\psi = \psi_1 \psi_2 \psi_3$

Drawing pictures, one observes that

$$\psi_3 \left(1_m \otimes \sigma_{r,n}^- \otimes 1_n\right) \psi_1$$
$$= (1_m \otimes \sigma_{r,n}^- \otimes 1_n)(1_{m+r} \otimes \overline{\mu}_{n,-})(1_m \otimes \sigma_{n,r}^+ \sigma_{r,n}^+ \otimes 1_n).$$

Thus,

$$\alpha_+ \sim (\beta \otimes 1_{2n})(1_m \otimes \sigma_{n,r}^- \otimes 1_n)(\gamma \otimes 1_{r+n})$$
$$\times (1_m \otimes \sigma_{r,n}^- \otimes 1_n)(1_{m+r} \otimes \overline{\mu}_{n,-})(1_m \otimes \sigma_{n,r}^+ \sigma_{r,n}^+ \otimes 1_n).$$

By Lemma 2.23, we can replace $\overline{\mu}_{n,-}$ with 1_n and simultaneously remove 1_n on the right of the other factors. This and the identity $\sigma_{r,n}^- \sigma_{n,r}^+ = 1_{r+n}$ give

$$\alpha_+ \sim (\beta \otimes 1_n)(1_m \otimes \sigma_{n,r}^-)(\gamma \otimes 1_r)(1_m \otimes \sigma_{r,n}^+) = \alpha_-. \qquad \square$$

Lemma 2.25. *Under the assumptions of Lemma 2.24, the M-equivalence class of the braid*

$$(1_n \otimes \beta)(\sigma_{r,n}^\varepsilon \otimes 1_m)(1_r \otimes \gamma)(\sigma_{n,r}^{-\varepsilon} \otimes 1_m)$$

does not depend on $\varepsilon = \pm$.

Proof. This follows from Lemma 2.24 by applying the involution $\mu \mapsto \overline{\mu}$ and using Lemma 2.21. $\qquad \square$

2.7.3 Proof of Lemma 2.11

The independence of the sign ε follows from Lemma 2.25, where the symbols $r, n, m, \varepsilon, \beta$, and γ should be replaced respectively with

$$n, \; m, \; t+r, \; -\varepsilon, \; (\alpha \otimes 1_t)(1_n \otimes \sigma_{t,r}^\nu)(\beta \otimes 1_r), \; (\gamma \otimes 1_r)(1_m \otimes \sigma_{r,t}^{-\nu})(\delta \otimes 1_t).$$

The independence of the sign ν follows from the fact that conjugate braids are M-equivalent and Lemma 2.24, where the symbols $n, m, \varepsilon, \beta, \gamma$ should be replaced respectively with

$$t, \; m+n, \; \nu, \; (1_n \otimes \delta)(\sigma^\varepsilon_{m,n} \otimes 1_r)(1_m \otimes \alpha), \; (1_m \otimes \beta)(\sigma^{-\varepsilon}_{n,m} \otimes 1_t)(1_n \otimes \gamma).$$

We now prove (2.3). By the first claim of the lemma, it suffices to consider the case $\varepsilon = \nu$. Consider the braid

$$\langle\!\langle \alpha, \beta, \gamma, \delta \,|\, \varepsilon \rangle\!\rangle = (\alpha \otimes \gamma)(1_n \otimes \sigma^\varepsilon_{m,r} \otimes 1_t)(1_n \otimes \theta^\varepsilon_m \otimes \sigma^\varepsilon_{t,r})(1_n \otimes \sigma^\varepsilon_{t,m} \otimes 1_r)$$
$$\times (\beta \otimes \delta)(1_n \otimes \sigma^{-\varepsilon}_{m,t} \otimes 1_r)(1_n \otimes \theta^{-\varepsilon}_m \otimes \sigma^{-\varepsilon}_{r,t})(1_n \otimes \sigma^{-\varepsilon}_{r,m} \otimes 1_t) \in B_{m+n+r+t}.$$

Note the obvious conjugacy

$$\langle\!\langle \alpha, \beta, \gamma, \delta \,|\, \varepsilon \rangle\!\rangle \sim_c \langle\!\langle \beta, \alpha, \delta, \gamma \,|\, -\varepsilon \rangle\!\rangle. \tag{2.11}$$

We claim that

$$\langle \alpha, \beta, \gamma, \delta \,|\, \varepsilon, \varepsilon \rangle \sim \langle\!\langle \alpha, \beta, \gamma, \delta \,|\, \varepsilon \rangle\!\rangle. \tag{2.12}$$

This will imply (2.3) for $\nu = \varepsilon$: applying (2.12), (2.11), and (2.4), we obtain

$$\langle \alpha, \beta, \gamma, \delta \,|\, \varepsilon, \varepsilon \rangle \sim \langle \beta, \alpha, \delta, \gamma \,|\, -\varepsilon, -\varepsilon \rangle \sim \langle \delta, \gamma, \beta, \alpha \,|\, \varepsilon, \varepsilon \rangle.$$

The case $\nu = -\varepsilon$ of (2.3) follows then from the first claim of the lemma.

A sequence of moves establishing (2.12) for $\varepsilon = +$ is given pictorially in Figure 2.27. (These moves can be described algebraically, which however is less instructive.) Here, instead of drawing braids we draw their closures. This is more economical in terms of space and does not hinder the argument since conjugate braids are M-equivalent.

The first and the last diagrams in Figure 2.27 represent the closures of the braids $\langle \alpha, \beta, \gamma, \delta \,|\, +, + \rangle$ and $\langle\!\langle \alpha, \beta, \gamma, \delta \,|\, + \rangle\!\rangle$, respectively. The first transformation in Figure 2.27 is a single move $M(\mu_m, -)$. (It would be more logical to write $+_+$ in the box but we write simply $+$.) The next two moves are isotopies in the class of closed braid diagrams (this amounts to conjugation of braids). Note that the box with $+$ followed by a box with $-$ is just the trivial braid; this splitting of the trivial braid is needed for the next move. The fourth move is the inverse to $M'(\overline{\mu}_m, -)$. The last move is an isotopy of closed braid diagrams. Since all these moves preserve the M-equivalence class of a braid, we obtain (2.12) for $\varepsilon = +$. The case $\varepsilon = -$ is treated similarly using the mirror image of Figure 2.27. $\qquad\qquad\square$

Exercise 2.7.1. Verify that the moves M_2, M_3 correspond to each other under the involution $\beta \mapsto \overline{\beta}$ on the set of braids.

Exercise 2.7.2. Let $\mu \in B_{n+k}$ with $n \geq 1, k \geq 0$ be an n-right-ghost braid. Verify that $1_r \otimes \mu \equiv 1_{r+n}$ for any $r \geq 0$ and $(\delta \otimes 1_k) \mu (\delta^{-1} \otimes 1_k) \equiv 1_n$ for any $\delta \in B_n$.

Solution. For any $\beta \in B_{m+r+n}$ with $m \geq 0$,

$$(\beta \otimes 1_k)(1_m \otimes 1_r \otimes \mu) = (\beta \otimes 1_k)(1_{m+r} \otimes \mu) \sim \beta.$$

For any $\beta \in B_{m+n}$ with $m \geq 0$,

$$(\beta \otimes 1_k)\big(1_m \otimes (\delta \otimes 1_k)\mu(\delta^{-1} \otimes 1_k)\big)$$
$$= (\beta \otimes 1_k)(1_m \otimes \delta \otimes 1_k)(1_m \otimes \mu)(1_m \otimes \delta^{-1} \otimes 1_k)$$
$$\sim_c (1_m \otimes \delta^{-1} \otimes 1_k)(\beta \otimes 1_k)(1_m \otimes \delta \otimes 1_k)(1_m \otimes \mu)$$
$$= \big((1_m \otimes \delta^{-1})\beta(1_m \otimes \delta) \otimes 1_k\big)(1_m \otimes \mu)$$
$$\sim (1_m \otimes \delta^{-1})\beta(1_m \otimes \delta) \sim_c \beta.$$

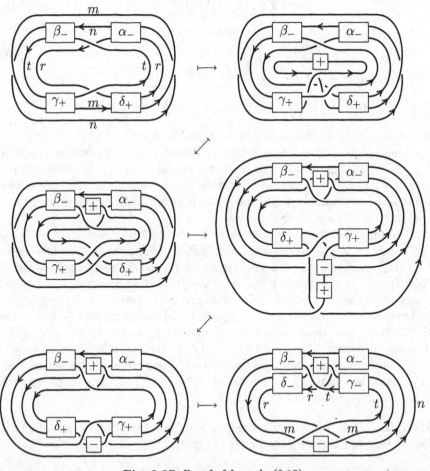

Fig. 2.27. Proof of formula (2.12)

Notes

The content of Section 2.1 is standard. Theorem 2.1 was first pointed out by Artin [Art25] without proof; see also [Mor78, Th. 1], and [BZ85, Prop. 10.16].

Theorem 2.3 is due to Alexander [Ale23a]. The algorithm of Section 2.4.3 transforming a link diagram into a closed braid diagram is due to Vogel [Vog90], who improved a previous construction by Yamada [Yam87]. Bendings were introduced by Vogel (under a different name). The height of a link diagram was introduced by Traczyk [Tra98], who also stated Lemmas 2.4–2.6. Our proof of Lemmas 2.5 and 2.6 is based on arguments from [Vog90, Sect. 5]. Corollary 2.7 is due to Yamada [Yam87]. Exercise 2.4.1 is due to Traczyk [Tra98].

Theorem 2.8 was announced by Markov [Mar36] in 1936. The first published proof appeared in the monograph [Bir74]. According to [Bir74, p. 49], this proof "is based on notes taken at a seminar at Princeton University in 1954. The speaker is unknown to us...." Different proofs were given by Bennequin [Ben83], Morton [Mor86], and Traczyk [Tra98]. The proof of Markov's theorem given above follows Traczyk [Tra98].

3

Homological Representations
of the Braid Groups

Braid groups, viewed as the groups of isotopy classes of self-homeomorphisms of punctured disks, naturally act on the homology of topological spaces obtained from the punctured disks by functorial constructions. We discuss here two such constructions and study the resulting linear representations of the braid groups: the Burau representation (Sections 3.1–3.3) and the Lawrence–Krammer–Bigelow representation (Sections 3.5–3.7). As an application of the Burau representation, we construct in Section 3.4 the one-variable Alexander–Conway polynomial of links in $\mathbf{R}^3$. As an application of the Lawrence–Krammer–Bigelow representation, we establish the linearity of B_n for all n (Section 3.5.4).

3.1 The Burau representation

For all $n \geq 1$, W. Burau introduced a linear representation of the braid group B_n by $n \times n$ matrices over the ring of Laurent polynomials

$$\Lambda = \mathbf{Z}[t, t^{-1}].$$

This representation has been extensively studied from various viewpoints. In this section we define the Burau representation and discuss its main properties.

3.1.1 Definition

Fix $n \geq 2$. For $i = 1, \ldots, n-1$, consider the following $n \times n$ matrix over the ring $\Lambda = \mathbf{Z}[t, t^{-1}]$:

$$U_i = \begin{pmatrix} I_{i-1} & 0 & 0 & 0 \\ 0 & 1-t & t & 0 \\ 0 & 1 & 0 & 0 \\ 0 & 0 & 0 & I_{n-i-1} \end{pmatrix},$$

C. Kassel, V. Turaev, *Braid Groups*, DOI: 10.1007/978-0-387-68548-9_3,
© Springer Science+Business Media, LLC 2008

where I_k denotes the unit $k \times k$ matrix. When $i = 1$, there is no unit matrix in the upper left corner of U_i. When $i = n - 1$, there is no unit matrix in the lower right corner of U_i. Substituting $t = 1$ in the definition of $U_1, \ldots, U_{n-1}$, we obtain permutation $n \times n$ matrices. One can therefore view $U_1, \ldots, U_{n-1}$ as one-parameter deformations of permutation matrices.

Each matrix U_i has a block-diagonal form with blocks being the unit matrices and the 2×2 matrix

$$U = \begin{pmatrix} 1-t & t \\ 1 & 0 \end{pmatrix}. \tag{3.1}$$

By the Cayley–Hamilton theorem, any 2×2 matrix M over the ring Λ satisfies $M^2 - \text{tr}(M)M + \det(M)I_2 = 0$, where $\text{tr}(M)$ is the trace of M and $\det(M)$ is the determinant of M. For $M = U$, this gives $U^2 - (1-t)U - tI_2 = 0$. Since the unit matrices also satisfy this equation,

$$U_i^2 - (1-t)U_i - tI_n = 0$$

for all i. This can be rewritten as $U_i(U_i - (1-t)I_n) = tI_n$. Hence, U_i is invertible over Λ and its inverse is computed by

$$U_i^{-1} = t^{-1}(U_i - (1-t)I_n) = \begin{pmatrix} I_{i-1} & 0 & 0 & 0 \\ 0 & 0 & 1 & 0 \\ 0 & t^{-1} & 1-t^{-1} & 0 \\ 0 & 0 & 0 & I_{n-i-1} \end{pmatrix}.$$

The block form of the matrices $U_1, \ldots, U_{n-1}$ implies that $U_i U_j = U_j U_i$ for all i, j with $|i - j| \geq 2$. We also have

$$U_i U_{i+1} U_i = U_{i+1} U_i U_{i+1}$$

for $i = 1, \ldots, n - 2$. To check this, it is enough to verify the equality

$$\begin{pmatrix} 1-t & t & 0 \\ 1 & 0 & 0 \\ 0 & 0 & 1 \end{pmatrix} \begin{pmatrix} 1 & 0 & 0 \\ 0 & 1-t & t \\ 0 & 1 & 0 \end{pmatrix} \begin{pmatrix} 1-t & t & 0 \\ 1 & 0 & 0 \\ 0 & 0 & 1 \end{pmatrix}$$

$$= \begin{pmatrix} 1 & 0 & 0 \\ 0 & 1-t & t \\ 0 & 1 & 0 \end{pmatrix} \begin{pmatrix} 1-t & t & 0 \\ 1 & 0 & 0 \\ 0 & 0 & 1 \end{pmatrix} \begin{pmatrix} 1 & 0 & 0 \\ 0 & 1-t & t \\ 0 & 1 & 0 \end{pmatrix}.$$

This equality is an exercise in matrix multiplication.

By Lemma 1.2, the formula $\psi_n(\sigma_i) = U_i$ with $i = 1, \ldots, n - 1$ defines a group homomorphism ψ_n from the braid group B_n with $n \geq 2$ to the group $\text{GL}_n(\Lambda)$ of invertible $n \times n$ matrices over Λ. This is the *Burau representation* of B_n. In particular, for $n = 2$, this representation is the homomorphism $B_2 \to \text{GL}_2(\Lambda)$, sending the generator σ_1 of $B_2 \cong \mathbf{Z}$ to the matrix (3.1).

By convention, the Burau representation ψ_1 of the (trivial) group B_1 is the trivial homomorphism $B_1 \to \mathrm{GL}_1(\Lambda)$.

Observe that $\det U_i = -t$ for all i. This implies that for any $\beta \in B_n$,

$$\det \psi_n(\beta) = (-t)^{\langle \beta \rangle},$$

where $\langle \beta \rangle \in \mathbf{Z}$ is the image of β under the homomorphism $B_n \to \mathbf{Z}$ sending the generators $\sigma_1, \ldots, \sigma_{n-1}$ to 1.

The Burau representations $\{\psi_n\}_{n \geq 1}$ are compatible with the natural inclusions $\iota : B_n \hookrightarrow B_{n+1}$: for any $n \geq 1$ and $\beta \in B_n$,

$$\psi_{n+1}(\iota(\beta)) = \begin{pmatrix} \psi_n(\beta) & 0 \\ 0 & 1 \end{pmatrix}. \tag{3.2}$$

3.1.2 Unitarity

The study of the Burau representation $\psi_n : B_n \to \mathrm{GL}_n(\Lambda)$ has to a great extent been focused on its kernel and image. We establish here a simple property of the image showing that it is contained in a rather narrow subgroup of $\mathrm{GL}_n(\Lambda)$. This property will not be used in the sequel.

Consider the involutive automorphism of the ring Λ, $\lambda \mapsto \overline{\lambda}$ for $\lambda \in \Lambda$, sending t to t^{-1}. For a matrix $A = (\lambda_{i,j})$ over Λ, set $\overline{A} = (\overline{\lambda}_{i,j})$ and let $A^T = (\lambda_{j,i})$ be the transpose of A. Let Ω_n be the lower triangular $n \times n$ matrix over Λ with all diagonal terms equal to 1 and all subdiagonal terms equal to $1 - t$:

$$\Omega_n = \begin{pmatrix} 1 & 0 & 0 & \cdots & 0 \\ 1-t & 1 & 0 & \cdots & 0 \\ 1-t & 1-t & 1 & \cdots & 0 \\ \vdots & \vdots & \vdots & \ddots & \vdots \\ 1-t & 1-t & 1-t & \cdots & 1 \end{pmatrix}.$$

Theorem 3.1. *For any $n \geq 1$ and $A \in \psi_n(B_n) \subset \mathrm{GL}_n(\Lambda)$,*

$$\overline{A}\, \Omega_n A^T = \Omega_n. \tag{3.3}$$

Proof. If (3.3) holds for a matrix A, then it holds for its inverse: multiplying (3.3) on the left by $\overline{A}^{-1}$ and on the right by $(A^T)^{-1}$, we obtain the same formula with A replaced by A^{-1}. If (3.3) holds for two matrices A_1, A_2, then it holds for their product:

$$\overline{A_1 A_2}\, \Omega_n (A_1 A_2)^T = \overline{A_1}\, \overline{A_2}\, \Omega_n A_2^T A_1^T = \overline{A_1}\, \Omega_n A_1^T = \Omega_n.$$

Now, since the matrices $U_1, \ldots, U_{n-1}$ generate the group $\psi_n(B_n)$, it is enough to prove (3.3) for $A = U_i$ with $i = 1, \ldots, n-1$. Present $A = U_i$ and Ω_n in the block form

$$A = \begin{pmatrix} I_{i-1} & 0 & 0 \\ 0 & U & 0 \\ 0 & 0 & I_{n-i-1} \end{pmatrix}, \quad \Omega_n = \begin{pmatrix} \Omega_{i-1} & 0 & 0 \\ K_{2,i-1} & \Omega_2 & 0 \\ K_{n-i-1,i-1} & K_{n-i-1,2} & \Omega_{n-i-1} \end{pmatrix},$$

where

$$U = \begin{pmatrix} 1-t & t \\ 1 & 0 \end{pmatrix}, \quad \Omega_2 = \begin{pmatrix} 1 & 0 \\ 1-t & 1 \end{pmatrix},$$

and $K_{p,q}$ is the $p \times q$ matrix with all entries equal to $1-t$. A direct computation gives

$$\overline{A}\,\Omega_n A^T = \begin{pmatrix} \Omega_{i-1} & 0 & 0 \\ \overline{U}\,K_{2,i-1} & \overline{U}\,\Omega_2 U^T & 0 \\ K_{n-i-1,i-1} & K_{n-i-1,2} U^T & \Omega_{n-i-1} \end{pmatrix}.$$

Note that $\overline{U} K_{2,i-1} = K_{2,i-1}$, since

$$\overline{U} \begin{pmatrix} 1-t \\ 1-t \end{pmatrix} = \begin{pmatrix} 1-t^{-1} & t^{-1} \\ 1 & 0 \end{pmatrix} \begin{pmatrix} 1-t \\ 1-t \end{pmatrix} = \begin{pmatrix} 1-t \\ 1-t \end{pmatrix}.$$

Similarly, $K_{n-i-1,2} U^T = K_{n-i-1,2}$, since

$$(1-t, 1-t)\, U^T = (1-t, 1-t) \begin{pmatrix} 1-t & 1 \\ t & 0 \end{pmatrix} = (1-t, 1-t).$$

A direct computation gives $\overline{U}\,\Omega_2 U^T = \Omega_2$. Substituting these formulas in the expression for $\overline{A}\,\Omega_n A^T$, we conclude that $\overline{A}\,\Omega_n A^T = \Omega_n$. $\quad\square$

Applying the involution $A \mapsto \overline{A}$ and the transposition to (3.3), we obtain $\overline{A}\,\overline{\Omega}_n^T A^T = \overline{\Omega}_n^T$. Therefore for any $A \in \psi_n(B_n)$ and $\lambda, \mu \in \Lambda$,

$$\overline{A}\,(\lambda \Omega_n + \mu \overline{\Omega}_n^T) A^T = \lambda \Omega_n + \mu \overline{\Omega}_n^T.$$

In particular, setting $\lambda = \mu = 1$, we obtain

$$\overline{A}\,\Theta_n A^T = \Theta_n, \tag{3.4}$$

where $\Theta_n = \Omega_n + \overline{\Omega}_n^T$ is the following $n \times n$ matrix:

$$\Theta_n = \begin{pmatrix} 2 & 1-t^{-1} & 1-t^{-1} & \cdots & 1-t^{-1} \\ 1-t & 2 & 1-t^{-1} & \cdots & 1-t^{-1} \\ 1-t & 1-t & 2 & \cdots & 1-t^{-1} \\ \vdots & \vdots & \vdots & \ddots & \vdots \\ 1-t & 1-t & 1-t & \cdots & 2 \end{pmatrix}.$$

The matrix Θ_n is "Hermitian" in the sense that $\overline{\Theta}_n^T = \Theta_n$.

Remark 3.2. Sending $t \in \Lambda$ to a complex number ζ of absolute value 1, we obtain a ring homomorphism $p_\zeta : \Lambda \to \mathbf{C}$. The involution $\lambda \mapsto \overline{\lambda}$ on the ring Λ corresponds under p_ζ to complex conjugation. Applying p_ζ to the entries of $n \times n$ matrices over Λ, we obtain a group homomorphism $\mathrm{GL}_n(\Lambda) \to \mathrm{GL}_n(\mathbf{C})$, also denoted by p_ζ. This gives a representation

$$P_\zeta = p_\zeta \psi_n : B_n \to \mathrm{GL}_n(\mathbf{C}).$$

Formula (3.4) implies that

$$\overline{P_\zeta(\beta)}\, p_\zeta(\Theta_n)\, P_\zeta(\beta)^T = p_\zeta(\Theta_n)$$

for all $\beta \in B_n$. For $\zeta = 1$, we have $p_\zeta(\Theta_n) = 2I_n$. Therefore the Hermitian matrix $p_\zeta(\Theta_n)$ is positive definite for all ζ sufficiently close to 1. For such ζ, the matrices in $P_\zeta(B_n) \subset \mathrm{GL}_n(\mathbf{C})$ are obtained by transposition and conjugation from unitary matrices.

3.1.3 The kernel of ψ_n

A homomorphism from a group to a group of matrices is said to be *faithful* if its kernel is trivial. The homomorphism ψ_1 is faithful, since $B_1 = \{1\}$. The homomorphism ψ_2 is also faithful. Indeed, the matrix $U = U_1 \in \mathrm{GL}_2(\Lambda)$, which is the image of the generator σ_1 of $B_2 \cong \mathbf{Z}$, satisfies

$$(1, -1)\, U = (-t, t) = -t\,(1, -1).$$

Hence, $(1, -1)\, U^k = (-t)^k\, (1, -1)$ for all $k \in \mathbf{Z}$ and we can conclude that U is of infinite order in $\mathrm{GL}_2(\Lambda)$. In Section 3.3.2 we shall show that $\mathrm{Ker}\,\psi_3 = \{1\}$. For $n \geq 4$, the question whether ψ_n is faithful, i.e., whether $\mathrm{Ker}\,\psi_n = \{1\}$, remained open for a long time. Note that $\mathrm{Ker}\,\psi_n \subset \mathrm{Ker}\,\psi_{n+1}$ under the inclusion $B_n \subset B_{n+1}$. Therefore, if $\mathrm{Ker}\,\psi_n \neq \{1\}$, then we have also $\mathrm{Ker}\,\psi_m \neq \{1\}$ for all $m \geq n$.

Theorem 3.3. $\mathrm{Ker}\,\psi_n \neq \{1\}$ *for* $n \geq 5$.

At the moment of writing (2007), it is unknown whether $\mathrm{Ker}\,\psi_4 = \{1\}$.

We point out explicit braids on five and six strings annihilated by the Burau representation. Set

$$\gamma = \sigma_4 \sigma_3^{-1} \sigma_2^{-1} \sigma_1^2 \sigma_2^{-1} \sigma_1^{-2} \sigma_2^{-2} \sigma_1^{-1} \sigma_4^{-5} \sigma_2 \sigma_3 \sigma_4^3 \sigma_2 \sigma_1^2 \sigma_2 \sigma_3^{-1} \in B_5.$$

Then the commutator

$$\rho = [\gamma \sigma_4 \gamma^{-1}, \sigma_4 \sigma_3 \sigma_2 \sigma_1^2 \sigma_2 \sigma_3 \sigma_4]$$

is a nontrivial element of $\mathrm{Ker}\,\psi_5 \subset B_5$. Here for elements a, b of a group,

$$[a, b] = a^{-1} b^{-1} ab.$$

The braid ρ is represented by a word of length 120 in the generators $\sigma_1^{\pm 1}, \sigma_2^{\pm 1}, \sigma_3^{\pm 1}, \sigma_4^{\pm 1}$ (observe that γ has length 26, while $\sigma_4^{-1}\gamma$ and $\gamma^{-1}\sigma_4$ have length 25). For $n = 6$ we can produce a shorter word representing an element of the kernel. Set

$$\gamma = \sigma_4 \sigma_5^{-2} \sigma_2^{-1} \sigma_1^3 \sigma_2^{-1} \sigma_5^{-1} \sigma_4 \in B_6 \,.$$

The commutator

$$\rho' = [\gamma \sigma_3 \gamma^{-1}, \sigma_3]$$

is a nontrivial element of $\mathrm{Ker}\,\psi_6 \subset B_6$. The braid ρ' is represented by a word of length 44 in the generators. That ρ, ρ' lie in the kernel of the Burau representation can in principle be verified by a direct computation. That they are nontrivial braids can be obtained using the solution of the word problem in B_n given in Section 1.5.1, or the normal form of braids discussed in Section 6.5.4, or the prime handle reduction of Section 7.5. These computations, however, shed no light on the geometric reasons forcing ρ, ρ' to lie in the kernel. These reasons will be discussed in Section 3.2.

Exercise 3.1.1. Show that $\mathrm{Ker}\,\psi_n \subset B_n$ is invariant under the involutive anti-automorphism $h : B_n \to B_n$ sending σ_i to itself for $i = 1, \ldots, n - 1$. (Hint: Verify that $U_i^T = D U_i D^{-1}$, where $i = 1, \ldots, n - 1$ and $D = D_n$ is the diagonal $n \times n$ matrix with diagonal terms $1, t, t^2, \ldots, t^{n-1}$. Deduce that

$$\psi_n(h(\beta)) = D^{-1} \psi_n(\beta)^T D$$

for all $\beta \in B_n$.)

3.2 Nonfaithfulness of the Burau representation

The aim of this section is to prove Theorem 3.3 for $n \geq 6$. The case $n = 5$ is somewhat subtler; for this case, we refer the reader to [Big99].

We begin with a study of homological representations of mapping class groups of surfaces.

3.2.1 Homological representations

Let Σ be a connected oriented surface (possibly with boundary $\partial \Sigma$). Recall that by self-homeomorphisms of Σ we mean orientation-preserving homeomorphisms $\Sigma \to \Sigma$ fixing the boundary pointwise. The isotopy classes of self-homeomorphisms of Σ form the mapping class group $\mathfrak{M}(\Sigma)$; see Section 1.6.1, where we take $M = \Sigma$, $Q = \emptyset$. A self-homeomorphism of Σ induces an automorphism of the homology group $H = H_1(\Sigma; \mathbf{Z})$. It is clear that isotopic self-homeomorphisms of Σ are homotopic and therefore induce the same automorphism of H. This defines a group homomorphism $\mathfrak{M}(\Sigma) \to \mathrm{Aut}(H)$, called the *homological representation* of $\mathfrak{M}(\Sigma)$.

Recall the *intersection form* $H \times H \to \mathbf{Z}$. This is a skew-symmetric bilinear form whose value $[\alpha] \cdot [\beta] \in \mathbf{Z}$ on the homology classes $[\alpha], [\beta] \in H$ represented by oriented loops α, β on Σ is the algebraic intersection number of these loops computed as follows. Deforming slightly α and β, we can assume that they meet transversely in a finite set of points that are not self-crossings of α or of β. Then

$$[\alpha] \cdot [\beta] = \sum_{p \in \alpha \cap \beta} \varepsilon_p ,$$

where $\varepsilon_p = +1$ if the tangent vectors of α, β at p form a positively oriented basis and $\varepsilon_p = -1$ otherwise. This sum does not depend on the choice of the loops α, β in their homology classes and defines a bilinear form $H \times H \to \mathbf{Z}$. The identity $[\alpha] \cdot [\beta] = -[\beta] \cdot [\alpha]$ shows that this intersection form is skew-symmetric. The action of $\mathfrak{M}(\Sigma)$ on H preserves the intersection form.

The homological representation has a more general "twisted" version, which comes up in the following setting. Suppose for concreteness that $\partial \Sigma \neq \emptyset$ and fix a base point $d \in \partial \Sigma$. Consider a surjective homomorphism φ from $\pi_1(\Sigma, d)$ onto a group G. Let $\widetilde{\Sigma} \to \Sigma$ be the covering corresponding to the kernel of φ. The group of covering transformations of $\widetilde{\Sigma}$ is identified with G. Pick an arbitrary point $\widetilde{d} \in \partial \widetilde{\Sigma}$ lying over d and consider the relative homology group $\widetilde{H} = H_1(\widetilde{\Sigma}, G\widetilde{d}; \mathbf{Z})$, where $G\widetilde{d}$ is the G-orbit of $\widetilde{d}$, i.e., the set of all points of $\widetilde{\Sigma}$ lying over d. The action of G on $\widetilde{\Sigma}$ induces a left action of G on $\widetilde{H}$ and turns $\widetilde{H}$ into a left module over the group ring $\mathbf{Z}[G]$. This module is free of rank $n = \mathrm{rk}\, H_1(\Sigma; \mathbf{Z})$. This follows from the fact that Σ deformation retracts onto a union of n simple closed loops on Σ meeting only at their common origin d (here we crucially use the assumption $\partial \Sigma \neq \emptyset$; cf. Figure 1.15, where Σ is the complement of n points in a disk). Let $\mathrm{Aut}(\widetilde{H})$ be the group of $\mathbf{Z}[G]$-linear automorphisms of $\widetilde{H}$. Clearly, $\mathrm{Aut}(\widetilde{H}) \cong \mathrm{GL}_n(\mathbf{Z}[G])$.

Any self-homeomorphism f of Σ fixes the boundary $\partial \Sigma$ pointwise and, in particular, fixes d. It induces therefore an automorphism $f_\#$ of the fundamental group $\pi_1(\Sigma, d)$. Let $\mathfrak{M}_\varphi(\Sigma, d)$ be the group of isotopy classes of self-homeomorphisms f of Σ such that $\varphi \circ f_\# = \varphi$. We construct a homomorphism $\mathfrak{M}_\varphi(\Sigma, d) \to \mathrm{Aut}(\widetilde{H})$ called the *twisted homological representation* of $\mathfrak{M}_\varphi(\Sigma, d)$. Every self-homeomorphism f of Σ representing an element of $\mathfrak{M}_\varphi(\Sigma, d)$ lifts uniquely to a homeomorphism $\widetilde{f} : \widetilde{\Sigma} \to \widetilde{\Sigma}$ fixing $\widetilde{d}$. The equality $\varphi \circ f_\# = \varphi$ ensures that $\widetilde{f}$ commutes with the action of G on $\widetilde{\Sigma}$. Therefore $\widetilde{f}$ fixes the set $G\widetilde{d}$ pointwise: $\widetilde{f}(g\widetilde{d}) = g\widetilde{f}(\widetilde{d}) = g\widetilde{d}$ for all $g \in G$. Let $\widetilde{f}_*$ be the automorphism of $\widetilde{H} = H_1(\widetilde{\Sigma}, G\widetilde{d}; \mathbf{Z})$ induced by $\widetilde{f}$. Since $\widetilde{f}$ commutes with the action of G, this automorphism is $\mathbf{Z}[G]$-linear. The map $f \mapsto \widetilde{f}_*$ defines a group homomorphism $\mathfrak{M}_\varphi(\Sigma, d) \to \mathrm{Aut}(\widetilde{H})$, which is the homological representation in question. The group $\widetilde{H}$ carries a natural intersection form preserved by $\mathfrak{M}_\varphi(\Sigma, d)$ but we shall not need it.

3.2.2 The homomorphism Ψ_n

We apply the general scheme of twisted homological representations to punctured disks. Fix $n \geq 1$. Let Q be the set $\{(1,0),(2,0),\ldots,(n,0)\} \subset \mathbf{R}^2$ and let D be a closed Euclidean disk in $\mathbf{R}^2$ containing Q in its interior. We provide D with the counterclockwise orientation as in Figure 1.15. Observe that for any point p in the interior of D, the group

$$H_1(D - \{p\}; \mathbf{Z}) \cong \mathbf{Z}$$

is generated by the homology class of a small loop encircling p counterclockwise. Each loop γ in $D - \{p\}$ represents k times this generator, where k is the *winding number* of γ around p. Set

$$\Sigma = D - Q$$

and fix a base point $d \in \partial\Sigma = \partial D$. Consider the group homomorphism φ from $\pi_1(\Sigma, d)$ to the infinite cyclic group $\{t^k\}_{k\in\mathbf{Z}}$ sending the homotopy class of a loop γ to $t^{-w(\gamma)}$, where $w(\gamma)$ is the *total winding number* of γ defined as the sum of its winding numbers around the points $(1,0),(2,0),\ldots,(n,0)$. The kernel of φ determines an infinite cyclic covering $\widetilde{\Sigma} \to \Sigma$. We identify its group of covering transformations with the infinite cyclic group $\{t^k\}_{k\in\mathbf{Z}}$. Pick a point $\widetilde{d} \in \partial\widetilde{\Sigma}$ over d and set

$$\widetilde{H} = H_1\Big(\widetilde{\Sigma}, \bigcup_{k\in\mathbf{Z}} t^k\widetilde{d}; \mathbf{Z}\Big).$$

Observe that any self-homeomorphism of D permuting the points of Q preserves the total winding number of loops in Σ. This is obvious for the small loops encircling the points of Q and holds for arbitrary loops, since their total winding numbers depend only on their homology classes in the group $H_1(\Sigma; \mathbf{Z}) \cong \mathbf{Z}^n$, which is generated by the homology classes of the small loops. Therefore the restriction to Σ defines a group homomorphism

$$\mathfrak{M}(D, Q) \to \mathfrak{M}_\varphi(\Sigma, d).$$

(It is actually an isomorphism but we do not need this.) Composing this homomorphism with the twisted homological representation $\mathfrak{M}_\varphi(\Sigma, d) \to \mathrm{Aut}(\widetilde{H})$ defined in Section 3.2.1, we obtain a group homomorphism

$$\Psi_n : \mathfrak{M}(D, Q) \to \mathrm{Aut}(\widetilde{H}).$$

The image of $f \in \mathfrak{M}(D, Q)$ under Ψ_n is the automorphism $\widetilde{f}_*$ of $\widetilde{H}$ induced by the lift $\widetilde{f} : \widetilde{\Sigma} \to \widetilde{\Sigma}$ of $f|_\Sigma : \Sigma \to \Sigma$ fixing $\widetilde{d}$.

In the next two subsections we show that $\mathrm{Ker}\,\Psi_n \neq \{1\}$ for $n \geq 6$. After that we show that Ψ_n is equivalent to the Burau representation ψ_n for all n. This will imply the nonfaithfulness of the latter for $n \geq 6$.

3.2.3 The kernel of Ψ_n

We give a construction of elements in $\text{Ker}\,\Psi_n$ using half-twists about spanning arcs as introduced in Section 1.6.2.

We say that two spanning arcs α, β on (D, Q) are *transversal* if they have no common endpoints and meet transversely at a finite number of points. For any transversal spanning arcs α, β on (D, Q), we define their *algebraic intersection* $\langle \alpha, \beta \rangle \in \Lambda = \mathbf{Z}[t, t^{-1}]$. Consider the open arcs $\alpha \cap \Sigma = \alpha - \partial\alpha$ and $\beta \cap \Sigma = \beta - \partial\beta$ on $\Sigma = D - Q$. Orient these arcs in an arbitrary way and pick arbitrary lifts $\widetilde{\alpha}, \widetilde{\beta} \subset \widetilde{\Sigma}$ of α, β with induced orientations. Now we can set

$$\langle \alpha, \beta \rangle = \sum_{k \in \mathbf{Z}} (t^k \widetilde{\alpha} \cdot \widetilde{\beta})\, t^k \in \Lambda, \tag{3.5}$$

where $t^k \widetilde{\alpha} \cdot \widetilde{\beta} \in \mathbf{Z}$ is the algebraic intersection number of the oriented arcs $t^k \widetilde{\alpha}$ and $\widetilde{\beta}$ on $\widetilde{\Sigma}$. Note that although the arcs $t^k \widetilde{\alpha}$ and $\widetilde{\beta}$ are not compact, they have only a finite number of intersections, and moreover, the sum on the right-hand side of (3.5) is finite. This is so because the covering projection $\widetilde{\Sigma} \to \Sigma$ maps $\widetilde{\beta}$ bijectively onto β and maps the set $(\bigcup_{k \in \mathbf{Z}} t^k \widetilde{\alpha}) \cap \widetilde{\beta}$ bijectively onto the finite set $\alpha \cap \beta$. This shows also that every point $p \in \alpha \cap \beta \subset \Sigma$ lifts to an intersection point of $t^k \widetilde{\alpha}$ with $\widetilde{\beta}$ for exactly one $k = k_p \in \mathbf{Z}$. Therefore,

$$\langle \alpha, \beta \rangle = \sum_{p \in \alpha \cap \beta} \varepsilon_p\, t^{k_p}, \tag{3.6}$$

where $\varepsilon_p = \pm 1$ is the intersection sign of α and β at p. As an exercise, the reader may verify that for any $p, q \in \alpha \cap \beta$, the difference $k_p - k_q$ is the total winding number of the loop in Σ going from p to q along α and then from q to p along β. The expression $\langle \alpha, \beta \rangle$ is defined only up to multiplication by ± 1 and a power of t depending on the choice of orientations on α, β and the choice of their lifts $\widetilde{\alpha}, \widetilde{\beta}$. This will not be important for us, since we are interested only in whether $\langle \alpha, \beta \rangle = 0$. Note that

$$\langle \beta, \alpha \rangle = \sum_{k \in \mathbf{Z}} (t^k \widetilde{\beta} \cdot \widetilde{\alpha})\, t^k = \sum_{k \in \mathbf{Z}} (\widetilde{\beta} \cdot t^{-k} \widetilde{\alpha})\, t^k$$

$$= -\sum_{k \in \mathbf{Z}} (t^{-k} \widetilde{\alpha} \cdot \widetilde{\beta})\, t^k = -\sum_{k \in \mathbf{Z}} (t^k \widetilde{\alpha} \cdot \widetilde{\beta})\, t^{-k}$$

$$= -\overline{\langle \alpha, \beta \rangle},$$

where the overbar denotes the ring involution on Λ sending t to t^{-1}. Hence, $\langle \alpha, \beta \rangle = 0 \Rightarrow \langle \beta, \alpha \rangle = 0$.

As we know, every spanning arc α on (D, Q) gives rise to a half-twist $\tau_\alpha : (D, Q) \to (D, Q)$ acting as the identity outside a disk neighborhood of α and mapping α onto itself via an orientation-reversing involution. Restricting τ_α to $\Sigma = D - Q$, we obtain a self-homeomorphism of Σ, denoted again by τ_α.

Lemma 3.4. *Let α, β be transversal spanning arcs on (D, Q). If $\langle \alpha, \beta \rangle = 0$, then $\Psi_n(\tau_\alpha \tau_\beta) = \Psi_n(\tau_\beta \tau_\alpha)$.*

Proof. To prove the lemma we compute the homological action of the half-twists. As a warmup, we compute the action of τ_α on $H = H_1(\Sigma; \mathbf{Z})$. Consider the loop α' on D drawn in Figure 3.1. This loop has a "figure-eight" shape and its only self-crossing lies on α. We orient α and α' so that $[\alpha] \cdot [\alpha'] = -2$, where $[\alpha] \in H_1(D, Q; \mathbf{Z})$ is the relative homology class of α and $[\alpha'] \in H$ is the homology class of α'. The dot $\cdot$ denotes the bilinear intersection form $H_1(D, Q; \mathbf{Z}) \times H \to \mathbf{Z}$ determined by the counterclockwise orientation of D.

The effect of the half-twist τ_α on an oriented curve transversal to α is to insert $(\alpha')^{\pm 1}$ at each crossing of α with this curve; see Figure 1.14. It is easy to check that for any $h \in H$,

$$(\tau_\alpha)_*(h) = h + ([\alpha] \cdot h) [\alpha'].$$

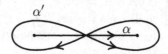

Fig. 3.1. The loop α' associated with a spanning arc α

The automorphism $\Psi_n(\tau_\alpha)$ of $\widetilde{H} = H_1(\widetilde{\Sigma}, \bigcup_{k \in \mathbf{Z}} t^k \widetilde{d}; \mathbf{Z})$ is defined by $\Psi_n(\tau_\alpha) = (\widetilde{\tau}_\alpha)_*$, where $\widetilde{\tau}_\alpha : \widetilde{\Sigma} \to \widetilde{\Sigma}$ is the lift of $\tau_\alpha : \Sigma \to \Sigma$ fixing $\widetilde{d}$. Observe that the loop α' on Σ associated to α has zero total winding number and therefore lifts to a loop $\widetilde{\alpha}'$ on $\widetilde{\Sigma}$. Consider an arbitrary oriented path γ in $\widetilde{\Sigma}$ with endpoints in $\bigcup_{k \in \mathbf{Z}} t^k \widetilde{d}$. The effect of $\widetilde{\tau}_\alpha$ on γ is to insert a lift of $(\alpha')^{\pm 1}$ at each crossing of γ with the preimage of α in $\widetilde{\Sigma}$. Thus $(\widetilde{\tau}_\alpha)_*$ acts on the relative homology class $[\gamma] \in \widetilde{H}$ by

$$(\widetilde{\tau}_\alpha)_*([\gamma]) = [\gamma] + \lambda_\gamma [\widetilde{\alpha}'],$$

where $\lambda_\gamma \in \Lambda$ is a Laurent polynomial whose coefficients are the algebraic intersection numbers of γ with lifts of α to $\widetilde{\Sigma}$. Since $\langle \alpha, \beta \rangle = 0$, any lift of α has algebraic intersection number zero with any lift of β to $\widetilde{\Sigma}$ and hence with any lift $\widetilde{\beta}'$ of β' to $\widetilde{\Sigma}$. Therefore, $\lambda_{\widetilde{\beta}'} = 0$ and $(\widetilde{\tau}_\alpha)_*([\widetilde{\beta}']) = [\widetilde{\beta}']$. Similarly, $(\widetilde{\tau}_\beta)_*([\gamma]) = [\gamma] + \mu_\gamma [\widetilde{\beta}']$ for all γ as above and some $\mu_\gamma \in \Lambda$. The equality $\langle \beta, \alpha \rangle = 0$ implies that $(\widetilde{\tau}_\beta)_*([\widetilde{\alpha}']) = [\widetilde{\alpha}']$. We conclude that for all γ,

$$(\widetilde{\tau}_\alpha \widetilde{\tau}_\beta)_*([\gamma]) = [\gamma] + \lambda_\gamma [\widetilde{\alpha}'] + \mu_\gamma [\widetilde{\beta}'] = (\widetilde{\tau}_\beta \widetilde{\tau}_\alpha)_*([\gamma]).$$

Therefore $(\widetilde{\tau}_\alpha \widetilde{\tau}_\beta)_* = (\widetilde{\tau}_\beta \widetilde{\tau}_\alpha)_*$. $\square$

To prove that $\operatorname{Ker}\Psi_n \neq \{1\}$, it remains to construct two spanning arcs α, β satisfying the conditions of Lemma 3.4 and such that $\tau_\alpha \tau_\beta \neq \tau_\beta \tau_\alpha$ in $\mathfrak{M}(D, Q)$. For $n = 6$, such spanning arcs α, β are drawn in Figure 3.2. To check the equality $\langle \alpha, \beta \rangle = 0$, one applies (3.6) and the computations after it (this is left as an exercise for the reader). To prove that τ_α and τ_β do not commute in $\mathfrak{M}(D, Q)$, one can use a brute-force computation using, for instance, the action of the mapping class group on $\pi_1(\Sigma, d)$. We give a geometric argument in the next subsection.

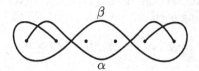

Fig. 3.2. Spanning arcs α, β for $n = 6$

3.2.4 Dehn twists

To show that two half-twists do not commute, we shall appeal to the theory of Dehn twists. We begin with the relevant definitions. Let Σ be an arbitrary oriented surface. By a *simple closed curve* on Σ, we mean the image of an embedding $S^1 \hookrightarrow \Sigma^\circ = \Sigma - \partial\Sigma$. (Note that simple closed curves are not assumed to be oriented.) A simple closed curve c on Σ gives rise to a self-homeomorphism t_c of Σ, called the *Dehn twist about c*. It is defined as follows. Set $I = [0, 1]$ and identify a cylinder neighborhood of c in Σ with $S^1 \times I$ so that $c = S^1 \times \{1/2\}$ and the product of the counterclockwise orientation on $S^1 = \{z \in \mathbf{C} \,|\, |z| = 1\}$ and the right-handed orientation on I corresponds to the given orientation on Σ. The Dehn twist $t_c : \Sigma \to \Sigma$ is the identity outside $S^1 \times I$ and sends any $(x, s) \in S^1 \times I$ to

$$(e^{2\pi i s}x, s) \in S^1 \times I.$$

It is clear that t_c is an orientation-preserving homeomorphism. Its isotopy class depends neither on the choice of the cylinder neighborhood of c nor on the choice of its identification with $S^1 \times I$. Note that if f is a self-homeomorphism of Σ, then $f(c)$ is a simple closed curve of Σ and $t_{f(c)} = f t_c f^{-1}$, where equality means isotopy in the class of self-homeomorphisms of Σ.

Two simple closed curves c, d on Σ are said to be *isotopic* if there is a self-homeomorphism of Σ that is isotopic to the identity and sends c onto d. It is clear that if c, d are isotopic, then $t_c = t_d$.

The question whether two Dehn twists commute (up to isotopy) has a simple geometric solution contained in the following lemma.

Lemma 3.5. *Let c, d be simple closed curves on an oriented surface Σ. The Dehn twists t_c, t_d commute if and only if c, d are isotopic to disjoint simple closed curves.*

Proof. If c, d are disjoint, then they have disjoint cylinder neighborhoods, so that the Dehn twists t_c, t_d obviously commute. If c, d are isotopic to disjoint simple closed curves c', d', then $t_c = t_{c'}$ commutes with $t_d = t_{d'}$. The proof of the converse is based on the techniques and results of [Tra79], which we now recall. For simple closed curves c, d on Σ, denote by $i(c, d)$ the minimum number of intersections of simple closed curves on Σ isotopic to c, d, respectively, and meeting each other transversely. Thus,

$$i(c, d) = \min_{c', d'} \operatorname{card}(c' \cap d') \geq 0,$$

where c', d' run over all pairs of simple closed curves on Σ isotopic to c, d, respectively, and such that c' meets d' transversely. In particular, $i(c, c) = 0$, since c is isotopic to a simple closed curve disjoint from c.

Proposition 1 on p. 68 of [Tra79] includes as a special case the following claim: if c, d, e are three simple closed curves on Σ, then

$$|i(t_c(d), e) - i(c, d)\, i(c, e)| \leq i(d, e).$$

Setting $e = d$, we obtain

$$i(t_c(d), d) = i(c, d)^2. \tag{3.7}$$

This implies that if c, c' are simple closed curves on Σ such that $t_c = t_{c'}$, then $i(c, d) = i(c', d)$ for any d.

Suppose now that two Dehn twists t_c, t_d commute. Then

$$t_d = t_c\, t_d\, t_c^{-1} = t_{t_c(d)}.$$

By the previous paragraph, $i(t_c(d), d) = i(d, d) = 0$. By (3.7), $i(c, d) = 0$. Hence, c, d are isotopic to disjoint simple closed curves.   □

The next lemma yields a necessary geometric condition for two simple closed curves on an oriented surface to be isotopic to curves with fewer intersections.

Lemma 3.6. *Let c, d be simple closed curves on Σ intersecting transversely at finitely many points. If c, d are isotopic to simple closed curves c', d' on Σ that are transversal and satisfy $\operatorname{card}(c' \cap d') < \operatorname{card}(c \cap d)$, then the curves c, d have a "digon," i.e., an embedded disk in Σ whose boundary consists of a subarc of c and a subarc of d and whose interior does not meet $c \cup d$; see Figure 3.3.*

For a proof, see [Tra79, pp. 46–48] or [PR00, Prop. 3.2].

The half-twists about arcs are related to the Dehn twists as follows. Suppose that $\Sigma = M - Q$, where M is an oriented surface and Q a finite subset of $M^\circ = M - \partial M$. Let α be a spanning arc on (M, Q). Consider a closed disk in M containing α in its interior and meeting Q only along the endpoints of α. Let $c = c(\alpha) \subset \Sigma$ be the boundary of this disk. This simple closed curve is determined by α up to isotopy in Σ. The Dehn twist $t_c : \Sigma \to \Sigma$ can be computed from the half-twist $\tau_\alpha : \Sigma \to \Sigma$ by

$$t_c = \tau_\alpha^2.$$

Indeed, both sides act as the identity outside a disk neighborhood of α as well as inside a smaller concentric disk neighborhood of α. In the annulus between these disks, both t_c and τ_α^2 act as the Dehn twist about the core circle of the annulus.

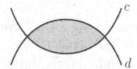

Fig. 3.3. A digon

We can now prove that the half-twists $\tau_\alpha, \tau_\beta \in \mathfrak{M}(D, Q)$ associated with the arcs α, β in Figure 3.2 do not commute. If they do, then their restrictions to the six-punctured disk $\Sigma = D - Q$ also commute. Then the Dehn twists

$$t_{c(\alpha)} = \tau_\alpha^2 : \Sigma \to \Sigma \quad \text{and} \quad t_{c(\beta)} = \tau_\beta^2 : \Sigma \to \Sigma$$

commute. By Lemmas 3.5 and 3.6, the curves $c(\alpha)$ and $c(\beta)$ must have a digon in Σ. Drawing these curves, one observes that they have 16 crossings and no digons in Σ. Hence τ_α, τ_β do not commute.

3.2.5 Equivalence of representations

The following theorem shows that the representation Ψ_n of B_n constructed in Section 3.2.2 is equivalent to the Burau representation ψ_n for all $n \geq 1$. Recall the isomorphism $\eta : B_n \to \mathfrak{M}(D, Q)$ defined in Section 1.6.3.

Theorem 3.7. *There is a group isomorphism* $\mu : \mathrm{GL}_n(\Lambda) \to \mathrm{Aut}(\widetilde{H})$, *where* $\widetilde{H} = H_1(\widetilde{\Sigma}, \bigcup_{k \in \mathbf{Z}} t^k \widetilde{d}; \mathbf{Z})$, *such that the following diagram is commutative:*

$$
\begin{array}{ccc}
B_n & \xrightarrow{\ \eta\ } & \mathfrak{M}(D, Q) \\
{\scriptstyle \psi_n}\downarrow & & \downarrow{\scriptstyle \Psi_n} \\
\mathrm{GL}_n(\Lambda) & \xrightarrow{\ \mu\ } & \mathrm{Aut}(\widetilde{H})
\end{array}
\qquad (3.8)
$$

Proof. We first compute the Λ-module $\widetilde{H} = H_1(\widetilde{\Sigma}, \bigcup_{k \in \mathbf{Z}} t^k \widetilde{d}; \mathbf{Z})$. Observe that Σ deformation retracts on the graph $\Gamma \subset \Sigma$ formed by one vertex d and n oriented loops $X_1, \ldots, X_n$ on Σ shown in Figure 1.15. The total winding numbers of these loops are equal to -1. The homomorphism φ sends the generators of $\pi_1(\Sigma, d)$ represented by these loops to t. The infinite cyclic covering $\widetilde{\Sigma}$ of Σ deformation retracts on an infinite graph $\widetilde{\Gamma} \subset \widetilde{\Sigma}$ with vertices $\{t^k \widetilde{d}\}_{k \in \mathbf{Z}}$ and oriented edges $\{t^k \widetilde{X}_i\}_{k \in \mathbf{Z}, i=1,\ldots,n}$, where each edge $t^k \widetilde{X}_i$ connects $t^k \widetilde{d}$ to $t(t^k \widetilde{d}) = t^{k+1} \widetilde{d}$ and is oriented from the former to the latter. The generator t acts on $\widetilde{\Gamma}$ by sending $t^k \widetilde{X}_i$ onto $t^{k+1} \widetilde{X}_i$. The cellular chain complex of the pair $(\widetilde{\Gamma}, \bigcup_{k \in \mathbf{Z}} t^k \widetilde{d})$ is 0 except in dimension 1, where it is equal to $\bigoplus_{i=1}^n \Lambda \widetilde{X}_i$. Therefore

$$\widetilde{H} = H_1\Big(\widetilde{\Sigma}, \bigcup_{k \in \mathbf{Z}} t^k \widetilde{d}; \mathbf{Z}\Big) = H_1\Big(\widetilde{\Gamma}, \bigcup_{k \in \mathbf{Z}} t^k \widetilde{d}; \mathbf{Z}\Big) = \bigoplus_{i=1}^n \Lambda[\widetilde{X}_i]$$

is a free Λ-module with basis $[\widetilde{X}_1], \ldots, [\widetilde{X}_n]$. We use this basis to identify $\mathrm{Aut}(\widetilde{H})$ with $\mathrm{GL}_n(\Lambda)$ in the standard way. The action of a matrix $(\lambda_{i,j}) \in \mathrm{GL}_n(\Lambda)$ on $\widetilde{H}$ sends each $[\widetilde{X}_j]$ to $\sum_i \lambda_{i,j}[\widetilde{X}_i]$.

We define a group isomorphism

$$\mu : \mathrm{GL}_n(\Lambda) \to \mathrm{GL}_n(\Lambda) = \mathrm{Aut}(\widetilde{H})$$

as the composition of the matrix transposition and inversion: $\mu(U) = (U^T)^{-1}$ for $U \in \mathrm{GL}_n(\Lambda)$. To check that the diagram (3.8) is commutative, we need to verify that for all $\beta \in B_n$,

$$\Psi_n \eta(\beta) = \mu \psi_n(\beta).$$

Since both sides are multiplicative with respect to β, it suffices to check this equality for a set of generators of B_n. We do it for the generators $\sigma_1^{-1}, \ldots, \sigma_{n-1}^{-1}$. Pick $i = 1, \ldots, n-1$. The homeomorphism $\eta(\sigma_i^{-1}) : D \to D$ exchanges the points $(i, 0), (i+1, 0) \in Q$ via a clockwise rotation of the arc $[i, i+1] \times \{0\}$ by an angle of π. This homeomorphism keeps X_k fixed for $k \neq i, i+1$, transforms X_i into a loop homotopic to the product $X_i X_{i+1} X_i^{-1}$, and transforms X_{i+1} into X_i. The lift of this homeomorphism to $\widetilde{\Sigma}$ keeps $\widetilde{X}_k$ fixed for $k \neq i, i+1$, transforms $\widetilde{X}_{i+1}$ into $\widetilde{X}_i$, and stretches $\widetilde{X}_i$ into the path

$$\widetilde{X}_i (t\widetilde{X}_{i+1}) (t\widetilde{X}_i)^{-1}.$$

The induced automorphism $\Psi_n \eta(\sigma_i^{-1})$ of $\widetilde{H}$ acts by

$$[\widetilde{X}_i] \mapsto (1-t)[\widetilde{X}_i] + t[\widetilde{X}_{i+1}], \qquad [\widetilde{X}_{i+1}] \mapsto [\widetilde{X}_i],$$

and $[\widetilde{X}_k] \mapsto [\widetilde{X}_k]$ for $k \neq i, i+1$. The matrix of this automorphism in the basis $[\widetilde{X}_1], \ldots, [\widetilde{X}_n]$ is precisely $U_i^T = \mu \psi_n(\sigma_i^{-1})$. $\square$

Remarks 3.8. (a) Similar methods, extended to arcs from the points of Q to the base point $d \in \partial D$, show that $\operatorname{Ker} \psi_5 \neq \{1\}$; see [Big99].

(b) Applying the construction of Section 3.2.1 to the natural projection $\pi_1(\Sigma, d) \to H_1(\Sigma)$, we obtain a matrix representation of the Torelli subgroup of $\mathfrak{M}(\Sigma)$ consisting of the self-homeomorphisms of Σ acting as the identity on $H_1(\Sigma)$. When Σ is the complement of n points in a 2-disk, this group is the pure braid group P_n and this representation is a version of the Gassner representation of P_n by $n \times n$ matrices over

$$\mathbf{Z}[H_1(\Sigma)] = \mathbf{Z}[t_1^{\pm 1}, \dots, t_n^{\pm 1}].$$

For more on the Gassner representation, see [Bir74], [Per06].

Exercise 3.2.1. Show that the arcs α, β in Figure 3.2 (where $n = 6$) can be computed by $\alpha = \eta(\gamma_1)(\alpha_3)$ and $\beta = \eta(\gamma_2)(\alpha_3)$, where α_3 is the spanning arc $[3, 4] \times \{0\}$ on (D, Q),

$$\gamma_1 = \sigma_1 \sigma_2^{-1} \sigma_5^{-1} \sigma_4 \quad \text{and} \quad \gamma_2 = \sigma_1^{-2} \sigma_2 \sigma_5^2 \sigma_4^{-1}.$$

Deduce that the commutator $[\gamma_1 \sigma_3 \gamma_1^{-1}, \gamma_2 \sigma_3 \gamma_2^{-1}]$ is a nontrivial element of $\operatorname{Ker} \psi_6$. This implies that the braid $\rho' = [\gamma_2^{-1} \gamma_1 \sigma_3 \gamma_1^{-1} \gamma_2, \sigma_3]$ introduced in Section 3.1.3 is a nontrivial element of $\operatorname{Ker} \psi_6$.

Exercise 3.2.2. Show that the isomorphism $B_n \cong \mathfrak{M}(D, Q)$ defined in Section 1.6.3 sends the center of B_n onto the infinite cyclic subgroup of $\mathfrak{M}(D, Q)$ generated by the Dehn twist about a simple closed curve in $D - Q$ obtained by pushing the circle ∂D inside $D - Q$.

Exercise 3.2.3. Show that if a simple closed curve c on a surface Σ bounds a disk in Σ, then the Dehn twist t_c is isotopic to the identity.

3.3 The reduced Burau representation

We show here that the Burau representation is reducible. As an application, we prove the faithfulness of ψ_3. Throughout this section, $\Lambda = \mathbf{Z}[t, t^{-1}]$.

3.3.1 Reduction of ψ_n

Recall the matrices

$$U_1, \dots, U_{n-1} \in \operatorname{GL}_n(\Lambda)$$

from Section 3.1.1. As above, the symbol I_k denotes the unit $k \times k$ matrix. The following theorem shows that the Burau representation is reducible.

Theorem 3.9. *Let $n \geq 3$ and $V_1, V_2, \ldots, V_{n-1}$ be the $(n-1) \times (n-1)$ matrices over Λ given by*

$$V_1 = \begin{pmatrix} -t & 0 & 0 \\ 1 & 1 & 0 \\ 0 & 0 & I_{n-3} \end{pmatrix}, \quad V_{n-1} = \begin{pmatrix} I_{n-3} & 0 & 0 \\ 0 & 1 & t \\ 0 & 0 & -t \end{pmatrix},$$

and for $1 < i < n-1$,

$$V_i = \begin{pmatrix} I_{i-2} & 0 & 0 & 0 & 0 \\ 0 & 1 & t & 0 & 0 \\ 0 & 0 & -t & 0 & 0 \\ 0 & 0 & 1 & 1 & 0 \\ 0 & 0 & 0 & 0 & I_{n-i-2} \end{pmatrix}.$$

Then for all $i = 1, \ldots, n-1$,

$$C^{-1} U_i C = \begin{pmatrix} V_i & 0 \\ *_i & 1 \end{pmatrix}, \tag{3.9}$$

where C is the $n \times n$ matrix

$$C = C_n = \begin{pmatrix} 1 & 1 & 1 & \cdots & 1 \\ 0 & 1 & 1 & \cdots & 1 \\ 0 & 0 & 1 & \cdots & 1 \\ \vdots & \vdots & \vdots & \ddots & \vdots \\ 0 & 0 & 0 & \cdots & 1 \end{pmatrix}$$

and $*_i$ is the row of length $n - 1$ equal to 0 if $i < n - 1$ and to $(0, \ldots, 0, 1)$ if $i = n - 1$.

Proof. For $i = 1, \ldots, n - 1$, set

$$V_i' = \begin{pmatrix} V_i & 0 \\ *_i & 1 \end{pmatrix}.$$

It suffices to prove that $U_i C = C V_i'$ for all i. Fix i and observe that for any $k = 1, \ldots, n$, the kth column of $U_i C$ is the sum of the first k columns of U_i. A direct computation shows that $U_i C$ is obtained from C by replacing the (i, i)th entry by $1 - t$ and replacing the $(i+1, i)$th entry by 1. Similarly, for any $\ell = 1, \ldots, n$, the ℓth row of $C V_i'$ is the sum of the last ℓ rows of V_i'. A direct computation shows that $C V_i'$ is obtained from C by the same modification as above. Hence $U_i C = C V_i'$. □

Since the matrices $U_1, \ldots, U_{n-1} \in \mathrm{GL}_n(\Lambda)$ satisfy the braid relations, so do the conjugate matrices $C^{-1} U_1 C, \ldots, C^{-1} U_{n-1} C$. Formula (3.9) implies that the matrices $V_1, \ldots, V_{n-1}$ also satisfy the braid relations. It is obvious that these matrices are invertible over Λ and therefore belong to $\mathrm{GL}_{n-1}(\Lambda)$.

By Lemma 1.2, the formula $\psi_n^r(\sigma_i) = V_i$ defines a group homomorphism $\psi_n^r : B_n \to \mathrm{GL}_{n-1}(\Lambda)$ for all $n \geq 3$. It is called the *reduced Burau representation*. For $n = 2$, we define the reduced Burau representation to be the homomorphism $\psi_2^r : B_2 \to \mathrm{GL}_1(\Lambda)$ sending σ_1 to the 1×1 matrix $(-t)$. This value is chosen so that formula (3.9) holds also for $n = 2$. This formula implies that for any $n \geq 2$ and any braid $\beta \in B_n$,

$$C^{-1}\psi_n(\beta)C = \begin{pmatrix} \psi_n^r(\beta) & 0 \\ *_\beta & 1 \end{pmatrix}, \tag{3.10}$$

where $*_\beta$ is a row of length $n-1$ over Λ depending on β. The following lemma shows how to compute this row from the matrix $\psi_n^r(\beta)$.

Lemma 3.10. *For* $i = 1, \ldots, n-1$, *let* a_i *be the* ith *row of the matrix* $\psi_n^r(\beta) - I_{n-1}$. *Then*

$$-(1 + t + \cdots + t^{n-1})*_\beta = \sum_{i=1}^{n-1} (1 + t + \cdots + t^i)\, a_i .$$

Proof. Consider the Λ-module Λ^n whose elements are identified with rows of length n over Λ. The group $\mathrm{GL}_n(\Lambda)$ acts on Λ^n on the right via the multiplication of rows by matrices. A direct verification shows that the vector

$$E = (1, t, t^2, \ldots, t^{n-1}) \in \Lambda^n$$

satisfies $EU_i = E$ for all i. Hence, $E\psi_n(\beta) = E$. Then the vector

$$F = EC = (1, 1+t, 1+t+t^2, \ldots, 1+t+\cdots+t^{n-1}) \in \Lambda^n$$

satisfies

$$F\begin{pmatrix} \psi_n^r(\beta) & 0 \\ *_\beta & 1 \end{pmatrix} = ECC^{-1}\psi_n(\beta)C = EC = F.$$

Subtracting $FI_n = F$, we obtain

$$F\begin{pmatrix} \psi_n^r(\beta) - I_{n-1} \\ *_\beta \end{pmatrix} = 0.$$

This equality means that the linear combination of the rows a_i of the matrix $\psi_n^r(\beta) - I_{n-1}$ with coefficients $1, 1+t, 1+t+t^2, \ldots, 1+t+\cdots+t^{n-2}$ is equal to $-(1 + t + \cdots + t^{n-1})*_\beta$. □

This lemma shows that no information is lost under the passage from the Burau representation to its reduced form. In particular, if $\psi_n^r(\beta) = I_{n-1}$, then $*_\beta = 0$ and $\psi_n(\beta) = I_n$. Therefore $\mathrm{Ker}\,\psi_n^r \subset \mathrm{Ker}\,\psi_n$. The opposite inclusion directly follows from (3.10). We conclude that $\mathrm{Ker}\,\psi_n^r = \mathrm{Ker}\,\psi_n$.

Remark 3.11. A homological interpretation of ψ_n^r is obtained by replacing in Section 3.2 the Λ-module $\widetilde{H} = H_1(\widetilde{\Sigma}, \bigcup_{k \in \mathbf{Z}} t^k \widetilde{d}; \mathbf{Z})$ by the Λ-module $\widetilde{H}^r = H_1(\widetilde{\Sigma}; \mathbf{Z})$. In the homological sequence of the pair $(\widetilde{\Sigma}, \bigcup_{k \in \mathbf{Z}} t^k \widetilde{d})$,

$$H_1\left(\bigcup_{k \in \mathbf{Z}} t^k \widetilde{d}; \mathbf{Z}\right) \to \widetilde{H}^r \to \widetilde{H} \to H_0\left(\bigcup_{k \in \mathbf{Z}} t^k \widetilde{d}; \mathbf{Z}\right) \to H_0(\widetilde{\Sigma}; \mathbf{Z}),$$

the leftmost term is zero because $\bigcup_{k \in \mathbf{Z}} t^k \widetilde{d}$ is a discrete space. Therefore the homomorphism $\widetilde{H}^r \to \widetilde{H}$ is an embedding, so that we can view $\widetilde{H}^r$ as a submodule of $\widetilde{H}$. Clearly,

$$H_0\left(\bigcup_{k \in \mathbf{Z}} t^k \widetilde{d}; \mathbf{Z}\right) = \Lambda, \quad H_0(\widetilde{\Sigma}; \mathbf{Z}) = \mathbf{Z},$$

and the homomorphism $H_0(\bigcup_{k \in \mathbf{Z}} t^k \widetilde{d}; \mathbf{Z}) \to H_0(\widetilde{\Sigma}; \mathbf{Z})$ is the homomorphism $\Lambda \to \mathbf{Z}$ sending t to 1. The kernel $(t-1)\Lambda$ of this homomorphism is a free Λ-module of rank 1. Therefore the quotient $\widetilde{H}/\widetilde{H}^r$ is a free Λ-module of rank one, so that $\widetilde{H} \cong \widetilde{H}^r \oplus \Lambda$. The action of $\mathfrak{M}(D, Q)$ on $\widetilde{H}$ preserves $\widetilde{H}^r$ and gives a homological interpretation of ψ_n^r. However, this action does not preserve the complementary module Λ. This is the geometric reason for the fact that the Burau representation can be reduced but is not a direct sum of its reduced form with a one-dimensional representation. As an exercise, the reader may verify that $\widetilde{H}^r \cong \Lambda^{n-1}$.

3.3.2 The faithfulness of ψ_3

We prove that the Burau representation ψ_3 is faithful. Consider the group homomorphism $\varphi : \mathrm{GL}_2(\Lambda) \to \mathrm{SL}_2(\mathbf{Z})$ obtained by the substitution $t \mapsto -1$. It transforms the reduced Burau matrices

$$V_1 = \begin{pmatrix} -t & 0 \\ 1 & 1 \end{pmatrix}, \quad V_2 = \begin{pmatrix} 1 & t \\ 0 & -t \end{pmatrix}$$

into the integral matrices

$$a_1 = \varphi(V_1) = \begin{pmatrix} 1 & 0 \\ 1 & 1 \end{pmatrix}, \quad a_2 = \varphi(V_2) = \begin{pmatrix} 1 & -1 \\ 0 & 1 \end{pmatrix}.$$

By Appendix A, the group $\mathrm{SL}_2(\mathbf{Z})$ is generated by the transpose matrices $A = a_1^T$, $B = a_2^T$ with defining relations $ABA = BAB$ and $(ABA)^4 = 1$. Hence, $\mathrm{SL}_2(\mathbf{Z})$ is generated by a_1, a_2 with defining relations $a_1 a_2 a_1 = a_2 a_1 a_2$ and $(a_1 a_2 a_1)^4 = 1$.

The homomorphism $\varphi \circ \psi_3^r : B_3 \to \mathrm{SL}_2(\mathbf{Z})$ sends the standard braid generators σ_1, σ_2 to a_1, a_2, respectively. It is clear that this homomorphism is surjective and its kernel is the normal subgroup generated by the braid $(\sigma_1 \sigma_2 \sigma_1)^4$.

Since this braid is central in B_3, the kernel in question is the cyclic group $\langle (\sigma_1\sigma_2\sigma_1)^4 \rangle \subset B_3$. Consequently,

$$\operatorname{Ker}\psi_3 \subset \operatorname{Ker}(\varphi \circ \psi_3^{\mathrm{r}}) = \langle (\sigma_1\sigma_2\sigma_1)^4 \rangle.$$

Observe that

$$V_1V_2V_1 = \begin{pmatrix} 0 & -t^2 \\ -t & 0 \end{pmatrix} \quad \text{and} \quad (V_1V_2V_1)^2 = \begin{pmatrix} t^3 & 0 \\ 0 & t^3 \end{pmatrix}.$$

Therefore, for any nonzero $k \in \mathbf{Z}$,

$$\psi_3^{\mathrm{r}}\big((\sigma_1\sigma_2\sigma_1)^{4k}\big) = (V_1V_2V_1)^{4k} = \begin{pmatrix} t^{6k} & 0 \\ 0 & t^{6k} \end{pmatrix} \neq I_2.$$

Hence $\operatorname{Ker}\psi_3 = \operatorname{Ker}\psi_3^{\mathrm{r}} = \{1\}$.

Exercise 3.3.1. Show that $\psi_n^{\mathrm{r}}\big((\sigma_1 \cdots \sigma_{n-1})^n\big) = t^n I_{n-1}$ for all $n \geq 2$. (Hint: Use the homological interpretation of ψ_n^{r}; observe that the element of $\mathfrak{M}(D,Q)$ corresponding to $(\sigma_1 \cdots \sigma_{n-1})^n$ is the Dehn twist about a circle in D concentric to ∂D.) Note that a similar equality does not hold for ψ_n, for instance, $\psi_2(\sigma_1^2) \neq t^2 I_2$.

3.4 The Alexander–Conway polynomial of links

We use here the reduced Burau representations $\psi_1^{\mathrm{r}}, \psi_2^{\mathrm{r}}, \ldots$ and the theory of Markov functions from Section 2.5.2 to construct the one-variable Alexander–Conway polynomial of links.

3.4.1 An example of a Markov function

We construct a Markov function with values in the Laurent polynomial ring $\mathbf{Z}[s, s^{-1}]$. The associated link invariant will be studied in the next subsection. Let

$$g : \Lambda = \mathbf{Z}[t, t^{-1}] \to \mathbf{Z}[s, s^{-1}]$$

be the ring homomorphism sending t to s^2. For a braid β on $n \geq 2$ strings, consider the following rational function in s with integral coefficients:

$$f_n(\beta) = (-1)^{n+1}\frac{s^{-\langle\beta\rangle}(s - s^{-1})}{s^n - s^{-n}}\, g\big(\det(\psi_n^{\mathrm{r}}(\beta) - I_{n-1})\big),$$

where $\langle\beta\rangle \in \mathbf{Z}$ is the image of β under the homomorphism $B_n \to \mathbf{Z}$ sending the generators $\sigma_1, \ldots, \sigma_{n-1}$ to 1. For example, for $n = 2$ and $k \in \mathbf{Z}$,

$$f_2(\sigma_1^k) = -s^{-k}(s + s^{-1})^{-1}\big((-s^2)^k - 1\big).$$

In particular, $f_2(\sigma_1) = f_2(\sigma_1^{-1}) = 1$. By definition, $f_1(B_1) = 1$.

Lemma 3.12. *The mappings* $\{f_n : B_n \to \mathbf{Z}[s, s^{-1}]\}_{n \geq 1}$ *form a Markov function.*

Proof. Pick a braid $\beta \in B_n$ with $n \geq 1$. A conjugation of β in B_n preserves both $\langle \beta \rangle$ and $\det(\psi_n^r(\beta) - I_{n-1})$ and therefore preserves $f_n(\beta)$. This implies the first condition in the definition of a Markov function.

Set $\beta_+ = \iota(\beta)\sigma_n \in B_{n+1}$, where ι is the natural inclusion $B_n \hookrightarrow B_{n+1}$. We verify now that $f_{n+1}(\beta_+) = f_n(\beta)$. For $n = 1$, we have $\beta = 1$, $\beta_+ = \sigma_1$, and $f_2(\beta_+) = f_2(\sigma_1) = 1 = f_1(\beta)$. Suppose that $n \geq 2$. We first observe the equalities

$$\frac{s^{-\langle \beta \rangle}(s - s^{-1})}{s^n - s^{-n}} = \frac{s^{n-1-\langle \beta \rangle}}{1 + s^2 + s^4 + \cdots + s^{2(n-1)}}$$

and

$$n - 1 - \langle \beta \rangle = (n+1) - 1 - \langle \beta_+ \rangle.$$

Therefore the desired formula $f_{n+1}(\beta_+) = f_n(\beta)$ is equivalent to the following formula:

$$(1 + t + \cdots + t^{n-1}) \det(\psi_{n+1}^r(\beta_+) - I_n)$$
$$= -(1 + t + \cdots + t^n) \det(\psi_n^r(\beta) - I_{n-1}). \quad (3.11)$$

By (3.2) and (3.10),

$$\psi_{n+1}(\iota(\beta)) = \begin{pmatrix} \psi_n(\beta) & 0 \\ 0 & 1 \end{pmatrix} = \begin{pmatrix} C_n & 0 \\ 0 & 1 \end{pmatrix} \begin{pmatrix} \psi_n^r(\beta) & 0 & 0 \\ *_\beta & 1 & 0 \\ 0 & 0 & 1 \end{pmatrix} \begin{pmatrix} C_n^{-1} & 0 \\ 0 & 1 \end{pmatrix}.$$

Therefore,

$$\begin{pmatrix} \psi_{n+1}^r(\beta_+) & 0 \\ *_{\beta_+} & 1 \end{pmatrix} = C_{n+1}^{-1} \, \psi_{n+1}(\beta_+) \, C_{n+1}$$

$$= C_{n+1}^{-1} \, \psi_{n+1}(\iota(\beta)) \, \psi_{n+1}(\sigma_n) \, C_{n+1}$$

$$= C_{n+1}^{-1} \begin{pmatrix} C_n & 0 \\ 0 & 1 \end{pmatrix} \begin{pmatrix} \psi_n^r(\beta) & 0 & 0 \\ *_\beta & 1 & 0 \\ 0 & 0 & 1 \end{pmatrix} \begin{pmatrix} C_n^{-1} & 0 \\ 0 & 1 \end{pmatrix} \begin{pmatrix} I_{n-1} & 0 & 0 \\ 0 & 1-t & t \\ 0 & 1 & 0 \end{pmatrix} C_{n+1}. \quad (3.12)$$

Observe that

$$C_n^{-1} = \begin{pmatrix} 1 & -1 & 0 & \cdots & 0 & 0 \\ 0 & 1 & -1 & \cdots & 0 & 0 \\ 0 & 0 & 1 & \cdots & 0 & 0 \\ \vdots & \vdots & \vdots & \ddots & \vdots & \\ 0 & 0 & 0 & \cdots & 1 & -1 \\ 0 & 0 & 0 & \cdots & 0 & 1 \end{pmatrix}.$$

A direct computation shows that the product of the first three (resp. the last three) matrices on the right-hand side of (3.12) is equal to

$$\begin{pmatrix} \psi_n^r(\beta) & 0 & 0 \\ *_\beta & 1 & -1 \\ 0 & 0 & 1 \end{pmatrix}, \quad \text{resp.} \quad \begin{pmatrix} I_{n-2} & 0 & 0 & 0 \\ 0 & 1 & t & 0 \\ 0 & 0 & 1-t & 1 \\ 0 & 0 & 1 & 1 \end{pmatrix}.$$

To multiply these two matrices we expand

$$\begin{pmatrix} \psi_n^r(\beta) & 0 & 0 \\ *_\beta & 1 & -1 \\ 0 & 0 & 1 \end{pmatrix} = \begin{pmatrix} X & Y & 0 & 0 \\ Z & T & 0 & 0 \\ P & Q & 1 & -1 \\ 0 & 0 & 0 & 1 \end{pmatrix},$$

where X is a square matrix over Λ of size $n-2$, Y is a column over Λ of height $n-2$, Z and P are rows over Λ of length $n-2$, and $T, Q \in \Lambda$. The formulas above give

$$\begin{pmatrix} \psi_{n+1}^r(\beta_+) & 0 \\ *_{\beta_+} & 1 \end{pmatrix} = \begin{pmatrix} X & Y & tY & 0 \\ Z & T & tT & 0 \\ P & Q & tQ-t & 0 \\ 0 & 0 & 1 & 1 \end{pmatrix}.$$

Hence,

$$\psi_{n+1}^r(\beta_+) - I_n = \begin{pmatrix} X - I_{n-2} & Y & tY \\ Z & T-1 & tT \\ P & Q & tQ-t-1 \end{pmatrix}.$$

To compute the determinant of this $n \times n$ matrix, we multiply the $(n-1)$st column by $-t$ and add the result to the nth column. This gives

$$\det(\psi_{n+1}^r(\beta_+) - I_n) = \det J,$$

where

$$J = \begin{pmatrix} X - I_{n-2} & Y & 0 \\ Z & T-1 & t \\ P & Q & -t-1 \end{pmatrix}.$$

Observe that

$$\psi_n^r(\beta) - I_{n-1} = \begin{pmatrix} X - I_{n-2} & Y \\ Z & T-1 \end{pmatrix} \quad \text{and} \quad *_\beta = \begin{pmatrix} P & Q \end{pmatrix}.$$

These formulas and Lemma 3.10 imply that adding the rows of J with coefficients

$$1, \ 1+t, \ 1+t+t^2, \ldots, \ 1+t+\cdots+t^{n-1},$$

we obtain a new bottom row whose first $n-1$ entries are equal to 0. The last, nth entry is equal to

$$(1+t+\cdots+t^{n-2})t + (1+t+\cdots+t^{n-1})(-t-1) = -(1+t+\cdots+t^n).$$

Therefore,

$$(1 + t + \cdots + t^{n-1}) \det(\psi_{n+1}^{\mathrm{r}}(\beta_+) - I_n)$$

$$= \det \begin{pmatrix} X - I_{n-2} & Y & 0 \\ Z & T - 1 & t \\ 0 & 0 & -(1 + t + \cdots + t^n) \end{pmatrix}.$$

This implies (3.11). Hence

$$f_{n+1}(\sigma_n \, \iota(\beta)) = f_{n+1}(\iota(\beta) \, \sigma_n) = f_{n+1}(\beta_+) = f_n(\beta).$$

A similar argument shows that $f_{n+1}(\sigma_n^{-1}\iota(\beta)) = f_n(\beta)$. This verifies the second condition in the definition of a Markov function.   □

For an oriented link $L \subset \mathbf{R}^3$, set $\widehat{f}(L) = f_n(\beta)$, where β is an arbitrary braid on n strings whose closure is isotopic to L. By Section 2.5.2 and the previous lemma, $\widehat{f}(L)$ is an isotopy invariant of L independent of the choice of β. We study this invariant in the next subsection.

3.4.2 The Alexander–Conway polynomial

The (one-variable) Alexander–Conway polynomial is a fundamental and historically the first polynomial invariant of oriented links in $\mathbf{R}^3$. This polynomial extends to a two-variable polynomial invariant of oriented links in $\mathbf{R}^3$, known as the Jones–Conway or HOMFLY-PT polynomial. The latter will be constructed in the context of Iwahori–Hecke algebras in Section 4.4.

We begin with an axiomatic definition of the Alexander–Conway polynomial. We shall say that three oriented links $L_+, L_-, L_0 \subset \mathbf{R}^3$ form a *Conway triple* if they coincide outside a 3-ball in $\mathbf{R}^3$ and look as in Figure 3.4 inside this ball. The *Alexander–Conway polynomial of links* is a mapping ∇ assigning to every oriented link $L \subset \mathbf{R}^3$ a Laurent polynomial $\nabla(L) \in \mathbf{Z}[s, s^{-1}]$ satisfying the following three axioms:

(i) $\nabla(L)$ is invariant under isotopy of L;
(ii) if L is a trivial knot, then $\nabla(L) = 1$;
(iii) for any Conway triple $L_+, L_-, L_0 \subset \mathbf{R}^3$,

$$\nabla(L_+) - \nabla(L_-) = (s^{-1} - s)\,\nabla(L_0).$$

The latter equality is known as the *Alexander–Conway skein relation*.

As an example of a computation using the skein relation, consider the Conway triple L_+, L_-, L_0 in Figure 3.5. Here L_+ (resp. L_-) is obtained from an oriented link $L \subset \mathbf{R}^3$ by adding a small positive (resp. negative) curl. Both links L_+ and L_- are isotopic to L, while L_0 is the disjoint union of L with a trivial knot. Axioms (i) and (iii) imply that $\nabla(L_0) = 0$. We conclude that ∇ annihilates all links obtained as a disjoint union of a nonempty link with a trivial knot. In particular, ∇ annihilates all trivial links with two or more components.

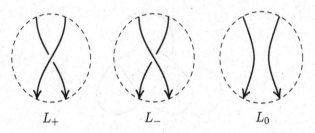

L_+ L_- L_0

Fig. 3.4. A Conway triple

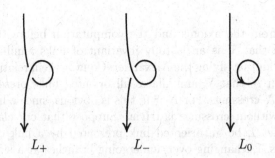

L_+ L_- L_0

Fig. 3.5. Example of a Conway triple

Theorem 3.13. *The Alexander–Conway polynomial of links exists and is unique. The invariant $\widehat{f}$ of links in $\mathbf{R}^3$ constructed in Section 3.4.1 coincides with the Alexander–Conway polynomial.*

Proof. We first prove the uniqueness: there is at most one mapping from the set of oriented links in $\mathbf{R}^3$ to $\mathbf{Z}[s, s^{-1}]$ satisfying axioms (i)–(iii). The proof requires the notion of an ascending link diagram, which we now introduce. An oriented link diagram $\mathcal{D}$ on $\mathbf{R}^2$ is *ascending* if it satisfies the following two conditions:

(a) the components of $\mathcal{D}$ can be indexed by $1, \ldots, m$ (where m is the number of the components) so that at every crossing of distinct components, the component with smaller index lies below the component with larger index;

(b) each component of $\mathcal{D}$ can be provided with a base point (not a crossing) such that starting from this point and moving along the component in the positive direction, we always reach the self-crossings of this component for the first time along the undergoing branch and for the second time along the overgoing branch.

An example of an ascending link diagram is given in Figure 3.6. It is a simple geometric exercise to see that the link presented by an ascending diagram is necessarily trivial.

Suppose now that there are two mappings from the set of oriented links in $\mathbf{R}^3$ to $\mathbf{Z}[s, s^{-1}]$ satisfying axioms (i)–(iii) of the Alexander–Conway polynomial. Let ∇ be their difference. We have to prove that $\nabla = 0$.

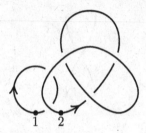

Fig. 3.6. An ascending link diagram

It is clear from the axioms and the computation before the statement of the theorem that ∇ is an isotopy invariant of links annihilating trivial knots and links and satisfying the Alexander–Conway skein relation. We prove by induction on N that ∇ annihilates all oriented links presented by link diagrams with N crossings. For $N = 0$, this is obvious, since a link presented by a diagram without crossings is trivial. Suppose that our claim holds for a certain N. Let L be an oriented link presented by a link diagram with $N + 1$ crossings. Exchanging over/undergoing branches at a single crossing, we obtain a diagram of another link, L'. The links L, L' together with the link L_0 obtained by smoothing the crossing in question form a Conway triple as in Figure 3.4. The link L_0 is presented by a link diagram with N crossings. By the induction assumption, $\nabla(L_0) = 0$. The skein relation gives $\nabla(L) = \nabla(L')$. Thus, the value of ∇ on L is not changed when overcrossings are traded for undercrossings. However, these operations can transform our diagram into an ascending one. Since ∇ annihilates the links presented by ascending diagrams, $\nabla(L) = 0$. This completes the induction step. Hence $\nabla = 0$.

To prove the remaining claims of the theorem, it is enough to show that the link invariant $\widehat{f}$ constructed in the previous subsection satisfies the axioms of the Alexander–Conway polynomial. By Corollary 2.9 and the results above, $\widehat{f}$ is a well-defined isotopy invariant of links. If L is a trivial knot, then L is the closure of a trivial braid on one string and therefore $\widehat{f}(L) = 1$. We verify now that $\widehat{f}$ satisfies the Alexander–Conway skein relation.

Given $n \geq 2$, $i \in \{1, \ldots, n - 1\}$, and two braids $\alpha, \beta \in B_n$, we see directly from the definitions that the closures of the braids $\alpha\sigma_i\beta$, $\alpha\sigma_i^{-1}\beta$, and $\alpha\beta$ form a Conway triple of links in $\mathbf{R}^3$. The proof of Alexander's theorem (Theorem 2.3) shows that conversely, an arbitrary Conway triple of links in $\mathbf{R}^3$ arises in this way from certain n, i, α, β. Thus, we need to prove the identity

$$f_n(\alpha\sigma_i\beta) - f_n(\alpha\sigma_i^{-1}\beta) = (s^{-1} - s)\, f_n(\alpha\beta).$$

Since f_n is invariant under conjugation in B_n and σ_i is a conjugate of σ_1 in B_n (see Exercise 1.1.4), we can assume without loss of generality that $i = 1$. Further conjugating by α, we can assume that $\alpha = 1$. Thus we need to prove that for any $\beta \in B_n$,

$$f_n(\sigma_1\beta) - f_n(\sigma_1^{-1}\beta) = (s^{-1} - s)\, f_n(\beta).$$

This reduces to the equality

$$s^{-1}g(D_+) - s\,g(D_-) = (s^{-1} - s)\,g(D_0),\qquad(3.13)$$

where

$$D_\pm = \det(\psi_n^{\mathrm{r}}(\sigma_1^{\pm 1}\beta) - I_{n-1}) \quad\text{and}\quad D_0 = \det(\psi_n^{\mathrm{r}}(\beta) - I_{n-1}).$$

Multiplying both sides of (3.13) by s, we reduce (3.13) to the equality

$$D_+ - tD_- = (1 - t)D_0.$$

To verify the latter, we expand

$$\psi_n^{\mathrm{r}}(\beta) = \begin{pmatrix} a & b & x \\ c & d & y \\ p & q & M \end{pmatrix},$$

where $a, b, c, d \in \Lambda$, x, y are rows over Λ of length $n-3$, p, q are columns over Λ of height $n - 3$, and M is an $(n - 3) \times (n - 3)$ matrix over Λ. By definition, $D_0 = \det A_0$, where

$$A_0 = \begin{pmatrix} a - 1 & b & x \\ c & d - 1 & y \\ p & q & M - I_{n-3} \end{pmatrix}.$$

Also

$$\psi_n^{\mathrm{r}}(\sigma_1\beta) = \begin{pmatrix} -t & 0 & 0 \\ 1 & 1 & 0 \\ 0 & 0 & I_{n-3} \end{pmatrix}\begin{pmatrix} a & b & x \\ c & d & y \\ p & q & M \end{pmatrix} = \begin{pmatrix} -ta & -tb & -tx \\ a+c & b+d & x+y \\ p & q & M \end{pmatrix}.$$

Subtracting I_{n-1}, then multiplying the first row by $-t^{-1}$, and finally subtracting the first row from the second one, we obtain $D_+ = -t\det A_+$, where

$$A_+ = \begin{pmatrix} a + t^{-1} & b & x \\ c - t^{-1} & d - 1 & y \\ p & q & M - I_{n-3} \end{pmatrix}.$$

Similarly,

$$\psi_n^{\mathrm{r}}(\sigma_1^{-1}\beta) = \begin{pmatrix} -t^{-1} & 0 & 0 \\ t^{-1} & 1 & 0 \\ 0 & 0 & I_{n-3} \end{pmatrix}\begin{pmatrix} a & b & x \\ c & d & y \\ p & q & M \end{pmatrix}$$

$$= \begin{pmatrix} -t^{-1}a & -t^{-1}b & -t^{-1}x \\ t^{-1}a+c & t^{-1}b+d & t^{-1}x+y \\ p & q & M - I_{n-3} \end{pmatrix}.$$

Subtracting I_{n-1}, then adding the first row to the second one, and finally multiplying the first row by $-t$, we obtain $D_- = -t^{-1} \det A_-$, where

$$A_- = \begin{pmatrix} a+t & b & x \\ c-1 & d-1 & y \\ p & q & M-I_{n-3} \end{pmatrix}.$$

The matrices A_0, A_+, A_- differ only in the first columns, which we denote by A_0^1, A_+^1, A_-^1, respectively. Clearly,

$$-tA_+^1 + A_-^1 = (1-t)A_0^1.$$

We conclude that $D_+ - tD_- = (1-t)D_0$.

The function $L \mapsto \widehat{f}(L)$ satisfies all conditions of the Alexander–Conway polynomial except that a priori it takes values in the field of rational functions in s rather than in its subring of Laurent polynomials $\mathbf{Z}[s, s^{-1}]$. However, applying the skein relation and an induction on the number of crossings of a link diagram as at the beginning of the proof, one observes that all the values of $\widehat{f}$ are integral polynomials in $s - s^{-1}$. In particular, all the values of $\widehat{f}$ are Laurent polynomials in s. □

3.5 The Lawrence–Krammer–Bigelow representation

We discuss a linear representation of B_n introduced by R. Lawrence and studied by D. Krammer and S. Bigelow. The definition of this representation is based on a study of a certain infinite covering of the configuration space of pairs of points on the punctured disk.

In this section we fix $n \geq 1$ and use the symbols D, $Q = \{(1,0), \ldots, (n,0)\}$, $\Sigma = D - Q$ introduced in Section 3.2.2.

3.5.1 The configuration spaces $\mathcal{F}$ and $\mathcal{C}$

Let $\mathcal{F}$ be the space of ordered pairs of distinct points in Σ. In other words, the space $\mathcal{F}$ is the complement of the diagonal $\{(x,x)\}_{x \in \Sigma}$ in $\Sigma \times \Sigma$. It is clear that $\mathcal{F}$ is a noncompact connected 4-dimensional manifold with boundary. It has a natural orientation obtained by squaring the counterclockwise orientation of Σ. In the notation of Section 1.4.1, we have

$$\mathcal{F} = \mathcal{F}_2(\Sigma) = \mathcal{F}_{n,2}(D).$$

The formula $(x,y) \mapsto (y,x)$ for distinct $x, y \in \Sigma$ defines an involution on $\mathcal{F}$. The quotient space $\mathcal{C}$ of this involution is the space of nonordered pairs of distinct points in Σ. Since the involution $(x,y) \mapsto (y,x)$ on $\mathcal{F}$ is orientation-preserving and fixed-point free, the space $\mathcal{C}$ is an oriented noncompact connected 4-dimensional manifold with boundary. Note for the record that the projection $\mathcal{F} \to \mathcal{C}$ is a 2-fold covering. In the notation of Section 1.4.3, we have $\mathcal{C} = \mathcal{C}_2(\Sigma) = \mathcal{C}_{n,2}(D)$.

In the sequel, a nonordered pair of distinct points $x, y \in \Sigma$ is denoted by $\{x, y\}$. Note that $\{x, y\} = \{y, x\} \in \mathcal{C}$. A (continuous) path $\xi : I \to \mathcal{C}$, where $I = [0, 1]$, can be written in the form $\xi = \{\xi_1, \xi_2\}$ for two (continuous) paths $\xi_1, \xi_2 : I \to \Sigma$. The equality $\xi = \{\xi_1, \xi_2\}$ means that $\xi(s) = \{\xi_1(s), \xi_2(s)\}$ for all $s \in I$. The path ξ is a loop if $\{\xi_1(0), \xi_2(0)\} = \{\xi_1(1), \xi_2(1)\}$, so that either

$$\xi_1(0) = \xi_1(1) \neq \xi_2(0) = \xi_2(1)$$

or

$$\xi_1(0) = \xi_2(1) \neq \xi_1(1) = \xi_2(0).$$

In the first case, the paths ξ_1, ξ_2 are loops on Σ. In the second case, the paths ξ_1, ξ_2 are not loops but their product $\xi_1\xi_2$ is well defined and is a loop on Σ.

We introduce two numerical invariants w and u of loops in $\mathcal{C}$. Consider a loop $\xi = \{\xi_1, \xi_2\}$ in $\mathcal{C}$ as above. If ξ_1, ξ_2 are loops, then $w(\xi) = w(\xi_1) + w(\xi_2)$, where $w(\xi_i)$ is the total winding number of ξ_i around $\{(1, 0), \ldots, (n, 0)\}$; see Section 3.2.2. If $\xi_1(1) = \xi_2(0)$, then the product path $\xi_1\xi_2$ is a loop on Σ and we set $w(\xi) = w(\xi_1\xi_2)$.

To define the second invariant $u(\xi)$, consider the map

$$s \mapsto \frac{\xi_1(s) - \xi_2(s)}{|\xi_1(s) - \xi_2(s)|} : I \to S^1 \subset \mathbf{C}. \tag{3.14}$$

This map sends $s = 0, 1$ either to the same numbers or to opposite numbers. Therefore, the map

$$s \mapsto \left(\frac{\xi_1(s) - \xi_2(s)}{|\xi_1(s) - \xi_2(s)|} \right)^2 : I \to S^1 \tag{3.15}$$

is a loop on S^1. The counterclockwise orientation of S^1 determines a generator of $H_1(S^1; \mathbf{Z}) \cong \mathbf{Z}$. The loop (3.15) on S^1 is homologous to k times the generator with $k \in \mathbf{Z}$, and we set $u(\xi) = k$. Note that $u(\xi)$ is even if ξ_1, ξ_2 are loops and odd otherwise. The invariants $w(\xi)$ and $u(\xi)$ are preserved under homotopy of ξ and are additive with respect to the multiplication of loops.

For example, consider the loop $\xi = \{\xi_1, \xi_2\}$, where ξ_1 is the constant loop in a point $z \in \Sigma$ and ξ_2 is an arbitrary loop in $\Sigma - \{z\}$. Then $w(\xi) = w(\xi_2)$ and $u(\xi) = 2v$, where v is the winding number of ξ_2 around z. In particular, if ξ_2 is a small loop encircling counterclockwise a point of Q and $z \in \partial\Sigma = \partial D$, then $w(\xi) = 1$ and $u(\xi) = 0$. To give another example, pick a small closed disk $B \subset \Sigma$ and two distinct points $a, b \in \partial B$. Let ξ_1 (resp. ξ_2) parametrize the arc on ∂B leading from a to b (resp. from b to a) counterclockwise. For the loop $\xi = \{\xi_1, \xi_2\}$, we have $w(\xi) = w(\xi_1\xi_2) = 0$ and $u(\xi) = 1$.

3.5.2 The covering space $\widetilde{C}$ and the module $\mathcal{H}$

We fix once for all two distinct points $d_1, d_2 \in \partial \Sigma = \partial D$ and take $c = \{d_1, d_2\}$ as the base point of C. The formula

$$\xi \mapsto q^{w(\xi)} t^{u(\xi)}$$

defines a group homomorphism φ from the fundamental group $\pi_1(C, c)$ to the multiplicative free abelian group with generators q, t. The examples in the previous subsection show that this homomorphism is surjective.

Let $\widetilde{C} \to C$ be the covering corresponding to the subgroup $\mathrm{Ker}\, \varphi$ of $\pi_1(C, c)$. The generators q and t act on $\widetilde{C}$ as commuting covering transformations, and $C = \widetilde{C}/(q, t)$. A loop ξ in C lifts to a loop in $\widetilde{C}$ if and only if $w(\xi) = u(\xi) = 0$.

The 2-fold covering $\mathcal{F} \to C$ is a quotient of the covering $\widetilde{C} \to C$, as we now explain. Observe that a loop $\xi = \{\xi_1, \xi_2\}$ on C lifts to a loop on $\mathcal{F}$ if and only if ξ_1, ξ_2 are loops on Σ. The latter holds if and only if $u(\xi)$ is even. Hence, the covering $\mathcal{F} \to C$ is determined by the subgroup of $\pi_1(C, c)$ formed by the homotopy classes of loops ξ with $u(\xi) \in 2\mathbf{Z}$. Therefore $\mathcal{F} = \widetilde{C}/(q, t^2)$ is the quotient of $\widetilde{C}$ by the group of homeomorphisms generated by q and t^2.

The action of q, t on $\widetilde{C}$ induces an action of q, t on the abelian group

$$\mathcal{H} = H_2(\widetilde{C}; \mathbf{Z}).$$

This turns $\mathcal{H}$ into a module over the commutative ring

$$R = \mathbf{Z}[q^{\pm 1}, t^{\pm 1}].$$

The module $\mathcal{H}$ can be explicitly computed using a deformation retraction of C onto a 2-dimensional CW-space; see [Big03], [PP02]. The computation shows that $\mathcal{H}$ is a free R-module of rank $n(n-1)/2$, that is,

$$\mathcal{H} \cong R^{n(n-1)/2}. \tag{3.16}$$

For more on the structure of $\mathcal{H}$, see Section 3.5.6.

3.5.3 An action of B_n on $\mathcal{H}$

As we know from Section 1.6, the braid group B_n is canonically isomorphic to the mapping class group $\mathfrak{M}(D, Q)$. In the remaining part of this chapter, we make no distinction between these two groups. We now construct an action of B_n on $\mathcal{H}$. Any self-homeomorphism f of the pair (D, Q) induces a homeomorphism $\widehat{f} : C \to C$ by

$$\widehat{f}(\{x, y\}) = \{f(x), f(y)\},$$

where x, y are distinct points of $\Sigma = D - Q$. Clearly, $\widehat{f}(c) = c$, so that we can consider the automorphism $\widehat{f}_{\#}$ of $\pi_1(C, c)$ induced by $\widehat{f}$.

Lemma 3.14. *We have $\varphi \circ \widehat{f}_{\#} = \varphi$.*

Proof. We need to prove that $w \circ \widehat{f}_{\#} = w$ and $u \circ \widehat{f}_{\#} = u$. The first equality is proven by the same argument as in Section 3.2.2. To prove the second equality, consider the inclusion of configuration spaces $\mathcal{C} = \mathcal{C}_2(\Sigma) \hookrightarrow \mathcal{C}_2(D)$ induced by the inclusion $\Sigma \hookrightarrow D$. The definition of the numerical invariant u for loops in $\mathcal{C}$ extends to loops in $\mathcal{C}_2(D)$ word for word and gives a homotopy invariant of loops in $\mathcal{C}_2(D)$. The Alexander–Tietze theorem stated in Section 1.6.1 implies that the self-homeomorphism of $\mathcal{C}_2(D)$ induced by f is homotopic to the identity. Hence, $u \circ \widehat{f}_{\#} = u$ and therefore $\varphi \circ \widehat{f}_{\#} = \varphi$.   □

The equality $\varphi \circ \widehat{f}_{\#} = \varphi$ implies that $\widehat{f}$ lifts uniquely to a map $\widetilde{f} : \widetilde{\mathcal{C}} \to \widetilde{\mathcal{C}}$ keeping fixed all points of $\widetilde{\mathcal{C}}$ lying over c. The same equality ensures that $\widetilde{f}$ commutes with the covering transformations of $\widetilde{\mathcal{C}}$. The map $\widetilde{f}$ is a homeomorphism with inverse $\widetilde{f^{-1}}$. Therefore the induced endomorphism $\widetilde{f}_*$ of $\mathcal{H} = H_2(\widetilde{\mathcal{C}}; \mathbf{Z})$ is an R-linear automorphism. Consider the mapping

$$B_n = \mathfrak{M}(D, Q) \to \mathrm{Aut}_R(\mathcal{H})$$

sending the isotopy class of f to $\widetilde{f}_* : \mathcal{H} \to \mathcal{H}$. This mapping is a group homomorphism. It is called the *Lawrence–Krammer–Bigelow representation* of B_n. A fundamental property of this representation is contained in the following theorem.

Theorem 3.15. *The Lawrence–Krammer–Bigelow representation of the braid group B_n is faithful for all $n \geq 1$.*

This theorem is proven in Sections 3.6 and 3.7. One can give explicit matrices describing the action of the generators $\sigma_1, \ldots, \sigma_{n-1} \in B_n$ on $\mathcal{H}$; see [Kra02], [Big01], [Bud05]. The proof of Theorem 3.15 given below uses neither these matrices nor the isomorphism (3.16).

3.5.4 The linearity of B_n

We say that a group G is *linear* if there is an injective group homomorphism $G \to \mathrm{GL}_N(\mathbf{R})$ for some integer $N \geq 1$. We state an important corollary of Theorem 3.15.

Theorem 3.16. *For all $n \geq 1$, the braid group B_n is linear.*

This theorem follows from Theorem 3.15 and the isomorphism (3.16). Indeed, choosing a basis of the R-module $\mathcal{H}$, we can identify $\mathrm{Aut}_R(\mathcal{H})$ with the matrix group $\mathrm{GL}_{n(n-1)/2}(R)$. The ring $R = \mathbf{Z}[q^{\pm 1}, t^{\pm 1}]$ can be embedded in the field of real numbers by assigning to q, t algebraically independent real values. This induces an embedding

$$\mathrm{GL}_{n(n-1)/2}(R) \hookrightarrow \mathrm{GL}_{n(n-1)/2}(\mathbf{R}).$$

Composing it with the Lawrence–Krammer–Bigelow representation, we obtain a faithful homomorphism $B_n \to \mathrm{GL}_{n(n-1)/2}(\mathbf{R})$.

We give another proof of Theorem 3.16 entirely avoiding the use of the isomorphism (3.16). This proof gives an embedding of B_n into $\mathrm{GL}_N(\mathbf{R})$ for $N = n(n+1)$. We begin with a simple algebraic lemma.

Lemma 3.17. *Let* $L = \mathbf{Z}[x_1^{\pm 1}, x_2^{\pm 1}]$ *be the ring of Laurent polynomials in the variables* x_1, x_2. *Let* C *be a free* L-*module of finite rank* $N \geq 1$. *For an arbitrary* L-*submodule* H *of* C, *the group* $\mathrm{Aut}_L(H)$ *of* L-*automorphisms of* H *embeds into* $\mathrm{GL}_N(\mathbf{R})$.

Proof. Let $Q = \mathbf{Q}(x_1, x_2)$ be the field of rational functions in the variables x_1, x_2 with rational coefficients. Clearly, Q is the field of fractions of L. Consider the Q-vector space $\overline{H} = Q \otimes_L H$. Since H is a submodule of a free L-module, it has no L-torsion, and hence the natural homomorphism $H \to \overline{H}$ sending $h \in H$ to $1 \otimes h$ is injective. Any L-automorphism of H extends uniquely to a Q-automorphism of $\overline{H}$. In this way, the group $\mathrm{Aut}_L(H)$ embeds into $\mathrm{GL}_m(Q)$, where $m = \dim_Q \overline{H}$. The field Q can be embedded in $\mathbf{R}$ by assigning to x_1, x_2 algebraically independent real values. This gives embeddings $\mathrm{Aut}_L(H) \subset \mathrm{GL}_m(Q) \subset \mathrm{GL}_m(\mathbf{R})$. Note that the inclusion $i : H \hookrightarrow C$ induces a homomorphism of Q-vector spaces $\overline{H} \to \overline{C}$, where $\overline{C} = Q \otimes_L C$. This homomorphism is injective: any element of its kernel can be multiplied by an element of L to give an element of $\mathrm{Ker}(i) = 0$. Therefore $m \leq \dim_Q \overline{C} = N$, so that $\mathrm{Aut}_L(H) \subset \mathrm{GL}_m(\mathbf{R}) \subset \mathrm{GL}_N(\mathbf{R})$. $\qquad\square$

Note that for any topological manifold M with boundary ∂M, the inclusion $M^\circ = M - \partial M \hookrightarrow M$ is a homotopy equivalence. The homotopy inverse $M \to M^\circ$ can be obtained by pushing M into M° using a cylinder neighborhood of ∂M in M.

We can now prove Theorem 3.16. It is clear that $\mathcal{F}^\circ = \mathcal{F} - \partial \mathcal{F}$ is the complement of the diagonal $\{(x,x)\}_{x \in \Sigma^\circ}$ in $\Sigma^\circ \times \Sigma^\circ$. By Lemma 1.26, assigning to any ordered pair of points the first point, we obtain a locally trivial fiber bundle $\mathcal{F}^\circ \to \Sigma^\circ$ whose fiber is the complement of a point in Σ°. The base Σ° of this bundle deformation retracts onto a wedge of n circles, while the fiber deformation retracts onto a wedge of $n + 1$ circles. This implies that $\mathcal{F}^\circ$ deformation retracts onto a 2-dimensional CW-complex, $X \subset \mathcal{F}^\circ$, with one zero-cell, $2n + 1$ one-cells, and $n(n+1)$ two-cells. Since the inclusion $\mathcal{F}^\circ \hookrightarrow \mathcal{F}$ is a homotopy equivalence, the inclusion $X \hookrightarrow \mathcal{F}$ also is a homotopy equivalence.

Recall from Section 3.5.2 that $\widetilde{C}$ can be viewed as the covering of $\mathcal{F}$ with the group of covering transformations $\mathbf{Z} \times \mathbf{Z}$ generated by q and t^2. The covering $\widetilde{C} \to \mathcal{F}$ restricts to a covering $\widetilde{X} \to X$ with the same group of covering transformations. Here $\widetilde{X}$ is the preimage of $X \subset \mathcal{F}$ in $\widetilde{C}$, and the inclusion $\widetilde{X} \subset \widetilde{C}$ is a homotopy equivalence. The cellular chain complex of $\widetilde{X}$ has the form $C_2 \to C_1 \to C_0$, where each C_i is a free module over the ring

$$R_0 = \mathbf{Z}[q^{\pm 1}, t^{\pm 2}] \subset R.$$

The rank of the R_0-module C_i is equal to the number of i-cells in X. Therefore

$$\mathcal{H} = H_2(\widetilde{C}; \mathbf{Z}) = H_2(\widetilde{X}; \mathbf{Z}) = \mathrm{Ker}(\partial : C_2 \to C_1)$$

is an R_0-submodule of C_2. We now apply Lemma 3.17, where we substitute

$$x_1 = q, \ x_2 = t^2, \ C = C_2, \ H = \mathcal{H}, \ \text{and} \ N = n(n+1) \,.$$

By this lemma, $\mathrm{Aut}_{R_0}(\mathcal{H})$ embeds into $\mathrm{GL}_N(\mathbf{R})$. Composing with the embeddings

$$B_n \hookrightarrow \mathrm{Aut}_R(\mathcal{H}) \subset \mathrm{Aut}_{R_0}(\mathcal{H}) \,,$$

we obtain the claim of the theorem. $\qquad\qquad\qquad\qquad\qquad\qquad\qquad$ □

3.5.5 A sesquilinear form on $\mathcal{H}$

The module $\mathcal{H}$ carries a natural R-valued sesquilinear form defined as follows. The orientation of C lifts to $\widetilde{C}$ and turns the latter into an oriented (four-dimensional) manifold. Consider the associated intersection form $\mathcal{H} \times \mathcal{H} \to \mathbf{Z}$. Its value $g_1 \cdot g_2$ on homology classes $g_1, g_2 \in \mathcal{H}$ is obtained by representing these classes by transversal 2-cycles G_1, G_2 in $\widetilde{C}$ and counting the intersections of G_1, G_2 with signs $\pm$ determined by the orientation of $\widetilde{C}$. The intersection form $\mathcal{H} \times \mathcal{H} \to \mathbf{Z}$ is symmetric and invariant under the action of orientation-preserving homeomorphisms $\widetilde{C} \to \widetilde{C}$. In particular, this form is invariant under the action of the covering transformations q, t.

Define a pairing

$$\langle \, , \, \rangle : \mathcal{H} \times \mathcal{H} \to R$$

by

$$\langle g_1, g_2 \rangle = \sum_{k,\ell \in \mathbf{Z}} (q^k t^\ell g_1 \cdot g_2)\, q^k t^\ell \,. \tag{3.17}$$

The sum on the right-hand side is finite, since the 2-cycles G_1, G_2 as above lie in compact subsets of $\widetilde{C}$ and therefore the cycles $q^k t^\ell G_1$ and G_2 are disjoint except for a finite set of pairs (k, ℓ).

The pairing (3.17) is invariant under the action of orientation-preserving homeomorphisms $\widetilde{C} \to \widetilde{C}$ commuting with the covering transformations q, t. In particular, it is preserved under the action of the braid group B_n on $\mathcal{H}$.

Lemma 3.18. *For any $g_1, g_2 \in \mathcal{H}$ and $r \in R$,*

$$\langle g_2, g_1 \rangle = \overline{\langle g_1, g_2 \rangle}, \quad \langle g_1, rg_2 \rangle = r \langle g_1, g_2 \rangle, \quad \langle rg_1, g_2 \rangle = \overline{r} \langle g_1, g_2 \rangle, \tag{3.18}$$

where $r \mapsto \overline{r}$ is the involutive automorphism of the ring R sending q to q^{-1} and t to t^{-1}.

Proof. We have

$$\langle g_2, g_1 \rangle = \sum_{k,\ell \in \mathbf{Z}} (q^k t^\ell g_2 \cdot g_1) \, q^k t^\ell$$

$$= \sum_{k,\ell \in \mathbf{Z}} (g_1 \cdot q^k t^\ell g_2) \, q^k t^\ell$$

$$= \sum_{k,\ell \in \mathbf{Z}} (q^{-k} t^{-\ell} g_1 \cdot g_2) \, q^k t^\ell$$

$$= \sum_{k,\ell \in \mathbf{Z}} (q^k t^\ell g_1 \cdot g_2) \, q^{-k} t^{-\ell}$$

$$= \overline{\langle g_1, g_2 \rangle}.$$

To verify the equalities $\langle g_1, r g_2 \rangle = r \langle g_1, g_2 \rangle$ and $\langle r g_1, g_2 \rangle = \bar{r} \langle g_1, g_2 \rangle$, it suffices to consider the case $r = q^i t^j$ with $i, j \in \mathbf{Z}$. We have

$$\langle g_1, q^i t^j g_2 \rangle = \sum_{k,\ell \in \mathbf{Z}} (q^k t^\ell g_1 \cdot q^i t^j g_2) \, q^k t^\ell$$

$$= q^i t^j \sum_{k,\ell \in \mathbf{Z}} (q^{k-i} t^{\ell-j} g_1 \cdot g_2) \, q^{k-i} t^{\ell-j}$$

$$= q^i t^j \langle g_1, g_2 \rangle$$

and

$$\langle q^i t^j g_1, g_2 \rangle = \sum_{k,\ell \in \mathbf{Z}} (q^{k+i} t^{\ell+j} g_1 \cdot g_2) \, q^k t^\ell$$

$$= q^{-i} t^{-j} \sum_{k,\ell \in \mathbf{Z}} (q^{k+i} t^{\ell+j} g_1 \cdot g_2) \, q^{k+i} t^{\ell+j}$$

$$= q^{-i} t^{-j} \langle g_1, g_2 \rangle. \qquad \square$$

According to Budney [Bud05], the form $\langle \, , \, \rangle : \mathcal{H} \times \mathcal{H} \to R$ is nonsingular in the sense that the determinant of its matrix with respect to a basis of $\mathcal{H}$ is nonzero. Moreover, replacing q, t with appropriate complex numbers, one obtains a negative definite Hermitian form; see [Bud05]. This gives an injective group homomorphism from B_n into the unitary group $U_{n(n-1)/2}$.

3.5.6 Remarks

We make a few remarks aimed at familiarizing the reader with the module $\mathcal{H}$. These remarks will not be used in the sequel.

It is quite easy to see that the module $\mathcal{H}$ is nontrivial and in fact rather big. Let $X, \widetilde{X}$ be the same spaces as in the proof of Theorem 3.16. Note that the ring $R_0 = \mathbf{Z}[q^{\pm 1}, t^{\pm 2}]$ embeds into the field $Q = \mathbf{Q}(q, t^2)$ of rational functions in the variables q, t^2. For an R_0-module H, denote the dimension of the Q-vector space $Q \otimes_{R_0} H$ by $\operatorname{rk} H$. We verify that $\operatorname{rk} \mathcal{H} \geq n(n-1)$. Indeed,

$$\operatorname{rk} H_0(\widetilde{X}; \mathbf{Z}) - \operatorname{rk} H_1(\widetilde{X}; \mathbf{Z}) + \operatorname{rk} \mathcal{H} = \chi(X) = n(n-1),$$

where $\chi(X)$ is the Euler characteristic of X. For every 0-cell x of $\widetilde{X}$ there is a path in $\widetilde{X}$ leading from x to qx, so that $(1-q)x$ is the boundary of a 1-chain. Hence $Q \otimes_R H_0(\widetilde{X}; \mathbf{Z}) = 0$ and $\operatorname{rk} H_0(\widetilde{X}; \mathbf{Z}) = 0$. Therefore, $\operatorname{rk} \mathcal{H} \geq n(n-1)$. The isomorphism (3.16) implies that $\mathcal{H} \cong R_0^{n(n-1)}$.

Specific elements of $\mathcal{H}$ may be derived from arbitrary disjoint spanning arcs α, β on (D, Q). Consider the associated loops $\alpha', \beta' : S^1 \to \Sigma$ as in Figure 3.1. Choosing these loops closely enough to α, β, we may assume that they do not meet. The formula

$$(s_1, s_2) \mapsto \{\alpha'(s_1), \beta'(s_2)\} \in \mathcal{C}$$

for $s_1, s_2 \in S^1$ defines an embedding of the torus $S^1 \times S^1$ into $\mathcal{C}$. The induced homomorphism of the fundamental groups sends $\pi_1(S^1 \times S^1)$ to the kernel of φ. Therefore this embedding lifts to an embedding of the torus into $\widetilde{\mathcal{C}}$. It can be shown that the fundamental class of the torus represents a nontrivial homology class in $\mathcal{H}$. Such classes, corresponding to various α, β, are permuted by the action of B_n on $\mathcal{H}$. A similar but subtler construction applies to pairs of spanning arcs on (D, Q) meeting at one common endpoint; it gives a mapping of an orientable closed surface of genus 2 to $\widetilde{\mathcal{C}}$; see [Big03]. Moreover, each spanning arc on (D, Q) gives rise to a mapping of an orientable closed surface of genus 3 to $\widetilde{\mathcal{C}}$; see [Big03] and Section 3.7.1. Applying these constructions to the arcs

$$[1, 2] \times \{0\}, \ [2, 3] \times \{0\}, \ldots, [n-1, n] \times \{0\}$$

on (D, Q) and to pairs of such arcs, one obtains $n(n-1)/2$ homology classes in $\mathcal{H}$ forming an R-basis of $\mathcal{H}$.

3.6 Noodles vs. spanning arcs

In this section we introduce and study so-called noodles on the n-punctured disk $\Sigma = D - Q$, where $n \geq 1$ and $Q = \{(1, 0), \ldots, (n, 0)\} \subset D$. Noodles will be used in a crucial way in the proof of Theorem 3.15 in Section 3.7.

3.6.1 Noodles

A *noodle* in Σ is an oriented embedded arc $N \subset \Sigma$ such that

$$\partial N = N \cap \partial \Sigma.$$

The boundary ∂N of a noodle N consists of the two endpoints of N lying on $\partial \Sigma = \partial D$. An example of a noodle is shown in Figure 3.7.

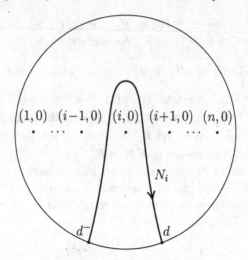

Fig. 3.7. The noodle N_i

We focus now on the intersections of a noodle N with spanning arcs on (D, Q). Let α be a spanning arc on (D, Q) intersecting N transversely at finitely many points. The intersection of N and α can be simplified using digons as in Section 3.2.4. A *digon* for N, α is an embedded disk in $\Sigma^\circ = \Sigma - \partial\Sigma$ whose boundary is formed by a subarc of N and a subarc of α and whose interior does not meet $N \cup \alpha$; cf. Figure 3.3, where c, d should be replaced by N, α. Each digon determines an obvious isotopy of α (rel $\partial\alpha$) decreasing the number of points in $N \cap \alpha$ by two. The following lemma shows that conversely, if there is such an isotopy, then the pair N, α has digons.

Lemma 3.19. *If there is an isotopy of α (rel $\partial\alpha$) decreasing the number of points in $N \cap \alpha$, then the pair N, α has at least one digon.*

Proof. We deduce this lemma from Lemma 3.6 by extending the arcs N and α to simple closed curves on a bigger surface. Pick closed disk neighborhoods $U_1, U_2 \subset D$ of the endpoints of α such that $U_1 \cap U_2 = U_i \cap N = \emptyset$ for $i = 1, 2$ and each circle ∂U_i meets α at exactly one point. Consider the punctured disk $D_- = D - (U_1^\circ \cup U_2^\circ)$. Clearly, $\partial D_- = \partial U_1 \cup \partial U_2 \cup \partial D$. We now form a new surface S by gluing the following three pieces: the punctured disk D_-, an annulus $A = S^1 \times [0, 1]$, and a punctured torus T obtained as the complement of a small open disk on $S^1 \times S^1$. (Instead of the torus, we can use any orientable surface of positive genus.) The surfaces D_-, A, and T are glued along homeomorphisms $\partial A \approx \partial U_1 \cup \partial U_2$ and $\partial T \approx \partial D$ chosen so that the resulting surface, S, is orientable. Connecting the points $\alpha \cap \partial U_1$, $\alpha \cap \partial U_2$ in A, we can extend the arc $\alpha \cap D_-$ to a simple closed curve $\widehat{\alpha}$ on S; see Figure 3.8. Similarly, the arc $N \subset D_-$ extends to a simple closed curve $\widehat{N}$ on S going once along a longitude of T.

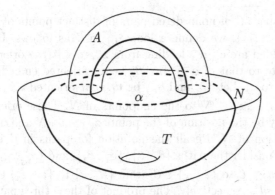

Fig. 3.8. The surface S

If there is an isotopy of α (rel $\partial\alpha$) decreasing the number of points in $N \cap \alpha$, then there is an isotopy of $\widehat{\alpha}$ in S decreasing the number of points in $\widehat{N} \cap \widehat{\alpha}$. By Lemma 3.6, the pair $\widehat{N}, \widehat{\alpha}$ has a digon on S. Such a digon cannot approach a branch of $\widehat{N}$ or $\widehat{\alpha}$ from different sides and therefore meets neither T nor A. Therefore such a digon lies on D_- and is a digon for N, α.   □

3.6.2 Algebraic intersection of noodles and arcs

The intersection of a noodle N and a spanning arc α can be measured in terms of a so-called *algebraic intersection* $\langle N, \alpha \rangle$. This is an element of the ring $\mathbf{Z}[q^{\pm 1}, t^{\pm 1}]$ defined up to multiplication by monomials $q^w t^u$ with $w \in \mathbf{Z}$ and $u \in 2\mathbf{Z} \subset \mathbf{Z}$. The algebraic intersection $\langle N, \alpha \rangle$ depends on a choice of orientation on α, which we fix from now on. As above, we endow Σ with counterclockwise orientation. The orientations of α and Σ allow us to speak of the "right" and "left" sides of α in Σ. Pushing α slightly to the left (keeping the endpoints), we obtain a "parallel" oriented spanning arc α^- on (D, Q) with the same starting and terminal endpoints as α and disjoint from α otherwise. Slightly deforming N, we can assume that N intersects α transversely in $m \geq 0$ points $z_1, \ldots, z_m$ (the numeration is arbitrary). We choose the parallel arc α^- very closely to α so that α^- meets N transversely in m points $z_1^-, \ldots, z_m^-$, where each pair z_i^-, z_i is joined by a short subarc of N lying in the narrow strip on Σ bounded by $\alpha^- \cup \alpha$; cf. Figure 3.9 below (the strip in question is shaded). For $i \in \{1, \ldots, m\}$, let $\varepsilon_i = \pm 1$ be the intersection sign of N and α at z_i (recall that both N and α are oriented). Thus, $\varepsilon_i = +1$ if N crosses α at z_i from left to right and $\varepsilon_i = -1$ otherwise. Denote the starting endpoint and the terminal endpoint of N by d^- and d, respectively. Fix arbitrary points

$$z^- \in \alpha^- - \partial\alpha^-, \quad z \in \alpha - \partial\alpha$$

and fix paths θ^-, θ in Σ leading respectively from d^- to z^- and from d to z and having disjoint images (these paths are allowed to meet N, α^-, and α elsewhere).

Recall the space $\mathcal{C}$ of nonordered pairs of distinct points of Σ. For every pair $i, j \in \{1, \ldots, m\}$, we define a loop $\xi_{i,j}$ in $\mathcal{C}$ as follows. Let β_i^- be an oriented embedded arc on α^- leading from z^- to z_i^- (the orientation of β_i^- may be opposite to that of α^-). Let β_j be an oriented embedded arc on α leading from z to z_j. Let $\gamma_{i,j}^-$ and $\gamma_{i,j}$ be *disjoint* oriented arcs in N leading from the points $z_i^-, z_j \in N$ to the endpoints of N. These oriented arcs are determined only by the position of the points z_i^-, z_j on N and do not depend on the orientation of N. Recall the notation for paths in $\mathcal{C}$ introduced in Section 3.5.1. Consider the paths $\{\theta^-, \theta\}$, $\{\beta_i^-, \beta_j\}$, and $\{\gamma_{i,j}^-, \gamma_{i,j}\}$ in $\mathcal{C}$. They lead from $\{d^-, d\} \in \mathcal{C}$ to $\{z^-, z\} \in \mathcal{C}$, from $\{z^-, z\}$ to $\{z_i^-, z_j\} \in \mathcal{C}$, and from $\{z_i^-, z_j\}$ to $\{d^-, d\}$, respectively. The product of these three paths

$$\xi_{i,j} = \{\theta^-, \theta\} \{\beta_i^-, \beta_j\} \{\gamma_{i,j}^-, \gamma_{i,j}\} \tag{3.19}$$

is a loop in $\mathcal{C}$ beginning and ending in $\{d^-, d\}$. Set

$$\langle N, \alpha \rangle = \sum_{i=1}^{m} \sum_{j=1}^{m} \varepsilon_i \, \varepsilon_j \, q^{w(\xi_{i,j})} \, t^{u(\xi_{i,j})} \in \mathbf{Z}[q^{\pm 1}, t^{\pm 1}],$$

where w and u are the integral invariants of loops in $\mathcal{C}$ introduced in Section 3.5.1. The expression on the right-hand side does not depend on the numeration of points in $N \cap \alpha$. Under a different choice of z^-, z, θ^-, θ, all loops $\xi_{i,j}$ are multiplied on the left by one and the same loop in $\mathcal{C}$ of the form $\{\xi_1, \xi_2\}$, where ξ_1, ξ_2 are loops in Σ. Then $\langle N, \alpha \rangle$ is multiplied by a monomial in $q^{\pm 1}, t^{\pm 2}$.

For example, if N is disjoint from α, then $\langle N, \alpha \rangle = 0$. If N crosses α in only one point, then $m = 1$ and $\langle N, \alpha \rangle = q^k t^\ell$ for some $k, \ell \in \mathbf{Z}$.

We state two fundamental properties of the algebraic intersections of noodles and arcs.

Lemma 3.20. *The algebraic intersection $\langle N, \alpha \rangle$ is invariant under isotopies of N and α in Σ constant on the endpoints.*

Proof. It suffices to fix N and to prove that $\langle N, \alpha \rangle$ is invariant under isotopies of the spanning arc α. A generic isotopy of α in Σ can be split into a finite sequence of local moves of three types:

(i) an isotopy of α in Σ keeping α transversal to N,

(ii) a move pushing a small subarc of α across a subarc of N,

(iii) an inverse to (ii).

It is clear from the definitions that the moves of type (i) do not change $\langle N, \alpha \rangle$. Any move of type (ii) adds two new intersection points z_{m+1}, z_{m+2} to the set $N \cap \alpha = \{z_1, \ldots, z_m\}$. Assume for concreteness that the subarc of N connecting z_{m+1} with z_{m+2} lies on the right of the arc α; see Figure 3.9. Clearly, the sign $\varepsilon_i = \pm 1$ is preserved under this move for $i = 1, \ldots, m$. For all

$i, j = 1, \ldots, m$, the loops $\xi_{i,j}$ computed before and after the move are homotopic to each other. Therefore such pairs (i, j) contribute the same expression to $\langle N, \alpha \rangle$ before and after the move. For $i = 1, \ldots, m + 2$, the loops $\xi_{i,m+1}$ and $\xi_{i,m+2}$ are homotopic and the obvious equality $\varepsilon_{m+1} = -\varepsilon_{m+2}$ implies that the contributions of the pairs $(i, m + 1)$, $(i, m + 2)$ cancel each other. Similarly, for any $i = 1, \ldots, m$, the loops $\xi_{m+1,i}$ and $\xi_{m+2,i}$ are homotopic and the contributions of the pairs $(m + 1, i)$, $(m + 2, i)$ cancel each other. Therefore $\langle N, \alpha \rangle$ is preserved under the move.   □

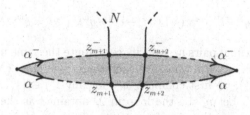

Fig. 3.9. Additional crossings

We say that a spanning arc α on (D, Q) can be *isotopped off* a noodle N if there is a continuous family of spanning arcs $\{\alpha_s\}_{s \in [0,1]}$ on (D, Q) such that $\alpha_0 = \alpha$ and α_1 is disjoint from N. Such a family $\{\alpha_s\}_s$ is called an *isotopy* of α. Note that the spanning arcs α_s necessarily have the same endpoints.

Lemma 3.21. *A spanning arc α can be isotopped off a noodle N if and only if $\langle N, \alpha \rangle = 0$.*

Proof. If there is an isotopy $\{\alpha_s\}_s$ of $\alpha = \alpha_0$ in Σ such that α_1 is disjoint from N, then $\langle N, \alpha \rangle = \langle N, \alpha_1 \rangle = 0$. The hard part of the lemma is the opposite implication. Applying a preliminary isotopy to α, we can assume that α intersects N transversely at a *minimal* number of points $z_1, \ldots, z_m$ with $m \geq 0$. We assume that $m \geq 1$ and show that $\langle N, \alpha \rangle \neq 0$.

We keep the notation introduced above in the definition of $\langle N, \alpha \rangle$. For any $i, j \in \{1, \ldots, m\}$, set $w_{i,j} = w(\xi_{i,j}) \in \mathbf{Z}$ and $u_{i,j} = u(\xi_{i,j}) \in \mathbf{Z}$. Then

$$\langle N, \alpha \rangle = \sum_{i=1}^{m} \sum_{j=1}^{m} \varepsilon_i \varepsilon_j q^{w_{i,j}} t^{u_{i,j}} . \tag{3.20}$$

Observe that

$$\varepsilon_i = (-1)^{u_{i,i}}$$

for all i. Indeed, if $\varepsilon_i = +1$, then N crosses the arc α at z_i from left to right and therefore the paths $\gamma_{i,i}^-$, $\gamma_{i,i}$ end respectively in d^-, d. Then $\xi_{i,i}$ has the form $\{\xi_1, \xi_2\}$, where ξ_1, ξ_2 are loops in Σ. In this case

$$u_{i,i} = u(\xi_{i,i}) = 0 \pmod 2.$$

Similarly, if $\varepsilon_i = -1$, then $u_{i,i} = 1 \pmod 2$. In both cases $\varepsilon_i = (-1)^{u_{i,i}}$.

We shall use the lexicographic order on monomials $q^w t^u$ with $w, u \in \mathbf{Z}$. More precisely, we write $q^w t^u \geq q^{w'} t^{u'}$ with $w, u, w', u' \in \mathbf{Z}$ if either $w > w'$ or $w = w'$ and $u \geq u'$. We say that an ordered pair (i, j) with $i, j \in \{1, \dots, m\}$ is *maximal* (for given N, α) if $q^{w_{i,j}} t^{u_{i,j}} \geq q^{w_{k,l}} t^{u_{k,l}}$ for all $k, l \in \{1, \dots, m\}$. A maximal pair necessarily exists because the lexicographic order on the monomials is total. A maximal pair may be nonunique. We claim that

$$\text{if } (i, j) \text{ is maximal, then } u_{i,i} = u_{j,j}. \tag{3.21}$$

This claim implies that every maximal pair (i, j) contributes the monomial

$$\varepsilon_i \varepsilon_j \, q^{w_{i,j}} t^{u_{i,j}} = (-1)^{u_{i,i}} (-1)^{u_{j,j}} q^{w_{i,j}} t^{u_{i,j}} = q^{w_{i,j}} t^{u_{i,j}}$$

to $\langle N, \alpha \rangle$. All maximal pairs necessarily contribute the same monomial, which then occurs in $\langle N, \alpha \rangle$ with a positive coefficient. Therefore $\langle N, \alpha \rangle \neq 0$.

To prove (3.21), we first compute $w_{i,j}$ for any $i, j \in \{1, \dots, m\}$ (not necessarily maximal). Let η_i^- be the loop in Σ obtained as the product of the path $\theta^- \beta_i^-$ with the path going from z_i^- to d^- along N. Let η_j be the loop in Σ obtained as the product of $\theta \beta_j$ with the path going from z_j to d along N. We claim that

$$w_{i,j} = w(\eta_i^-) + w(\eta_j). \tag{3.22}$$

Indeed, if the path $\gamma_{i,j}^-$ appearing in (3.19) ends at d^-, then the path $\gamma_{i,j}$ ends at d, the paths $\theta^- \beta_i^- \gamma_{i,j}^-$ and $\theta \beta_j \gamma_{i,j}$ are loops, and $w_{i,j}$ is the sum of their total winding numbers. Formula (3.22) follows in this case from the equalities $\eta_i^- = \theta^- \beta_i^- \gamma_{i,j}^-$ and $\eta_j = \theta \beta_j \gamma_{i,j}$. Assume that $\gamma_{i,j}^-$ ends at d. Then $\gamma_{i,j}$ ends at d^-,

$$\eta_i^- = \theta^- \beta_i^- \gamma_{i,j}^- N^{-1}, \quad \eta_j = \theta \beta_j \gamma_{i,j} N,$$

where N is viewed as a path from d^- to d. By definition, $w_{i,j}$ is the total winding number of the loop $\theta^- \beta_i^- \gamma_{i,j}^- \theta \beta_j \gamma_{i,j}$. This loop is homotopic in Σ to the loop

$$\theta^- \beta_i^- \gamma_{i,j}^- N^{-1} N \theta \beta_j \gamma_{i,j} N N^{-1} = \eta_i^- N \eta_j N^{-1}.$$

The loop $\eta_i^- N \eta_j N^{-1}$ is homologous to $\eta_i^- \eta_j$ in Σ. Hence (3.22).

Inspecting the loops η_i^- and η_i, we observe that the difference between their homology classes $[\eta_i^-], [\eta_i] \in H_1(\Sigma; \mathbf{Z})$ is represented by the loop going from d to d^- along N^{-1}, then from d^- to z^- along θ^-, then from z^- to z along a path lying in the strip between α^- and α, and finally from z to d along θ^{-1}. Therefore the difference $[\eta_i^-] - [\eta_i] \in H_1(\Sigma; \mathbf{Z})$ does not depend on i. This implies that the number

$$W = w(\eta_i^-) - w(\eta_i) \in \mathbf{Z}$$

does not depend on i. Formula (3.22) implies that for all $i, j = 1, \dots, m$,

$$w_{i,j} = w(\eta_i) + w(\eta_j) + W. \tag{3.23}$$

Suppose that the pair (i,j) is maximal. Then $w_{i,j}$ is maximal among all the integers $w_{k,l}$. By (3.23), both numbers $w(\eta_i)$ and $w(\eta_j)$ must be maximal among all the integers $w(\eta_k)$. Then

$$w(\eta_i) = w(\eta_j) \quad \text{and} \quad w_{i,i} = w_{i,j}.$$

The maximality of (i,j) implies that $u_{i,i} \leq u_{i,j}$. We claim that $u_{i,i} = u_{i,j}$. For $i = j$, this is obvious and we assume that $i \neq j$.

Suppose, seeking a contradiction, that $u_{i,i} < u_{i,j}$. Let μ be the (embedded) subarc of α connecting z_i and z_j. Let ν be the (embedded) subarc of N connecting z_i and z_j. We orient μ from z_i to z_j and ν from z_j to z_i. The product $\mu\nu$ is a loop on Σ based at z_i. We distinguish two cases.

Case 1: The arc ν approaches α at z_i from the right (in other words, ν does not pass through z_i^-). Then the loop $\mu\nu$ does not pass through z_i^- and we can consider its winding number, $v \in \mathbf{Z}$, around z_i^-. We claim that $v > 0$. To see this, we compute v as follows. As was already observed, $2v = u(\{z_i^-, \mu\nu\})$, where u is the invariant of loops in $\mathcal{C}$ defined in Section 3.5.1 and z_i^- stands for the constant path in the point z_i^-. Observe that $\beta_j \sim \beta_i\mu$, where $\sim$ denotes the homotopy of paths in $\Sigma - \{z_i^-\}$ relative to the endpoints. The assumption that ν does not pass through z_i^- implies that

$$\gamma_{i,i}^- = \gamma_{i,j}^- \quad \text{and} \quad \gamma_{i,i} = \nu^{-1}\gamma_{i,j};$$

see Figure 3.10. Then

$$\xi_{i,j} = \{\theta^-, \theta\}\{\beta_i^-, \beta_j\}\{\gamma_{i,j}^-, \gamma_{i,j}\} \sim \{\theta^-, \theta\}\{\beta_i^-, \beta_i\}\{z_i^-, \mu\nu\}\{\gamma_{i,i}^-, \gamma_{i,i}\}.$$

The latter loop is homologous in $\mathcal{C}$ to the loop

$$\{\theta^-, \theta\}\{\beta_i^-, \beta_i\}\{\gamma_{i,i}^-, \gamma_{i,i}\}\{z_i^-, \mu\nu\} = \xi_{i,i}\{z_i^-, \mu\nu\}.$$

Therefore,

$$2v = u(\{z_i^-, \mu\nu\}) = u(\xi_{i,j}) - u(\xi_{i,i}) = u_{i,j} - u_{i,i}.$$

The assumption $u_{i,i} < u_{i,j}$ implies that $v > 0$.

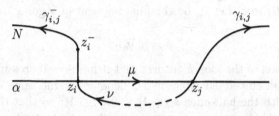

Fig. 3.10. Case 1: the paths $\gamma_{i,j}^-$ and $\gamma_{i,j}$

We can now bring one more loop into the picture. Consider the short subarc of N connecting z_i to z_i^- in the strip between α and α^-. Pick a loop ρ in a small neighborhood of this subarc such that

(i) ρ begins and ends in z_i;
(ii) ρ does not meet z_i^- and winds clockwise v times around z_i^-;
(iii) ρ has $v - 1$ transversal self-crossings;
(iv) ρ meets $\mu\nu$ only at z_i (see Figure 3.11).

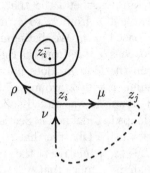

Fig. 3.11. The loop ρ for $v = 3$

Note that the winding number of the loop $\mu\nu\rho$ around z_i^- is equal to 0. Hence, this loop lifts to an appropriate covering of the complement of $\{z_i^-\}$. We now describe this lift in more detail.

Let $D_\bullet = D - \{z_i^-\}$ and $p : \widehat{D}_\bullet \to D_\bullet$ be the universal (infinite cyclic) covering. Let $\widehat{\mu} : [0, 1] \to \widehat{D}_\bullet$ be an arbitrary lift of μ (so that $p\widehat{\mu} = \mu$). There is a unique lift $\widehat{\nu} : [0, 1] \to \widehat{D}_\bullet$ of ν such that $\widehat{\nu}(0) = \widehat{\mu}(1)$. Consider also the unique lift $\widehat{\rho} : [0, 1] \to \widehat{D}_\bullet$ of ρ such that $\widehat{\rho}(0) = \widehat{\nu}(1)$. By abuse of notation, we shall denote the paths $\mu, \nu, \rho, \widehat{\mu}, \widehat{\nu}, \widehat{\rho}$ and their images by the same letters. Since the winding number of $\mu\nu\rho$ around z_i^- is zero, the path $\widehat{\mu}\widehat{\nu}\widehat{\rho}$ is a loop. Our choice of ρ ensures that $\widehat{\rho}$ is an embedded arc in $\widehat{D}_\bullet$ meeting $\widehat{\mu}\widehat{\nu}$ only at the endpoints. However, the embedded arcs $\widehat{\mu}$ and $\widehat{\nu}$ in $\widehat{D}_\bullet$ may meet in several points besides their common endpoint $\widehat{\mu}(1) = \widehat{\nu}(0)$. Let a be the first point of $\widehat{\mu}$ that lies also on $\widehat{\nu}$ (possibly $a = \widehat{\mu}(1)$). Let $\widehat{\mu}_a$ be the initial segment of $\widehat{\mu}$ going from $\widehat{\mu}(0)$ to a. Let $\widehat{\nu}_a$ be the final segment of $\widehat{\nu}$ going from a to $\widehat{\nu}(1)$. Set

$$\delta = \widehat{\mu}_a \widehat{\nu}_a \widehat{\rho}.$$

The construction of the loop δ ensures that it has no self-crossings. This loop parametrizes an embedded circle in $\widehat{D}_\bullet$ denoted by the same symbol δ. We identify $\widehat{D}_\bullet$ with the half-open strip $\mathbf{R} \times [0, 1) \subset \mathbf{R}^2$ so that the orientation in $\widehat{D}_\bullet$ induced by the counterclockwise orientation in $D_\bullet$ is identified with the counterclockwise orientation in $\mathbf{R}^2$. The Jordan curve theorem implies that δ bounds an embedded disk $B \subset \widehat{D}_\bullet$.

We verify now that the loop δ encircles B counterclockwise. Let C be the component of $D_\bullet - \rho$ surrounding z_i^-. We check first that $C \cap p(B) = \emptyset$. Indeed, suppose that there is a point $b \in B$ such that $p(b) \in C$. We can connect the point $p(b)$ to any other point b' of C by an arc in C. This arc lifts to an arc in $\widehat{D}_\bullet$ beginning in b. The latter arc never meets δ, since its projection to $D_\bullet$ never meets μ, ν, or ρ. Hence this lifted arc lies in the interior $B^\circ = B - \partial B$ of B, and its terminal endpoint projects to b'. Thus, $C \subset p(B)$. Since B is compact, so is $p(B)$. On the other hand, it is clear that C is not contained in a compact subset of $D_\bullet$. This contradiction shows that $C \cap p(B) = \emptyset$. Observe now that C lies on the right of ρ. If B lies on the right of $\widehat{\rho} \subset \delta$, then necessarily $C \cap p(B) \neq \emptyset$, a contradiction. Thus, B lies on the left of $\widehat{\rho}$ and of δ. Hence, δ goes counterclockwise around B.

We claim that $B \cap p^{-1}(Q) = \emptyset$. Indeed, being a compact subset of $\widehat{D}_\bullet$, the disk B may contain only a finite number of points of the (discrete) set $p^{-1}(Q) \subset \widehat{D}_\bullet$. Observe that the paths μ, ν, ρ lie in $\Sigma = D - Q$ and do not meet Q. Therefore $\partial B \cap p^{-1}(Q) = \emptyset$, so that $B \cap p^{-1}(Q) \subset B^\circ$. The loop $\delta = \partial B$ is homologous in $B - p^{-1}(Q)$ to the sum of small loops encircling the points of $B \cap p^{-1}(Q)$ counterclockwise. The latter loops are projected by p homeomorphically onto small loops encircling certain points of Q counterclockwise. Therefore,

$$\mathrm{card}(B \cap p^{-1}(Q)) = w(p \circ \delta),$$

where $w(p \circ \delta)$ is the total winding number of the loop $p \circ \delta$ in Σ around the points of Q. We have

$$p \circ \delta = \mu_a \nu_a \rho,$$

where $\mu_a = p(\widehat{\mu}_a)$ is the initial segment of μ going from z_i to $p(a)$ along α, and $\nu_a = p(\widehat{\nu}_a)$ is the final segment of ν going from $p(a)$ to z_i along N. Then $p(a) \in N \cap \alpha$, so that $p(a) = z_k$ for some $k = 1, \ldots, n$. Since ρ is contractible in Σ, the loop $\mu_a \nu_a \rho$ is homotopic to $\mu_a \nu_a$ in Σ and

$$w(\mu_a \nu_a \rho) = w(\mu_a \nu_a).$$

Recall the loops η_k, η_i in Σ based at the terminal endpoint d of N. The difference between their homology classes $[\eta_k], [\eta_i] \in H_1(\Sigma; \mathbf{Z})$ depends neither on the choice of the path θ nor on the choice of its terminal endpoint $z \in \alpha$. Taking $z = z_i$, one immediately deduces from the definition of η_k, η_i that $[\eta_k] - [\eta_i] = [\mu_a \nu_a]$. Therefore,

$$w(\mu_a \nu_a) = w(\eta_k) - w(\eta_i).$$

To sum up, we have

$$\mathrm{card}(B \cap p^{-1}(Q)) = w(p \circ \delta) = w(\mu_a \nu_a \rho) = w(\mu_a \nu_a) = w(\eta_k) - w(\eta_i).$$

Since $w(\eta_i)$ is maximal, $\mathrm{card}(B \cap p^{-1}(Q)) \leq 0$. Hence $B \cap p^{-1}(Q) = \emptyset$.

We shall need a few simple facts concerning the covering $p : \widehat{D}_\bullet \to D_\bullet$. The group of covering transformations of p is an infinite cyclic group generated by the covering transformation $g : \widehat{D}_\bullet \to \widehat{D}_\bullet$ corresponding to the loop encircling z_i counterclockwise. The set $p^{-1}(N)$ consists of an infinite number of disjoint closed intervals in $\widehat{D}_\bullet$ with boundary on $\partial\widehat{D}_\bullet$. These intervals can be numerated by integers so that the action of g shifts the index by 1. This implies that any nontrivial covering transformation $\widehat{D}_\bullet \to \widehat{D}_\bullet$ maps each component of $p^{-1}(N)$ to a different component of $p^{-1}(N)$. The same facts hold for the set $p^{-1}(\alpha) \subset \widehat{D}_\bullet$ with the only difference that its components are closed intervals lying in the interior of $\widehat{D}_\bullet$.

We claim that under our assumptions the pair N, α has a digon. This would imply that the intersection $N \cap \alpha$ is not minimal. The latter contradicts our choice of α in its isotopy class. Therefore, the assumption $u_{i,i} < u_{i,j}$ must have been false, so that $u_{i,i} = u_{i,j}$.

We now construct a digon for N, α. Suppose first that

$$B^\circ \cap p^{-1}(N) \neq \emptyset \quad \text{or} \quad B^\circ \cap p^{-1}(\alpha) \neq \emptyset$$

(or both). Observe that the circle $\delta = \partial B$ is formed by three embedded arcs: the arc $\widehat{\mu}_a$ lying on $p^{-1}(\alpha)$, the arc $\widehat{\nu}_a$ lying on $p^{-1}(N)$, and the arc $\widehat{\rho}$ meeting the set $p^{-1}(N) \cup p^{-1}(\alpha)$ only in its two endpoints. Note that the boundary of the one-manifold $p^{-1}(N)$ is contained in $\partial\widehat{D}_\bullet$ and lies therefore outside of B. If $B^\circ \cap p^{-1}(N) \neq \emptyset$, then $B^\circ \cap p^{-1}(N)$ is a finite set of disjoint embedded arcs with endpoints on $\widehat{\mu}_a$. At least one of these arcs bounds together with a subarc of $\widehat{\mu}_a$ a disk $D_1 \subset B$ whose interior does not meet $p^{-1}(N)$. If $B^\circ \cap p^{-1}(N) = \emptyset$, then we set $D_1 = B$. Similarly, the boundary of $p^{-1}(\alpha) \subset \widehat{D}_\bullet$ is contained in $p^{-1}(Q)$ and lies outside of B. If the interior D_1° of D_1 meets $p^{-1}(\alpha)$, then they meet along a finite number of disjoint embedded arcs with endpoints on $p^{-1}(N) \cap \partial D_1$. At least one of these arcs bounds together with a subarc of $p^{-1}(N) \cap \partial D_1$ an embedded disk $D_2 \subset D_1$ whose interior does not meet $p^{-1}(\alpha)$. If $D_1^\circ \cap p^{-1}(\alpha) = \emptyset$, then we set $D_2 = D_1$. In any case, the boundary of D_2 is formed by an arc on $p^{-1}(N)$ and an arc on $p^{-1}(\alpha)$, while the interior D_2° of D_2 does not meet $p^{-1}(N \cup \alpha)$. Then

$$D_2^\circ \cap g(\partial D_2) = \emptyset,$$

for any nontrivial covering transformation $g : \widehat{D}_\bullet \to \widehat{D}_\bullet$ of the covering $p : \widehat{D}_\bullet \to D_\bullet$. The properties of the sets $p^{-1}(N)$ and $p^{-1}(\alpha)$ mentioned above imply that $\partial D_2 \cap g(\partial D_2) = \emptyset$. This implies that either $D_2 \cap g(D_2) = \emptyset$ or D_2 is contained in the interior of the disk $g(D_2)$. In the latter case, $g^{-1}(D_2) \subset D_2^\circ$, which contradicts the fact that D_2° does not meet $p^{-1}(N \cup \alpha)$. We conclude that $D_2 \cap g(D_2) = \emptyset$. Thus, the disk D_2 does not meet its images under nontrivial covering transformations of the covering $p : \widehat{D}_\bullet \to D_\bullet$. Hence, the restriction of p to D_2 is injective. This implies that $p(D_2)$ is a digon for N, α in Σ.

It remains to construct a digon for the pair N, α when $B° \cap p^{-1}(N \cup \alpha) = \emptyset$. The set $p^{-1}(\rho)$ consists of v copies of the line $\mathbf{R}$ embedded in $\widehat{D}_\bullet$; these lines meet each other at an infinite number of points (see Figure 3.12, where $v = 3$). The arcs μ_a, ν_a lie in the component of $D_\bullet - \rho$ adjacent to $\partial D_\bullet \approx S^1$ except for the points $\mu_a(0) = \nu_a(1) = z_i$. Therefore the arcs $\widehat{\mu}_a, \widehat{\nu}_a$ lie in the component of $\widehat{D}_\bullet - p^{-1}(\rho)$ adjacent to $\partial \widehat{D}_\bullet \approx \mathbf{R}$ except for the points $\widehat{\mu}_a(0) = \widehat{\mu}(0)$ and $\widehat{\nu}_a(1) = \widehat{\nu}(1)$ lying on $p^{-1}(z_i) \subset p^{-1}(\rho)$. Clearly, $\widehat{\nu}_a(1) = g^v(\widehat{\mu}_a(0))$, where $g : \widehat{D}_\bullet \to \widehat{D}_\bullet$ is the generator of the group of covering transformations chosen above and $v > 0$ is the winding number of the loop $\mu\nu$ around z_i^-. The disk B bounded by $\delta = \widehat{\mu}_a \widehat{\nu}_a \widehat{\rho}$ has to include the area between the arc $\widehat{\mu}_a \widehat{\nu}_a$ and $p^{-1}(\rho)$ (this area is shaded in Figure 3.12). Observing Figure 3.12, one immediately concludes that for $v \geq 2$, this area must meet $g(\widehat{\mu}_a \widehat{\nu}_a)$. This contradicts the assumption $B° \cap p^{-1}(N \cup \alpha) = \emptyset$. It follows that $v = 1$, so that $p^{-1}(\rho)$ is just a line and B is the area between this line and the arc $\widehat{\mu}_a \widehat{\nu}_a$. Then B projects injectively to $D_\bullet$, the loop ρ bounds a small disk containing z_i^-, and the union of this disk with $p(B)$ is a digon for N, α. This completes the proof of the equality $u_{i,i} = u_{i,j}$ in Case 1.

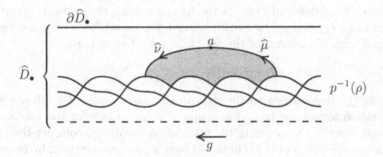

Fig. 3.12. The case $v = 3$

Case 2: The arc ν approaches α at z_i from the left (in other words, ν passes through z_i^-). Let us slightly push the arc ν near z_i^- to $\Sigma - \{z_i^-\}$ so that z_i^- lies on the left side of the resulting arc. Denote by ν' this new arc, also leading from z_j to z_i. The loop $\mu\nu'$ does not pass through z_i^- and we can consider its winding number, v, around z_i^-. We claim that $v > 0$. Observe first that the point z_i^- splits ν into two subarcs ν_1 and ν_2, where ν_1 leads from z_j to z_i^- and ν_2 leads from z_i^- to z_i. We have $\gamma_{i,j}^- = \nu_2 \gamma_{i,i}$ and $\gamma_{i,i}^- = \nu_1^{-1} \gamma_{i,j}$; see Figure 3.13. As in Case 1, we have $\beta_j \sim \beta_i \mu$. Therefore,

$$\xi_{i,j} = \{\theta^-, \theta\}\{\beta_i^-, \beta_j\}\{\gamma_{i,j}^-, \gamma_{i,j}\} \sim \{\theta^-, \theta\}\{\beta_i^-, \beta_i\}\{\nu_2, \mu\nu_1\}\{\gamma_{i,i}^-, \gamma_{i,i}\}.$$

The latter loop is homologous in $\mathcal{C}$ to the loop

$$\{\theta^-, \theta\}\{\beta_i^-, \beta_i\}\{\gamma_{i,i}^-, \gamma_{i,i}\}\{\nu_2, \mu\nu_1\} = \xi_{i,i}\{\nu_2, \mu\nu_1\}.$$

It is easy to deduce from the definitions and the construction of ν' that
$u(\{\nu_2, \mu\nu_1\}) = u(\{z_i^-, \mu\nu'\}) - 1 = 2v - 1$. Therefore,

$$2v - 1 = u(\{\nu_2, \mu\nu_1\}) = u(\xi_{i,j}) - u(\xi_{i,i}) = u_{i,j} - u_{i,i}.$$

The assumption $u_{i,i} < u_{i,j}$ implies that $v > 0$. The rest of the proof of the
equality $u_{i,i} = u_{i,j}$ goes as in Case 1 with the difference that instead of ν one
should everywhere use ν'.

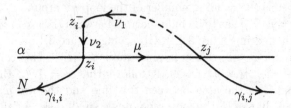

Fig. 3.13. Case 2: the paths $\gamma_{i,i}$ and $\gamma_{i,j}$

Analogous arguments prove that $u_{j,j} = u_{i,j}$ for any maximal pair (i,j).
This can also be deduced from the results above using the following symmetry
for the loops $\xi_{i,j}$ defined by (3.19), where i,j is an arbitrary (not necessarily
maximal) pair of elements of the set $\{1, \ldots, m\}$. Let us write

$$\xi_{i,j} = \xi_{i,j}(N, \alpha, z, z^+, \theta^-, \theta),$$

stressing the dependence on the data in the parentheses. We will use simi-
lar notation for $w_{i,j} = w(\xi_{i,j})$ and $u_{i,j} = u(\xi_{i,j})$. Consider the noodle $-N$
obtained from N by reversing the orientation. Similarly, consider the span-
ning arcs $-\alpha, -\alpha^-$ on (D, Q) obtained from α, α^-, respectively, by reversing
the orientation. It is clear that $-\alpha$ lies on the left of $-\alpha^-$, so that we can
set $(-\alpha^-)^- = -\alpha$. The noodle $-N$ crosses $-\alpha^-$ and $(-\alpha^-)^- = -\alpha$ in the
same points as before and we numerate them in the same way, except that z_i
becomes z_i^- and vice versa (for all i). It follows from the definitions that

$$\xi_{i,j}(N, \alpha, z^-, z, \theta^-, \theta) = \xi_{j,i}(-N, -\alpha^-, z, z^-, \theta, \theta^-)$$

for all i, j. This implies similar formulas for $w_{i,j}$ and $u_{i,j}$. Now, if the pair
(i,j) is maximal for (N, α), then the pair (j,i) is maximal for $(-N, -\alpha^-)$ and
by the results above,

$$
\begin{aligned}
u_{i,j}(N, \alpha, z^-, z, \theta^-, \theta) &= u_{j,i}(-N, -\alpha^+, z, z^-, \theta, \theta^-) \\
&= u_{j,j}(-N, -\alpha^+, z, z^-, \theta, \theta^-) \\
&= u_{j,j}(N, \alpha, z^-, z, \theta^-, \theta).
\end{aligned}
$$

We conclude that $u_{i,i} = u_{i,j} = u_{j,j}$ for any maximal pair (i,j). This
proves (3.21) and the lemma. $\square$

3.7 Proof of Theorem 3.15

The proof begins with two constructions. From each spanning arc α we derive a vector in $\mathcal{H}$ and from each noodle N we derive an oriented surface in $\tilde{\mathcal{C}}$. Then we compute the algebraic intersection $\langle N, \alpha \rangle$ in terms of these vectors and surfaces. This computation is used in the final subsection to finish the proof.

3.7.1 Homology classes associated with spanning arcs

Fix an oriented spanning arc α on (D, Q), where $Q = \{(1,0),(2,0),\ldots,(n,0)\}$. Pick disjoint closed disk neighborhoods

$$U_1, U_2, \ldots, U_n \subset D^\circ = D - \partial D$$

of the points $(1,0),(2,0),\ldots,(n,0)$, respectively. We shall always assume that α meets the disk neighborhoods U_i of its endpoints along certain radii and does not meet the other U_i. Let U be the set of all nonordered pairs $\{x, y\} \in \mathcal{C}$ such that at least one of the points $x, y \in \Sigma = D-Q$ lies in $\bigcup_{i=1}^{n} U_n$. Let $\tilde{U} \subset \tilde{\mathcal{C}}$ be the preimage of U under the covering map $\tilde{\mathcal{C}} \to \mathcal{C}$. It is clear that $\tilde{U}$ is invariant under the action of the covering transformations q, t on $\tilde{\mathcal{C}}$. This action turns the integral homology of $\tilde{U}$ and the relative integral homology of the pair $(\tilde{\mathcal{C}}, \tilde{U})$ into modules over the ring $\mathbf{Z}[q^{\pm 1}, t^{\pm 1}]$. We now associate with α a subset of $H_2(\tilde{\mathcal{C}}, \tilde{U}; \mathbf{Z})$ consisting of so-called α-classes.

Consider a parallel oriented spanning arc α^- as in Section 3.6.2. Recall that $\alpha \cup \alpha^-$ bounds a narrow strip in Σ and $\alpha \cap \alpha^- = \partial\alpha = \partial\alpha^-$. Consider the set $S_\alpha \subset \mathcal{C}$ consisting of all pairs $\{x, y\}$, where $x \in \alpha^- - \partial\alpha^-$ and $y \in \alpha - \partial\alpha$. Thus, $S_\alpha = (\alpha^- - \partial\alpha^-) \times (\alpha - \partial\alpha)$. Since S_α is simply connected, the embedding $S_\alpha \hookrightarrow \mathcal{C}$ lifts to an embedding $S_\alpha \hookrightarrow \tilde{\mathcal{C}}$. Fix such a lift and denote its image by $\tilde{S}_\alpha$. We regard S_α and $\tilde{S}_\alpha$ as open squares via

$$\tilde{S}_\alpha \approx S_\alpha = (\alpha^- - \partial\alpha^-) \times (\alpha - \partial\alpha)\,.$$

The surfaces S_α and $\tilde{S}_\alpha$ have a natural orientation obtained by multiplying the orientations in α^- and α. Pick subarcs $s \subset \alpha - \partial\alpha$ and $s^- \subset \alpha^- - \partial\alpha^-$ whose endpoints and complements in α, α^- lie in $\bigcup_{i=1}^{n} U_n$. Then $S = s^- \times s$ is a concentric closed subsquare of $\tilde{S}_\alpha$ whose boundary and complement in $\tilde{S}_\alpha$ lie in $\tilde{U}$. The oriented surface S represents an element of $H_2(\tilde{\mathcal{C}}, \tilde{U}; \mathbf{Z})$ independent of the choice of s, s^-. This element is denoted by $[S]$. Under a different choice of $\tilde{S}_\alpha$, it is multiplied by a monomial in q, t.

The image of $[S]$ under the boundary homomorphism

$$H_2(\tilde{\mathcal{C}}, \tilde{U}; \mathbf{Z}) \to H_1(\tilde{U}; \mathbf{Z})$$

is represented by the oriented circle $\partial S \subset \tilde{U}$. The following lemma shows that the homology class $[\partial S] \in H_1(\tilde{U}; \mathbf{Z})$ is annihilated by $(q-1)^2(qt+1)$.

Lemma 3.22. *We have* $(q-1)^2(qt+1)\,[\partial S] = 0$ *in* $H_1(\widetilde{U}; \mathbf{Z})$.

Proof. Let $(p_1, 0)$, $(p_2, 0)$ be the endpoints of α, where $p_1, p_2 \in \{1, 2, \ldots, n\}$. For brevity, we shall denote the point $(p_i, 0)$ simply by p_i, where $i = 1, 2$. For $i = 1, 2$, pick a point $u_i \in U_{p_i}$ lying in the strip between α^- and α. Consider the points $A, A', B, B' \in \Sigma$ and the eight paths

$$\alpha_1, \alpha_2, \alpha_3, \beta_1, \beta_2, \beta_3, \gamma_1, \gamma_2$$

in Σ drawn in Figure 3.14. The paths $\alpha_1, \alpha_2, \alpha_3, \beta_1, \beta_2, \beta_3$ are embedded arcs, while γ_i is a loop in U_{p_i} encircling p_i and based at u_i for $i = 1, 2$. It is understood that α goes along a radius of U_{p_1} from p_1 to A, then along α_2 from A to A', and then along a radius of U_{p_2} from A' to p_2 (the radii in question are not drawn in Figure 3.14). The arc α^- goes along a radius of U_{p_1} from p_1 to B', then along the path β_2^{-1} inverse to β_2, and then along a radius of U_{p_2} from B to p_2. One should think of α_2 (resp. of β_2) as being long and almost entirely exhausting α (resp. α^-), while the radii of U_{p_1}, U_{p_2} and the arcs $\alpha_1, \beta_3 \subset U_{p_1}$, $\alpha_3, \beta_1 \subset U_{p_2}$ are short.

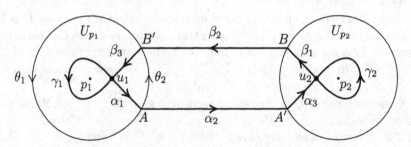

Fig. 3.14. The arcs $\alpha_1, \alpha_2, \alpha_3, \beta_1, \beta_2, \beta_3, \gamma_1, \gamma_2$

Consider the following loops in U based at $e = \{u_1, u_2\} = \{u_2, u_1\} \in U$:

$$a_1 = \{\gamma_1, u_2\}, \quad a_2 = \{u_1, \gamma_2\},$$
$$b_1 = \{\alpha_1, \beta_1\beta_2\beta_3\}\{\alpha_2\alpha_3, u_1\}, \quad b_2 = \{\alpha_1\alpha_2\alpha_3, \beta_1\}\{u_2, \beta_2\beta_3\},$$

where u_1, u_2 stand for the constant paths in the points u_1, u_2. Note that both loops b_1, b_2 are homotopic in $\mathcal{C}$ to the loop

$$\{\alpha_1\alpha_2\alpha_3, \beta_1\beta_2\beta_3\}.$$

(This certainly does not imply that b_1, b_2 are homotopic in U.) The homotopy classes of the loops a_1, a_2, b_1, b_2 in the fundamental group $\pi = \pi_1(U, e)$ will be denoted by the same symbols a_1, a_2, b_1, b_2. The symbol $\sim$ will denote homotopy in U for loops in U based at e. For any $x, y \in \pi$, set

$$x^y = y^{-1}xy \in \pi \quad \text{and} \quad [x, y] = x^{-1}x^y = x^{-1}y^{-1}xy \in \pi.$$

Observe the following relations in π:

$$[a_1, a_2] = 1, \quad [a_1, b_1 a_1 b_1] = 1, \quad [a_2, b_2 a_2 b_2] = 1. \tag{3.24}$$

The first relation is obvious, since

$$a_1 a_2 \sim \{\gamma_1, \gamma_2\} \sim a_2 a_1.$$

The relations $[a_1, b_1 a_1 b_1] = 1$ and $[a_2, b_2 a_2 b_2] = 1$ are proven similarly, and we shall prove only the first one. Consider the oriented arcs θ_1, θ_2 on ∂U_{p_1} as shown in Figure 3.14. These arcs lead from B' to A and from A to B' respectively, and their product $\theta_1 \theta_2$ is a loop parametrizing ∂U_{p_1}. We claim that

$$b_1 a_1 b_1 \sim \{u_1, \beta_1 \beta_2 \theta_1 \alpha_2 \alpha_3\}. \tag{3.25}$$

This will imply that

$$\begin{aligned}
a_1 b_1 a_1 b_1 &\sim \{\gamma_1, u_2\}\{u_1, \beta_1 \beta_2 \theta_1 \alpha_2 \alpha_3\} \\
&\sim \{\gamma_1, \beta_1 \beta_2 \theta_1 \alpha_2 \alpha_3\} \\
&\sim \{u_1, \beta_1 \beta_2 \theta_1 \alpha_2 \alpha_3\}\{\gamma_1, u_2\} \\
&\sim b_1 a_1 b_1 a_1.
\end{aligned}$$

Hence $[a_1, b_1 a_1 b_1] = 1$. We now prove (3.25). Observe first that

$$b_1 a_1 \sim \{\alpha_1, \beta_1 \beta_2 \beta_3\}\{\alpha_2 \alpha_3, \gamma_1\}$$

and

$$b_1 \sim \{u_1, \beta_1 \beta_2\}\{\alpha_1 \alpha_2 \alpha_3, \beta_3\} = \{\beta_1 \beta_2, u_1\}\{\beta_3, \alpha_1 \alpha_2 \alpha_3\}.$$

Therefore,

$$b_1 a_1 b_1 \sim \{\alpha_1, \beta_1 \beta_2 \beta_3\}\{\alpha_2 \alpha_3 \beta_1 \beta_2, \gamma_1\}\{\beta_3, \alpha_1 \alpha_2 \alpha_3\}.$$

The path $\alpha_2 \alpha_3 \beta_1 \beta_2$ is homotopic in $\Sigma - \gamma_1$ to θ_2. (By a homotopy of paths we always mean a homotopy keeping the endpoints of the paths fixed.) Hence,

$$\begin{aligned}
b_1 a_1 b_1 &\sim \{\alpha_1, \beta_1 \beta_2 \beta_3\}\{\theta_2, \gamma_1\}\{\beta_3, \alpha_1 \alpha_2 \alpha_3\} \\
&\sim \{\alpha_1, \beta_1 \beta_2\}\{A, \beta_3\}\{\theta_2, \gamma_1\}\{B', \alpha_1\}\{\beta_3, \alpha_2 \alpha_3\} \\
&\sim \{\alpha_1, \beta_1 \beta_2\}\{\theta_2, \beta_3 \gamma_1 \alpha_1\}\{\beta_3, \alpha_2 \alpha_3\}.
\end{aligned}$$

Observe that the path $\beta_3 \gamma_1 \alpha_1$ is homotopic in U_{p_1} to θ_1. Therefore

$$b_1 a_1 b_1 \sim \{\alpha_1, \beta_1 \beta_2\}\{\theta_2, \theta_1\}\{\beta_3, \alpha_2 \alpha_3\}.$$

Since the product $\alpha_1 \theta_2 \beta_3$ is homotopic to the constant path u_1, we obtain

$$b_1 a_1 b_1 \sim \{u_1, \beta_1 \beta_2 \theta_1 \alpha_2 \alpha_3\},$$

which proves (3.25).

We define the following four elements of π:

$$a = a_2^{-1}a_1, \quad b = b_2^{-1}b_1, \quad c_1 = [a_1, b_1], \quad c_2 = [a_2, b_2].$$

Then

$$a^{a_1} = a, \quad c_1^{b_1 a_1} c_1 = 1 = c_2^{b_2 a_2} c_2, \quad c_2 b a^{b_1} = a b^{a_1} c_1. \qquad (3.26)$$

To see this, rewrite all four relations via a_1, a_2, b_1, b_2. The first three relations are consequences of (3.24); in the last one, both sides are equal to $a_2^{-1} b_2^{-1} a_1 b_1$.

Pick a lift $\widetilde{e} \in \widetilde{C}$ of $e = \{u_1, u_2\}$. The group $\widetilde{\pi} = \pi_1(\widetilde{U}, \widetilde{e})$ is the subgroup of $\pi = \pi_1(U, e)$ formed by the homotopy classes of loops ξ in U such that $w(\xi) = u(\xi) = 0$. We claim that $a, b, c_1, c_2 \in \widetilde{\pi}$. Indeed, for $i = 1, 2$, we have $w(a_i) = w(\gamma_i) = 1$ and

$$w(b_i) = w(\alpha_1 \alpha_2 \alpha_3 \beta_1 \beta_2 \beta_3) = 0.$$

It follows from the definitions that $u(a_1) = u(a_2) = 0$ and $u(b_1) = u(b_2) = 1$. Hence $w(a) = u(a) = 0$ and $w(b) = u(b) = 0$, so that $a, b \in \widetilde{\pi}$. The commutator of any two elements of π belongs to $\widetilde{\pi}$, so that $c_1, c_2 \in \widetilde{\pi}$.

The image of any $x \in \widetilde{\pi}$ under the natural projection $\widetilde{\pi} \to H_1(\widetilde{U}; \mathbf{Z})$ will be denoted by $[x]$. It is clear that if $x \in \widetilde{\pi}$ and $y \in \pi$, then $x^y \in \widetilde{\pi}$. We claim that for all $x \in \widetilde{\pi}$ and $y \in \pi$,

$$[x^y] = q^{-w(y)} t^{-u(y)} [x], \qquad (3.27)$$

where we use the R-module structure on $H_1(\widetilde{U}; \mathbf{Z})$. To see this, present x, y by loops ξ, η in U, based at e. Then $x^y \in \pi$ is represented by the loop $\eta^{-1}\xi\eta$ in U. This loop lifts to a path $\mu_1 \mu_2 \mu_3$ in $\widetilde{U}$, where the path μ_1 is the lift of η^{-1} beginning at $\widetilde{e}$ and ending at the point

$$e' = q^{w(y^{-1})} t^{u(y^{-1})} \widetilde{e} = q^{-w(y)} t^{-u(y)} \widetilde{e},$$

the path μ_2 is the lift of ξ beginning at e', and μ_3 is the lift of η beginning at the terminal endpoint of μ_2. Since ξ represents $x \in \widetilde{\pi}$, the path μ_2 is a loop beginning and ending at e'. The path μ_3, being the lift of η beginning at e', must be the inverse of μ_1. Therefore the path $\mu_1 \mu_2 \mu_3$ is a loop and its homology class in $H_1(\widetilde{U}; \mathbf{Z})$ is equal to the homology class of μ_2. The latter is equal to $q^{-w(y)} t^{-u(y)} [x]$.

Applying (3.27), we obtain $[a^{a_1}] = q^{-1}[a]$ and

$$[c_1^{b_1 a_1} c_1] = q^{-1} t^{-1} [c_1] + [c_1], \quad [c_2^{b_2 a_2} c_2] = q^{-1} t^{-1} [c_2] + [c_2],$$
$$[c_2 b a^{b_1}] = [c_2] + [b] + t^{-1} [a], \quad [a b^{a_1} c_1] = [a] + q^{-1} [b] + [c_1].$$

Together with (3.26), this gives the following relations in $H_1(\widetilde{U}; \mathbf{Z})$:

$$(q-1)[a] = 0, \quad (qt+1)[c_1] = 0 = (qt+1)[c_2],$$
$$(q^{-1} - 1)[b] = (t^{-1} - 1)[a] + [c_2] - [c_1].$$

Combining these relations, we obtain

$$(q-1)^2(qt+1)[b] = 0.\qquad(3.28)$$

To compute the homology class $[S] \in H_2(\widetilde{C}, \widetilde{U}; \mathbf{Z})$, we need to choose the arcs $s \subset \alpha$ and $s^- \subset \alpha^-$ used in the definition of S. We take $s = \alpha_2$ and $s^- = \beta_2$. The endpoints of these arcs and their complements in α^-, α lie in $U_{p_1} \cup U_{p_2} \subset \bigcup_{i=1}^n U_i$, as required. The circle $\partial S \subset \widetilde{U}$ is parametrized by a loop in $\widetilde{U}$ that is a lift of the following loop $b' \subset U$ based at $\{A, B\}$:

$$b' = \{A, \beta_2\}\{\alpha_2, B'\}\{A', \beta_2\}^{-1}\{\alpha_2, B\}^{-1}.$$

We claim that b' is homotopic to the following loop b'' in U also based at $\{A, B\}$:

$$b'' = \{A, \beta_2\beta_3\}\{\alpha_2\alpha_3, u_1\}\{u_2, \beta_2\beta_3\}^{-1}\{\alpha_2\alpha_3, B\}^{-1}.\qquad(3.29)$$

To see this, observe the obvious equalities of paths (up to homotopy in U)

$$\{A, \beta_3\}\{\alpha_2\alpha_3, u_1\} = \{\alpha_2\alpha_3, \beta_3\} = \{\alpha_2, B'\}\{\alpha_3, \beta_3\}.$$

Therefore

$$\{\alpha_2\alpha_3, u_1\} = \{A, \beta_3\}^{-1}\{\alpha_2, B'\}\{\alpha_3, \beta_3\}.$$

A similar argument shows that

$$\{u_2, \beta_2\beta_3\}^{-1} = \{\alpha_3, \beta_3\}^{-1}\{A', \beta_2\}^{-1}\{\alpha_3, B\}.$$

Substituting these expressions in (3.29) and observing that

$$\{A, \beta_2\beta_3\} = \{A, \beta_2\}\{A, \beta_3\} \quad \text{and} \quad \{\alpha_2\alpha_3, B\}^{-1} = \{\alpha_3, B\}^{-1}\{\alpha_2, B\}^{-1},$$

we conclude that b' is homotopic to b''. Observe now that

$$b_1 = \{\alpha_1, \beta_1\beta_2\beta_3\}\{\alpha_2\alpha_3, u_1\} \sim \{\alpha_1, \beta_1\}\{A, \beta_2\beta_3\}\{\alpha_2\alpha_3, u_1\},$$
$$b_2 = \{\alpha_1\alpha_2\alpha_3, \beta_1\}\{u_2, \beta_2\beta_3\} \sim \{\alpha_1, \beta_1\}\{\alpha_2\alpha_3, B\}\{u_2, \beta_2\beta_3\}.$$

Therefore the loop b'' is homotopic to the loop

$$\{\alpha_1, \beta_1\}^{-1}b_1 b_2^{-1}\{\alpha_1, \beta_1\}$$

in U. The latter loop is freely homotopic in U to $b_1 b_2^{-1}$. Since $b_1 b_2^{-1}$ is conjugate to $b = b_2^{-1}b_1$ in π, the loops b'' and b are freely homotopic in U. We conclude that b' is freely homotopic to b in U. Since b' lifts to a loop ∂S in $\widetilde{U}$, any homotopy of b' lifts to a homotopy of ∂S in $\widetilde{U}$. Hence, ∂S is freely homotopic to a lift of b to $\widetilde{U}$. Now the claim of the lemma directly follows from (3.28). $\square$

Lemma 3.22 and the exact homology sequence of the pair $(\widetilde{C}, \widetilde{U})$,

$$\cdots \to H_2(\widetilde{U}; \mathbf{Z}) \to \mathcal{H} \to H_2(\widetilde{C}, \widetilde{U}; \mathbf{Z}) \to H_1(\widetilde{U}; \mathbf{Z}) \to \cdots,$$

imply that the homology class $(q-1)^2(qt+1)[S] \in H_2(\widetilde{C}, \widetilde{U}; \mathbf{Z})$ is the image of a certain $v \in \mathcal{H}$ under the inclusion homomorphism $\mathcal{H} \to H_2(\widetilde{C}, \widetilde{U}; \mathbf{Z})$. Any such $v \in \mathcal{H}$ is called an α-*class with respect to the disks* $U_1, \ldots, U_n$ or, shorter, an α-*class*. An α-class can be represented by a 2-cycle in $\widetilde{C}$ obtained by gluing the 2-chain $(q-1)^2(qt+1)S$ with a 2-chain in $\widetilde{U}$ bounded by $(q-1)^2(qt+1)\,\partial S$.

It is clear that the α-class is determined by α only up to addition of elements of the image of the homomorphism $H_2(\widetilde{U}; \mathbf{Z}) \to \mathcal{H}$ induced by the inclusion $\widetilde{U} \hookrightarrow \widetilde{C}$ and up to multiplication by monomials in q, t (the latter is due to the indeterminacy in the choice of $\widetilde{S}_\alpha$). This describes completely the indeterminacy in the construction of an α-class. Indeed, it is easy to check that the set of α-classes does not depend on the choice of the arcs $s \subset \alpha - \partial\alpha$ and $s^- \subset \alpha^- - \partial\alpha^-$ used in the definition of the surface S. (To see this, observe that the surfaces S determined by s, s^- and by a pair of bigger arcs differ by an annulus in $\widetilde{U}$.) We show now that the set of α-classes is independent of the choice of the disks $U_1, \ldots, U_n$.

Lemma 3.23. *The set of α-classes in $\mathcal{H}$ does not depend on the choice of the disks* $U_1, \ldots, U_n$.

Proof. Let $\{U_i\}_{i=1}^n$ and $\{U_i'\}_{i=1}^n$ be two systems of closed disk neighborhoods of the points of $Q = \{(1,0), (2,0), \ldots, (n,0)\}$ in D° as at the beginning of this subsection. Let $\widetilde{U}$ and $\widetilde{U}'$ be the subsets of $\widetilde{C}$ associated with these systems of disks as above. Suppose first that $U_i' \subset U_i$ for all i. We can view U_i and U_i' as concentric disks with center $(i,0)$. By the assumptions, the arc α either does not meet the disk U_i or meets it along a radius whose intersection with U_i' is the radius of the latter. Contracting each U_i into U_i' along the radii, we obtain an isotopy $\{F_s : D \to D\}_{s \in I}$ of D into itself such that $F_0 = \mathrm{id}$, F_s fixes $\partial D \cup Q$ pointwise and fixes α setwise for all $s \in I$, and $F_1(U_i) = U_i'$ for all i. The induced homeomorphisms $\{\widetilde{F}_s : \widetilde{C} \to \widetilde{C}\}_{s \in I}$ form an isotopy of $\widetilde{C}$ into itself such that $\widetilde{F}_1(\widetilde{U}) = \widetilde{U}'$.

Observe now that any self-homeomorphism f of (D, Q) transforms α into a spanning arc $f(\alpha)$ on (D, Q), and the orientation of α induces an orientation of $f(\alpha)$ via f. It is clear from the definitions that the induced homomorphism $\widetilde{f}_* : \mathcal{H} \to \mathcal{H}$ sends the set of α-classes with respect to the disks $\{U_i\}_i$ onto the set of $f(\alpha)$-classes with respect to the disks $\{f(U_i)\}_i$. Applying this to $f = F_1$ and observing that $f(\alpha) = \alpha$, $f(U_i) = U_i'$ for all i, and $\widetilde{f}_* = \mathrm{id}$ (because $\widetilde{f} = \widetilde{F}_1$ is isotopic—and hence homotopic—to $\widetilde{F}_0 = \mathrm{id}$), we conclude that the set of α-classes with respect to the disks $\{U_i\}_{i=1}^n$ coincides with the set of α-classes with respect to the disks $\{U_i'\}_{i=1}^n$. The general case is obtained by transitivity using a third system of disks $\{U_i''\}_{i=1}^n$ such that $U_i'' \subset U_i \cap U_i'$ for all $i = 1, \ldots, n$. $\square$

3.7.2 Surfaces associated with noodles

For a noodle N on D, the set

$$F = F_N = \{\{x,y\} \in \mathcal{C} \mid x,y \in N^\circ = N - \partial N\}$$

is a surface in $\mathcal{C}^\circ = \mathcal{C} - \partial\mathcal{C}$ homeomorphic to the open triangle

$$\{(x_1, x_2) \in (0,1)^2 \mid x_1 < x_2\}.$$

The surface F is therefore homeomorphic to the plane $\mathbf{R}^2$. Since F is contractible, it lifts to a surface

$$\widetilde{F} = \widetilde{F}_N \subset \widetilde{\mathcal{C}}^\circ = \widetilde{\mathcal{C}} - \partial\widetilde{\mathcal{C}}$$

also homeomorphic to $\mathbf{R}^2$. It is clear that $\widetilde{\mathcal{C}}^\circ$ is an open oriented smooth four-dimensional manifold and $\widetilde{F}$ is a smooth two-dimensional submanifold.

Lemma 3.24. *The surface $\widetilde{F}$ is a closed subset of $\widetilde{\mathcal{C}}^\circ$.*

Proof. Pick an arbitrary point $a \in \widetilde{\mathcal{C}}^\circ - \widetilde{F}$. Let $\{x,y\} \in \mathcal{C}$ be the projection of a to $\mathcal{C}$, where x,y are distinct points of Σ. The inclusion $a \in \widetilde{\mathcal{C}}^\circ$ implies that $x,y \in \Sigma^\circ$. If $x \notin N$ or $y \notin N$, then x and y have disjoint open neighborhoods $U_x, U_y \subset \Sigma^\circ$, respectively, such that at least one of them does not meet N. (Here we use the obvious fact that N is a closed subset of Σ.) Then $U_x \times U_y$ is a neighborhood of the point $\{x,y\}$ in $\mathcal{C}^\circ - F$ and the preimage of $U_x \times U_y$ in $\widetilde{\mathcal{C}}^\circ$ is an open neighborhood of a contained in $\widetilde{\mathcal{C}}^\circ - \widetilde{F}$. If $x,y \in N$, then x and y have disjoint open disk neighborhoods $U_x, U_y \subset \Sigma^\circ$, respectively, such that both U_x and U_y meet N along an open interval. Then $U_x \times U_y$ is an open neighborhood of the point $\{x,y\} \in F$ homeomorphic to an open 4-ball and meeting F along an open 2-disk. The preimage of this neighborhood in $\widetilde{\mathcal{C}}^\circ$ consists of disjoint open 4-balls. One of them meets $\widetilde{F}$ along an open 2-disk and the others do not meet $\widetilde{F}$. The point $a \in \widetilde{\mathcal{C}}^\circ - \widetilde{F}$ lying over $\{x,y\} \in F$ has to lie in one of those open 4-balls that do not meet $\widetilde{F}$. We conclude that in all cases, the point a has an open neighborhood in $\widetilde{\mathcal{C}}^\circ$ disjoint from $\widetilde{F}$. Thus, the set $\widetilde{\mathcal{C}}^\circ - \widetilde{F}$ is open in $\widetilde{\mathcal{C}}^\circ$ and the set $\widetilde{F}$ is closed in $\widetilde{\mathcal{C}}^\circ$. $\qquad\square$

Note one important consequence of this lemma: the intersection of $\widetilde{F}$ with any compact subset of $\widetilde{\mathcal{C}}^\circ$ is compact. We use this property to define an integral intersection number of $\widetilde{F}$ with an arbitrary element of $\mathcal{H}$ as follows. We first orient F: at a point $\{x,y\} = \{y,x\} \in F$ such that $x \in N^\circ$ is closer to the starting endpoint on N than $y \in N^\circ$, the orientation of F is the product of the orientations of N at x and y in this order. This orientation of F lifts to $\widetilde{F}$ in the obvious way so that $\widetilde{F}$ becomes oriented. Since, as was observed in Section 3.5.4, the inclusion $\widetilde{\mathcal{C}}^\circ \hookrightarrow \widetilde{\mathcal{C}}$ is a homotopy equivalence,

$$\mathcal{H} = H_2(\widetilde{\mathcal{C}}; \mathbf{Z}) = H_2(\widetilde{\mathcal{C}}^\circ; \mathbf{Z}).$$

The rest of the definition is quite standard. To define the intersection number $\widetilde{F} \cdot v \in \mathbf{Z}$ for $v \in \mathcal{H}$, we pick a 2-cycle V in $\widetilde{C}^\circ$ representing v. By the remarks above, V meets $\widetilde{F} \approx \mathbf{R}^2$ along a compact subset, which necessarily lies inside a closed 2-disk in $\widetilde{F}$. We can slightly deform V in $\widetilde{C}^\circ$ to make it transversal to this disk, keeping V disjoint from the rest of $\widetilde{F}$. The set $\widetilde{F} \cap V$ is then discrete and compact. It is therefore finite, so that one can count its points with signs $\pm$ determined by the orientation of $\widetilde{C}$, $\widetilde{F}$, and V. A standard argument from the theory of homological intersections shows that the resulting integer $\widetilde{F} \cdot v = \widetilde{F} \cdot V$ depends only on v. Specifically, any two 2-cycles V_1, V_2 in $\widetilde{C}^\circ$ representing v differ by the boundary of a 3-chain in $\widetilde{C}^\circ$; such a chain can be made transversal to $\widetilde{F}$ and then its intersection with $\widetilde{F}$ is a compact oriented 1-manifold. The fact that this 1-manifold has the same numbers of inputs and outputs implies that $\widetilde{F} \cdot V_1 = \widetilde{F} \cdot V_2$.

In analogy with formula (3.17), we set for any $v \in \mathcal{H}$,

$$\langle \widetilde{F}, v \rangle = \sum_{k, \ell \in \mathbf{Z}} (q^k t^\ell \widetilde{F} \cdot v)\, q^k t^\ell. \tag{3.30}$$

Here $q^k t^\ell \widetilde{F}$ is the image of $\widetilde{F}$ under the covering transformations $q^k t^\ell$ of the covering $\widetilde{C} \to C$. Note that when k, ℓ run over $\mathbf{Z}$, the surface $q^k t^\ell \widetilde{F}$ runs over all possible lifts of F to $\widetilde{C}$. A priori, the sum on the right-hand side of (3.30) may be infinite; Lemma 3.25 below shows that it is finite.

The same computations as in the proof of Lemma 3.18 show that under a different choice of the lift $\widetilde{F}$ of F, the expression $\langle \widetilde{F}, v \rangle$ is multiplied by a monomial in $q^{\pm 1}, t^{\pm 1}$.

Lemma 3.25. *Let $r \mapsto r^*$ be the involution of the ring $R = \mathbf{Z}[q^{\pm 1}, t^{\pm 1}]$ sending q to q and t to $-t$. Let N be a noodle on D and let α be an oriented spanning arc on (D, Q). Then for any α-class $v \in \mathcal{H}$,*

$$\langle \widetilde{F}_N, v \rangle = -(q - 1)^2 (qt + 1) \langle N, \alpha \rangle^*, \tag{3.31}$$

where $\langle N, \alpha \rangle \in R$ is the algebraic intersection defined in Section 3.6.2.

Proof. Note that the left-hand side of (3.31) is defined up to multiplication by monomials in $q^{\pm 1}, t^{\pm 1}$, while the right-hand side is defined up to multiplication by monomials in $q^{\pm 1}, t^{\pm 2}$. The equality is understood in the sense that the sides have a common representative. Then all representatives of the right-hand side represent also the left-hand side.

Pushing the endpoints of N along ∂D, we can deform N into a noodle N' with starting point d_1 and terminal point d_2, where $d_1, d_2 \in \partial D$ are the points used in the construction of $\widetilde{C}$. The surfaces F_N and $F_{N'}$ differ only in a subset of a cylinder neighborhood of ∂C in C. We can choose the lifts $\widetilde{F}_N$ and $\widetilde{F}_{N'}$ so that they differ only in a subset of a cylinder neighborhood of $\partial \widetilde{C}$ in $\widetilde{C}$. Since v can be represented by a 2-cycle in the complement of such a neighborhood, $\langle \widetilde{F}_N, v \rangle = \langle \widetilde{F}_{N'}, v \rangle$. It follows from the definitions that $\langle N, \alpha \rangle = \langle N', \alpha \rangle$. Thus, without loss of generality we can assume that the starting point of N is d_1 and the terminal point of N is d_2.

It is enough to prove (3.31) for a specific choice of $\widetilde{F} = \widetilde{F}_N \subset \widetilde{\mathcal{C}}$. Fix a lift $\widetilde{c} \in \widetilde{\mathcal{C}}$ of $c = \{d_1, d_2\} \in \mathcal{C}$. For $\widetilde{F}$, we take the lift of $F = F_N$ containing $\widetilde{c}$.

We need to specify a lift $\widetilde{S}_\alpha \subset \widetilde{\mathcal{C}}$ of the surface S_α defined in Section 3.7.1. To this end, fix points $z^- \in \alpha^-$, $z \in \alpha$ and fix paths θ^-, θ in $\Sigma = D - Q$ having disjoint images and leading from d_1 to z^- and from d_2 to z, respectively. Consider the path $\{\theta^-, \theta\}$ in $\mathcal{C}$ leading from $c = \{d_1, d_2\}$ to $\{z^-, z\}$. Let Θ be the lift of this path to $\widetilde{\mathcal{C}}$ starting at $\widetilde{c}$. The path Θ terminates at a point $\Theta(1)$ lying over $\{z^-, z\} \in S_\alpha$. We choose for $\widetilde{S}_\alpha \subset \widetilde{\mathcal{C}}$ the lift of S_α containing $\Theta(1)$. The surfaces S_α and $\widetilde{S}_\alpha$ are oriented as in Section 3.7.1.

Assume that N intersects α (resp. α^-) transversely in m points $z_1, \ldots, z_m$ (resp. $z_1^-, \ldots, z_m^-$) as in Section 3.6.2. Then F intersects S_α transversely in the points $\{z_i^-, z_j\}$, where $i, j = 1, \ldots, m$. Therefore for any $k, \ell \in \mathbf{Z}$, the image of $\widetilde{F}$ under the covering transformation $q^k t^\ell$ meets $\widetilde{S}_\alpha$ transversely in at most m^2 points. Adding the corresponding intersection signs, we obtain an integer, denoted by $q^k t^\ell(\widetilde{F}) \cdot \widetilde{S}_\alpha \in \mathbf{Z}$. Set

$$\sigma = \sum_{k, \ell \in \mathbf{Z}} (q^k t^\ell \widetilde{F} \cdot \widetilde{S}_\alpha) \, q^k t^\ell \in R.$$

The sum on the right-hand side is finite (it has at most m^2 terms).

We compute σ as follows. Observe that for every pair $i, j \in \{1, \ldots, m\}$, there are unique integers $k_{i,j}, \ell_{i,j} \in \mathbf{Z}$ such that $q^{k_{i,j}} t^{\ell_{i,j}} \widetilde{F}$ intersects $\widetilde{S}_\alpha$ at a point lying over $\{z_i^-, z_j\} \in \mathcal{C}$. Let $\varepsilon_{i,j} = \pm 1$ be the corresponding intersection sign. Then

$$\sigma = \sum_{i=1}^m \sum_{j=1}^m \varepsilon_{i,j} \, q^{k_{i,j}} t^{\ell_{i,j}}.$$

We now express the right-hand side in terms of the loops $\xi_{i,j}$ and other data introduced in Section 3.6.2 (where $d^- = d_1$ and $d = d_2$). We claim that

$$q^{k_{i,j}} t^{\ell_{i,j}} = \varphi(\xi_{i,j}),$$

or in other words, that $k_{i,j} = w(\xi_{i,j})$ and $\ell_{i,j} = u(\xi_{i,j})$ for all i, j. Indeed, we can lift $\xi_{i,j}$ to a path $\Theta\beta\gamma$ in $\widetilde{\mathcal{C}}$ beginning at $\widetilde{c}$, where Θ, β, γ are lifts of $\{\theta^-, \theta\}, \{\beta_i^-, \beta_j\}, \{\gamma_{i,j}^-, \gamma_{i,j}\}$, respectively. By the choice of $\widetilde{S}_\alpha$, the point $\Theta(1) = \beta(0)$ lies on $\widetilde{S}_\alpha$. Then the path β lies entirely on $\widetilde{S}_\alpha$. The path $\Theta\beta\gamma$, being a lift of the loop $\xi_{i,j}$, ends at

$$\gamma(1) = \varphi(\xi_{i,j})(\widetilde{c}) \in \varphi(\xi_{i,j})\widetilde{F}_N.$$

Hence, the lift γ of $\{\gamma_{i,j}^-, \gamma_{i,j}\}$ lies on $\varphi(\xi_{i,j})\widetilde{F}$ and the point $\gamma(0) = \beta(1)$ lies over $\{z_i^-, z_j\}$ and belongs to $\varphi(\xi_{i,j})\widetilde{F} \cap \widetilde{S}_\alpha$. This proves our claim.

We now claim that for all i, j,

$$\varepsilon_{i,j} = -(-1)^{u(\xi_{i,j})} \varepsilon_i \varepsilon_j,$$

where ε_i (resp. ε_j) is the intersection sign of N and α at z_i (resp. at z_j). Observe first that $\varepsilon_{i,j}$ is the intersection sign of the surfaces F_N and S_α at the point $\{z_i^-, z_j\} \in C$. Let x^- (resp. x) be a positive tangent vector of N at z_i^- (resp. at z_j). Let y^- (resp. y) be a positive tangent vector of α^- at z_i^- (resp. of α at z_j). Assume for concreteness that the point z_i^- lies closer to d_1 along N than z_j. Then the orientation of F_N at the point $\{z_i^-, z_j\}$ is determined by the pair of vectors (x^-, x). The orientation of S_α at $\{z_i^-, z_j\}$ is determined by the pair of vectors (y^-, y). The distinguished orientation of C at $\{z_i^-, z_j\}$ is equal to $\varepsilon_i \varepsilon_j$ times the orientation of C determined by the following tuple of four tangent vectors:

$$(x^-, y^-, x, y).$$

Then

$$\varepsilon_{i,j} = -\varepsilon_i \varepsilon_j = -(-1)^{u(\xi_{i,j})} \varepsilon_i \varepsilon_j,$$

since in the case at hand the paths $\gamma_{i,j}^-$ and $\gamma_{i,j}$ end at d_1 and d_2, respectively, and the integer $u(\xi_{i,j})$ is even. The case in which z_j lies closer to d_1 along N than z_i^- is treated similarly.

To sum up,

$$\sigma = \sum_{i=1}^{m} \sum_{j=1}^{m} -(-1)^{u(\xi_{i,j})} \varepsilon_i \varepsilon_j \, q^{w(\xi_{i,j})} t^{u(\xi_{i,j})} = -\langle N, \alpha \rangle^*.$$

We can now prove (3.31). Let $U_1, \ldots, U_n$ and $\widetilde{U}$ be as in Section 3.7.1. Choosing the disks $U_1, \ldots, U_n$ small enough, we can assume that they do not meet N. Then

$$q^k t^\ell \widetilde{F} \cap \widetilde{U} = \emptyset \tag{3.32}$$

for all $k, \ell \in \mathbf{Z}$. Recall that the α-class v is represented by a sum of a 2-chain in $\widetilde{U}$ and a 2-chain $(q-1)^2(qt+1)S$. By (3.32), the 2-chain in $\widetilde{U}$ does not contribute to $\langle \widetilde{F}, v \rangle$, so that we can safely replace v by $(q-1)^2(qt+1)S$. By definition, $S \subset \widetilde{S}_\alpha$ is a subsurface of $\widetilde{S}_\alpha$ such that $\widetilde{S}_\alpha - S \subset \widetilde{U}$. Therefore, a similar argument shows that in the computation of $\langle \widetilde{F}, v \rangle$, we can replace S by $\widetilde{S}_\alpha$. Using the same computations as in the proof of Lemma 3.18, we obtain the equalities

$$\langle \widetilde{F}, v \rangle = (q-1)^2(qt+1) \sum_{k,l \in \mathbf{Z}} (q^k t^l \widetilde{F} \cdot \widetilde{S}_\alpha) q^k t^l$$

$$= (q-1)^2(qt+1) \sigma$$

$$= -(q-1)^2(qt+1) \langle N, \alpha \rangle^*. \qquad \square$$

Lemma 3.26. *If a self-homeomorphism f of (D, Q) represents an element of the kernel $\mathrm{Ker}(B_n \to \mathrm{Aut}_R(\mathcal{H}))$, then $\langle N, f(\alpha) \rangle = \langle N, \alpha \rangle$ for any noodle N and any oriented spanning arc α on (D, Q).*

Proof. As was already observed above, the homomorphism $\widetilde{f}_* : \mathcal{H} \to \mathcal{H}$ transforms any α-class $v \in \mathcal{H}$ into an $f(\alpha)$-class. Formula (3.31) and the assumption $\widetilde{f}_* = \mathrm{id}$ imply that

$$
\begin{aligned}
-(q-1)^2(qt+1)\,\langle N, f(\alpha) \rangle^* &= \langle \widetilde{F}, \widetilde{f}_*(v) \rangle \\
&= \langle \widetilde{F}, v \rangle \\
&= -(q-1)^2(qt+1)\,\langle N, \alpha \rangle^*.
\end{aligned}
$$

Therefore, $\langle N, f(\alpha) \rangle = \langle N, \alpha \rangle$. $\square$

3.7.3 End of the proof

Pick an arbitrary element of the kernel $\mathrm{Ker}(B_n \to \mathrm{Aut}_R(\mathcal{H}))$. By Corollary 1.34, it can be represented by a smooth self-homeomorphism f of the disk D permuting the points of the set $Q = \{(1,0), \ldots, (n,0)\}$. We shall prove that f is isotopic to the identity map (rel $Q \cup \partial D$). This will imply that

$$
\mathrm{Ker}(B_n \to \mathrm{Aut}_R(\mathcal{H})) = \{1\}.
$$

We begin with the following assertion.

Claim 3.27. *A spanning arc α on (D, Q) can be isotopped off a noodle N if and only if $f(\alpha)$ can be isotopped off N.*

To see this, orient α in an arbitrary way and endow $f(\alpha)$ with the orientation induced via f. Lemma 3.26 implies that $\langle N, f(\alpha) \rangle = 0$ if and only if $\langle N, \alpha \rangle = 0$. Now, Lemma 3.21 implies that α can be isotopped off N if and only if $f(\alpha)$ can be isotopped off N.

We shall apply Claim 3.27 to the following arcs and noodles. Denote by α_i the arc $[i, i+1] \times 0 \subset D$ and denote by N_i the noodle shown in Figure 3.7, where $i = 1, \ldots, n-1$. We shall assume that the noodles $N_1, \ldots, N_{n-1}$ are pairwise disjoint (then their endpoints lie consecutively on ∂D). It is clear that the arc α_i is disjoint from the noodle N_j for all $j \neq i, i+1$. Claim 3.27 implies that the spanning arc $f(\alpha_i)$ can be isotopped off N_j for $j \neq i, i+1$. Therefore, the arc $f(\alpha_i)$ may end only at the points $(i, 0)$ and $(i+1, 0)$. In other words, $f(\alpha_i)$ has the same endpoints as α_i for all i. For $n \geq 3$, this implies that f induces the identity permutation on Q. We assume that $n \geq 3$, postponing the cases $n = 1$ and $n = 2$ to the end of the proof.

As was just explained, we can isotop the spanning arc $f(\alpha_1)$ off N_3. This isotopy extends to an isotopy of the homeomorphism f (rel $Q \cup \partial D$), so that we can assume from the very beginning that the arc $f(\alpha_1)$ does not meet N_3.

Similarly, $f(\alpha_1)$ can be isotopped off N_4. By Section 3.6.1, this can be done by a sequence of isotopies eliminating digons for the pair $(N_4, f(\alpha_1))$. Since N_4 and $f(\alpha_1)$ do not meet N_3, neither do the digons in question. Hence the isotopies along these digons do not create intersections of $f(\alpha_1)$ with N_3. Repeating this argument, we can ensure that $f(\alpha_1)$ is disjoint from all the noodles N_i with $i = 3, 4, \ldots, n - 1$. Drawing these (disjoint) noodles, one easily observes that all spanning arcs in their complement are isotopic to α_1. Then, applying one more isotopy, we can arrange that $f(\alpha_1) = \alpha_1$. Note that all self-homeomorphisms of a closed interval keeping the endpoints fixed are isotopic to the identity. Therefore we can further isotop f so that it becomes the identity on α_1. Applying a similar procedure to α_2, we can ensure that $f|_{\alpha_2} = \text{id}$ while keeping $f|_{\alpha_1} = \text{id}$. Continuing in this way, we can isotop f so that it preserves the interval $[1, n] \times \{0\}$ pointwise. Applying a further isotopy, we can ensure that $f = \text{id}$ in an open neighborhood of this interval in D. In other words, $f = \text{id}$ outside an annular neighborhood A of ∂D in $\Sigma = D - Q$.

We identify A with $\partial D \times [0, 1]$, so that $\partial D \subset \partial A$ is identified with $\partial D \times \{0\}$. The (smooth) homeomorphism $f|_A : A \to A$ must be isotopic (rel ∂A) to the kth power of the Dehn twist about the circle $\partial D \times \{1/2\} \subset A$ for some $k \in \mathbf{Z}$; see, for instance, [Iva02, Lemma 4.1.A]. Thus, f is isotopic to g^k, where g is the self-homeomorphism of D acting as the Dehn twist on A and as the identity on $D - A$.

We claim that the homeomorphism g acts on $\mathcal{H}$ via multiplication by the monomial $q^{2n} t^b$ for some $b \in \mathbf{Z}$. (In fact, $b = 2$ but we shall not need it.) Then $\tilde{f}_* : \mathcal{H} \to \mathcal{H}$ is multiplication by $q^{2nk} t^{bk}$. For $k \neq 0$, this cannot be the identity map: if it is, then

$$(q^{2nk} t^{bk} - 1)\mathcal{H} = 0$$

and the linearity of the function

$$\mathcal{H} \to \mathbf{Z}[q^{\pm 1}, t^{\pm 1}], \quad v \mapsto \langle \widetilde{F}_N, v \rangle$$

implies that this function is identically zero for any noodle N. By Lemma 3.25 we must have $\langle N, \alpha \rangle = 0$ for all N, α. The latter is not true, as was observed before the statement of Lemma 3.20. This contradiction shows that $k = 0$, so that f is isotopic to the identity.

To compute the action of g on $\mathcal{H}$, consider the homeomorphism $\hat{g} : \mathcal{C} \to \mathcal{C}$ defined by $\hat{g}(\{x, y\}) = \{g(x), g(y)\}$ for distinct $x, y \in \Sigma$; cf. Section 3.5.3. Consider the lift $\tilde{g} : \widetilde{\mathcal{C}} \to \widetilde{\mathcal{C}}$ of $\hat{g}$ keeping fixed all points lying over the base point $c = \{d_1, d_2\} \in \mathcal{C}$. Since $g = \text{id}$ outside A, we have $\hat{g} = \text{id}$ outside the set $\{(x, y) \in \mathcal{C} \mid x \in A \text{ or } y \in A\}$. Let $\widetilde{A} \subset \widetilde{\mathcal{C}}$ be the preimage of this set under the covering projection $\widetilde{\mathcal{C}} \to \mathcal{C}$. The homeomorphism $\tilde{g}$ has to act on $\widetilde{\mathcal{C}} - \widetilde{A}$ as a covering transformation $q^a t^b$ for some $a, b \in \mathbf{Z}$. The set $\widetilde{A}$ is a tubular neighborhood of $\partial \widetilde{\mathcal{C}}$ in $\widetilde{\mathcal{C}}$ and therefore any 2-cycle in $\widetilde{\mathcal{C}}$ can be deformed into $\widetilde{\mathcal{C}} - \widetilde{A}$. Hence, $\tilde{g}$ acts on $\mathcal{H}$ as multiplication by $q^a t^b$.

We now verify that $a = 2n$. For $i = 1, 2$, define a path $\delta_i : I \to A$ by $\delta_i(s) = d_i \times s$, where $s \in I = [0, 1]$ and $d_1, d_2 \in \partial D$ are the points used in the construction of $\widetilde{C}$. Set $\delta = \{\delta_1, \delta_2\} : I \to C$ and let $\widetilde{\delta} : I \to \widetilde{C}$ be an arbitrary lift of δ. The point $\widetilde{\delta}(0)$ lies over c and therefore $\widetilde{g}(\widetilde{\delta}(0)) = \widetilde{\delta}(0)$. The point $\widetilde{\delta}(1)$ lies in the closure of $\widetilde{C} - \widetilde{A}$ and therefore $\widetilde{g}(\widetilde{\delta}(1)) = q^a t^b \widetilde{\delta}(1)$. Therefore the path $\widetilde{g} \circ \widetilde{\delta} : I \to \widetilde{C}$ leads from $\widetilde{\delta}(0)$ to $q^a t^b \widetilde{\delta}(1)$. Multiplying by $\widetilde{\delta}^{-1}$, we obtain the path $\widetilde{\delta}^{-1}(\widetilde{g} \circ \widetilde{\delta})$ leading from $\widetilde{\delta}(1)$ to $q^a t^b \widetilde{\delta}(1)$ in $\widetilde{C}$. By the definition of the covering $\widetilde{C} \to C$, the integer a must be the value of the invariant w on the loop obtained by projecting the latter path to C. This loop is nothing but

$$\delta^{-1}(\widehat{g} \circ \delta) = \{\delta_1^{-1}(g \circ \delta_1), \delta_2^{-1}(g \circ \delta_2)\}.$$

Hence,

$$a = w(\delta^{-1}(\widehat{g} \circ \delta)) = w(\delta_1^{-1}(g \circ \delta_1)) + w(\delta_2^{-1}(g \circ \delta_2)).$$

It remains to observe that $w(\delta_i^{-1}(g \circ \delta_i)) = n$ for $i = 1, 2$. This completes the proof in the case $n \geq 3$.

The remaining cases $n = 1, 2$ are easy. For $n = 1$, there is nothing to prove, since $B_1 = \{1\}$. The group B_2 is infinite cyclic, and the square of a generator is the Dehn twist as in the previous paragraphs, which, as we have just explained, represents an element of infinite order in $\mathrm{Aut}_R(\mathcal{H})$.

Notes

The Burau representation ψ_n was introduced by Burau [Bur36]. A version of Theorem 3.1 was first obtained by Squier [Squ84], who used a different, more complicated, matrix in the role of Θ_n. The matrix Θ_n in Theorem 3.1 was pointed out by Perron [Per06].

The representations ψ_2, ψ_3 were long known to be faithful; see [Bir74]. Moody [Moo91] first proved that ψ_n is nonfaithful for $n \geq 9$. Long and Paton [LP93] extended Moody's argument to $n \geq 6$. Bigelow [Big99] proved that ψ_5 is nonfaithful. Our exposition in Section 3.2 follows the ideas and techniques of these papers. The examples in Section 3.1.3 are taken from [Big99]. The proof of Lemma 3.5 was suggested to the authors by Nikolai Ivanov; see also [PR00, Prop. 3.7]. Theorem 3.7 is folklore. The reducibility of ψ_n is well known; see [Bir74].

The Alexander–Conway polynomial is a refinement, due to J. H. Conway, of the Alexander polynomial of links; see [Lic97] for an exposition. Burau computed the Alexander polynomial of the closure of a braid from its Burau matrix; see [Bir74]. The refinement of this result to the Alexander–Conway polynomial (Section 3.3 and the second claim of Theorem 3.13) is due to V. Turaev (unpublished).

The Lawrence–Krammer–Bigelow representation is one of a family of representations introduced by Lawrence [Law90]. Her work was inspired by a study of the Jones polynomial of links and was concerned with representations of Hecke algebras arising from the actions of braids on the homology of configuration spaces. Theorem 3.15 was proven independently and from different viewpoints by Krammer [Kra02] and Bigelow [Big01] after Krammer proved it for $n = 4$ in [Kra00]. The theory of noodles (Section 3.6) and the proof of Theorem 3.15 given in Section 3.7 are due to Bigelow [Big01]. (In loc. cit. Bigelow also uses the concept of a "fork" introduced by Krammer in [Kra00]. Here we have avoided the use of forks.) For more on this and related topics, see the surveys [Big02], [Tur02], [BB05].

Symmetric Groups and Iwahori–Hecke Algebras

The study of the braid group B_n naturally leads to the so-called Iwahori–Hecke algebra H_n. This algebra is a finite-dimensional quotient of the group algebra of B_n depending on two parameters q and z. Our interest in the Iwahori–Hecke algebras is due to their connections to braids and links and to their beautiful representation theory discussed in the next chapter.

As an application of the theory of Iwahori–Hecke algebras, we introduce the two-variable Jones–Conway polynomial of oriented links in Euclidean 3-space. This polynomial, known also as HOMFLY or HOMFLY-PT, extends both the Alexander–Conway link polynomial introduced in the previous chapter and the famous Jones link polynomial.

For $q = 1$ and $z = 0$, the Iwahori–Hecke algebra H_n is the group algebra of the symmetric group $\mathfrak{S}_n$. For arbitrary values of the parameters, H_n shares a number of properties of the group algebra of $\mathfrak{S}_n$. We begin therefore by recalling basic properties of $\mathfrak{S}_n$.

4.1 The symmetric groups

The symmetric group $\mathfrak{S}_n$ with $n \geq 1$ is the group of all permutations of the set $\{1, 2, \ldots, n\}$. The group law of $\mathfrak{S}_n$ is the composition of permutations, and the neutral element is the identity permutation that fixes all elements of $\{1, 2, \ldots, n\}$.

4.1.1 A presentation of $\mathfrak{S}_n$ by generators and relations

Fix an integer $n \geq 1$. For integers i, j such that $1 \leq i < j \leq n$, we denote by $\tau_{i,j}$ the permutation exchanging i and j and leaving the other elements of $\{1, 2, \ldots, n\}$ fixed. Such a permutation is called a *transposition*. There are $n(n-1)/2$ transpositions in $\mathfrak{S}_n$.

C. Kassel, V. Turaev, *Braid Groups*, DOI: 10.1007/978-0-387-68548-9_4,
© Springer Science+Business Media, LLC 2008

When $j = i + 1$, we write s_i for $\tau_{i,j}$. The transpositions $s_1, \ldots, s_{n-1}$ are called *simple transpositions*. It is an easy exercise to check that the simple transpositions satisfy the following relations for all $i, j = 1, \ldots, n - 1$:

$$\begin{aligned} s_i s_j &= s_j s_i && \text{if } |i - j| \geq 2\,, \\ s_i s_j s_i &= s_j s_i s_j && \text{if } |i - j| = 1\,, \\ s_i^2 &= 1\,. \end{aligned} \tag{4.1}$$

Let G_n denote the group with generators $\dot{s}_1, \ldots, \dot{s}_{n-1}$ and relations obtained from (4.1) by replacing each s_i with $\dot{s}_i$. The group G_1 is trivial. The group G_2 has a single generator $\dot{s}_1$ subject to the unique relation $\dot{s}_1^2 = 1$; it follows that G_2 is a cyclic group of order 2. For each n, there is a canonical group homomorphism $G_n \to G_{n+1}$ sending $\dot{s}_i \in G_n$ to $\dot{s}_i \in G_{n+1}$ for $i = 1, \ldots, n - 1$.

Theorem 4.1. *For all $n \geq 1$, there is a group homomorphism*

$$\varphi : G_n \to \mathfrak{S}_n$$

such that $\varphi(\dot{s}_i) = s_i$ for all $i = 1, \ldots, n - 1$. The homomorphism φ is an isomorphism.

The definition of G_n and relations (4.1) directly imply the existence (and the uniqueness) of φ. The bijectivity of φ will be proved in Section 4.1.2 using Lemmas 4.2 and 4.3 below.

Theorem 4.1 provides the standard presentation of the group $\mathfrak{S}_n$ by generators and relations. As an application, we can define the sign of a permutation. By definition of G_n, there is a unique group homomorphism $\chi : G_n \to \{\pm 1\}$ such that $\chi(\dot{s}_i) = -1$ for all $i = 1, \ldots, n - 1$. The *sign* $\varepsilon(w) \in \{\pm 1\}$ of a permutation $w \in \mathfrak{S}_n$ is defined by

$$\varepsilon(w) = \chi(\varphi^{-1}(w))\,.$$

Clearly, $\varepsilon(s_i) = \chi(\dot{s}_i) = -1$ for all $i = 1, \ldots, n - 1$.

Lemma 4.2. *For any $n \geq 1$, every element of G_n can be written as a word in the letters $\dot{s}_1, \ldots, \dot{s}_{n-1}$ with $\dot{s}_{n-1}$ appearing at most once.*

Proof. We proceed by induction on n. The statement holds for $n = 1$ and $n = 2$ in view of the computation of G_1 and G_2 above. We suppose that the lemma holds for $n - 1 \geq 2$ and prove it for n. Since $\dot{s}_i^2 = 1$ or, equivalently, $\dot{s}_i^{-1} = \dot{s}_i$ for all $i = 1, \ldots, n - 1$, any element of G_n can be written as a word in the letters $\dot{s}_1, \ldots, \dot{s}_{n-1}$.

Let $w = w_1 \dot{s}_{n-1} w_2 \dot{s}_{n-1} w_3$ be an element of G_n in which $\dot{s}_{n-1}$ appears at least twice. We may assume that $\dot{s}_{n-1}$ does not appear in w_2. Hence w_2 belongs to the image of G_{n-1} in G_n under the canonical homomorphism $G_{n-1} \to G_n$.

By the induction hypothesis, we can write w_2 as a word in $\dot{s}_1, \ldots, \dot{s}_{n-2}$ in which $\dot{s}_{n-2}$ appears at most once.

If $\dot{s}_{n-2}$ does not appear in w_2, then w_2 is a word in $\dot{s}_1, \ldots, \dot{s}_{n-3}$. Now, $\dot{s}_{n-1}\dot{s}_i = \dot{s}_i\dot{s}_{n-1}$ for all $i \leq n-3$. Therefore, w_2 commutes with $\dot{s}_{n-1}$ and

$$w = w_1\dot{s}_{n-1}w_2\dot{s}_{n-1}w_3 = w_1w_2\dot{s}_{n-1}^2w_3 = w_1w_2w_3 \, .$$

We thus have reduced the number of occurrences of $\dot{s}_{n-1}$ in w by two.

If $\dot{s}_{n-2}$ appears exactly once in w_2, then $w_2 = w'\dot{s}_{n-2}w''$, where both w' and w'' are words in $\dot{s}_1, \ldots, \dot{s}_{n-3}$. Clearly, w' and w'' commute with $\dot{s}_{n-1}$ and

$$\begin{aligned} w &= w_1\dot{s}_{n-1}w_2\dot{s}_{n-1}w_3 \\ &= w_1\dot{s}_{n-1}w'\dot{s}_{n-2}w''\dot{s}_{n-1}w_3 \\ &= w_1w'\dot{s}_{n-1}\dot{s}_{n-2}\dot{s}_{n-1}w''w_3 \, . \end{aligned}$$

Using the relation $\dot{s}_{n-1}\dot{s}_{n-2}\dot{s}_{n-1} = \dot{s}_{n-2}\dot{s}_{n-1}\dot{s}_{n-2}$, we obtain

$$w = w_1w'\dot{s}_{n-2}\dot{s}_{n-1}\dot{s}_{n-2}w''w_3 \, .$$

We have thus reduced the number of occurrences of $\dot{s}_{n-1}$ in w by one. Iterating this procedure, we arrive at the desired conclusion. $\qquad\square$

We define the following subsets of G_n:

$$\begin{aligned} \dot{\Sigma}_1 &= \{1, \dot{s}_1\}, \\ \dot{\Sigma}_2 &= \{1, \dot{s}_2, \dot{s}_2\dot{s}_1\}, \\ \dot{\Sigma}_3 &= \{1, \dot{s}_3, \dot{s}_3\dot{s}_2, \dot{s}_3\dot{s}_2\dot{s}_1\}, \\ &\;\;\vdots \\ \dot{\Sigma}_{n-1} &= \{1, \dot{s}_{n-1}, \dot{s}_{n-1}\dot{s}_{n-2}, \ldots, \dot{s}_{n-1}\dot{s}_{n-2}\cdots\dot{s}_2\dot{s}_1\} \, . \end{aligned}$$

Observe that card $\dot{\Sigma}_i = i + 1$ for all $i = 1, \ldots, n-1$.

Lemma 4.3. *Any element of G_n can be written as a product $w_1w_2\cdots w_{n-1}$, where $w_i \in \dot{\Sigma}_i$ for $i = 1, \ldots, n-1$.*

Proof. We prove the lemma by induction on n. For $n = 1$ and $n = 2$, the assertion is obvious. We suppose that it holds for $n - 1 \geq 2$ and prove it for n. By Lemma 4.2 it suffices to treat an element $w \in G_n$ represented by a word in $\dot{s}_1, \ldots, \dot{s}_{n-1}$ in which $\dot{s}_{n-1}$ appears exactly once: $w = w_1\dot{s}_{n-1}w_2$, where w_1 and w_2 are words in $\dot{s}_1, \ldots, \dot{s}_{n-2}$. By the induction hypothesis, $w_2 = u_1u_2\cdots u_{n-2}$, where $u_i \in \dot{\Sigma}_i$ for $i = 1, \ldots, n-2$. Since $\dot{s}_{n-1}\dot{s}_i = \dot{s}_i\dot{s}_{n-1}$ for $i \leq n-3$, the elements of $\dot{\Sigma}_i$ with $i \leq n-3$ commute with $\dot{s}_{n-1}$. Hence,

$$w = w_1\dot{s}_{n-1}w_2 = w_1\dot{s}_{n-1}u_1u_2\cdots u_{n-2} = w_1u_1u_2\cdots u_{n-3}\dot{s}_{n-1}u_{n-2} \, .$$

The element $w_1u_1u_2\cdots u_{n-3}$ comes from G_{n-1} and can be expanded as $v_1v_2\cdots v_{n-2}$ with $v_i \in \dot{\Sigma}_i$ for $i = 1, \ldots, n-2$, whereas $\dot{s}_{n-1}u_{n-2} \in \dot{\Sigma}_{n-1}$. $\quad\square$

4.1.2 Proof of Theorem 4.1

It is well known (and easy to prove) that the simple transpositions $s_1, \ldots, s_{n-1}$ generate $\mathfrak{S}_n$. Therefore, the homomorphism $\varphi : G_n \to \mathfrak{S}_n$ is surjective. Hence, $\operatorname{card} G_n \geq \operatorname{card} \mathfrak{S}_n = n!$. On the other hand, consider the map φ' sending $(w_1, w_2, \ldots, w_{n-1}) \in \dot{\Sigma}_1 \times \dot{\Sigma}_2 \times \cdots \times \dot{\Sigma}_{n-1}$ to $w_1 w_2 \cdots w_{n-1} \in G_n$. Lemma 4.3 implies that φ' is surjective. Hence,

$$\operatorname{card} G_n \leq \prod_{i=1}^{n-1} \operatorname{card} \dot{\Sigma}_i = n! \,.$$

Therefore, $\operatorname{card} G_n = \operatorname{card} \mathfrak{S}_n$. Hence $\varphi : G_n \to \mathfrak{S}_n$ is a bijection.   □

As observed in this proof, the mapping $\varphi' : \dot{\Sigma}_1 \times \dot{\Sigma}_2 \times \cdots \times \dot{\Sigma}_{n-1} \to G_n$ is surjective. Since $\operatorname{card}(G_n) = n! = \operatorname{card}(\dot{\Sigma}_1 \times \dot{\Sigma}_2 \times \cdots \times \dot{\Sigma}_{n-1})$, this mapping is a bijection. We thus obtain the following corollary of Theorem 4.1.

Corollary 4.4. *Consider the following subsets of $\mathfrak{S}_n$:*

$$\Sigma_1 = \{1, s_1\},$$
$$\Sigma_2 = \{1, s_2, s_2 s_1\},$$
$$\Sigma_3 = \{1, s_3, s_3 s_2, s_3 s_2 s_1\},$$
$$\vdots$$
$$\Sigma_{n-1} = \{1, s_{n-1}, s_{n-1} s_{n-2}, \ldots, s_{n-1} s_{n-2} \cdots s_2 s_1\}.$$

For any $w \in \mathfrak{S}_n$, there is a unique element

$$(w_1, w_2, \ldots, w_{n-1}) \in \Sigma_1 \times \Sigma_2 \times \cdots \times \Sigma_{n-1}$$

such that $w = w_1 w_2 \cdots w_{n-1}$.

4.1.3 Reduced expressions and length of a permutation

Since $s_i^{-1} = s_i$ for $i = 1, \ldots, n-1$ and $s_1, \ldots, s_{n-1}$ generate $\mathfrak{S}_n$, any permutation $w \in \mathfrak{S}_n$ can be expanded as a product $w = s_{i_1} s_{i_2} \cdots s_{i_r}$, where $i_1, i_2, \ldots, i_r \in \{1, 2, \ldots, n-1\}$. If r is minimal among all such expansions of w, then we say that $s_{i_1} s_{i_2} \cdots s_{i_r}$ is a *reduced expression* for w and that $s_{i_1} s_{i_2} \cdots s_{i_r}$ is a *reduced word*. A permutation may have many different reduced expressions.

We define the *length* $\lambda(w)$ of a permutation w as the length r of a reduced expression $s_{i_1} s_{i_2} \cdots s_{i_r}$ for w. Observe the following:

(a) If $s_{i_1} s_{i_2} \cdots s_{i_r}$ is a reduced expression for w, then

$$s_{i_r}^{-1} \cdots s_{i_2}^{-1} s_{i_1}^{-1} = s_{i_r} \cdots s_{i_2} s_{i_1}$$

is a reduced expression for w^{-1}. It follows that $\lambda(w^{-1}) = \lambda(w)$ for any w.

(b) If $s_{i_1} s_{i_2} \ldots s_{i_r}$ is a reduced word, then for all indices $1 \leq p < q \leq r$ the truncated word $s_{i_p} s_{i_{p+1}} \cdots s_{i_q}$ is reduced.

(c) The neutral element $1 \in \mathfrak{S}_n$ is the only element of length zero, whereas the simple transpositions are the only elements of length one.

(d) The sign of a permutation w can be computed from its length by

$$\varepsilon(w) = (-1)^{\lambda(w)}. \tag{4.2}$$

Lemma 4.5. *For any $w \in \mathfrak{S}_n$ and any $s_i \in S$,*

$$\lambda(s_i w) = \lambda(w) \pm 1 \quad and \quad \lambda(w s_i) = \lambda(w) \pm 1.$$

Proof. By definition of the length, $\lambda(s_i w) \leq \lambda(w) + 1$. Replacing in this formula w by $s_i w$, we obtain

$$\lambda(w) = \lambda(s_i^2 w) \leq \lambda(s_i w) + 1.$$

Therefore, $\lambda(w) - 1 \leq \lambda(s_i w) \leq \lambda(w) + 1$. By (4.2), since

$$\varepsilon(s_i w) = \varepsilon(s_i) \varepsilon(w) = -\varepsilon(w),$$

we cannot have $\lambda(s_i w) = \lambda(w)$. Therefore, $\lambda(s_i w) = \lambda(w) \pm 1$.

We derive $\lambda(w s_i) = \lambda(w) \pm 1$ from the previous equality by replacing w and $w s_i$ with their inverses. $\qquad\square$

4.1.4 Inversions and the exchange theorem

Given a permutation $w \in \mathfrak{S}_n$, we define an *inversion* of w to be a pair of integers (i, j) such that $1 \leq i < j \leq n$ and $w(i) > w(j)$. We write $I(w)$ for the set of transpositions $\tau_{i,j}$ of $\mathfrak{S}_n$ such that (i, j) is an inversion of w. By definition, the cardinality of $I(w)$ is equal to the number of inversions of w.

It is clear that $I(1) = \emptyset$ and $I(s_i) = \{s_i\}$ for $i = 1, \ldots, n - 1$. Note also that $\tau_{i,j} \in I(\tau_{i,j})$ for any transposition $\tau_{i,j} \in \mathfrak{S}_n$.

In order to formulate the next lemma, recall the *symmetric difference* $A \bigtriangleup B$ of two subsets A and B of a given set G; it is defined by

$$A \bigtriangleup B = (A \cup B) - (A \cap B).$$

The symmetric difference is an associative, commutative composition law on the set of subsets of G, with the empty set as the neutral element. When G is a group,

$$g^{-1}(A \bigtriangleup B) g = (g^{-1} A g) \bigtriangleup (g^{-1} B g), \tag{4.3}$$

for all $g \in G$, where for $A \subset G$ and $g \in G$, we set

$$g^{-1} A g = \{g^{-1} a g \,|\, a \in A\}.$$

In the proof of the next lemma, we use the following elementary fact:

$$A \bigtriangleup \{a\} = \begin{cases} A \cup \{a\} & \text{if } a \notin A, \\ A - \{a\} & \text{if } a \in A. \end{cases} \tag{4.4}$$

Lemma 4.6. *We have* $I(vw) = w^{-1}I(v)w \,\Delta\, I(w)$ *for all* $v, w \in \mathfrak{S}_n$.

Proof. We prove the lemma by induction on $\lambda(w)$.

(a) If $\lambda(w) = 0$, then $w = 1$ and

$$w^{-1}I(v)w \,\Delta\, I(w) = I(v) \,\Delta\, \emptyset = I(v) = I(vw).$$

(b) If $\lambda(w) = 1$, then $w = s_k$ for some $k = 1, \ldots, n-1$. We have to prove
that for all $v \in \mathfrak{S}_n$,

$$I(vs_k) = s_k^{-1}I(v)s_k \,\Delta\, \{s_k\}. \tag{4.5}$$

Let us first check that

$$I(vs_k) - \{s_k\} = s_k^{-1}I(v)s_k - \{s_k\}. \tag{4.6}$$

Indeed, a transposition $\tau_{i,j}$ belongs to $I(vs_k) - \{s_k\}$ if and only if $\tau_{i,j} \neq s_k$
and $(vs_k)(i) > (vs_k)(j)$. Since $s_k(i) < s_k(j)$, these conditions hold if and only
if $\tau_{i,j} \neq s_k$ and $(s_k(i), s_k(j))$ is an inversion of v. In turn, this is equivalent
to $\tau_{i,j} \neq s_k$ and $s_k\tau_{i,j}s_k^{-1} = \tau_{s_k(i),s_k(j)} \in I(v)$. The latter conditions are
equivalent to $\tau_{i,j} \in s_k^{-1}I(v)s_k - \{s_k\}$. This proves (4.6).

Next, observe that the inclusion $s_k \in I(v)$ holds if and only if $s_k \notin I(vs_k)$.
Indeed, $v(k) > v(k+1)$ is equivalent to

$$(vs_k)(k) = v(k+1) < v(k) = (vs_k)(k+1).$$

We can now prove (4.5). If $s_k \in I(v)$, then by the observation above,
$s_k \notin I(vs_k)$. Therefore, by (4.6) and (4.4),

$$\begin{aligned}
I(vs_k) &= I(vs_k) - \{s_k\} \\
&= s_k^{-1}I(v)s_k - \{s_k\} \\
&= s_k^{-1}I(v)s_k \,\Delta\, \{s_k\}.
\end{aligned}$$

If $s_k \notin I(v)$, then $s_k \in I(vs_k)$ and, by (4.6) and (4.4),

$$\begin{aligned}
I(vs_k) &= \big(I(vs_k) - \{s_k\}\big) \cup \{s_k\} \\
&= \big(s_k^{-1}I(v)s_k - \{s_k\}\big) \cup \{s_k\} \\
&= s_k^{-1}I(v)s_k \cup \{s_k\} \\
&= s_k^{-1}I(v)s_k \,\Delta\, \{s_k\}.
\end{aligned}$$

(c) If $\lambda(w) > 1$, then $w = us_k$, where $u \in \mathfrak{S}_n$ and $\lambda(u) = \lambda(w) - 1$. We
have

$$\begin{aligned}
I(vw) &= I(vus_k) \\
&= s_k^{-1}I(vu)s_k \,\Delta\, \{s_k\} \\
&= s_k^{-1}\big(u^{-1}I(v)u \,\Delta\, I(u)\big)s_k \,\Delta\, \{s_k\} \\
&= \big(s_k^{-1}u^{-1}I(v)us_k \,\Delta\, s_k^{-1}I(u)s_k\big) \,\Delta\, \{s_k\} \\
&= s_k^{-1}u^{-1}I(v)us_k \,\Delta\, \big(s_k^{-1}I(u)s_k \,\Delta\, \{s_k\}\big) \\
&= w^{-1}I(v)w \,\Delta\, I(w).
\end{aligned}$$

The second and sixth equalities follow from the case $\lambda(w) = 1$, the third one from the induction hypothesis, the fourth one from (4.3), and the fifth one from the associativity of Δ. $\qquad\square$

Lemma 4.7. Let $T = \{\tau_{i,j} \mid 1 \le i < j \le n\} \subset \mathfrak{S}_n$. For any $w \in \mathfrak{S}_n$,
  (a) $\lambda(w) = \operatorname{card} I(w)$;
  (b) $\lambda(w) \le n(n-1)/2$ and $\lambda(w) = n(n-1)/2$ if and only if $I(w) = T$;
  (c) $I(w) = \{\tau \in T \mid \lambda(w\tau) < \lambda(w)\}$;
  (d) $\lambda(ws_i) = \lambda(w) + 1$ if and only if $w(i) < w(i+1)$.

Proof. (a) Let $r = \lambda(w)$ and $w = s_{i_1} s_{i_2} \cdots s_{i_r}$ be a reduced expression for w. A repeated application of Lemma 4.6 shows that

$$I(w) = I(s_{i_1} s_{i_2} \cdots s_{i_r}) = \{t_1\} \Delta \cdots \Delta \{t_r\},$$

where $t_1, \ldots, t_r \in \mathfrak{S}_n$ are the transpositions defined by

$$t_k = (s_{i_{k+1}} \cdots s_{i_r})^{-1} s_{i_k} (s_{i_{k+1}} \cdots s_{i_r}) \tag{4.7}$$

for $1 \le k \le r-1$. In particular, $t_r = s_{i_r}$. Observe that

$$
\begin{aligned}
wt_k &= s_{i_1} s_{i_2} \cdots s_{i_{k-1}} s_{i_k} s_{i_{k+1}} \cdots s_{i_r} (s_{i_{k+1}} \cdots s_{i_r})^{-1} s_{i_k} (s_{i_{k+1}} \cdots s_{i_r}) \\
&= s_{i_1} \cdots s_{i_{k-1}} \widehat{s_{i_k}} s_{i_{k+1}} \cdots s_{i_r},
\end{aligned} \tag{4.8}
$$

where the hat over s_{i_k} indicates that it has been removed. We claim that the transpositions $t_1, \ldots, t_r$ are all distinct. Indeed, suppose that $t_p = t_q$ for some $p < q$. A computation similar to the one above shows that

$$w = wt_p^2 = wt_p t_q = s_{i_1} \cdots \widehat{s_{i_p}} \cdots \widehat{s_{i_q}} \cdots s_{i_r}.$$

Then $\lambda(w) < r$, a contradiction. Consequently, $I(w)$ is the disjoint union of the singletons $\{t_1\}, \ldots, \{t_r\}$, and $I(w)$ has $r = \lambda(w)$ elements.
  (b) We have $\lambda(w) = \operatorname{card} I(w) \le \operatorname{card} T = n(n-1)/2$.
  (c) We saw in the proof of (a) that $I(w) = \{t_1, \ldots, t_r\}$ and

$$wt_k = s_{i_1} \cdots \widehat{s_{i_k}} \cdots s_{i_r}$$

for all $k = 1, \ldots, r$. Therefore, $\lambda(wt_k) < \lambda(w)$. This shows that $\lambda(w\tau) < \lambda(w)$ for any $\tau \in I(w)$.
  If $\tau \in T$ does not belong to $I(w)$, then $\tau = \tau^{-1}\tau\tau \notin \tau^{-1}I(w)\tau$, whereas $\tau \in I(\tau)$. Therefore,

$$\tau \in \tau^{-1}I(w)\tau \, \Delta \, I(\tau) = I(w\tau).$$

By the previous argument,

$$\lambda(w) = \lambda(w\tau^2) < \lambda(w\tau).$$

  (d) By Lemma 4.5 and (c), the equality $\lambda(ws_i) = \lambda(w) + 1$ holds if and only if $s_i \notin I(w)$, which is equivalent to $w(i) < w(i+1)$. $\qquad\square$

We now state the so-called *exchange theorem*.

Theorem 4.8. *Let* $s_{i_1} \cdots s_{i_r}$ *be a reduced expression for* $w \in \mathfrak{S}_n$, *where* $r = \lambda(w)$. *If* $\lambda(ws_j) < \lambda(w)$ *for some* $j \in \{1, \ldots, n-1\}$, *then there is* $k \in \{1, \ldots, r\}$ *such that* $ws_j = s_{i_1} \cdots \widehat{s_{i_k}} \cdots s_{i_r}$. *If* $\lambda(s_j w) < \lambda(w)$ *for some* $j \in \{1, \ldots, n-1\}$, *then there is* $k \in \{1, \ldots, r\}$ *such that* $s_j w = s_{i_1} \cdots \widehat{s_{i_k}} \cdots s_{i_r}$.

Proof. We saw in the proof of Lemma 4.7 (a) that if $t_1, \ldots, t_r$ are the transpositions defined by (4.7), then $I(w) = \{t_1, \ldots, t_r\}$. If $\lambda(ws_j) < \lambda(w)$, then $s_j \in I(w)$ by Lemma 4.7 (c). Therefore, $s_j = t_k$ for some $k \in \{1, \ldots, r\}$. By (4.8),

$$ws_j = wt_k = s_{i_1} \cdots \widehat{s_{i_k}} \cdots s_{i_r}.$$

The second claim is deduced from the first one by replacing w with w^{-1}. $\square$

Corollary 4.9. *Let* $w \in \mathfrak{S}_n$. *If* $\lambda(ws_j) < \lambda(w)$ *for some* $j \in \{1, \ldots, n-1\}$, *then there is a reduced expression for* w *ending with* s_j. *If* $\lambda(s_j w) < \lambda(w)$ *for some* $j \in \{1, \ldots, n-1\}$, *then there is a reduced expression for* w *beginning with* s_j.

This is a direct corollary of the previous theorem: if $\lambda(ws_j) < \lambda(w)$, then $ws_j = s_{i_1} \cdots \widehat{s_{i_k}} \cdots s_{i_r}$ and $w = s_{i_1} \cdots \widehat{s_{i_k}} \cdots s_{i_r} s_j$ is a reduced expression for w, since its length is equal to $r = \lambda(w)$. The second claim is proven similarly.

We conclude with a lemma needed in the proof of Lemma 4.18 below.

Lemma 4.10. *If* $\lambda(s_i w s_j) = \lambda(w)$ *and* $\lambda(s_i w) = \lambda(w s_j)$ *for* $w \in \mathfrak{S}_n$ *and some* $i, j \in \{1, \ldots, n-1\}$, *then* $s_i w = w s_j$ *and* $s_i w s_j = w$.

Proof. (a) Suppose first that $\lambda(s_i w) = \lambda(w s_j) > \lambda(s_i w s_j) = \lambda(w)$. By Lemma 4.6,

$$I(s_i w) = w^{-1} I(s_i) w \, \Delta \, I(w) = \{w^{-1} s_i w\} \, \Delta \, I(w).$$

Since $\lambda(s_i w s_j) < \lambda(s_i w)$ and $\lambda(w s_j) > \lambda(w)$, Lemma 4.7(c) implies that s_j belongs to $I(s_i w)$, but not to $I(w)$. Therefore $s_j = w^{-1} s_i w$; hence $s_i w = w s_j$ and $s_i w s_j = w s_j^2 = w$.

(b) If $\lambda(s_i w) = \lambda(w s_j) < \lambda(s_i w s_j) = \lambda(w)$, then we apply a similar argument, using that $I(w) = I(s_i(s_i w)) = \{w^{-1} s_i w\} \, \Delta \, I(s_i w)$. $\square$

4.1.5 Equivalence of reduced expressions

For $n \geq 1$, let M_n be the set of all finite sequences of integers from the set $\{1, \ldots, n-1\}$, including the empty sequence. We equip M_n with the associative product given by concatenation. In this way, M_n becomes a monoid with the empty sequence as the neutral element.

On M_n we consider the equivalence relation $\sim$ generated by the following two families of relations:

$$S_1\,(i,j)\,S_2 \sim S_1\,(j,i)\,S_2 \qquad\qquad (4.9)$$

for all $S_1, S_2 \in M_n$ and all $i,j \in \{1,\ldots,n-1\}$ such that $|i-j| \geq 2$, and

$$S_1\,(i,j,i)\,S_2 \sim S_1\,(j,i,j)\,S_2 \qquad\qquad (4.10)$$

for all $S_1, S_2 \in M_n$ and all $i,j \in \{1,\ldots,n-1\}$ such that $|i-j| = 1$. Observe that equivalent sequences have the same length. The equivalence relation $\sim$ has been devised so that

$$(i_1,\ldots,i_k) \sim (j_1,\ldots,j_k) \in M_n \implies s_{i_1}\cdots s_{i_k} = s_{j_1}\cdots s_{j_k} \in \mathfrak{S}_n\,.$$

Lemma 4.11. *If $s_{i_1}\cdots s_{i_k}$ and $s_{j_1}\cdots s_{j_k}$ are reduced expressions for the same permutation $w \in \mathfrak{S}_n$, then $(i_1,\ldots,i_k) \sim (j_1,\ldots,j_k)$ in M_n.*

This lemma shows that for any $w \in \mathfrak{S}_n$, we can pass from one reduced expression for w to any other reduced expression for w using only the relations

$$s_i s_j = s_j s_i \qquad \text{if } |i-j| \geq 2,$$
$$s_i s_j s_i = s_j s_i s_j \qquad \text{if } |i-j| = 1\,.$$

Proof. We prove the lemma by induction on k. If $k = 0$, then $w = 1$ has only one reduced expression. If $k = 1$, then $w = s_i$ for some i and w has only one reduced expression.

Assume that $k \geq 2$. From the equality $s_{i_1}\cdots s_{i_k} = s_{j_1}\cdots s_{j_k}$ we deduce that $s_{i_2}\cdots s_{i_k} = s_{i_1} s_{j_1}\cdots s_{j_k}$. Since $s_{i_2}\cdots s_{i_k}$ is reduced,

$$\lambda(s_{i_1} s_{j_1}\cdots s_{j_k}) = \lambda(s_{i_2}\cdots s_{i_k}) = k-1 < k = \lambda(s_{j_1}\cdots s_{j_k})\,.$$

Therefore, by Theorem 4.8, there is an integer p with $1 \leq p \leq k$ such that

$$s_{i_2}\cdots s_{i_k} = s_{i_1} s_{j_1}\cdots s_{j_k} = s_{j_1}\cdots \widehat{s_{j_p}}\cdots s_{j_k}\,. \qquad (4.11)$$

Since $s_{i_2}\cdots s_{i_k}$ and $s_{j_1}\cdots \widehat{s_{j_p}}\cdots s_{j_k}$ represent the same permutation and have the same length $k-1$ and since $s_{i_2}\cdots s_{i_k}$ is reduced, $s_{j_1}\cdots \widehat{s_{j_p}}\cdots s_{j_k}$ is also reduced. By the induction hypothesis, $(i_2,\ldots,i_k) \sim (j_1,\ldots,\widehat{j_p},\ldots,j_k)$. Hence,

$$(i_1,i_2,\ldots,i_k) \sim (i_1,j_1,\ldots,\widehat{j_p},\ldots,j_k)\,. \qquad (4.12)$$

The word $s_{j_1}\cdots s_{j_p}$, being a part of the reduced word $s_{j_1}\cdots s_{j_k}$, is reduced. The second equality in (4.11) implies that $s_{i_1} s_{j_1}\cdots s_{j_{p-1}}$ and $s_{j_1}\cdots s_{j_p}$ are equal in $\mathfrak{S}_n$. Since these words have the same length p and one of them is reduced, so is the other one. If $p < k$, then we apply the induction hypothesis to these words, obtaining $(i_1,j_1,\ldots,j_{p-1}) \sim (j_1,\ldots,j_p)$. From this and (4.12), we obtain

$$\begin{aligned}
(i_1,i_2,\ldots,i_k) &\sim (i_1,j_1,\ldots,\widehat{j_p},\ldots,j_k)\\
&= (i_1,j_1,\ldots,j_{p-1})(j_{p+1},\ldots,j_k)\\
&\sim (j_1,\ldots,j_p)(j_{p+1},\ldots,j_k)\\
&= (j_1,\ldots,j_p,j_{p+1},\ldots,j_k)\,,
\end{aligned}$$

which was to be proven.

If $p = k$, then (4.12) becomes $(i_1, i_2, \ldots, i_k) \sim (i_1, j_1, \ldots, j_{k-1})$. This equivalence implies that

$$s_{i_1} s_{j_1} \cdots s_{j_{k-1}} = s_{i_1} s_{i_2} \cdots s_{i_k} = s_{j_1} s_{j_2} \cdots s_{j_k} \, .$$

Summarizing our argument, we see that to prove the implication

$$s_{i_1} \cdots s_{i_k} = s_{j_1} \cdots s_{j_k} \implies (i_1, \ldots, i_k) \sim (j_1, \ldots, j_k),$$

it is enough to prove the implication

$$s_{i_1} s_{j_1} \cdots s_{j_{k-1}} = s_{j_1} \cdots s_{j_k} \implies (i_1, j_1, \ldots, j_{k-1}) \sim (j_1, \ldots, j_k). \quad (4.13)$$

We now start the argument all over again with the reduced expressions $s_{j_1} \cdots s_{j_k} = s_{i_1} s_{j_1} \cdots s_{j_{k-1}}$. Proceeding as above, we show that in order to prove (4.13), it is enough to prove the implication

$$s_{j_1} s_{i_1} s_{j_1} \cdots s_{j_{k-2}} = s_{i_1} s_{j_1} s_{j_2} \cdots s_{j_{k-1}}$$
$$\implies (j_1, i_1, j_1, \ldots, j_{k-2}) \sim (i_1, j_1, j_2, \ldots, j_{k-1}). \quad (4.14)$$

We first prove (4.14) when $|i_1 - j_1| \geq 2$. Then $s_{i_1} s_{j_1} = s_{j_1} s_{i_1}$ and

$$s_{i_1} s_{j_1} s_{j_1} \cdots s_{j_{k-2}} = s_{j_1} s_{i_1} s_{j_1} \cdots s_{j_{k-2}} = s_{i_1} s_{j_1} s_{j_2} \cdots s_{j_{k-1}} \, .$$

Multiplying on the left by $s_{j_1} s_{i_1}$ in $\mathfrak{S}_n$, we obtain

$$s_{j_1} \cdots s_{j_{k-2}} = s_{j_2} \cdots s_{j_{k-1}} \, .$$

Both sides are reduced expressions of length $k - 2$. By the induction assumption, $(j_1, \ldots, j_{k-2}) \sim (j_2, \ldots, j_{k-1})$. From this and (4.9), we obtain

$$(j_1, i_1, j_1, \ldots, j_{k-2}) = (j_1, i_1)(j_1, \ldots, j_{k-2})$$
$$\sim (i_1, j_1)(j_2, \ldots, j_{k-1})$$
$$= (i_1, j_1, j_2, \ldots, j_{k-1}),$$

which was to be proven.

If $|i_1 - j_1| = 1$, then we proceed again as above and reduce the proof of (4.14) to showing that the equality

$$s_{i_1} s_{j_1} s_{i_1} s_{j_1} \cdots s_{j_{k-3}} = s_{j_1} s_{i_1} s_{j_1} s_{j_2} \cdots s_{j_{k-2}}$$

implies that $(i_1, j_1, i_1, j_1, \ldots, j_{k-3}) \sim (j_1, i_1, j_1, j_2, \ldots, j_{k-2})$. This and the equality $s_{j_1} s_{i_1} s_{j_1} = s_{i_1} s_{j_1} s_{i_1}$ imply that

$$s_{j_1} s_{i_1} s_{j_1} s_{j_1} \cdots s_{j_{k-3}} = s_{i_1} s_{j_1} s_{i_1} s_{j_1} \cdots s_{j_{k-3}} = s_{j_1} s_{i_1} s_{j_1} s_{j_2} \cdots s_{j_{k-2}} \, ,$$

which, after left multiplication by $s_{j_1} s_{i_1} s_{j_1}$, gives

$$s_{j_1} \cdots s_{j_{k-3}} = s_{j_2} \cdots s_{j_{k-2}} \, .$$

Since both sides of this equality are reduced expressions of length $k-3$, we can apply the induction hypothesis and obtain $(j_1, \ldots, j_{k-3}) \sim (j_2, \ldots, j_{k-2})$. From this and (4.10),

$$
\begin{aligned}
(i_1, j_1, i_1, j_1, \ldots, j_{k-3}) &= (i_1, j_1, i_1)(j_1, \ldots, j_{k-3}) \\
&\sim (j_1, i_1, j_1)(j_2, \ldots, j_{k-2}) \\
&= (j_1, i_1, j_1, j_2, \ldots, j_{k-2}),
\end{aligned}
$$

which was to be proven. □

The following theorem is useful for defining maps from the symmetric groups to monoids.

Theorem 4.12. *For any monoid M and any $x_1, \ldots, x_{n-1} \in M$ satisfying the relations*

$$
\begin{aligned}
x_i x_j &= x_j x_i &&\text{if } |i-j| \geq 2, \\
x_i x_j x_i &= x_j x_i x_j &&\text{if } |i-j| = 1,
\end{aligned}
$$

there is a set-theoretic map $\rho : \mathfrak{S}_n \to M$ defined by

$$\rho(w) = x_{i_1} \cdots x_{i_k},$$

for any $w \in \mathfrak{S}_n$ and any reduced expression $w = s_{i_1} \cdots s_{i_k}$.

Proof. Define a monoid homomorphism $\rho' : M_n \to M$ by

$$\rho'(i_1, \ldots, i_k) = x_{i_1} \cdots x_{i_k}$$

for all $(i_1, \ldots, i_k) \in M_n$. We claim that $\rho'(S) = \rho'(S')$ for all $S, S' \in M_n$ such that $S \sim S'$. Indeed, by definition of the equivalence $\sim$, it suffices to prove the claim when $S = S_1 (i,j) S_2$ (resp. $S = S_1 (i,j,i) S_2$) and $S' = S_1 (j,i) S_2$ (resp. $S' = S_1 (j,i,j) S_2$) for $S_1, S_2 \in M_n$ and $i, j \in \{1, \ldots, n-1\}$ such that $|i-j| \geq 2$ (resp. $|i-j| = 1$). By the assumptions of the theorem,

$$
\begin{aligned}
\rho'(S_1 (i,j) S_2) &= \rho'(S_1) \, x_i x_j \, \rho'(S_2) \\
&= \rho'(S_1) \, x_j x_i \, \rho'(S_2) \\
&= \rho'(S_1 (j,i) S_2)
\end{aligned}
$$

if $|i-j| \geq 2$, and

$$
\begin{aligned}
\rho'(S_1 (i,j,i) S_2) &= \rho'(S_1) \, x_i x_j x_i \, \rho'(S_2) \\
&= \rho'(S_1) \, x_j x_i x_j \, \rho'(S_2) \\
&= \rho'(S_1 (j,i,j) S_2)
\end{aligned}
$$

if $|i-j| = 1$.

To prove the theorem, we need only check that ρ is well defined, i.e., if $s_{i_1} \cdots s_{i_k}$ and $s_{j_1} \cdots s_{j_k}$ are reduced expressions for $w \in \mathfrak{S}_n$, then

$$x_{i_1} \cdots x_{i_k} = x_{j_1} \cdots x_{j_k}.$$

By Lemma 4.11, $(i_1, \ldots, i_k) \sim (j_1, \ldots, j_k)$ in M_n. By the claim above,

$$x_{i_1} \cdots x_{i_k} = \rho'(i_1, \ldots, i_k) = \rho'(j_1, \ldots, j_k) = x_{j_1} \cdots x_{j_k}. \qquad \square$$

4.1.6 The longest element of $\mathfrak{S}_n$

Let $w_0 \in \mathfrak{S}_n$ be the permutation $i \mapsto n + 1 - i$ for all $i \in \{1, \ldots, n-1\}$:

$$w_0 = \begin{pmatrix} 1 & 2 & \cdots & n-1 & n \\ n & n-1 & \cdots & 2 & 1 \end{pmatrix}. \qquad (4.15)$$

It is clear that w_0 is the only permutation $w \in \mathfrak{S}_n$ such that $w(i) > w(j)$ for all $i, j \in \{1, \ldots, n-1\}$ with $i < j$. In other words, $w = w_0$ if and only if the set $I(w)$ consists of all transpositions. By Lemma 4.7 (a), $\lambda(w_0) = n(n-1)/2$ and $\lambda(w) < n(n-1)/2$ for $w \neq w_0$. Because of this, w_0 is called the *longest element* of $\mathfrak{S}_n$. We record two other properties of w_0 (the second one will be used in Section 6.5.2).

Lemma 4.13. *If $w \in \mathfrak{S}_n$ satisfies $\lambda(ws_i) < \lambda(w)$ for all $i \in \{1, \ldots, n-1\}$, then $w = w_0$.*

Proof. By Lemma 4.7 (c), $s_i \in I(w)$ for all i. Then $w(i) > w(i+1)$ for all i. The only permutation satisfying these inequalities is w_0. $\qquad \square$

Lemma 4.14. *For any $u, v \in \mathfrak{S}_n$ such that $uv = w_0$,*

$$\lambda(u) + \lambda(v) = \lambda(w_0).$$

Proof. The lemma trivially holds for $u = w_0$ and $v = 1$.

We claim that for any $u \in \mathfrak{S}_n$, $u \neq w_0$, there is a sequence $s_{i_1}, \ldots, s_{i_r}$ of simple transpositions such that $us_{i_1} \cdots s_{i_r} = w_0$ and $\lambda(us_{i_1} \cdots s_{i_r}) = \lambda(u) + r$. Before we prove the claim, let us show that it implies the lemma for u and $v = u^{-1}w_0 = s_{i_1} \cdots s_{i_r}$. Clearly, $\lambda(v) \leq r$ and

$$\lambda(w_0) = \lambda(uv) \leq \lambda(u) + \lambda(v) \leq \lambda(u) + r = \lambda(us_{i_1} \cdots s_{i_r}) = \lambda(w_0).$$

Therefore, $\lambda(u) + \lambda(v) = \lambda(w_0)$.

Let us now establish the claim. Since $u \neq w_0$, by Lemma 4.13, there is s_{i_1} such that $\lambda(us_{i_1}) \geq \lambda(u)$. By Lemma 4.5, we have $\lambda(us_{i_1}) = \lambda(u) + 1$. If $\lambda(us_{i_1}) = \lambda(w_0)$, then $us_{i_1} = w_0$, since w_0 is the unique element of $\mathfrak{S}_n$ of maximal length, and we are done. If $\lambda(us_{i_1}) < \lambda(w_0)$, then again by Lemma 4.13, we can find s_{i_2} such that $\lambda(us_{i_1}s_{i_2}) \geq \lambda(us_{i_1})$. Then

$$\lambda(us_{i_1}s_{i_2}) = \lambda(us_{i_1}) + 1 = \lambda(u) + 2.$$

If $\lambda(us_{i_1}s_{i_2}) = \lambda(w_0)$, then $us_{i_1}s_{i_2} = w_0$ and we are done. If not, we continue as above until we find the required sequence $s_{i_1}, \ldots, s_{i_r}$. $\qquad \square$

Exercise 4.1.1. Using Theorem 4.8, prove that if $\lambda(s_{i_1} \cdots s_{i_r}) < r$, then there are $p, q \in \{1, \ldots, r\}$ such that $p < q$ and

$$s_{i_1} \cdots s_{i_r} = s_{i_1} \cdots \widehat{s_{i_p}} \cdots \widehat{s_{i_q}} \cdots s_{i_r},$$

where $\widehat{s_{i_p}}$ and $\widehat{s_{i_q}}$ are removed on the right-hand side.

Exercise 4.1.2. Deduce Theorem 4.1 from Theorem 4.12, using the latter to construct a left inverse $\mathfrak{S}_n \to G_n$ of $\varphi : G_n \to \mathfrak{S}_n$.

Exercise 4.1.3. (a) Show that $w_{k,\ell} = s_k s_{k-1} \cdots s_\ell$ is a reduced word for each pair (k, ℓ) such that $1 \leq \ell \leq k \leq n - 1$.

(b) Prove that the word $w_{k_1,\ell_1} w_{k_2,\ell_2} \cdots w_{k_r,\ell_r}$ obtained by concatenating words as in (a) is reduced for $k_1 < k_2 < \cdots < k_r$.

Exercise 4.1.4. For any integer $k \geq 1$, set $[k]_q = 1 + q + \cdots + q^{k-1} \in \mathbf{Z}[q]$. Show that

$$\sum_{w \in \mathfrak{S}_n} q^{\lambda(w)} = [1]_q [2]_q [3]_q \cdots [n]_q.$$

(Hint: Use Exercise 4.1.3 and Corollary 4.4.)

Exercise 4.1.5. Prove that for any $w \in \mathfrak{S}_n$ and any $i = 1, \ldots, n-1$, the equality $\lambda(s_i w) = \lambda(w) + 1$ holds if and only if $w^{-1}(i) < w^{-1}(i+1)$.

4.2 The Iwahori–Hecke algebras

4.2.1 Presentation by generators and relations

We fix an integer $n \geq 1$ and a commutative ring R together with two elements $q, z \in R$. We assume that q is invertible in R.

Definition 4.15. *The Iwahori–Hecke algebra $H_n = H_n^R(q, z)$ is the unital associative R-algebra generated by $T_1, \ldots, T_{n-1}$ subject to the relations*

$$T_i T_j = T_j T_i \tag{4.16}$$

for $i, j = 1, 2, \ldots, n-1$ such that $|i - j| \geq 2$,

$$T_i T_{i+1} T_i = T_{i+1} T_i T_{i+1} \tag{4.17}$$

for $i = 1, 2, \ldots, n-2$, and

$$T_i^2 = z T_i + q\mathbf{1} \tag{4.18}$$

for $i = 1, \ldots, n-1$. By definition, $H_1 = H_1^R(q, z) = R$.

Any element of H_n is a linear combination of monomials $T_{i_1} T_{i_2} \cdots T_{i_r}$, including the empty monomial, which we identify with the unit 1 of H_n. By (4.18), each generator T_i is invertible in H_n with inverse

$$T_i^{-1} = q^{-1}(T_i - z\mathbf{1}). \tag{4.19}$$

Therefore, each monomial $T_{i_1} T_{i_2} \cdots T_{i_r}$ is invertible in H_n.

By Theorem 4.1, for $q = 1$, $z = 0$, we have $H_n^R(q, z) \cong R[\mathfrak{S}_n]$.

4.2.2 The one-parameter Iwahori–Hecke algebras

Many authors consider the one-parameter Iwahori–Hecke algebra $H_n^R(q)$, which by definition is $H_n^R(q, z)$ with $z = q - 1$. The algebra $H_n^R(q)$ is the unital associative R-algebra generated by $T_1, \ldots, T_{n-1}$ subject to relations (4.16), (4.17), and

$$T_i^2 = (q - 1)T_i + q1. \qquad (4.20)$$

There is essentially no loss of generality in considering the one-parameter Iwahori–Hecke algebras rather than the two-parameter ones. Indeed, the two-parameter algebra $H_n^R(q, z)$ is isomorphic to an algebra of the form $H_n^R(q')$ possibly after extending the ring of scalars. To see this, consider the presentation of $H_n^R(q, z)$ by generators and relations exhibited in Section 4.2.1. For $i = 1, \ldots, n-1$, set $T_i' = u^{-1}T_i$ for some invertible element u. Clearly, $T_1', \ldots, T_{n-1}'$ satisfy (4.16) and (4.17). From (4.18) we obtain

$$(T_i')^2 = u^{-1}zT_i' + u^{-2}q$$

for $i = 1, \ldots, n-1$. Let R' be the smallest ring containing R and a root u of the quadratic polynomial $X^2 + zX - q$: if R contains a root of this polynomial, then $R' = R$; otherwise, R' is a quadratic extension of R. Then the map $T_i \mapsto uT_i'$ $(i = 1, \ldots, n - 1)$ induces an algebra isomorphism $H_n^{R'}(q, z) \cong H_n^{R'}(u^{-2}q)$.

4.2.3 Basis of H_n

We return to the two-parameter Iwahori–Hecke algebra $H_n = H_n^R(q, z)$. We now show that H_n is a free R-module on a basis indexed by the elements of the symmetric group $\mathfrak{S}_n$. Recall the notation from Section 4.1: the symbol s_i denotes the simple transposition exchanging i and $i + 1$ for $i = 1, \ldots, n - 1$, and $\lambda(w)$ denotes the length of $w \in \mathfrak{S}_n$.

Lemma 4.16. *(a) For each $w \in \mathfrak{S}_n$, there is a unique $T_w \in H_n$ such that $T_w = T_{i_1} \cdots T_{i_r}$ whenever $w = s_{i_1} \cdots s_{i_r}$ is a reduced expression for w.*
(b) For $w \in \mathfrak{S}_n$ and any simple transposition s_i,

$$T_i T_w = \begin{cases} T_{s_i w} & \text{if } \lambda(s_i w) > \lambda(w), \\ q T_{s_i w} + z\, T_w & \text{if } \lambda(s_i w) < \lambda(w). \end{cases}$$

Observe that if $w = 1 \in \mathfrak{S}_n$, then $T_w = 1 \in H_n$.

Proof. (a) This follows from (4.16), (4.17), and Theorem 4.12.
   (b) Let $s_{i_1} \cdots s_{i_r}$ be a reduced expression for w. If $\lambda(s_i w) > \lambda(w)$, then $s_i s_{i_1} \cdots s_{i_r}$ is a reduced expression for $s_i w$. Therefore, $T_{s_i w} = T_i T_w$.
   If $\lambda(s_i w) < \lambda(w)$, then we may assume by Corollary 4.9 that $s_{i_1} = s_i$. Then $s_i w$ has $s_{i_2} \cdots s_{i_r}$ as a reduced expression. Hence, by (4.18),

$$T_i T_w = T_i T_{i_1} \cdots T_{i_r} = T_i^2 T_{i_2} \cdots T_{i_r}$$
$$= z T_i T_{i_2} \cdots T_{i_r} + q T_{i_2} \cdots T_{i_r}$$
$$= z T_w + q T_{s_i w} .$$ □

Theorem 4.17. *The R-module H_n is free of rank n! with basis $\{T_w \mid w \in \mathfrak{S}_n\}$.*

Proof. Let H be the R-submodule of H_n spanned by the vectors T_w ($w \in \mathfrak{S}_n$). By Lemma 4.16 (b), H is a left ideal of H_n. Since $1 = T_1 \in H$, we have $H = H_n$. To prove the theorem, it remains to show that the vectors T_w ($w \in \mathfrak{S}_n$) are linearly independent over R. To this end, we construct an action of H_n on a free R-module of rank $n!$.

Let V be the free R-module with a basis $\{e_w\}_{w \in \mathfrak{S}_n}$ indexed by the elements $w \in \mathfrak{S}_n$. We define $2n - 2$ homomorphisms $\{L_i, R_i : V \to V\}_{i=1}^{n-1}$ as follows. For $i = 1, \ldots, n-1$, set

$$L_i(e_w) = \begin{cases} e_{s_i w} & \text{if } \lambda(s_i w) > \lambda(w) , \\ q e_{s_i w} + z e_w & \text{if } \lambda(s_i w) < \lambda(w) , \end{cases} \tag{4.21}$$

and

$$R_i(e_w) = \begin{cases} e_{w s_i} & \text{if } \lambda(w s_i) > \lambda(w) , \\ q e_{w s_i} + z e_w & \text{if } \lambda(w s_i) < \lambda(w) . \end{cases} \tag{4.22}$$

To complete the proof of Theorem 4.17, we need the following two lemmas.

Lemma 4.18. *We have $L_i R_j = R_j L_i$ for all $i, j = 1, \ldots, n-1$.*

Proof. It suffices to check that $L_i R_j(e_w) = R_j L_i(e_w)$ for all $w \in \mathfrak{S}_n$. We distinguish six cases depending on the lengths of w, $s_i w$, $w s_j$, and $s_i w s_j$. In the following proof we use (4.21) and (4.22) repeatedly.

(i) If $\lambda(w) < \lambda(s_i w) = \lambda(w s_j) < \lambda(s_i w s_j)$, then

$$L_i R_j(e_w) = L_i(e_{w s_j}) = e_{s_i w s_j} = R_j(e_{s_i w}) = R_j L_i(e_w) .$$

(ii) If $\lambda(w) > \lambda(s_i w) = \lambda(w s_j) > \lambda(s_i w s_j)$, then

$$L_i R_j(e_w) = q L_i(e_{w s_j}) + z L_i(e_w)$$
$$= q(q e_{s_i w s_j} + z e_{w s_j}) + z(q e_{s_i w} + z e_w)$$
$$= q(q e_{s_i w s_j} + z e_{s_i w}) + z(q e_{w s_j} + z e_w)$$
$$= q R_j(e_{s_i w}) + z R_j(e_w)$$
$$= R_j L_i(e_w) .$$

(iii) If $\lambda(w) < \lambda(s_i w) = \lambda(w s_j) > \lambda(s_i w s_j)$, then we necessarily have $\lambda(s_i w s_j) = \lambda(w)$. Applying Lemma 4.10, we obtain $s_i w = w s_j$. Then

$$L_i R_j(e_w) = L_i(e_{w s_j}) = q e_{s_i w s_j} + z e_{w s_j}$$
$$= q e_{s_i w s_j} + z e_{s_i w}$$
$$= R_j(e_{s_i w}) = R_j L_i(e_w) .$$

(iv) The case $\lambda(w) > \lambda(s_iw) = \lambda(ws_j) < \lambda(s_iws_j)$ is treated like (iii).

(v) If $\lambda(ws_j) < \lambda(w) < \lambda(s_iw) > \lambda(s_iws_j) = \lambda(w)$, then

$$L_iR_j(e_w) = q\,L_i(e_{ws_j}) + z\,L_i(e_w)$$
$$= qe_{s_iws_j} + ze_{s_iw}$$
$$= R_j(e_{s_iw}) = R_jL_i(e_w).$$

(vi) The case $\lambda(s_iw) < \lambda(w) < \lambda(ws_j) > \lambda(s_iws_j) = \lambda(w)$ is treated like (v).   □

Lemma 4.19. *For any reduced expression $s_{i_1}s_{i_2}\ldots s_{i_r}$ representing $w \in \mathfrak{S}_n$, set $R = R_{i_r}\ldots R_{i_2}R_{i_1} \in \mathrm{End}_R(V)$ and $L = L_{i_1}L_{i_2}\ldots L_{i_r} \in \mathrm{End}_R(V)$. Then*

$$e_w = R(e_1) = L(e_1).$$

Proof. The equality $e_w = R(e_1)$ is proved by induction on $r = \lambda(w)$. For $r = 1$, this equality follows from the definition of $R = R_{i_1}$. For $r \geq 2$, set $w' = s_{i_1}s_{i_2}\ldots s_{i_{r-1}}$ and $R' = R_{i_{r-1}}\ldots R_{i_2}R_{i_1}$ and suppose that $R'(e_1) = e_{w'}$. Since $\lambda(w) = \lambda(w's_{i_r}) > \lambda(w')$, by (4.22),

$$R(e_1) = R_{i_r}(R'(e_1)) = R_{i_r}(e_{w'}) = e_{w's_{i_r}} = e_w.$$

The identity $e_w = L(e_1)$ is proved similarly.   □

Lemma 4.20. *The endomorphisms $L_1, \ldots, L_{n-1}$ of the R-module V satisfy relations (4.16), (4.17), and (4.18) in which T_i is replaced by L_i.*

Proof. (a) If $\lambda(s_iw) > \lambda(w)$, then

$$L_i^2(e_w) = L_i(e_{s_iw}) = qe_w + ze_{s_iw} = z\,L_i(e_w) + qe_w.$$

If $\lambda(s_iw) < \lambda(w)$, then

$$L_i^2(e_w) = L_i(qe_{s_iw} + ze_w) = z\,L_i(e_w) + qe_w.$$

(b) Let $s_{i_1}s_{i_2}\ldots s_{i_r}$ be a reduced word for $w \in \mathfrak{S}_n$ and $R = R_{i_r}\ldots R_{i_2}R_{i_1}$. For i and j such that $|i - j| = 1$,

$$L_iL_jL_i(e_w) = L_iL_jL_iR(e_1) = RL_iL_jL_i(e_1)$$
$$= R(e_{s_is_js_i}) = R(e_{s_js_is_j})$$
$$= RL_jL_iL_j(e_1) = L_jL_iL_jR(e_1)$$
$$= L_jL_iL_j(e_w).$$

We have used Lemma 4.18 for the second and sixth equalities and Lemma 4.19 for the first, third, fifth, and seventh equalities, whereas the fourth equality follows from the relation $s_is_js_i = s_js_is_j$ in $\mathfrak{S}_n$.

(c) The equalities $L_iL_j = L_jL_i$ for $|i - j| \geq 2$ are proved similarly using the relations $s_is_j = s_js_i$.   □

By Lemma 4.20, there is an algebra homomorphism $H_n \to \mathrm{End}_R(V)$ sending T_i to L_i for $i = 1, \ldots, n-1$. In other words, H_n acts on V by

$$T_i\, v = L_i(v)$$

for all $v \in V$ and $i = 1, \ldots, n-1$. Lemma 4.19 implies that $T_w\, e_1 = e_w$ for all $w \in \mathfrak{S}_n$.

We can now prove the linear independence of the elements T_w ($w \in \mathfrak{S}_n$) of H_n. Suppose that there is an additive relation

$$\sum_{w \in \mathfrak{S}_n} a_w T_w = 0,$$

where $a_w \in R$ for all $w \in \mathfrak{S}_n$. Applying both sides to $e_1 \in V$, we obtain

$$0 = \sum_{w \in \mathfrak{S}_n} a_w T_w\, e_1 = \sum_{w \in \mathfrak{S}_n} a_w\, e_w \in V.$$

Since the set $\{e_w\}_{w \in \mathfrak{S}_n}$ is a basis of V, we have $a_w = 0$ for all w. This establishes the linear independence of the elements $T_w \in H_n$, and completes the proof of Theorem 4.17.   □

4.2.4 Consequences of Theorem 4.17

We record two useful consequences of Theorem 4.17. Observe first that there is an algebra homomorphism $\iota : H_n \to H_{n+1}$ sending each generator T_i of H_n ($i = 1, \ldots, n-1$) to the generator T_i of H_{n+1}. The homomorphism ι turns H_{n+1} into a left and right H_n-module by $ha = \iota(h)a$ and $ah = a\iota(h)$ for $h \in H_n$, $a \in H_{n+1}$.

Proposition 4.21. *The homomorphism $\iota : H_n \to H_{n+1}$ is injective. As a left H_n-module, H_{n+1} is free of rank $n + 1$ with basis*

$$\{1, T_n, T_n T_{n-1}, \ldots, T_n T_{n-1} \cdots T_2 T_1\}.$$

Proof. By definition of T_w, we have $\iota(T_w) = T_w$ for all $w \in \mathfrak{S}_n$, where on the right-hand side w is considered as an element of $\mathfrak{S}_{n+1}$. By Theorem 4.17, ι sends a basis of H_n to a subset of a basis of H_{n+1}. Therefore ι is injective.

As a consequence of Corollary 4.4, any $w \in \mathfrak{S}_{n+1} - \mathfrak{S}_n$ can be written uniquely as $w = w' s_n s_{n-1} \cdots s_k$ for some $w' \in \mathfrak{S}_n$ and an integer k such that $1 \le k \le n$. We claim that

$$T_w = T_{w'} T_n T_{n-1} \cdots T_k. \tag{4.23}$$

Indeed, since w', considered as an element of $\mathfrak{S}_{n+1}$, fixes $n + 1$, we have

$$w'(n) < w'(n+1) = n + 1.$$

Therefore, by Lemma 4.7(d), $\lambda(w') < \lambda(w's_n)$. More generally, for each ℓ such that $1 \le \ell \le n$,

$$
\begin{aligned}
(w's_ns_{n-1}\cdots s_{\ell+1})(\ell) &= (w's_ns_{n-1}\cdots s_{\ell+2})(\ell) = \cdots \\
&= (w's_ns_{n-1})(\ell) = (w's_n)(\ell) = w'(\ell) \\
&< n+1 = w'(n+1) \\
&= (w's_n)(n) = (w's_ns_{n-1})(n-1) \\
&= \cdots = (w's_ns_{n-1}\cdots s_{\ell+1})(\ell+1).
\end{aligned}
$$

Therefore, by Lemma 4.7 (d),

$$
\lambda(w's_ns_{n-1}\cdots s_{\ell+1}) < \lambda(w's_ns_{n-1}\cdots s_{\ell+1}s_\ell).
$$

It follows then by induction that if $s_{i_1}\cdots s_{i_r}$ is an arbitrary reduced expression for $w' \in \mathfrak{S}_n$, then $s_{i_1}\cdots s_{i_r}s_ns_{n-1}\cdots s_k$ is a reduced expression for w. Therefore, by definition of T_w and $T_{w'}$,

$$
T_w = T_{i_1}\cdots T_{i_r}T_nT_{n-1}\cdots T_k = T_{w'}T_nT_{n-1}\cdots T_k,
$$

which proves (4.23).

Since the elements T_w with $w \in \mathfrak{S}_{n+1}$ span H_{n+1} as an R-module, (4.23) implies that the elements $\{1, T_n, T_{n-1}T_{n-1}, \ldots, T_nT_{n-1}\cdots T_2T_1\}$ generate H_{n+1} as a left H_n-module. Their linear independence over H_n follows from the linear independence of the elements T_w $(w \in \mathfrak{S}_{n+1})$ over R. $\square$

Proposition 4.22. *For any $n \ge 2$, there is an isomorphism of R-modules*

$$
\varphi : H_n \oplus (H_n \otimes_{H_{n-1}} H_n) \to H_{n+1}
$$

given for any $a \in H_n$ and any finite family $\{b_i, c_i\}_i \subset H_n$ by

$$
\varphi\left(a + \sum_i b_i \otimes c_i\right) = \iota(a) + \sum_i b_iT_nc_i.
$$

Proof. Since H_{n-1} is generated by $T_1, \ldots, T_{n-2}$ and $T_iT_n = T_nT_i$ for $i \le n-2$,

$$
\varphi(bh \otimes c) = bhT_nc = bT_nhc = \varphi(b \otimes hc)
$$

for all $h \in H_{n-1}$, $b, c \in H_n$. This shows that φ is well defined. Clearly, φ is a morphism of left H_n-modules.

By Proposition 4.21, H_n is a free left H_{n-1}-module with basis

$$
\{1, T_{n-1}, T_{n-1}T_{n-2}, \ldots, T_{n-1}T_{n-2}\cdots T_2T_1\}.
$$

Therefore, $H_n \oplus (H_n \otimes_{H_{n-1}} H_n)$ is a free left H_n-module with basis

$$
\{1\} \amalg \{1 \otimes 1, 1 \otimes T_{n-1}, 1 \otimes T_{n-1}T_{n-2}, \ldots, 1 \otimes T_{n-1}T_{n-2}\cdots T_2T_1\}.
$$

The map φ sends this basis to the set

$$\{1\} \amalg \{T_n, T_n T_{n-1}, T_n T_{n-1} T_{n-2}, \ldots, T_n T_{n-1} T_{n-2} \cdots T_2 T_1\},$$

which by Proposition 4.21 is a basis of the left H_n-module H_{n+1}. This implies that φ is an isomorphism. □

Exercise 4.2.1. Show that the assignment $T_i \mapsto -q T_i^{-1}$ $(i = 1, \ldots, n-1)$ defines an algebra automorphism of H_n.

Exercise 4.2.2. Prove that any algebra homomorphism $H_n \to R$ sends all T_i $(i = 1, \ldots, n-1)$ to one and the same root of the polynomial $X^2 - zX - q$.

Exercise 4.2.3 (The Hecke algebra associated to $\mathrm{GL}_n(\mathbf{F}_q)$). Let $\mathbf{F}_q$ be a finite field of cardinality q and $G = \mathrm{GL}_n(\mathbf{F}_q)$. We denote by $\mathbf{C}(G)$ the complex vector space of functions from G to $\mathbf{C}$. For any $g \in G$, define

$$\delta_g \in \mathbf{C}(G)$$

to be the function vanishing everywhere except on g, where its value is 1. Given $f, f' \in \mathbf{C}(G)$, let $f * f'$ be the element of $\mathbf{C}(G)$ given by

$$(f * f')(g) = \sum_{h \in G} f(h) \, f'(h^{-1} g)$$

for all $g \in G$. For $f \in \mathbf{C}(G)$, set

$$\varepsilon(f) = \sum_{g \in G} f(g) \in \mathbf{C}.$$

(a) Show that $\{\delta_g\}_{g \in G}$ is a basis of $\mathbf{C}(G)$, the operation $*$ is associative, and $\varepsilon(f * f') = \varepsilon(f) \varepsilon(f')$ for all $f, f' \in \mathbf{C}(G)$.

(b) Let $B \subset G$ be the subgroup of upper triangular matrices. Define $\mathbf{C}(B\backslash G/B)$ to be the subspace of $\mathbf{C}(G)$ consisting of the functions f such that $f(bg) = f(gb) = f(g)$ for all $g \in G$, $b \in B$. Show that $\mathbf{C}(B\backslash G/B)$ is closed under $*$ and the following function is a unit (with respect to $*$):

$$\delta_0 = \frac{1}{\mathrm{card}(B)} \sum_{g \in B} \delta_g \in \mathbf{C}(B\backslash G/B).$$

(c) For any permutation $w \in \mathfrak{S}_n$, consider the set BwB of all elements of G of the form bwb', where $b, b' \in B$ and w is identified with the corresponding permutation matrix. Define δ_w to be the function on G whose value is $1/\mathrm{card}(B)$ on BwB and 0 elsewhere. Show that $\{\delta_w\}_{w \in \mathfrak{S}_n}$ is a basis of $\mathbf{C}(B\backslash G/B)$. Hint: Use the *Bruhat decomposition*

$$G = \coprod_{w \in \mathfrak{S}_n} BwB.$$

(d) Prove that for all $w, w' \in \mathfrak{S}_n$ and $g \in G$,

$$\left(\delta_w * \delta_{w'}\right)(g) = \frac{1}{\text{card}(B)^2} \, \text{card}\left(BwB \cap gB(w')^{-1}B\right).$$

(e) Compute $\text{card}(Bs_i B)$ for each simple transposition $s_i \in \mathfrak{S}_n$. Show that the function $\delta_{s_i} * \delta_{s_i}$ is zero outside $B \cup Bs_i B$, and for any $g \in B$,

$$\left(\delta_{s_i} * \delta_{s_i}\right)(g) = q \, \delta_0 \,.$$

Using $\varepsilon : \mathbf{C}(B\backslash G/B) \to \mathbf{C}$, deduce that

$$\delta_{s_i} * \delta_{s_i} = (q-1) \, \delta_{s_i} + q \, \delta_0 \,.$$

(f) Prove that for all s_i and $w \in \mathfrak{S}_n$ such that $\lambda(s_i w) > \lambda(w)$,

$$\delta_{s_i} * \delta_w = \delta_{s_i w} \,.$$

(g) Conclude that the algebra $\mathbf{C}(B\backslash G/B)$ is isomorphic to the Iwahori–Hecke algebra $H_n^{\mathbf{C}}(q)$.

4.3 The Ocneanu traces

As in the previous section, we fix a commutative ring R together with two elements $q, z \in R$. We now assume that both q and z are invertible in R. The aim of this section is to construct for all $n \geq 1$ a trace $\tau_n : H_n \to R$ on $H_n = H_n^R(q, z)$. This trace will be instrumental in the construction of a two-variable polynomial invariant of links in the next section.

We proceed by induction on n. For $n = 1$, we define $\tau_1 : H_1 = R \to R$ to be the identity map. For $n = 2$, we define $\tau_2 : H_2 \to R$ on the basis $\{1, T_1\}$ by

$$\tau_2(1) = \frac{1-q}{z} \quad \text{and} \quad \tau_2(T_1) = 1 \,. \tag{4.24}$$

Suppose that $\tau_n : H_n \to R$ is defined for some $n \geq 2$. We define the trace $\tau_{n+1} : H_{n+1} \to R$ using the isomorphism $\varphi : H_n \oplus (H_n \otimes_{H_{n-1}} H_n) \to H_{n+1}$ of Proposition 4.22 as follows. Set

$$\tau_{n+1}(\varphi(a)) = \frac{1-q}{z} \, \tau_n(a)$$

and

$$\tau_{n+1}(\varphi(b \otimes c)) = \tau_n(bc)$$

for all $a, b, c \in H_n$. Induction on n shows that $\tau_n : H_n \to R$ is R-linear, that is, $\tau_n(ra) = r\tau_n(a)$ for all $r \in R$, $a \in H_n$. The linear form τ_n is called the *Ocneanu trace* on H_n.

Proposition 4.23. *For all $n \geq 1$ and all $a, b \in H_n$,*

(i) $\tau_n(ab) = \tau_n(ba)$,
(ii) $\tau_{n+1}(T_n a) = \tau_{n+1}(T_n^{-1} a) = \tau_n(a)$.

Proof. (i) We prove the relation $\tau_n(ab) = \tau_n(ba)$ by induction on n. It holds for $n = 1$ and $n = 2$ because H_1 and H_2 are commutative. We now suppose that this relation holds for τ_n and prove it for τ_{n+1}. Since H_{n+1} is generated by $T_1, \ldots, T_n$ and φ is onto, it is enough to show that

$$\tau_{n+1}(\omega T_i) = \tau_{n+1}(T_i \omega)$$

for all ω in the image of φ and all $i = 1, \ldots, n$.

(a) If $\omega = \varphi(a)$ for some $a \in H_n$, then by definition of τ_{n+1},

$$\tau_{n+1}(\omega T_i) = \begin{cases} \frac{1-q}{z} \tau_n(a T_i) & \text{if } i < n, \\ \tau_n(a) & \text{if } i = n, \end{cases}$$

and

$$\tau_{n+1}(T_i \omega) = \begin{cases} \frac{1-q}{z} \tau_n(T_i a) & \text{if } i < n, \\ \tau_n(a) & \text{if } i = n. \end{cases}$$

The relation $\tau_{n+1}(\omega T_i) = \tau_{n+1}(T_i \omega)$ follows from the induction hypothesis.

(b) Suppose that $\omega = \varphi(a \otimes b) = a T_n b$ for some $a, b \in H_n$. If $i < n$, then

$$\begin{aligned}
\tau_{n+1}(\omega T_i) = \tau_{n+1}(a T_n b T_i) &= \tau_n(ab T_i) \\
&= \tau_n(T_i ab) = \tau_{n+1}(T_i a T_n b) \\
&= \tau_{n+1}(T_i \omega),
\end{aligned}$$

where the third equality follows from the induction hypothesis.

When $i = n$, we have to check the equality

$$\tau_{n+1}(a T_n b T_n) = \tau_{n+1}(T_n a T_n b).$$

There are four cases to consider.

(b$_1$) If a and b belong to H_{n-1}, then they commute with T_n and the relation is obvious.

(b$_2$) Let $a \in H_{n-1}$ and $b = b' T_{n-1} b''$, where $b', b'' \in H_{n-1}$. Observe that $a, b',$ and b'' commute with T_n. We have

$$\begin{aligned}
\tau_{n+1}(a T_n b T_n) &= \tau_{n+1}(a T_n b' T_{n-1} b'' T_n) \\
&= \tau_{n+1}(ab' T_n T_{n-1} T_n b'') \\
&= \tau_{n+1}(ab' T_{n-1} T_n T_{n-1} b'') \\
&= \tau_n(ab' T_{n-1}^2 b'') \\
&= z\tau_n(ab' T_{n-1} b'') + q\tau_n(ab' b'') \\
&= z\tau_n(ab) + q\frac{1-q}{z} \tau_{n-1}(ab' b'').
\end{aligned}$$

On the other hand,

$$
\begin{aligned}
\tau_{n+1}(T_n a T_n b) &= \tau_{n+1}(T_n^2 ab) \\
&= z\tau_{n+1}(T_n ab) + q\tau_{n+1}(ab) \\
&= z\tau_n(ab) + q\frac{1-q}{z}\tau_n(ab'T_{n-1}b'') \\
&= z\tau_n(ab) + q\frac{1-q}{z}\tau_{n-1}(ab'b''),
\end{aligned}
$$

which is the same expression.

(b$_3$) The case $b \in H_{n-1}$ and $a = a'T_{n-1}a''$ with $a', a'' \in H_{n-1}$ is treated similarly.

(b$_4$) Suppose that $a = a'T_{n-1}a''$ and $b = b'T_{n-1}b''$ with $a', a'', b', b'' \in H_{n-1}$. Then

$$
\begin{aligned}
\tau_{n+1}(aT_n bT_n) &= \tau_{n+1}(aT_n b'T_{n-1}b''T_n) \\
&= \tau_{n+1}(ab'T_n T_{n-1}T_n b'') \\
&= \tau_{n+1}(ab'T_{n-1}T_n T_{n-1}b'') \\
&= \tau_n(ab'T_{n-1}^2 b'') \\
&= z\tau_n(ab'T_{n-1}b'') + q\tau_n(ab'b'') \\
&= z\tau_n(ab) + q\tau_n(a'T_{n-1}a''b'b'') \\
&= z\tau_n(ab) + q\tau_{n-1}(a'a''b'b'').
\end{aligned}
$$

On the other hand,

$$
\begin{aligned}
\tau_{n+1}(T_n aT_n b) &= \tau_{n+1}(T_n a'T_{n-1}a''T_n b) \\
&= \tau_{n+1}(a'T_n T_{n-1}T_n a''b) \\
&= \tau_{n+1}(a'T_{n-1}T_n T_{n-1}a''b) \\
&= \tau_n(a'T_{n-1}^2 a''b) \\
&= z\tau_n(a'T_{n-1}a''b) + q\tau_n(a'a''b) \\
&= z\tau_n(ab) + q\tau_n(a'a''b'T_{n-1}b'') \\
&= z\tau_n(ab) + q\tau_{n-1}(a'a''b'b''),
\end{aligned}
$$

which proves the desired relation.

(ii) By definition of τ_{n+1},

$$
\tau_{n+1}(T_n a) = \tau_{n+1}(\varphi(1 \otimes a)) = \tau_n(a)
$$

for all $a \in H_n$. Since $T_n^{-1} = q^{-1}T_n - q^{-1}z1$, we obtain

$$
\begin{aligned}
\tau_{n+1}(T_n^{-1}a) &= q^{-1}\tau_{n+1}(T_n a) - q^{-1}z\tau_{n+1}(a) \\
&= q^{-1}\tau_n(a) - \frac{1-q}{z}q^{-1}z\tau_n(a) \\
&= \tau_n(a).
\end{aligned}
$$

$\square$

Exercise 4.3.1. Show that on the basis

$$\{1, T_1, T_2, T_1 T_2, T_2 T_1, T_1 T_2 T_1\}$$

of H_3 the trace $\tau_3 : H_3 \to R$ is computed by

$$\tau_3(1) = \frac{(1-q)^2}{z^2}, \quad \tau_3(T_1) = \tau_3(T_2) = \frac{1-q}{z},$$

$$\tau_3(T_1 T_2) = \tau_3(T_2 T_1) = 1, \quad \tau_3(T_1 T_2 T_1) = z + \frac{q(1-q)}{z}.$$

4.4 The Jones–Conway polynomial

We now use the theory of Iwahori–Hecke algebras presented above to construct a two-parameter polynomial invariant of oriented links in $\mathbf{R}^3$. Recall from Section 2.5.2 the notion of a Markov function on the braid groups. We build an explicit Markov function as follows. Let R be a commutative ring with distinguished invertible elements q, z. Let $H_n = H_n^R(q, z)$ be the corresponding Iwahori–Hecke algebra with $n \geq 1$ and let $H_n^\times$ be the group of invertible elements of H_n. Consider the group homomorphism $\omega_n : B_n \to H_n^\times$ sending σ_i to T_i for $i = 1, \dots, n-1$. Composing ω_n with the Ocneanu trace $\tau_n : H_n \to R$ constructed in Section 4.3, we obtain a mapping $\tau_n \circ \omega_n : B_n \to R$. The following is an immediate consequence of Proposition 4.23.

Proposition 4.24. *The family* $\{\tau_n \circ \omega_n : B_n \to R\}_{n \geq 1}$ *is a Markov function.*

We now state the main theorem of this section. In the statement we use the notion of a Conway triple of links; see Section 3.3.

Theorem 4.25. *For any oriented link* $L \subset \mathbf{R}^3$ *and any braid* $\beta \in B_n$ *whose closure is isotopic to* L, *the element*

$$I_L(q, z) = \tau_n(\omega_n(\beta)) \in R$$

depends only on the isotopy class of L. *For the trivial knot* O,

$$I_O(q, z) = 1.$$

For any Conway triple (L_+, L_-, L_0) *of oriented links in* $\mathbf{R}^3$,

$$I_{L_+}(q, z) - q I_{L_-}(q, z) = z I_{L_0}(q, z).$$

Proof. The first assertion follows from the theory of Markov functions in Section 2.5.2 and Proposition 4.24. The trivial knot can be realized as the closure of the trivial braid $1 \in B_1 = \{1\}$. Therefore,

$$I_O(q, z) = \tau_1(\omega_1(1)) = \tau_1(1) = 1.$$

Let us check that $A = I_{L_+}(q, z) - qI_{L_-}(q, z) - zI_{L_0}(q, z)$ is zero for any Conway triple of oriented links (L_+, L_-, L_0). As observed in Section 3.4.2, such a triple can be isotopped to a Conway triple (L'_+, L'_-, L'_0), where

$$L'_+ = \widehat{\alpha\sigma_i\beta}, \quad L'_- = \widehat{\alpha\sigma_i^{-1}\beta}, \quad L'_0 = \widehat{\alpha\beta}$$

for some $\alpha, \beta \in B_n$ and $1 \le i \le n - 1$. Using (4.18), we obtain

$$A = \tau_n\big(\omega_n(\alpha\sigma_i\beta)\big) - q\tau_n\big(\omega_n(\alpha\sigma_i^{-1}\beta)\big) - z\tau_n\big(\omega_n(\alpha\beta)\big)$$

$$= \tau_n\big(\omega_n(\alpha)T_i\,\omega_n(\beta)\big) - q\tau_n\big(\omega_n(\alpha)T_i^{-1}\omega_n(\beta)\big) - z\tau_n\big(\omega_n(\alpha)\,\omega_n(\beta)\big)$$

$$= \tau_n\Big(\omega_n(\alpha)\big(T_i - qT_i^{-1} - z1\big)\omega_n(\beta)\Big) = 0. \qquad \square$$

Corollary 4.26. *There is an isotopy invariant $L \mapsto P_L(x, y)$ of oriented links in $\mathbf{R}^3$ with values in $\mathbf{Z}[x, x^{-1}, y, y^{-1}]$ such that its value on the trivial knot O is 1 and for any Conway triple of oriented links (L_+, L_-, L_0),*

$$xP_{L_+}(x, y) - x^{-1}P_{L_-}(x, y) = yP_{L_0}(x, y)$$

(the skein relation). Such a link invariant $L \mapsto P_L(x, y)$ is unique.

Proof. Let $R = \mathbf{Z}[x, x^{-1}, y, y^{-1}]$ be the ring of Laurent polynomials in two variables x, y with integer coefficients. Set $P_L(x, y) = I_L(q, z) \in R$, where $I_L(q, z)$ is the link invariant provided by Theorem 4.25 for $q = x^{-2}$ and $z = x^{-1}y$. Clearly, $P_O(x, y) = I_O(q, z) = 1$. If (L_+, L_-, L_0) is a Conway triple, then by Theorem 4.25,

$$xP_{L_+}(x, y) - x^{-1}P_{L_-}(x, y) - yP_{L_0}(x, y)$$

$$= x\Big(P_{L_+}(x, y) - x^{-2}P_{L_-}(x, y) - x^{-1}yP_{L_0}(x, y)\Big)$$

$$= x\Big(I_{L_+}(q, z) - qI_{L_-}(q, z) - zI_{L_0}(q, z)\Big) = 0.$$

The uniqueness of $P_L(x, y)$ is proved in the same way as the uniqueness of the Alexander–Conway polynomial in Theorem 3.13. $\qquad \square$

We call $P_L(x, y)$ the *Jones–Conway polynomial* of L. In the literature it is also called the *HOMFLY polynomial*, the *HOMFLY-PT polynomial*, or the *two-variable Jones polynomial*.

Observe that $P_L(x, y)$ extends the Alexander–Conway polynomial $\nabla(L)$ introduced in Section 3.4.2. Namely, $\nabla(L) = P_L(1, s^{-1} - s)$. This follows directly from the uniqueness in Theorem 3.13.

Setting $x = t^{-1}$ and $y = t^{1/2} - t^{-1/2}$ in $P_L(x, y)$, one obtains the one-variable *Jones polynomial*

$$V_L(t) = P_L(t^{-1}, t^{1/2} - t^{-1/2}) \in \mathbf{Z}[t^{1/2}, t^{-1/2}]$$

satisfying the skein relation

$$t^{-1}V_{L_+}(t) - tV_{L_-}(t) = (t^{1/2} - t^{-1/2})\, V_{L_0}(t).$$

Exercise 4.4.1. Define the mirror image $\widetilde{L}$ of an oriented link L in $\mathbf{R}^3$ as the image of L under the reflection in a plane in $\mathbf{R}^3$. Prove that

$$P_{\widetilde{L}}(x,y) = P_L(x^{-1}, -y).$$

Exercise 4.4.2. Compute the polynomial P_L for the knots and links shown in Figure 2.1 and endowed with all possible orientations.

4.5 Semisimple algebras and modules

This section is a brief exposition of the theory of finite-dimensional semisimple algebras over a field. Fix a field K. By an *algebra* we mean an associative K-algebra with unit $1 \neq 0$. An algebra is *finite-dimensional* if it is finite-dimensional as a vector space over K.

4.5.1 Semisimple modules

Let A be an algebra. By an A-*module*, we mean a left A-module, that is, a K-vector space M together with a K-bilinear map $A \times M \to M$, $(a, m) \mapsto am$ such that $a(bm) = (ab)m$ and $1m = m$ for all a, $b \in A$, and $m \in M$. The map $a \mapsto (m \mapsto am)$ $(a \in A, m \in M)$ defines an algebra homomorphism $A \to \mathrm{End}_K(M)$ with values in the algebra of K-linear endomorphisms of M. Conversely, any algebra homomorphism $\chi : A \to \mathrm{End}_K(M)$ gives rise to an A-module structure on M by $am = \chi(a)(m)$ for $a \in A$ and $m \in M$.

By a *finite-dimensional* A-module we mean an A-module that is finite-dimensional as a vector space over K.

A *homomorphism of* A-*modules* $f : M \to M'$ is a K-linear map such that $f(am) = af(m)$ for all $a \in A$ and $m \in M$. We write $\mathrm{Hom}_A(M, M')$ for the vector space of all homomorphisms of A-modules $M \to M'$. We also set $\mathrm{End}_A(M) = \mathrm{Hom}_A(M, M)$.

If M' is a linear subspace of an A-module M such that $am' \in M'$ for all $a \in A$ and $m' \in M'$, then we say that M' is a A-*submodule* or, for short, a submodule of M. In this case, the embedding $M' \hookrightarrow M$ is a homomorphism of A-modules.

Definition 4.27. *(a) An A-module M is simple if M has no A-submodules except 0 and M.*

(b) An A-module is semisimple if it is isomorphic to a direct sum of a finite number of simple A-modules.

(c) An A-module M is completely reducible if for any A-submodule M' of M there is an A-submodule M'' such that $M = M' \oplus M''$.

Note that a simple A-module is semisimple, and if an A-module M is completely reducible, then any short exact sequence of A-modules

$$0 \to M' \to M \to M'' \to 0$$

splits, i.e., $M \cong M' \oplus M''$.

Proposition 4.28. *Let M be a finite-dimensional A-module. The following assertions are equivalent.*

(i) M is semisimple.
(ii) M is completely reducible.
(iii) $M = \sum_{i \in I} M_i$ is a sum of simple submodules M_i.

Proof. We first prove the implication (ii) $\Rightarrow$ (iii). Assume that M is nonzero and completely reducible. Since M is finite-dimensional over K, it must have nonzero submodules of minimal dimension as vector spaces over K; such submodules are necessarily simple. Consider the sum $M' \subset M$ of all simple submodules of M. This is a nonzero submodule of M. We are done if $M' = M$. If not, since M is completely reducible, there is a nonzero submodule $M'' \subset M$ such that $M = M' \oplus M''$. A nonzero submodule of M'' of minimal dimension is a simple submodule of M that is not in M'. This contradicts the definition of M'. Therefore, $M' = M$.

We next prove the implication (iii) $\Rightarrow$ (i). Suppose that $M = \sum_{i \in I} M_i$ is a sum of simple submodules M_i. Let $I' \subset I$ be a maximal subset such that $M_i \neq 0$ for $i \in I'$ and the sum $\sum_{i \in I'} M_i$ is direct. Such a subset I' exists and is finite because M is finite-dimensional. Let $M' = \sum_{i \in I'} M_i$. We claim that $M' = M$, which implies that M is a direct sum of a finite number of simple submodules. To prove the claim, it suffices to check that $M_k \subset M'$ for any $k \in I - I'$. Clearly, $M_k \cap M'$ is a submodule of M_k. Since M_k is simple, either $M_k \cap M' = 0$ or $M_k \cap M' = M_k$. If $M_k \cap M' = 0$, then the sum $\sum_{i \in I' \cup \{k\}} M_i$ is direct. This contradicts the maximality of I'. Therefore, $M_k \cap M' = M_k$, which implies that $M_k \subset M'$.

We finally prove the implication (i) $\Rightarrow$ (ii). Suppose that $M = \bigoplus_{i \in I} M_i$ is a direct sum of simple submodules, where I is a finite indexing set. Let M' be a submodule of M. Consider a maximal subset $I' \subset I$ such that the sum $M' + \sum_{i \in I'} M_i$ is direct. Reasoning analogous to that in the previous paragraph shows that $M' + \sum_{i \in I'} M_i = M$. Set $M'' = \sum_{i \in I'} M_i$. Then $M' \oplus M'' = M$, which proves that M is completely reducible. $\square$

Proposition 4.29. *Let M be a finite-dimensional semisimple A-module. Any A-submodule and any quotient A-module of M is semisimple.*

Proof. Let M_0 be a submodule of M. Let M_0' be the sum of all simple submodules of M_0. Since by Proposition 4.28, M is completely reducible, $M = M_0' \oplus M''$ for some submodule M'' of M. Together with $M_0' \subset M_0$, this implies that $M_0 = M_0' \oplus (M_0 \cap M'')$. If $M_0 \cap M'' \neq 0$, then this module contains a nonzero simple submodule, which is then contained in M_0'. This is impossible. Therefore, $M_0 \cap M'' = 0$ and $M_0 = M_0'$ is a sum of simple submodules. Using Proposition 4.28, we conclude that M_0 is semisimple.

Consider the quotient of M by a submodule M'. By Proposition 4.29, there is a submodule $M'' \subset M$ such that $M = M' \oplus M''$. By the previous paragraph, M'' is semisimple; hence so is $M/M' \cong M''$. $\square$

Recall that a *division ring* is a ring in which each nonzero element is invertible. A left module over a division ring D is called a left D-vector space. Any left D-vector space V has a basis, and two bases of V have the same cardinality, so that the concept of the dimension $\dim_D V$ of V makes sense (these results can be proved in the same way as the corresponding ones for vector spaces over a field).

The following proposition is called *Schur's lemma*.

Proposition 4.30. *(a) Let M and M' be simple A-modules. If M and M' are not isomorphic as A-modules, then $\operatorname{Hom}_A(M, M') = 0$.*

(b) The ring $\operatorname{End}_A(M)$ of A-module endomorphisms of a nonzero simple A-module M is a division ring.

(c) If the ground field K is algebraically closed and M is a nonzero finite-dimensional simple A-module, then $\dim_K \operatorname{End}_A(M) = 1$.

Proof. (a) Let $f \in \operatorname{Hom}_A(M, M')$. The kernel $\operatorname{Ker}(f)$ of f is a submodule of M. Since M is simple, $\operatorname{Ker}(f) = M$ or $\operatorname{Ker}(f) = 0$. In the first case, $f = 0$. In the second case, f is injective. Its image $f(M)$ is a submodule of M'. By the simplicity of the latter, $f(M) = M'$ or $f(M) = 0$. If $f(M) = M'$, then f is an isomorphism $M \to M'$. Thus, $f = 0$ or f is an isomorphism. The latter contradicts the assumptions. Hence, $f = 0$.

(b) By the proof of (a), any nonzero $f \in \operatorname{End}_A(M)$ is bijective. It is easy to check that the inverse f^{-1} of f is a homomorphism of A-modules. Hence, f is invertible in $\operatorname{End}_A(M)$.

(c) For any scalar $\lambda \in K$, the endomorphism $m \mapsto \lambda m$ lies in $\operatorname{End}_A(M)$. Conversely, let $f \in \operatorname{End}_A(M)$. Since K is algebraically closed and M is a finite-dimensional K-vector space, f has a nonzero eigenspace for some eigenvalue $\lambda \in K$. The eigenspace $\operatorname{Ker}(f - \lambda \operatorname{id}_M)$, being a nonzero submodule of the simple module M, must be equal to M. Hence, $f = \lambda \operatorname{id}_M$. In conclusion, $\operatorname{End}_A(M) \cong K$. $\square$

Corollary 4.31. *If K is algebraically closed and M, M' are isomorphic nonzero finite-dimensional simple A-modules, then $\dim_K \operatorname{Hom}_A(M, M') = 1$.*

Let Λ be the set of isomorphism classes of nonzero finite-dimensional simple A-modules. For each $\lambda \in \Lambda$, fix a simple A-module V_λ in the isomorphism class λ. For any integer $d \geq 1$, we denote by V_λ^d the direct sum of d copies of V_λ. We agree that $V_\lambda^d = 0$ if $d = 0$. The following proposition, known as the *Krull–Schmidt theorem*, asserts that the decomposition of a semisimple module into a direct sum of simple modules is unique.

Proposition 4.32. *If for some families $\{d(\lambda)\}_{\lambda \in \Lambda}$, $\{e(\lambda)\}_{\lambda \in \Lambda}$ of nonnegative integers, there is an A-module isomorphism*

$$\bigoplus_{\lambda \in \Lambda} V_\lambda^{d(\lambda)} \cong \bigoplus_{\lambda \in \Lambda} V_\lambda^{e(\lambda)},$$

then $d(\lambda) = e(\lambda)$ for all $\lambda \in \Lambda$.

Proof. Pick $\lambda_0 \in \Lambda$ and set $D = \mathrm{End}_A(V_{\lambda_0})$. By Proposition 4.30 (a),

$$\mathrm{Hom}\Big(\bigoplus_{\lambda \in \Lambda} V_\lambda^{d(\lambda)}, V_{\lambda_0}\Big) \cong \prod_{\lambda \in \Lambda} \mathrm{Hom}_A(V_\lambda^{d(\lambda)}, V_{\lambda_0})$$

$$\cong \prod_{\lambda \in \Lambda} \mathrm{Hom}_A(V_\lambda, V_{\lambda_0})^{d(\lambda)}$$

$$\cong \mathrm{Hom}_A(V_{\lambda_0}, V_{\lambda_0})^{d(\lambda_0)}$$

$$\cong D^{d(\lambda_0)}.$$

Similarly,

$$\mathrm{Hom}\Big(\bigoplus_{\lambda \in \Lambda} V_\lambda^{e(\lambda)}, V_{\lambda_0}\Big) \cong D^{e(\lambda_0)}.$$

The assumptions imply that $D^{d(\lambda_0)} \cong D^{e(\lambda_0)}$. By Proposition 4.30 (b), D is a division ring. Therefore, taking the dimensions over D, we obtain

$$d(\lambda_0) = \dim_D D^{d(\lambda_0)} = \dim_D D^{e(\lambda_0)} = e(\lambda_0). \qquad \square$$

4.5.2 Simple algebras

Definition 4.33. *An algebra A is simple if A is finite-dimensional and the only two-sided ideals of A are 0 and A.*

We give a typical example of a simple algebra.

Proposition 4.34. *Let V be a finite-dimensional left vector space over a division ring D. Then the algebra $\mathrm{End}_D(V)$ is simple.*

Proof. Pick a basis $\{v_1, \ldots, v_d\}$ of V. We have $V = Dv_1 \oplus \cdots \oplus Dv_d$. For $i, j \in \{1, \ldots, d\}$, define $f_{i,j} \in A = \mathrm{End}_D(V)$ by $f_{i,j}(v_k) = \delta_{j,k} v_i$ for all $k = 1, \ldots, d$ (here $\delta_{j,k}$ is the Kronecker symbol whose value is 1 if $j = k$, and 0 otherwise). One checks easily that $\{f_{i,j}\}_{i,j \in \{1,\ldots,d\}}$ is a basis of A considered as a vector space over D, and that $f_{i,j} \circ f_{k,\ell} = \delta_{j,k} f_{i,\ell}$ for all i, j, k, ℓ.

Let I be a nonzero two-sided ideal of A and let $f \in I$ be a nonzero element. Write $f = \sum_{i,j} a_{i,j} f_{i,j}$, where $a_{i,j} \in D$ for all $i, j \in \{1, \ldots, d\}$. Suppose that $a_{k,\ell} \neq 0$ for some $k, \ell \in \{1, \ldots, d\}$. Then

$$f_{k,k} \circ f \circ f_{\ell,\ell} = \sum_{i,j=1}^{d} a_{i,j} f_{k,k} \circ f_{i,j} \circ f_{\ell,\ell} = a_{k,\ell} f_{k,\ell}$$

belongs to I. It follows that $f_{k,\ell} \in I$. The relation $f_{i,j} = f_{i,k} \circ f_{k,\ell} \circ f_{\ell,j}$ implies that $f_{i,j} \in I$ for all $i, j = 1, \ldots, d$. Consequently, $I = A$. $\qquad \square$

The following is a converse to the previous proposition. It is a version of *Wedderburn's theorem.*

Proposition 4.35. *For any simple algebra A, there is a division ring D and a finite-dimensional D-vector space V such that $A \cong \mathrm{End}_D(V)$.*

Proof. Pick a left ideal $V \subset A$ of A of minimal positive dimension over K (possibly, $V = A$). The ideal V is an A-module, and by the minimality condition, it is a simple module. By Proposition 4.30 (b), $D = \mathrm{End}_A(V)$ is a division ring. We conclude using the next lemma with $I = V$. $\qquad\square$

Lemma 4.36. *Let A be an algebra having no two-sided ideals besides 0 and A. For any nonzero left ideal $I \subset A$, there is an algebra isomorphism $A \cong \mathrm{End}_D(I)$, where $D = \mathrm{End}_A(I)$ and I is viewed as a (left) D-module via the action of D on I defined by $(f, x) \mapsto f(x)$ for all $f \in D$ and $x \in I$.*

In this lemma we impose no conditions on the dimension of A, which may be finite or infinite.

Proof. For $a \in A$, define L_a (resp. R_a) to be the left (resp. the right) multiplication by a in A. By definition,

$$\mathrm{L}_a(b) = ab \quad \text{and} \quad \mathrm{R}_a(b) = ba \qquad (4.25)$$

for all $b \in A$. We have

$$\mathrm{L}_a \circ \mathrm{L}_b = \mathrm{L}_{ab} \quad \text{and} \quad \mathrm{R}_a \circ \mathrm{R}_b = \mathrm{R}_{ba} \qquad (4.26)$$

for all $a, b \in A$. Since I is a left ideal of A, we have $\mathrm{L}_a(I) \subset I$ for all $a \in A$, which implies that $\mathrm{L}_a \in \mathrm{End}_K(I)$. Since

$$\mathrm{L}_a(f(x)) = af(x) = f(ax) = f(\mathrm{L}_a(x))$$

for all $f \in D$ and $x \in I$, the endomorphism L_a belongs to $\mathrm{End}_D(I)$.

Let $\mathrm{L} : A \to \mathrm{End}_D(I)$ be the map sending $a \in A$ to $\mathrm{L}_a \in \mathrm{End}_D(I)$. Since $\mathrm{L}_a \circ \mathrm{L}_b = \mathrm{L}_{ab}$ for all $a, b \in A$ and $\mathrm{L}_1 = \mathrm{id}_I$, the map L is an algebra homomorphism. Let us show that L is an isomorphism. The kernel of L is a two-sided ideal of A. Since $\mathrm{L} \neq 0$, the assumptions of the lemma imply that the kernel of L must be zero. This proves the injectivity of L.

The proof of the surjectivity of L is a little bit more complicated; it goes as follows. If $x \in I$, then $\mathrm{R}_x(I) \subset I$. We claim that $\mathrm{R}_x \in D = \mathrm{End}_A(I)$. Indeed, for all $a \in A$, $x, y \in I$,

$$a\,\mathrm{R}_x(y) = a(yx) = (ay)x = \mathrm{R}_x(ay).$$

If $u \in \mathrm{End}_D(I)$, then for all $x, y \in I$,

$$u(yx) = u(\mathrm{R}_x(y)) = \mathrm{R}_x u(y) = u(y)x.$$

In particular, for any $a \in A$, $x, y \in I$,

$$u(yax) = u(y(ax)) = u(y)ax.$$

In other words, for all $a \in A$ and $y \in I$,

$$u \circ \mathrm{L}_{ya} = \mathrm{L}_{u(y)a} . \tag{4.27}$$

Now, IA is a nonzero two-sided ideal of A. By the assumptions of the lemma, $IA = A$. Equation (4.27) then implies that $u \circ \mathrm{L}_b \in \mathrm{L}(A) \subset \mathrm{End}_D(I)$ for all $u \in \mathrm{End}_D(I)$ and $b \in A$. This shows that the image of L is a left ideal of $\mathrm{End}_D(I)$. Since $\mathrm{id}_I = \mathrm{L}_1$ is in the image, the latter is equal to the whole algebra $\mathrm{End}_D(I)$, and the map L is surjective.   □

4.5.3 Modules over a simple algebra

We now prove that any simple algebra has a unique (up to isomorphism) nonzero simple module.

Proposition 4.37. *Let A be a simple algebra. Any nonzero left ideal I of A of minimal dimension is a simple A-module, and any nonzero simple A-module is isomorphic to I.*

Proof. Let I be a nonzero left ideal of A of minimal dimension. Any A-sub-module I' of I is a left ideal of A. By the minimality hypothesis on I, we must have $I' = 0$ or $I' = I$. Therefore, I is a simple A-module; it is finite-dimensional, since A is finite-dimensional.

Let M be a nonzero simple A-module. Set

$$I_0 = \{a \in A \,|\, am = 0 \ \text{for all} \ m \in M\} .$$

It is easy to check that I_0 is a two-sided ideal of A and $I_0 \neq A$, since $1 \in A$ does not annihilate M. Since A is simple, $I_0 = 0$. We have $IM \neq 0$; otherwise, we would have $I \subset I_0 = 0$. Therefore, there is $m \in M$ such that $Im \neq 0$. Consider the homomorphism of A-modules $I \to M, x \mapsto xm$. This homomorphism is nonzero and connects two simple A-modules. By Proposition 4.30 (a), the homomorphism $I \to M$ is an isomorphism.   □

Corollary 4.38. *Every simple algebra has a nonzero simple module. It is finite-dimensional and unique up to isomorphism.*

Proof. Every finite-dimensional algebra has a nonzero left ideal of minimal dimension. Therefore both claims follow directly from Proposition 4.37.   □

Proposition 4.39. *Any finite-dimensional module over a simple algebra is semisimple.*

Proof. Let A be a simple algebra. Consider A as a (left) module over itself. Let us first prove that this A-module is semisimple.

By Proposition 4.35, we may assume that $A = \mathrm{End}_D(V)$ for some division ring D and some finite-dimensional D-vector space V. Pick a basis $\{v_1, \ldots, v_d\}$ of V over D. The map $A = \mathrm{End}_D(V) \to V^d$ defined by

$$f \in A \mapsto (f(v_1), \ldots, f(v_d)) \in V^d$$

is a homomorphism of A-modules. This homomorphism is clearly injective. Since $\dim_D A = d^2 = \dim_D V^d$, it is an isomorphism. To establish that the A-module $A \cong V^d$ is a finite direct sum of simple A-modules, it suffices to check that V is a simple A-module. Let $V' \subset V$ be a nonzero A-submodule. Take a nonzero vector $v' \in V'$. For each $i = 1, \ldots, d$, we can construct $f_i \in A$ such that $f_i(v') = v_i$. It follows that $Av' = V$, hence $V' = V$.

If M is an arbitrary finite-dimensional A-module, then M necessarily has a finite number of generators over A; therefore, M is a quotient of the free A-module A^r of finite rank r (this is the direct sum of r copies of A). We have proved above that the A-module A is semisimple. Therefore, so is A^r. The semisimplicity of M follows from Proposition 4.29. □

We now state an important consequence of these propositions. Let M be a simple module over a simple algebra A. By Corollary 4.38, M is finite-dimensional. We know from Proposition 4.30 (b) that $D = \mathrm{End}_A(M)$ is a division ring. Since M is finite-dimensional, so is D. The dimensions of A, M, and D over the ground field K are related as follows.

Corollary 4.40. *With the notation above, $A \cong \mathrm{End}_D(M)$ and*

$$\dim_K A = \frac{(\dim_K M)^2}{\dim_K D}.$$

Proof. The division ring $D = \mathrm{End}_A(M)$ acts on M, turning M into a left D-vector space of finite dimension over $K \subset D$. Such a vector space has a finite basis over D of cardinality, say d. Lemma 4.36 and Proposition 4.37 imply that $A \cong \mathrm{End}_D(M)$ is isomorphic to the matrix algebra $M_d(D)$. Hence,

$$\dim_K A = \dim_K M_d(D) = \dim_D M_d(D) \dim_K D = d^2 \dim_K D.$$

We conclude by observing that $\dim_K M = \dim_D M \dim_K D = d \dim_K D$. □

4.5.4 The radical of a finite-dimensional algebra

Let A be a finite-dimensional algebra over K. Choosing a basis of A, we can identify $\mathrm{End}_K(A)$ with the matrix algebra $M_n(K)$, where $n = \dim_K A$. The trace of matrices induces a linear form $\mathrm{Tr} : \mathrm{End}_K(A) \to K$. It is easy to check that Tr is independent of the chosen basis.

Using the endomorphisms $\mathrm{R}_a \in \mathrm{End}_K(A)$ $(a \in A)$ of (4.25), we define a bilinear form $\langle \, , \, \rangle : A \times A \to K$ by

$$\langle a, b \rangle = \mathrm{Tr}(\mathrm{R}_b \circ \mathrm{R}_a) = \mathrm{Tr}(\mathrm{R}_{ab}) \tag{4.28}$$

for all $a, b \in A$. The bilinear form $\langle \, , \, \rangle$ is called the *trace form* of A.

Lemma 4.41. *For all a, b, $c \in A$,*

$$\langle a, b \rangle = \langle b, a \rangle \quad and \quad \langle ab, c \rangle = \langle ab, c \rangle = \langle b, ca \rangle.$$

Proof. The equality $\langle ab, c \rangle = \langle a, bc \rangle$ follows from the formula $\mathrm{R}_{(ab)c} = \mathrm{R}_{a(bc)}$. The proof of the equality $\langle a, b \rangle = \langle b, a \rangle$ relies on a well-known property of the trace, namely $\mathrm{Tr}(f \circ g) = \mathrm{Tr}(g \circ f)$ for all $f, g \in \mathrm{End}_K(A)$. We have

$$\langle a, b \rangle = \mathrm{Tr}(\mathrm{R}_b \circ \mathrm{R}_a) = \mathrm{Tr}(\mathrm{R}_a \circ \mathrm{R}_b) = \langle b, a \rangle.$$

Finally, using the previous equalities,

$$\langle a, bc \rangle = \langle bc, a \rangle = \langle b, ca \rangle. \qquad \square$$

The kernel $J(A)$ of the trace form, i.e., the vector space

$$J(A) = \{ a \in A \mid \langle a, b \rangle = 0 \text{ for all } b \in A \},$$

is called the *radical* of A.

Lemma 4.42. *The radical $J(A)$ is a two-sided ideal of A.*

Proof. Let $a \in A$ and $b \in J(A)$. We have to check that $ab, ba \in J(A)$. Using Lemma 4.41, for all $c \in A$,

$$\langle ab, c \rangle = \langle b, ca \rangle = 0 \quad and \quad \langle ba, c \rangle = \langle b, ac \rangle = 0. \qquad \square$$

Let $\{A_\lambda\}_{\lambda \in \Lambda}$ be a family of algebras over K with units $1_\lambda \in A_\lambda$. The *product algebra* $A = \prod_{\lambda \in \Lambda} A_\lambda$ is the vector space $\prod_{\lambda \in \Lambda} A_\lambda$ with coordinatewise addition and multiplication

$$(a_\lambda)_\lambda + (b_\lambda)_\lambda = (a_\lambda + b_\lambda)_\lambda \quad and \quad (a_\lambda)_\lambda \cdot (b_\lambda)_\lambda = (a_\lambda b_\lambda)_\lambda$$

for all $a_\lambda, b_\lambda \in A_\lambda$ with $\lambda \in \Lambda$. The vector $(1_\lambda)_\lambda$ is the unit of A. For each $\lambda \in \Lambda$, there is a natural inclusion $A_\lambda \hookrightarrow A$ sending $a \in A_\lambda$ to the family $(a_\mu)_{\mu \in \Lambda} \in A$, where $a_\mu = a$ for $\mu = \lambda$ and $a_\mu = 0$ for $\mu \neq \lambda$. We shall identify A_λ with its image in A. Under this identification, A_λ is a two-sided ideal of A and $A_\lambda A_\mu = 0$ for $\lambda \neq \mu$.

If the algebras $\{A_\lambda\}_{\lambda \in \Lambda}$ are finite-dimensional and the indexing set Λ is finite, then the algebra $\prod_{\lambda \in \Lambda} A_\lambda$ is finite-dimensional and we can compute its radical as follows.

Proposition 4.43. *If A is the product of a finite family $\{A_\lambda\}_{\lambda \in \Lambda}$ of finite-dimensional algebras, then*

$$J(A) = \prod_{\lambda \in \Lambda} J(A_\lambda).$$

Proof. Under the assumptions, A can be identified with the direct sum $\bigoplus_{\lambda \in \Lambda} A_\lambda$. It follows from the definition of the product in A that each right multiplication $R_a \in \mathrm{End}_K(A)$, where $a = (a_\lambda)_\lambda \in A$, is the direct sum over $\lambda \in \Lambda$ of the right multiplications R_{a_λ}. Therefore, the trace form $\langle \, , \, \rangle$ of A is the sum of the trace forms $\langle \, , \, \rangle_\lambda$ of the algebras $\{A_\lambda\}_{\lambda \in \Lambda}$, that is,

$$\langle (a_\lambda)_\lambda, (b_\lambda)_\lambda \rangle = \sum_{\lambda \in \Lambda} \langle a_\lambda, b_\lambda \rangle_\lambda$$

for all $(a_\lambda)_\lambda, (b_\lambda)_\lambda \in A$. It follows that $\prod_{\lambda \in \Lambda} J(A_\lambda) \subset J(A)$. We now prove the converse inclusion. Let $(a_\lambda)_\lambda \in J(A)$ and $b_\mu \in A_\mu$ for some $\mu \in \Lambda$. Considering b_μ as an element of A via the natural inclusion $A_\mu \hookrightarrow A$, we obtain

$$\langle a_\mu, b_\mu \rangle_\mu = \langle (a_\lambda)_\lambda, b_\mu \rangle = 0 \, .$$

Since this holds for all $b_\mu \in A_\mu$, we have $a_\mu \in J(A_\mu)$. Therefore, $(a_\lambda)_\lambda$ belongs to $\prod_{\lambda \in \Lambda} J(A_\lambda)$. $\qquad\square$

Recall that an ideal I of A is *nilpotent* if there is $N \geq 1$ such that $I^N = 0$, i.e., if $a_1 \cdots a_N = 0$ for all $a_1, \ldots, a_N \in I$.

Proposition 4.44. *Any nilpotent left ideal of a finite-dimensional algebra A is contained in $J(A)$.*

Proof. Let I be a nilpotent left ideal of A. To prove that $I \subset J(A)$, we have to check that $\langle a, b \rangle = 0$ for all $a \in I$ and $b \in A$. Set $c = ba \in I$. The ideal I being nilpotent, $c^N = 0$ for some $N \geq 1$. Hence, $(R_c)^N = R_{c^N} = 0$. In other words, R_c is a nilpotent endomorphism of A. Consequently, its trace vanishes. Therefore, by Lemma 4.41 and formula (4.28),

$$\langle a, b \rangle = \langle b, a \rangle = \mathrm{Tr}(R_c) = 0 \, . \qquad\qquad\square$$

4.5.5 Semisimple algebras

Definition 4.45. *An algebra A is semisimple if it is finite-dimensional and $J(A) = 0$.*

Equivalently, an algebra is semisimple if it is finite-dimensional and its trace form is nondegenerate.

Proposition 4.46. *A finite-dimensional algebra A is semisimple if and only if for some basis $\{a_1, \ldots, a_n\}$ of A,*

$$\det(\langle a_i, a_j \rangle_{i,j=1,\ldots,n}) \neq 0 \, .$$

Proof. The nondegeneracy of a symmetric bilinear form $\langle \, , \, \rangle$ on a finite-dimensional vector space with basis $\{a_1, \ldots, a_n\}$ is equivalent to the non-vanishing of the determinant $\det(\langle a_i, a_j \rangle_{i,j=1,\ldots,n})$. $\qquad\square$

Examples 4.47. (i) If G is a finite group and if the characteristic of K does not divide $\operatorname{card} G$, then the group algebra $K[G]$ is semisimple (this assertion is known as *Maschke's theorem*). Indeed, the set G is a basis of $K[G]$, and one checks easily that for all $g, h \in G$,

$$\langle g, h \rangle = \begin{cases} \operatorname{card} G & \text{if } gh = 1, \\ 0 & \text{if } gh \neq 1. \end{cases}$$

From this one deduces that the trace form of $K[G]$ is nondegenerate.

(ii) Let $A = \prod_{\lambda \in \Lambda} A_\lambda$ be the product of a finite family of algebras. It is clear that A is finite-dimensional if and only if all A_λ are finite-dimensional. It follows from this fact and Proposition 4.43 that A is semisimple if and only if all A_λ are semisimple.

(iii) All simple algebras over a field of characteristic zero are semisimple. This follows from Proposition 4.35 and Exercise 4.5.5 below.

Warning. A simple algebra over a field K of characteristic $p > 0$ is not necessarily semisimple. For instance, the algebra $M_p(K)$ of $p \times p$ matrices over K is simple by Proposition 4.34, but its trace form is zero; hence $J(M_p(K)) = M_p(K)$ (see Exercise 4.5.5).

4.5.6 A structure theorem for semisimple algebras

By a *subalgebra* of an algebra A over K, we mean a nonzero K-vector space $A' \subset A$ such that $ab \in A'$ for all $a, b \in A'$ and there is an element $1' \in A'$ such that $1'a = a1' = a$ for all $a \in A'$. Then $1' \neq 0$ and the multiplication in A restricted to A' turns the latter into an algebra with unit $1'$. Clearly, $1'$ coincides with the unit 1 of A if and only if $1 \in A'$.

Lemma 4.48. *Let A be an algebra and $\{A_\lambda\}_{\lambda \in \Lambda}$ a finite family of subalgebras of A such that $A = \bigoplus_{\lambda \in \Lambda} A_\lambda$ and $A_\lambda A_\mu = 0$ for any distinct $\lambda, \mu \in \Lambda$. Then the following hold:*

(a) A_λ is a two-sided ideal of A for any $\lambda \in \Lambda$.

(b) For each $\lambda \in \Lambda$, let $1_\lambda \in A_\lambda$ be defined from the expansion

$$1 = \sum_{\lambda \in \Lambda} 1_\lambda.$$

Then 1_λ belongs to the center of A and is the unit of A_λ.

(c) The map $f : \prod_{\lambda \in \Lambda} A_\lambda \to A$ defined by

$$f((a_\lambda)_\lambda) = \sum_{\lambda \in \Lambda} a_\lambda \in A$$

is an algebra isomorphism.

Proof. (a) We have

$$AA_\mu = \bigoplus_{\lambda \in \Lambda} A_\lambda A_\mu = A_\mu A_\mu \subset A_\mu .$$

The inclusion $A_\mu A \subset A_\mu$ is proved in a similar way.
  (b) If $a \in A$, then

$$\sum_{\lambda \in \Lambda} a 1_\lambda = a1 = a = 1a = \sum_{\lambda \in \Lambda} 1_\lambda a .$$

Since A_λ is a two-sided ideal of A, we have $a1_\lambda, 1_\lambda a \in A_\lambda$. By the uniqueness of expansions in a direct sum, $a1_\lambda = 1_\lambda a$. Thus, 1_λ is a central element of A.
  Since $A_\lambda A_\mu = 0$ for $\lambda \neq \mu$, for each $a_\mu \in A_\mu$,

$$a_\mu = 1a_\mu = \sum_{\lambda \in \Lambda} 1_\lambda a_\mu = 1_\mu a_\mu .$$

Similarly, $a_\mu = a_\mu 1_\mu$. Thus, 1_μ is the unit of A_μ.
  (c) It is clear that f is bijective. If $a = (a_\lambda)_\lambda$, $b = (b_\lambda)_\lambda \in \prod_{\lambda \in \Lambda} A_\lambda$, then

$$f(a)f(b) = \left(\sum_{\lambda \in \Lambda} a_\lambda \right) \left(\sum_{\lambda \in \Lambda} b_\lambda \right) = \sum_{\lambda \in \Lambda} a_\lambda b_\lambda = f(ab) .$$

The second equality follows from the hypothesis $A_\lambda A_\mu = 0$ for $\lambda \neq \mu$. This shows that f is an algebra isomorphism.   □

We now state the main structure theorem for semisimple algebras.

Theorem 4.49. *For any semisimple algebra A, there is a finite family of simple subalgebras $\{A_\lambda\}_{\lambda \in \Lambda}$ of A such that $A = \bigoplus_{\lambda \in \Lambda} A_\lambda$ and $A_\lambda A_\mu = 0$ for any distinct $\lambda, \mu \in \Lambda$. Such a family of subalgebras is unique.*

Proof. We proceed by induction on the dimension of A over K. If $\dim_K A = 1$, then A is necessarily simple.
  Assume that $\dim_K A > 1$ and that the theorem holds for all semisimple algebras of dimension $< \dim_K A$. Let $I \subset A$ be an arbitrary nonzero two-sided ideal of minimal dimension. Clearly, I contains no other nonzero two-sided ideals of A. Therefore, if $I = A$, then A is simple and the theorem is proved. If $I \neq A$, then set

$$I^\perp = \left\{ a \in A \mid \langle a, b \rangle = 0 \text{ for all } b \in I \right\},$$

where we use the trace form (4.28). It follows from Lemma 4.41 that $I^\perp$ is a two-sided ideal of A. Since the trace form is nondegenerate and $0 \neq I \neq A$, we have $0 \neq I^\perp \neq A$ and

$$\dim_K I + \dim_K I^\perp = \dim_K A .$$

The intersection $I \cap I^\perp$ is a two-sided ideal of A contained in I. By the minimality of I, we have either $I \cap I^\perp = I$ or $I \cap I^\perp = 0$. The equality $I \cap I^\perp = I$ is equivalent to the inclusion $I \subset I^\perp$ and is equivalent to the vanishing of the trace form on I. We claim that the latter is impossible. Indeed, I being minimal, the two-sided ideal $I^2 \subset I$ is either 0 or I. The equality $I^2 = 0$ would imply that A contains a nonzero nilpotent left ideal, which is impossible by Proposition 4.44. Therefore, $I^2 = I$, so that any $z \in I$ expands as $z = \sum_i x_i y_i$, where $x_i, y_i \in I$. If the trace form vanishes on I, then

$$\langle 1, z \rangle = \sum_i \langle 1, x_i y_i \rangle = \sum_i \langle x_i, y_i \rangle = 0 \,.$$

Therefore, $1 \in I^\perp$ and the two-sided ideal $I^\perp$ is equal to A, a contradiction.

We have thus proved that $I \cap I^\perp = 0$. Since $\dim_K I + \dim_K I^\perp = \dim_K A$, we obtain $A = I \oplus I^\perp$. The product ideals $II^\perp$ and $I^\perp I$, being contained in $I \cap I^\perp = 0$, must be equal to 0. As in Lemma 4.48 (b), the projections of the unit of A to I and $I^\perp$ are the units of I and $I^\perp$, respectively. Thus, I and $I^\perp$ are subalgebras of A and $A = I \times I^\perp$.

Any two-sided ideal $J \subset I$ of the algebra I is automatically a two-sided ideal of A. Since I is minimal, $J = 0$ or $J = I$. Hence, the algebra I is simple. The equality $A = I \times I^\perp$ and Proposition 4.43 imply that $J(I^\perp) \subset J(A)$. Since $J(A)$ vanishes, so does $J(I^\perp)$, which proves that the algebra $I^\perp$ is semisimple.

Since $\dim_K I^\perp < \dim_K A$, we may apply the induction hypothesis to $I^\perp$. We obtain a finite family $\{A_\lambda\}_{\lambda \in \Lambda'}$ of simple subalgebras of $I^\perp$ such that

$$I^\perp = \bigoplus_{\lambda \in \Lambda'} A_\lambda$$

and $A_\lambda A_\mu = 0$ for any distinct $\lambda, \mu \in \Lambda'$. We obtain the desired family $\{A_\lambda\}_{\lambda \in \Lambda}$ of simple subalgebras of A by setting $\Lambda = \Lambda' \amalg \{\lambda_0\}$ with $A_{\lambda_0} = I$.

In order to prove the uniqueness of the family $\{A_\lambda\}_{\lambda \in \Lambda}$, consider an arbitrary nonzero two-sided ideal J of A. We have

$$J = JA = \bigoplus_{\lambda \in \Lambda} JA_\lambda \,.$$

Each product ideal JA_λ is a two-sided ideal of A_λ. Since A_λ is a simple algebra, JA_λ is equal to 0 or to A_λ. Consequently, there is a nonempty set $\Lambda_0 \subset \Lambda$ such that $J = \bigoplus_{\lambda \in \Lambda_0} A_\lambda$. This shows that J is a subalgebra of A. Moreover, J is simple as an algebra if and only if Λ_0 consists of a single element λ_0, and then $J = A_{\lambda_0}$. We conclude that the family $\{A_\lambda\}_{\lambda \in \Lambda}$ consists of all nonzero two-sided ideals of A that are simple as algebras. This proves the uniqueness claim of the theorem. $\square$

Lemma 4.48 and Theorem 4.49 have the following consequences.

Corollary 4.50. *Any semisimple algebra is a product of simple algebras.*

Corollary 4.51. *Let J be a two-sided ideal of a semisimple algebra A. Then J and the quotient algebra A/J are semisimple algebras.*

Proof. Consider the splitting $A = \bigoplus_{\lambda \in \Lambda} A_\lambda$ of A as in Theorem 4.49. By Example 4.47 (ii), each A_λ is semisimple. We have seen in the proof of Theorem 4.49 that there is a set $\Lambda_0 \subset \Lambda$ such that $J = \bigoplus_{\lambda \in \Lambda_0} A_\lambda$. By Lemma 4.48,

$$J = \prod_{\lambda \in \Lambda_0} A_\lambda \quad \text{and} \quad A/J \cong \prod_{\lambda \in \Lambda - \Lambda_0} A_\lambda.$$

We conclude by using Example 4.47 (ii), which tells us that finite products of semisimple algebras are semisimple. □

4.5.7 Modules over a semisimple algebra

Let us first determine the simple modules over a semisimple algebra.

Proposition 4.52. *Let A be a semisimple algebra and $\{A_\lambda\}_{\lambda \in \Lambda}$ the family of simple subalgebras of A provided by Theorem 4.49. For any nonzero simple A-module M, there is a unique $\lambda \in \Lambda$ such that $M = A_\lambda M$. Moreover, M is a simple A_λ-module and $A_\mu M = 0$ for all $\mu \neq \lambda$.*

Proof. Let M be a nonzero simple A-module. Each $A_\lambda M$ is an A-submodule of M. We can write M as a sum of these submodules:

$$M = AM = \sum_{\lambda \in \Lambda} A_\lambda M. \tag{4.29}$$

Since $M \neq 0$, there is $\lambda \in \Lambda$ such that $A_\lambda M \neq 0$. By the simplicity of M, we have $A_\lambda M = M$. We claim that $A_\mu M = 0$ for $\mu \neq \lambda$. Indeed, $m \in M$ can be expanded as $m = \sum_i a_i m_i$ with $a_i \in A_\lambda$ and $m_i \in M$. If $a \in A_\mu$ with $\mu \neq \lambda$, then $am = \sum_i a a_i m_i = 0$, since $A_\mu A_\lambda = 0$. We next claim that the A_λ-module M is simple. Indeed, let N be a nonzero A_λ-submodule of M. Letting A_μ with $\mu \neq \lambda$ act on N as 0, we turn N into an A-submodule of M. Since M is simple as an A-module, $N = M$. □

Theorem 4.53. *Any finite-dimensional module over a semisimple algebra is semisimple.*

Proof. Consider a finite-dimensional module M over a semisimple algebra A. Expand A as a product of simple subalgebras $\{A_\lambda\}_{\lambda \in \Lambda}$ as in Theorem 4.49. Each vector space $A_\lambda M \subset M$ is a finite-dimensional module over A_λ. It follows from Proposition 4.39 that $A_\lambda M$ is a semisimple A_λ-module. Since a simple A_λ-module is simple as an A-module (where all A_μ with $\mu \neq \lambda$ act as 0), each $A_\lambda M$ is a semisimple A-module. Formula (4.29) implies that M is a sum of simple submodules. By Proposition 4.28, the A-module M is semisimple. □

We now summarize the representation theory of a (finite-dimensional) semisimple algebra A. Let $\{A_\lambda\}_{\lambda \in \Lambda}$ be the set of all nonzero two-sided ideals of A that are simple as algebras. This set is finite. For each $\lambda \in \Lambda$, there is a unique up to isomorphism simple A_λ-module V_λ. We view V_λ as an A-module by $A_\mu V_\lambda = 0$ for $\mu \neq \lambda$. Then the A-modules $\{V_\lambda\}_{\lambda \in \Lambda}$ are simple, and every simple A-module is isomorphic to exactly one of them. Moreover, for any finite-dimensional A-module M, there are a unique function $d_M : \Lambda \to \mathbf{N}$ and an isomorphism of A-modules

$$M \cong \bigoplus_{\lambda \in \Lambda} V_\lambda^{d_M(\lambda)} . \tag{4.30}$$

The function d_M is called the *dimension vector* of M.

Let D_λ be the division ring $\mathrm{End}_A(V_\lambda)$. The following theorem is a consequence of Corollary 4.40, Lemma 4.48, and Theorem 4.49.

Theorem 4.54. *With the notation above,*

$$A \cong \prod_{\lambda \in \Lambda} \mathrm{End}_{D_\lambda}(V_\lambda)$$

and

$$\dim_K A = \sum_{\lambda \in \Lambda} \frac{(\dim_K V_\lambda)^2}{\dim_K D_\lambda} .$$

Corollary 4.55. *If the ground field K is algebraically closed, then the algebra homomorphism*

$$A \to \prod_{\lambda \in \Lambda} \mathrm{End}_K(V_\lambda)$$

obtained as the product over Λ of the algebra homomorphisms $A \to \mathrm{End}_K(V_\lambda)$ induced by the action of A on V_λ is an isomorphism. Moreover,

$$\dim_K A = \sum_{\lambda \in \Lambda} (\dim_K V_\lambda)^2 .$$

Proof. Applying Proposition 4.30 (c) to the simple A-module V_λ, we obtain $\dim_K D_\lambda = 1$. Thus, $D_\lambda = K$. The corollary is then a reformulation of Theorem 4.54. □

Exercise 4.5.1. Let A be a finite-dimensional algebra with radical $J = J(A)$.

(a) Show that the quotient algebra A/J is semisimple.
(b) Prove that $1 + x$ is invertible in A for any $x \in J$.

Exercise 4.5.2. Let V be a finite-dimensional vector space over a field K. Show that every K-linear automorphism of the algebra $A = \mathrm{End}_K(V)$ is the conjugation by an element of A. (Hint: An automorphism of A defines a new A-module structure on V; now use the fact that A has only one isomorphism class of simple modules.)

Exercise 4.5.3. Let A be a semisimple algebra and M a finite-dimensional A-module. Show that there is an algebra isomorphism

$$\text{End}_A(M) \cong \prod_{\lambda \in \Lambda(A)} M_{d_M(\lambda)}\big(\text{End}_A(V_\lambda)\big).$$

Exercise 4.5.4. Let K be an algebraically closed field and A a semisimple K-algebra. Prove that there is an isomorphism of A-modules

$$A \cong \bigoplus_{\lambda \in \Lambda(A)} V_\lambda^{d_\lambda},$$

where $d_\lambda = \dim_K V_\lambda$.

Exercise 4.5.5. Let D be a division ring. For $1 \le i, j \le n$, let $E_{i,j} \in M_n(D)$ be the matrix whose entries are all zero except the (i,j) entry, which is 1.

(a) Verify that the trace of the right multiplication by $E_{i,j}$ in the matrix algebra $M_n(D)$ is n if $i = j$ and is 0 otherwise.
(b) Prove that the trace form of $M_n(D)$ is given for all $a, b \in M_n(D)$ by

$$\langle a, b \rangle = n \, \text{Tr}(ab).$$

(c) Deduce that $M_n(D)$ is semisimple if and only if n is invertible in D.

Exercise 4.5.6. Let K be a field of characteristic $p > 0$ and G the cyclic group of order p. Show that $(g - 1)^p = 0 \in K[G]$ for all $g \in G$. Deduce that the group algebra $K[G]$ contains a nonzero nilpotent ideal and is not semisimple.

Exercise 4.5.7. Let A be a finite-dimensional algebra over a field K of characteristic zero. Prove that all elements of the radical of A are nilpotent. (An element a of A is *nilpotent* if $a^N = 0$ for some integer $N \ge 1$.)
  Solution. Set $d = \dim_K A$. For each $a \in J(A)$ and $n \ge 1$,

$$\text{Tr}((\text{R}_a)^n) = \text{Tr}(\text{R}_{a^n}) = \langle a, a^{n-1} \rangle = 0.$$

If $\lambda_1, \ldots, \lambda_d$ are the eigenvalues of R_a in an algebraic closure of K, the previous equalities imply that
$$\lambda_1^n + \cdots + \lambda_d^n = 0$$
for all $n \ge 1$. By Newton's formulas (which require the ground field to be of characteristic zero), all elementary symmetric polynomials in $\lambda_1, \ldots, \lambda_d$ vanish. This implies that the characteristic polynomial of R_a is a monomial of degree d and hence $(\text{R}_a)^d = 0$. Therefore, $a^d = (\text{R}_a)^d(1) = 0$.

Exercise 4.5.8. Let K be a field of characteristic zero. Prove that a finite-dimensional K-algebra A that does not contain nonzero nilpotent left ideals is semisimple.

Solution. It suffices to prove that $J = J(A) = 0$. Assume that $J \neq 0$ and pick a nonzero left ideal $I \subset J$ of minimal dimension over K. By assumption, the ideal I is not nilpotent, and in particular, $I^2 \neq 0$. Hence, there is $x \in I$ such that $Ix \neq 0$. By the minimality of I and the inclusion $Ix \subset I$, we have $Ix = I$. Hence, there is $e \in I$ such that $ex = x$. It follows that

$$x = ex = e(ex) = e^2 x.$$

We thus obtain $(e - e^2)x = 0$. The left ideal

$$I' = \{y \in I \mid yx = 0\}$$

is a proper subideal of I, since $Ix \neq 0$. By the minimality of I, we must have $I' = 0$. Since $e - e^2 \in I'$, we have $e = e^2$. Hence,

$$e = e^2 = e^3 = \cdots.$$

Now, by Exercise 4.5.7, the element $e \in I \subset J$ is nilpotent (this is where we use the characteristic-zero assumption). From these two facts we deduce that $e = 0$. Hence, $x = ex = 0$ and $Ix = 0$, which contradicts the choice of x. Therefore, $J = 0$.

Exercise 4.5.9. An element e of an algebra A is an *idempotent* if $e = e^2$.

(a) Show that if $e \in A$ is an idempotent, then so is $f = 1 - e$.
(b) Suppose that an idempotent $e \in A$ is central, that is, e commutes with all elements of A. Set $f = 1 - e$. Prove that Ae and Af are two-sided ideals of A, that viewed as algebras Ae and Af have e and f as respective units, and that the map $Ae \times Af \to A, (a, b) \mapsto a + b$ is an algebra isomorphism.
(c) Show that the unique nonzero central idempotent of a simple algebra is its unit.

Exercise 4.5.10. A nonzero central idempotent e of an algebra A is *primitive* if it not expressible as a sum of two nonzero central idempotents whose product is zero.

(a) Prove that if e is a primitive central idempotent of A, then there are no algebras A_1, A_2 such that $Ae \cong A_1 \times A_2$.
(b) Let A be a product of $r < \infty$ simple algebras. Show that there is a unique set $\{e_1, \ldots, e_r\}$ of primitive central idempotents of A such that $e_k e_\ell = 0$ for all distinct $k, \ell \in \{1, \ldots, r\}$ and $e_1 + \cdots + e_r = 1$.

Exercise 4.5.11. Let A be a finite-dimensional algebra. Prove the following:

(a) The sum of two nilpotent left ideals of A is a nilpotent left ideal.
(b) Any nonnilpotent left ideal of A contains a nonzero idempotent.
(c) The sum J of all nilpotent left ideals of A is a two-sided ideal.
(d) If the ground field has characteristic zero, then J is the radical of A.

4.6 Semisimplicity of the Iwahori–Hecke algebras

We return to the Iwahori–Hecke algebras $H_n^R(q)$ of Section 4.2.2, where n is a positive integer, R is a commutative ring, and q is an invertible element of R.

Let us first analyze the behavior of $H_n^R(q)$ under a change of scalars. Let $f : R \to S$ be a homomorphism of commutative rings. Given an integer $n \geq 1$ and an invertible element $q \in R$, we have the R-algebra $H_n^R(q)$ and the S-algebra $H_n^S(\tilde{q})$, where $\tilde{q} = f(q) \in S$.

By Theorem 4.17, $H_n^R(q)$ is a free R-module of rank $n!$. We may therefore identify $\operatorname{End}_R(H_n^R(q))$ with the matrix algebra $M_{n!}(R)$. This allows us to define the R-bilinear trace form of $H_n^R(q)$,

$$\langle\,,\,\rangle_R : H_n^R(q) \times H_n^R(q) \to R,$$

by formula (4.28), where $R_c \in \operatorname{End}_R(H_n^R(q))$ is the right multiplication by c for any $c \in H_n^R(q)$. Similarly, we define the S-bilinear trace form of $H_n^S(\tilde{q})$,

$$\langle\,,\,\rangle_S : H_n^S(\tilde{q}) \times H_n^S(\tilde{q}) \to S.$$

Proposition 4.56. *There is an isomorphism of S-algebras*

$$\varphi : S \otimes_R H_n^R(q) \xrightarrow{\cong} H_n^S(\tilde{q})$$

such that

$$\langle \varphi(s \otimes x), \varphi(s' \otimes x') \rangle_S = ss' f(\langle x, x' \rangle_R) \tag{4.31}$$

for all $s, s' \in S$, $x, x' \in H_n^R(q)$.

Proof. Set $\varphi(s \otimes T_i) = s T_i \in H_n^S(\tilde{q})$ for $s \in S$ and $i = 1, \ldots, n-1$. It is easy to check that this defines a homomorphism of S-algebras

$$\varphi : S \otimes_R H_n^R(q) \to H_n^S(\tilde{q}).$$

By Theorem 4.17, $H_n^R(q)$ is a free R-module with basis $\{T_w \mid w \in \mathfrak{S}_n\}$. Similarly, $H_n^S(\tilde{q})$ is a free S-module with the same basis. It is clear that φ sends the basis $\{1 \otimes T_w \mid w \in \mathfrak{S}_n\}$ of $S \otimes_R H_n^R(q)$ to the basis $\{T_w \mid w \in \mathfrak{S}_n\}$ of $H_n^S(\tilde{q})$. Therefore, φ is an isomorphism.

By S-bilinearity, to prove (4.31), it is enough to check it for $s = s' = 1$, $x = T_w$, and $x' = T_{w'}$, where $w, w' \in \mathfrak{S}_n$. We have

$$\begin{aligned}
\langle \varphi(1 \otimes T_w), \varphi(1 \otimes T_{w'}) \rangle_S &= \langle T_w, T_{w'} \rangle_S \\
&= \operatorname{Tr}(R_{T_w T_{w'}} : H_n^S(\tilde{q}) \to H_n^S(\tilde{q})) \\
&= f(\operatorname{Tr}(R_{T_w T_{w'}} : H_n^R(q) \to H_n^R(q))) \\
&= f(\langle T_w, T_{w'} \rangle_R).
\end{aligned}$$

Here we used the fact that the structure constants for the multiplication of the basis elements $T_w \in H_n^S(\tilde{q})$ are the images under f of the corresponding structure constants of the basis elements $T_w \in H_n^R(q)$. □

Recall that an element of a ring is *algebraic* if it is the root of a nonzero polynomial with coefficients in $\mathbf{Z}$. We now state the main result of this section.

Theorem 4.57. *Let K be a field whose characteristic does not divide $n!$. The algebra $H_n^K(q)$ is semisimple for all $q \in K - \{0\}$ except a finite number of algebraic elements of $K - \{0, 1\}$.*

Note without proof a more precise result by Wenzl [Wen88]: $H_n^K(q)$ is semisimple provided q is not a root of unity of order d with $2 \leq d \leq n$.

Proof. If $q = 1$, then $H_n^K(q) \cong K[\mathfrak{S}_n]$ is semisimple by Example 4.47 (i).

Now suppose that $q \neq 1$. By definition, $H_n^K(q)$ is semisimple if and only if its trace form $\langle \, , \, \rangle_K$ is nondegenerate. Consider the basis $\{T_w\}_{w \in \mathfrak{S}_n}$ of $H_n^K(q)$. By Proposition 4.46, $H_n^K(q)$ is semisimple if and only if

$$\det(\langle T_w, T_{w'} \rangle_K)_{w, w' \in \mathfrak{S}_n} \neq 0 \,.$$

Let $R = \mathbf{Z}[q_0, q_0^{-1}]$ be the ring of Laurent polynomials in one variable q_0, and let $i : R \to K$ be the ring homomorphism such that $i(q_0) = q$. The R-algebra $H_n^R(q_0)$ carries a trace form $\langle \, , \, \rangle_R$, which by Proposition 4.56 is related to the trace form of $H_n^K(q)$ by

$$\langle T_w, T_{w'} \rangle_K = i\big(\langle T_w, T_{w'} \rangle_R \big)$$

for all $w, w' \in \mathfrak{S}_n$. Therefore,

$$\det(\langle T_w, T_{w'} \rangle_K)_{w, w' \in \mathfrak{S}_n} = i(D(q_0)) \,,$$

where

$$D(q_0) = \det(\langle T_w, T_{w'} \rangle_R)_{w, w' \in \mathfrak{S}_n} \in R \,.$$

In other words, $\det(\langle T_w, T_{w'} \rangle_K)_{w, w' \in \mathfrak{S}_n} \in K$ is the value of the Laurent polynomial $D(q_0)$ at $q_0 = q$.

We claim that $D(q_0) \neq 0$. To prove this claim, consider the ring homomorphism $\pi : R \to \mathbf{Q}$ sending q_0 to 1. By Proposition 4.56, there is an isomorphism of $\mathbf{Q}$-algebras $\mathbf{Q} \otimes_R H_n^R(q_0) \cong H_n^{\mathbf{Q}}(1)$. This isomorphism sends the basis $\{1 \otimes T_w\}_{w \in \mathfrak{S}_n}$ of $\mathbf{Q} \otimes_R H_n^R(q_0)$ to the basis $\{T_w\}_{w \in \mathfrak{S}_n}$ of $H_n^{\mathbf{Q}}(1)$. The trace form $\langle \, , \, \rangle_R$ of $H_n^R(q_0)$ is related to the trace form $\langle \, , \, \rangle_{\mathbf{Q}}$ of $H_n^{\mathbf{Q}}(1)$ by

$$\langle T_w, T_{w'} \rangle_{\mathbf{Q}} = \pi\big(\langle T_w, T_{w'} \rangle_R \big)$$

for all $w, w' \in \mathfrak{S}_n$. Hence,

$$\det(\langle T_w, T_{w'} \rangle_{\mathbf{Q}})_{w, w' \in \mathfrak{S}_n} = \pi(D(q_0)) \,.$$

Since $H_n^{\mathbf{Q}}(1) \cong \mathbf{Q}[\mathfrak{S}_n]$ is semisimple, it follows from Proposition 4.46 that $\det(\langle T_w, T_{w'} \rangle_{\mathbf{Q}})_{w, w' \in \mathfrak{S}_n} \neq 0$. This shows that $D(q_0) \neq 0$.

In conclusion, the K-algebra $H_n^K(q)$ is semisimple if and only if the value of the Laurent polynomial $D(q_0)$ at $q_0 = q$ is nonzero, i.e., if and only if q is not a root of $D(q_0)$ in K. We finally observe that a nonzero Laurent polynomial has finitely many roots and that its roots are algebraic. $\qquad\square$

Exercise 4.6.1. Let $R = \mathbf{Z}[q_0, q_0^{-1}]$. Compute the trace form on $H_n^R(q_0)$ for $n = 2$ and show that the corresponding Laurent polynomial $D(q_0)$ (defined in the previous proof) is equal to $(q_0 + 1)^2$.

Notes

The presentation (4.1) of the symmetric group is due to E. H. Moore [Moo97]. In Section 4.1 we followed [Mat99, Sect. 1.1]. The results of this section will be extended to Coxeter groups in Section 6.6.

Following an idea of André Weil, Shimura [Shi59] defined an "algebra of transformations" in connection with the Hecke operators of number theory. This algebra is defined as the convolution algebra of B-bi-invariant functions on a group G, where B is a subgroup of G such that $[B : B \cap xBx^{-1}] < \infty$ for all $x \in G$. In [Iwa64] Iwahori called Shimura's algebra of transformations a "Hecke ring" and gave it a presentation by generators and relations in the case that G is a Chevalley group over a finite field $\mathbf{F}_q$ and B is a Borel subgroup of G.

The Iwahori–Hecke algebra of Definition 4.15 is Shimura's algebra associated to the Chevalley group $G = \mathrm{GL}_n(\mathbf{F}_q)$ (for details, see Exercise 4.2.3 above, [Bou68, Chap. 4, Sect. 2, Exercises 22–24], [GHJ89, Sect. 2.10], [GP00, Sect. 8.4]).

In Sections 4.2–4.4 we have essentially followed [HKW86, Sects. 4–6]. The trace constructed in Section 4.3 is due to Ocneanu (see [FYHLMO85] and [Jon87, Sect. 5]). The existence of the two-variable Jones–Conway polynomial constructed in Section 4.4 was proved by Freyd, Yetter, Hoste, Lickorish, Millett, Ocneanu, Przytycki, and Traczyk soon after Vaughan Jones discovered the Jones polynomial in summer 1984; see [Jon85], [Jon87], [FYHLMO85], [PT87]. The discovery of the Jones polynomial and its generalizations laid the foundations for quantum topology; see [Tur94], [Kas95], [KRT97].

The content of Section 4.5 is standard and can be found in textbooks such as Bourbaki [Bou58], Curtis and Reiner [CR62], Pierce [Pie88], Drozd and Kirichenko [DK94], Benson [Ben98], Lang [Lan02]. Note that in positive characteristic our definition of a semisimple algebra is more restrictive than the definition given in these references. Lemma 4.36 is due to M. Rieffel.

5

Representations of the Iwahori–Hecke Algebras

In this chapter we study the linear representations of the one-parameter Iwahori–Hecke algebras of Section 4.2.2. Our aim is to classify their finite-dimensional representations over an algebraically closed field of characteristic zero in terms of partitions and Young diagrams. As an application, we prove that the reduced Burau representation introduced in Section 3.3 is irreducible. We end the chapter by a discussion of the Temperley–Lieb algebras.

5.1 The combinatorics of partitions and tableaux

We introduce the language of partitions, which is commonly used to describe the irreducible representations of the symmetric groups. We shall use this language in Section 5.3 to construct simple modules over the Iwahori–Hecke algebras.

5.1.1 Partitions

A *partition* of a nonnegative integer n is a finite sequence $\lambda = (\lambda_1, \lambda_2, \ldots, \lambda_p)$ of positive integers satisfying

$$\lambda_1 \geq \lambda_2 \geq \cdots \geq \lambda_p \quad \text{and} \quad |\lambda| = \lambda_1 + \lambda_2 + \cdots + \lambda_p = n.$$

We write $\lambda \vdash n$ to indicate that λ is a partition of n. The integers $\lambda_1, \lambda_2, \ldots, \lambda_p$ are called the *parts* of λ, and p is called the *number of parts*. By definition, $n = 0$ has a unique partition, namely the empty sequence $\emptyset$.

Let $\lambda = (\lambda_1, \lambda_2, \ldots, \lambda_p)$ be a partition with p parts. Setting $\lambda_k = 0$ for all $k > p$, we can identify λ with an infinite sequence $(\lambda_k)_{k \geq 1}$ of integers indexed by $k = 1, 2, \ldots$. This sequence is eventually zero, in the sense that $\lambda_k = 0$ for all sufficiently large k, and nonincreasing: $\lambda_k \geq \lambda_{k+1}$ for all k. Any eventually zero nonincreasing sequence $(\lambda_k)_{k \geq 1}$ of integers arises in this way from the partition $\lambda = (\lambda_1, \lambda_2, \ldots, \lambda_p)$ of $n = \sum_{k \geq 1} \lambda_k$. Here $p = \max\{k \mid \lambda_k \neq 0\}$. In particular, the empty partition $\emptyset$ corresponds to the constant zero sequence.

C. Kassel, V. Turaev, *Braid Groups*, DOI: 10.1007/978-0-387-68548-9_5,
© Springer Science+Business Media, LLC 2008

5.1.2 Diagrams

It is convenient to represent a partition $\lambda = (\lambda_1, \lambda_2, \ldots, \lambda_p)$ of $n \geq 0$ by its *diagram* $D(\lambda)$ (also called *Ferrers diagram* or *Young diagram*), which is defined as the set

$$D(\lambda) = \{(r, s) \mid 1 \leq r \leq p \text{ and } 1 \leq s \leq \lambda_r\}.$$

In particular, the diagram of the empty partition is the empty set. It follows from the definitions that $D(\lambda) = D(\lambda')$ if and only if $\lambda = \lambda'$.

We can represent $D(\lambda)$ graphically as a left-justified collection of boxes in the plane $\mathbf{R}^2$, each of them centered at the corresponding point $(r, s) \in \mathbf{R}^2$, with λ_1 boxes in the first row, λ_2 boxes in the second row, and so on until the last, pth, row, which contains λ_p boxes. The total number of boxes is equal to $|\lambda| = n$. In the figures we use the convention that the r-axis points downward and the s-axis points to the right. For instance, Figure 5.1 represents $D(\lambda)$ for $\lambda = (3, 2, 2, 1)$.

Fig. 5.1. The diagram of the partition $(3, 2, 2, 1)$

5.1.3 Operations on partitions

We define several operations on partitions needed for the sequel. Given two partitions $\lambda = (\lambda_k)_{k \geq 1}$ and $\lambda' = (\lambda'_k)_{k \geq 1}$ (possibly of different integers), we define sequences of integers $\lambda \wedge \lambda'$ and $\lambda \vee \lambda'$ by

$$(\lambda \wedge \lambda')_k = \min(\lambda_k, \lambda'_k) \quad \text{and} \quad (\lambda \vee \lambda')_k = \max(\lambda_k, \lambda'_k)$$

for all $k \geq 1$. These two sequences are nonincreasing and eventually zero, and thus define partitions. It is clear that

$$D(\lambda \wedge \lambda') = D(\lambda) \cap D(\lambda') \quad \text{and} \quad D(\lambda \vee \lambda') = D(\lambda) \cup D(\lambda').$$

The *conjugate* of a partition $\lambda \vdash n$ is the partition $\lambda^T \vdash n$ whose diagram is the set $\{(r, s) \mid (s, r) \in D(\lambda)\}$. In other words, the diagram of λ^T is obtained from the diagram of λ by exchanging its rows and columns. For instance, if $\lambda = (3, 2, 2, 1)$, then $\lambda^T = (4, 3, 1)$ (see Figure 5.2).

Fig. 5.2. The diagram of the conjugate partition of $(3, 2, 2, 1)$

5.1.4 Tableaux

A *tableau* T consists of a partition $\lambda \vdash n$ together with a bijection

$$D(\lambda) \to \{1, 2, \ldots, n\},$$

called the *labeling* and usually denoted by the same letter T. The values of the labeling are called the *labels* of the corresponding boxes. The partition λ is called the *shape* of T. Figure 5.3 shows two tableaux of shape $(3, 2, 2, 1)$.

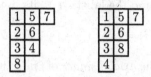

Fig. 5.3. Two tableaux of shape $(3, 2, 2, 1)$

Composing the labeling of a tableau T having n labels with a permutation σ of $\{1, 2, \ldots, n\}$, we obtain the labeling of another tableau σT of the same shape. In particular, $s_i T$ is the tableau T in which the labels i and $i + 1$ are switched. It is clear that $\sigma = \sigma' \iff \sigma T = \sigma' T$, and that two tableaux of the same shape can be obtained from each other by a unique permutation of the labels. Consequently, the number of tableaux of shape $\lambda \vdash n$ is equal to $n!$.

5.1.5 Standard tableaux

A tableau T of shape $\lambda \vdash n$ is said to be *standard* if its labeling increases from left to right in each row and from top to bottom in each column, i.e., if the labeling $T : D(\lambda) \to \{1, 2, \ldots, n\}$ satisfies

$$T(r, s) \leq T(r', s')$$

for all $(r, s), (r', s') \in D(\lambda)$ such that $r \leq r'$ and $s \leq s'$. For instance, the right tableau in Figure 5.3 is standard, whereas the left tableau is not (since the label 4 sits below the label 6).

Let $\mathcal{T}_\lambda$ be the set of standard tableaux of shape λ and $f^\lambda = \operatorname{card} \mathcal{T}_\lambda$. Exchanging rows and columns yields a bijection between $\mathcal{T}_\lambda$ and $\mathcal{T}_{\lambda^T}$, where λ^T is the conjugate partition of λ. Therefore, $f^{\lambda^T} = f^\lambda$.

Exercises 5.2.2 and 5.2.3 below provide explicit formulas for f^λ for certain partitions λ. A general formula for f^λ, called the *hook length formula*, is given in Exercise 5.2.6.

The following property of the numbers $\{f^\lambda\}_\lambda$ will play a key role in the classification of the simple modules over the Iwahori–Hecke algebras.

Theorem 5.1. *For all $n \geq 1$,*

$$\sum_{\lambda \vdash n} (f^\lambda)^2 = n! \,.$$

A proof of this theorem will be given in Section 5.2.4.

5.1.6 The axial distance

Let T be a tableau with n boxes. Suppose that the label $i \in \{1, \ldots, n-1\}$ sits in the box (r, s) of T and the label $i+1$ sits in the box (r', s') of T. Set

$$d_T(i) = (s' - r') - (s - r) \in \mathbf{Z}. \tag{5.1}$$

The integer $s-r$ is called the *axial distance* of i in T (it is the algebraic distance of the box (r, s) to the diagonal $\{(x, x) \mid x \in \mathbf{R}\}$ in $\mathbf{R}^2$). The integer $d_T(i)$ is then the difference between the axial distances of $i+1$ and i. We record some important properties of $d_T(i)$ in the following lemma.

Lemma 5.2. *Let T be a tableau with n boxes and $i, j \in \{1, \ldots, n-1\}$.*
   (a) Then

$$d_{s_j T}(i) = \begin{cases} -d_T(i) & \text{if } j = i, \\ d_T(i) & \text{if } |i - j| \geq 2. \end{cases}$$

   (b) Assume that $i \neq n-1$ and set $d = d_T(i)$, $e = d_T(i+1)$. Then

$$d_{s_i T}(i) = -d_{s_i s_{i+1} T}(i+1) = d_{s_i s_{i+1} s_i T}(i+1) = -d,$$
$$d_{s_{i+1} T}(i+1) = -d_{s_{i+1} s_i T}(i) = d_{s_i s_{i+1} s_i T}(i) = -e,$$
$$d_{s_i T}(i+1) = d_{s_{i+1} T}(i) = -d_{s_i s_{i+1} T}(i) = -d_{s_{i+1} s_i T}(i+1) = d + e.$$

Proof. (a) Let (r, s) be the box of T with label i and (r', s') the box of T with label $i+1$. Then (r, s) is the box of $s_i T$ with label $i+1$ and (r', s') is the box of $s_i T$ with label i. It follows that

$$d_{s_i T}(i) = (s - r) - (s' - r') = -d_T(i).$$

If $j = i$, then $d_{s_j T}(i) = d_{s_i T}(i) = -d_T(i)$. If $|i - j| \geq 2$, then T and $s_j T$ have the same boxes with labels i and $i + 1$. Therefore,

$$d_{s_j T}(i) = d_T(i).$$

(b) Suppose that the labels i, $i+1$, $i+2$ sit in the boxes (r,s), (r',s'), (r'',s'') of T, respectively. Then $d = (s'-r')-(s-r)$ and $e = (s''-r'')-(s'-r')$. The equalities $d_{s_i T}(i) = -d$ and $d_{s_{i+1}T}(i+1) = -e$ are consequences of (a). Since the labels $i+1$ and $i+2$ sit in the boxes (r,s) and (r'',s'') of $s_i T$,

$$
\begin{aligned}
d_{s_i T}(i+1) &= (s'' - r'') - (s - r) \\
&= \big((s'' - r'') - (s' - r')\big) + \big((s' - r') - (s - r)\big) \\
&= d + e.
\end{aligned}
$$

The labels i, $i+1$ sit in the boxes (r,s), (r'',s'') of $s_{i+1}T$, respectively. Therefore,

$$
d_{s_{i+1}T}(i) = (s'' - r'') - (s - r) = d + e.
$$

The computations of $d_{s_i s_{i+1}T}(j)$, $d_{s_{i+1}s_i T}(j)$, and $d_{s_i s_{i+1}s_i T}(j)$ with $j = i, i+1$ are similar. $\qquad\square$

When T is standard, we have the following additional information.

Lemma 5.3. *Let T be a standard tableau with n boxes.*
   (a) *If the labels i and $i+1$ of T sit in the same row, then $d_T(i) = 1$.*
   (b) *If i and $i+1$ sit in the same column, then $d_T(i) = -1$.*
   (c) *If i and $i+1$ sit neither in the same column nor in the same row, then $|d_T(i)| \geq 2$.*
   (d) *In all cases, $|d_T(i)| \leq n - 1$.*

Proof. Let (r,s) and (r',s') be the boxes of T with labels i and $i+1$, respectively.

   (a) If i and $i+1$ sit in the same row, then they necessarily occupy adjacent boxes, so that $r' = r$ and $s' = s+1$; therefore, $d_T(i) = 1$.

   (b) If i and $i+1$ sit in the same column, then $r' = r+1$ and $s' = s$; therefore, $d_T(i) = -1$.

   (c) Suppose that i and $i+1$ sit neither in the same column nor in the same row. If $r' > r$, then necessarily $s' < s$. Otherwise, consider the label k sitting in the box (r',s) of T. Since T is standard, $i < k < i+1$, which is impossible. Therefore,

$$
d_T(i) = (s' - s) - (r' - r) \leq -1 - 1 = -2.
$$

If $r' < r$, then for the same reason as above, we must have $s' > s$. In this case,

$$
d_T(i) = (s' - s) - (r' - r) \geq 1 + 1 = 2.
$$

In both cases, $|d_T(i)| \geq 2$.

   (d) The biggest value that $|d_T(i)|$ can reach occurs when one of the labels i or $i+1$ sits in the lowest box of the first column and the other one sits in the rightmost box of the first row. If the shape of T is the partition $(\lambda_1, \lambda_2, \ldots, \lambda_p)$ of n, then

$$
|d_T(i)| \leq (\lambda_1 - 1) + (p - 1) \leq \lambda_1 - 1 + \lambda_2 + \cdots + \lambda_p = n - 1. \qquad\square
$$

Exercise 5.1.1. (a) Define a binary relation $\leq$ on the set of all partitions by $\lambda \leq \lambda'$ if $D(\lambda) \subset D(\lambda')$. Show that $\leq$ is a partial order.
(b) Prove that for any partitions λ, λ', we have
    (i) $\lambda \wedge \lambda' \leq \lambda \leq \lambda \vee \lambda'$ and $\lambda \wedge \lambda' \leq \lambda' \leq \lambda \vee \lambda'$,
    (ii) if a partition μ satisfies $\mu \leq \lambda$ and $\mu \leq \lambda'$, then $\mu \leq \lambda \wedge \lambda'$,
    (iii) if a partition ν satisfies $\lambda \leq \nu$ and $\lambda' \leq \nu$, then $\lambda \vee \lambda' \leq \nu$.

5.2 The Young lattice

We provide the necessary background and then prove Theorem 5.1.

5.2.1 Corners

A *corner* of the diagram $D(\lambda)$ of a partition λ (or simply, a corner of λ) is a box centered on $(r, s) \in D(\lambda)$ such that neither $(r, s+1)$ nor $(r+1, s)$ belongs to $D(\lambda)$. In Figure 5.4 the three corners of $(3, 2, 2, 1)$ are marked.

It is clear that every nonempty partition has at least one corner, and distinct corners sit in distinct rows and in distinct columns. Moreover, every partition λ is determined by the set of its corners: the diagram of λ consists of the corners and the boxes lying to the left of a corner or above a corner.

Fig. 5.4. The corners of the partition $(3, 2, 2, 1)$

If (r, s) is a corner of $D(\lambda)$, then $\lambda_r > \lambda_{r+1}$. If we set

$$\mu_k = \begin{cases} \lambda_k & \text{if } k \neq r, \\ \lambda_k - 1 & \text{if } k = r, \end{cases}$$

then the sequence $(\mu_k)_k$ is nonincreasing and thus defines a partition μ of $n-1$, where $n = |\lambda|$. Clearly, $D(\mu) = D(\lambda) - \{(r, s)\}$. We say that μ is obtained from λ by *removing a corner*, which we symbolize by $\mu \hookrightarrow \lambda$. Figure 5.5 shows the three diagrams obtained by removing a corner from $D(3, 2, 2, 1)$.

Observe also that if $\lambda \vdash n$ and $\mu \vdash (n - 1)$ satisfy $D(\mu) \subset D(\lambda)$, then $\mu \hookrightarrow \lambda$, that is, μ is obtained from λ by removing a corner.

Lemma 5.4. *Let λ, λ' be distinct partitions of the same positive integer. Then there is at most one partition μ such that $\mu \hookrightarrow \lambda$ and $\mu \hookrightarrow \lambda'$, and there is at most one partition ν such that $\lambda \hookrightarrow \nu$ and $\lambda' \hookrightarrow \nu$. Moreover, there is μ such that $\mu \hookrightarrow \lambda$ and $\mu \hookrightarrow \lambda'$ if and only if there is ν such that $\lambda \hookrightarrow \nu$ and $\lambda' \hookrightarrow \nu$.*

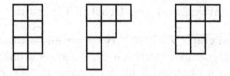

Fig. 5.5. The diagrams obtained by removing a corner of $D(3, 2, 2, 1)$

Proof. Let μ be such that $\mu \hookrightarrow \lambda$ and $\mu \hookrightarrow \lambda'$. Then $D(\mu) \subset D(\lambda) \cap D(\lambda')$ and card $D(\mu) = n - 1$, where $n = |\lambda| = |\lambda'| \geq 1$. Since $\lambda \neq \lambda'$, we have card $D(\lambda) \cap D(\lambda') < n$. It follows that

$$D(\mu) = D(\lambda) \cap D(\lambda') = D(\lambda \wedge \lambda'),$$

where $\lambda \wedge \lambda'$ is the partition defined in Section 5.1.3. Hence, $\mu = \lambda \wedge \lambda'$ and μ is necessarily unique. Note also that

$$\text{card } D(\lambda \vee \lambda') = \text{card}\big(D(\lambda) \cup D(\lambda')\big)$$
$$= \text{card } D(\lambda) + \text{card } D(\lambda') - \text{card}\big(D(\lambda) \cap D(\lambda')\big)$$
$$= 2n - (n - 1) = n + 1.$$

Hence, $\nu = \lambda \vee \lambda'$ is a partition of $n + 1$ such that $\lambda \hookrightarrow \nu$ and $\lambda' \hookrightarrow \nu$.

A similar argument shows that if ν is such that $\lambda \hookrightarrow \nu$ and $\lambda' \hookrightarrow \nu$, then necessarily $\nu = \lambda \vee \lambda'$ and the partition $\mu = \lambda \wedge \lambda'$ satisfies $\mu \hookrightarrow \lambda$ and $\mu \hookrightarrow \lambda'$. The conclusion of the lemma follows immediately. $\qquad\square$

Lemma 5.5. *Let $\lambda = (\lambda_1, \lambda_2, \ldots, \lambda_p)$ be an arbitrary partition. Suppose that there are ℓ partitions μ such that $\mu \hookrightarrow \lambda$. Then there are $\ell + 1$ partitions ν such that $\lambda \hookrightarrow \nu$.*

Proof. It is clear that (r, s) is a corner of λ if and only if $\lambda_r > \lambda_{r+1}$ (here we identify λ with an infinite nonincreasing eventually zero sequence of integers). Thus, the number ℓ of partitions μ such that $\mu \hookrightarrow \lambda$, which is the number of corners of λ, is equal to the number of all $r \geq 1$ such that $\lambda_r > \lambda_{r+1}$.

If a partition ν satisfies $\lambda \hookrightarrow \nu$, then

$$\nu_k = \begin{cases} \lambda_k & \text{if } k \neq r, \\ \lambda_k + 1 & \text{if } k = r \end{cases} \tag{5.2}$$

for some integer $r \geq 1$. If $r \geq 2$, then the assumption that ν is a partition implies that

$$\nu_{r-1} = \lambda_{r-1} \geq \nu_r = \lambda_r + 1,$$

from which it follows that $\lambda_{r-1} > \lambda_r$. Conversely, if $\lambda_{r-1} > \lambda_r$ for some $r \geq 2$, then (5.2) defines a partition ν such that $\lambda \hookrightarrow \nu$. The number of such partitions is equal to the number of all $r \geq 2$ such that $\lambda_{r-1} > \lambda_r$, hence to the number ℓ of all $r \geq 1$ such that $\lambda_r > \lambda_{r+1}$. But there is an additional ν such that $\lambda \hookrightarrow \nu$, namely the one given by $\nu_1 = \lambda_1 + 1$ and $\nu_k = \lambda_k$ for $k \geq 2$. In conclusion, the number of partitions ν such that $\lambda \hookrightarrow \nu$ is equal to $\ell + 1$. $\qquad\square$

5.2.2 The Young lattice and the Bratteli diagrams

Consider the oriented graph $\mathcal{Y}$ whose vertices are all partitions of nonnegative integers (including the empty partition $\emptyset$). There is a unique oriented edge $\mu \to \lambda$ in $\mathcal{Y}$ for each μ obtained from λ by removing a corner. This edge is also recorded by $\mu \hookrightarrow \lambda$. The graph $\mathcal{Y}$ is called the *Young lattice*.

For each $n \geq 0$, let $\mathcal{Y}_n$ be the finite oriented subgraph of $\mathcal{Y}$ whose vertices are the partitions λ with $|\lambda| \leq n$; any edge of $\mathcal{Y}$ between two vertices of $\mathcal{Y}_n$ is by definition an edge of $\mathcal{Y}_n$. The graphs $\mathcal{Y}_0, \mathcal{Y}_1, \mathcal{Y}_2, \ldots$ are called *Bratteli diagrams*. The Young lattice $\mathcal{Y}$ is the union of these graphs. Figure 5.6 represents $\mathcal{Y}_5$; this graph has 18 vertices and 25 edges.

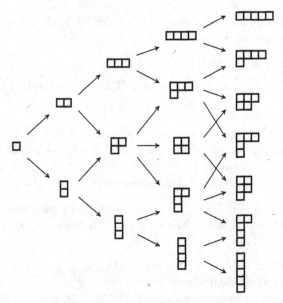

Fig. 5.6. The Bratteli diagram $\mathcal{Y}_5$

Lemma 5.6. *For any partition λ, the number $f^\lambda = \operatorname{card} \mathcal{T}_\lambda$ is equal to the number of oriented paths from $\emptyset$ to λ in $\mathcal{Y}$.*

Proof. The box with the largest label n in a standard tableau T of shape $\lambda \vdash n$ is necessarily a corner. Removing this box, we obtain a partition $\lambda^{(n-1)}$ of $n-1$ and a standard tableau of shape $\lambda^{(n-1)}$. Removing the corner labeled $n-1$ from the latter, we obtain a partition $\lambda^{(n-2)}$ of $n-2$ and a standard tableau of shape $\lambda^{(n-2)}$. Iterating this process until no boxes are left, we obtain an oriented path in $\mathcal{Y}$:

$$\emptyset = \lambda^{(0)} \hookrightarrow \lambda^{(1)} \hookrightarrow \cdots \hookrightarrow \lambda^{(n-1)} \hookrightarrow \lambda^{(n)} = \lambda. \qquad (5.3)$$

This path is uniquely determined by T.

Conversely, starting from an arbitrary oriented path (5.3) in $\mathcal{Y}$ from $\emptyset$ to $\lambda \dashv n$, we obtain a standard tableau of shape λ whose label i, where $i \in \{1, \ldots, n\}$, sits in the box added to $\lambda^{(i-1)}$ to obtain $\lambda^{(i)}$.

In this way, we obtain mutually inverse bijections between the set T_λ of standard tableaux of shape λ and the set of oriented paths from $\emptyset$ to λ in $\mathcal{Y}$. In particular, $f^\lambda = \operatorname{card} T_\lambda$ is the number of oriented paths from $\emptyset$ to λ in $\mathcal{Y}$. $\square$

The argument given in the proof of Lemma 5.6 shows that each graph $\mathcal{Y}_n$ and the Young lattice $\mathcal{Y} = \bigcup_n \mathcal{Y}_n$ are connected.

5.2.3 The operators D and U

Let $\mathbf{Z}[\mathcal{Y}]$ be the free abelian group with basis $\{\lambda\}$ indexed by all vertices of $\mathcal{Y}$. Define linear maps $D, U : \mathbf{Z}[\mathcal{Y}] \to \mathbf{Z}[\mathcal{Y}]$ by the following formulas: for $\lambda \dashv n \geq 1$, set

$$D(\lambda) = \sum_{\mu \hookrightarrow \lambda} \mu \quad \text{and} \quad U(\lambda) = \sum_{\lambda \hookrightarrow \nu} \nu.$$

Recall that for any partition λ of n, if $\mu \hookrightarrow \lambda$, then μ is a partition of $n-1$, and similarly if $\lambda \hookrightarrow \nu$, then ν is a partition of $n+1$. By definition, $D(\emptyset) = 0$ and $U(\emptyset) = \nu_0$, where $\nu_0 = (1)$ is the only partition of 1.

Let us record a combinatorial property of the operators D, U relating them to the integers f^λ. We use the following notation: for $k \geq 1$, let D^k (resp. U^k) be the composition of k copies of D (resp. of U). We also define D^0 and U^0 to be the identity map id of $\mathbf{Z}[\mathcal{Y}]$.

Lemma 5.7. *For any partition $\lambda \dashv n \geq 0$,*

$$D^n(\lambda) = f^\lambda \emptyset \quad \text{and} \quad U^n(\emptyset) = \sum_{\lambda \dashv n} f^\lambda \lambda.$$

Proof. It follows from the definitions that for each $k \geq 1$,

$$D^k(\lambda) = \sum_{\lambda^{(n-1)} \hookrightarrow \lambda} \sum_{\lambda^{(n-2)} \hookrightarrow \lambda^{(n-1)}} \cdots \sum_{\lambda^{(n-k)} \hookrightarrow \lambda^{(n-k+1)}} \lambda^{(n-k)}$$

$$= \sum_{\lambda^{(n-k)} \hookrightarrow \lambda^{(n-k+1)} \hookrightarrow \cdots \hookrightarrow \lambda^{(n-1)} \hookrightarrow \lambda} \lambda^{(n-k)} = \sum_{\mu \dashv (n-k)} f^\lambda_\mu \mu,$$

where f^λ_μ is the number of oriented paths in $\mathcal{Y}$ from μ to λ. For $k = n$, there is only one partition μ of $n - k = 0$, namely $\mu = \emptyset$, and then by Lemma 5.6, $f^\lambda_\mu = f^\lambda$. This yields the required formula for $D^n(\lambda)$.

A similar argument shows that for any partition $\mu \dashv m \geq 0$,

$$U^n(\mu) = \sum_{\lambda \dashv (m+n)} f^\lambda_\mu \lambda.$$

Applying this equality to $m = 0$ and $\mu = \emptyset$, we obtain the desired formula for $U^n(\emptyset)$. $\square$

The operators D and U enjoy the following remarkable property.

Lemma 5.8 (The Heisenberg relation). *We have* $DU - UD = \mathrm{id}$.

Proof. Let $\lambda \dashv n \geq 1$. By Lemma 5.4,

$$
\begin{aligned}
(DU)(\lambda) &= \sum_{\lambda \hookrightarrow \nu} D(\nu) = \sum_{\lambda \hookrightarrow \nu} \left(\sum_{\lambda' \hookrightarrow \nu} \lambda' \right) \\
&= a_\lambda^+ \lambda + \sum_{\lambda' \in A^+(\lambda)} \lambda',
\end{aligned}
\tag{5.4}
$$

where a_λ^+ is the number of partitions $\nu \dashv (n+1)$ such that $\lambda \hookrightarrow \nu$ and $A^+(\lambda)$ is the set of all $\lambda' \dashv n$ distinct from λ for which there is a (necessarily unique) partition $\nu \dashv (n+1)$ such that $\lambda \hookrightarrow \nu$ and $\lambda' \hookrightarrow \nu$.

Using the same lemma, we obtain

$$
\begin{aligned}
(UD)(\lambda) &= \sum_{\mu \hookrightarrow \lambda} U(\mu) = \sum_{\mu \hookrightarrow \lambda} \left(\sum_{\mu \hookrightarrow \lambda'} \lambda' \right) \\
&= a_\lambda^- \lambda + \sum_{\lambda' \in A^-(\lambda)} \lambda',
\end{aligned}
\tag{5.5}
$$

where a_λ^- is the number of partitions $\mu \dashv (n-1)$ such that $\mu \hookrightarrow \lambda$ and $A^-(\lambda)$ is the set of all $\lambda' \dashv n$ distinct from λ for which there is a (necessarily unique) partition $\mu \dashv (n-1)$ such that $\mu \hookrightarrow \lambda$ and $\mu \hookrightarrow \lambda'$.

The sets $A^+(\lambda)$ and $A^-(\lambda)$ coincide by Lemma 5.4, and $a_\lambda^+ = a_\lambda^- + 1$ by Lemma 5.5. Combining (5.4) and (5.5), we obtain

$$
(DU - UD)(\lambda) = \lambda.
$$

The same holds for $\lambda = \emptyset$, since $(DU - UD)(\emptyset) = D(\nu_0) = \emptyset$. □

Let us deduce the following more general formula: for each $n \geq 1$,

$$
DU^n - U^n D = n U^{n-1}.
\tag{5.6}
$$

This is proved by induction on n. If $n = 1$, then (5.6) coincides with the identity of Lemma 5.8. For $n \geq 2$, by the induction hypothesis and Lemma 5.8,

$$
\begin{aligned}
DU^n &= (DU^{n-1})U \\
&= \left(U^{n-1}D + (n-1)U^{n-2} \right) U \\
&= U^{n-1}DU + (n-1)U^{n-1} \\
&= U^{n-1}(UD + \mathrm{id}) + (n-1)U^{n-1} \\
&= U^n D + n U^{n-1}.
\end{aligned}
$$

5.2.4 Proof of Theorem 5.1

As an immediate consequence of Lemma 5.7, we obtain

$$(D^n U^n)(\emptyset) = \left(\sum_{\lambda \vdash n} (f^\lambda)^2 \right) \emptyset .$$

In order to prove Theorem 5.1, it therefore suffices to check that we also have

$$(D^n U^n)(\emptyset) = n! \, \emptyset .$$

We prove this equality by induction on n. The case $n = 0$ is trivial. For $n \geq 1$,

$$\begin{aligned}
(D^n U^n)(\emptyset) &= \left(D^{n-1}(DU^n) \right)(\emptyset) \\
&= \left(D^{n-1}(U^n D + n U^{n-1}) \right)(\emptyset) \\
&= (D^{n-1} U^n)(D(\emptyset)) + n \, (D^{n-1} U^{n-1})(\emptyset) \\
&= n(n-1)! \, \emptyset = n! \, \emptyset .
\end{aligned}$$

The second equality follows from (5.6), whereas the fourth equality follows from the induction hypothesis and from $D(\emptyset) = 0$. $\square$

Remark 5.9. The identity of Lemma 5.8 shows that $\mathbf{Z}[\mathcal{Y}]$ is a module over the *Weyl algebra* $\mathbf{Z}\langle D, U \rangle / (DU - UD - 1)$. Another classical example of a module over this algebra is given by the polynomials in one variable t, on which D acts by the derivation d/dt and U acts by the multiplication by t.

Exercise 5.2.1. Compute f^λ for all partitions $\lambda \vdash n$ with $n \leq 5$. (Hint: Use the Bratteli diagram $\mathcal{Y}_5$ in Figure 5.6.)

Exercise 5.2.2. Let $\lambda = (\lambda_1, \lambda_2, \ldots, \lambda_p)$ be a partition such that $p \geq 1$ and $\lambda_2 = \cdots = \lambda_p = 1$. Show that

$$f^\lambda = \binom{\lambda_1 + p - 2}{\lambda_1 - 1} .$$

(Hint: Use induction on $\lambda_1 + p$.)

Exercise 5.2.3. (a) Let $\lambda = (\lambda_1, \lambda_2)$ be a partition with two parts. Show that

$$f^\lambda = \binom{\lambda_1 + \lambda_2}{\lambda_2} - \binom{\lambda_1 + \lambda_2}{\lambda_2 - 1} = \frac{\lambda_1 - \lambda_2 + 1}{\lambda_1 + 1} \binom{\lambda_1 + \lambda_2}{\lambda_2} .$$

(b) Prove that

$$\sum_{\substack{\lambda_1 \geq \lambda_2 \geq 1 \\ \lambda_1 + \lambda_2 = n}} \left(f^{(\lambda_1, \lambda_2)} \right)^2 = \frac{1}{n+1} \binom{2n}{n} .$$

(Hint: Use the identity

$$\sum_{i=0}^{k} \binom{r}{i}\binom{s}{k-i} = \binom{r+s}{k},$$

where r, s, k are positive integers with $k \leq r + s$.)

Exercise 5.2.4. (a) Show that there is a unique family $g(\lambda_1, \ldots, \lambda_p)$ of integers, where $\lambda_1, \ldots, \lambda_p$ are arbitrary nonnegative integers with $p \geq 1$, such that

(i) $g(\lambda_1, \ldots, \lambda_p) = 0$ unless $\lambda_1 \geq \cdots \geq \lambda_p$,
(ii) $g(0) = 1$ and if $\lambda_p = 0$, then $g(\lambda_1, \ldots, \lambda_{p-1}, \lambda_p) = g(\lambda_1, \ldots, \lambda_{p-1})$,
(iii) if $\lambda_1 \geq \cdots \geq \lambda_p \geq 1$, then

$$g(\lambda_1, \ldots, \lambda_p) = \sum_{i=1}^{p} g(\lambda_1, \ldots, \lambda_i - 1, \ldots, \lambda_p).$$

(b) Prove that for any partition $\lambda = (\lambda_1, \lambda_2, \ldots, \lambda_p)$ of n, we have $f^\lambda = g(\lambda_1, \ldots, \lambda_p)$. (Hint: Giving a standard tableau with n boxes is the same as giving one with $n - 1$ boxes and saying where to put the nth box.)

Exercise 5.2.5. Let $x_1, \ldots, x_p$ be indeterminates and let $\Delta(x_1, \ldots, x_p)$ be the polynomial defined by

$$\Delta(x_1, \ldots, x_p) = \prod_{1 \leq i < j \leq p} (x_i - x_j)$$

if $p \geq 2$, and by $\Delta(x_1) = 1$ if $p = 1$.
    (a) Show that

$$\sum_{i=1}^{p} x_i \Delta(x_1, \ldots, x_i + y, \ldots, x_p) = \left(x_1 + \cdots + x_p + \frac{p(p-1)}{2} y \right) \Delta(x_1, \ldots, x_p).$$

(Hint: The left-hand side is a homogeneous polynomial, antisymmetric in $x_1, \ldots, x_p$.)
    (b) Show that the integers $g(\lambda_1, \ldots, \lambda_p)$ of Exercise 5.2.4 satisfy

$$\frac{g(\lambda_1, \ldots, \lambda_p)}{(\lambda_1 + \cdots + \lambda_p)!} = \frac{\Delta(\lambda_1 + p - 1, \lambda_2 + p - 2, \ldots, \lambda_p)}{(\lambda_1 + p - 1)! \, (\lambda_2 + p - 2)! \cdots \lambda_p!},$$

provided $\lambda_1 + p - 1 \geq \lambda_2 + p - 2 \geq \cdots \geq \lambda_p$.

Exercise 5.2.6 (Hook length formula). Let $D = D(\lambda)$ be the diagram of a partition λ. For $(i, j) \in D$, the *hook* $H_{i,j}$ consists of the box (i, j) together with the boxes of D lying below (i, j) in the same column or lying to the right of (i,j) in the same row. The number $h_{i,j}$ of boxes in $H_{i,j}$ is called the *hook length*, and it is computed by

$$h_{i,j} = \lambda_i + \lambda_j^T - i - j + 1,$$

where λ^T is the conjugate partition of λ.

(a) Prove that

$$\prod_{(i,j)\in D} h_{i,j} = \frac{(\lambda_1 + p - 1)!\,(\lambda_2 + p - 2)! \cdots \lambda_p!}{\Delta(\lambda_1 + p - 1, \lambda_2 + p - 2, \ldots, \lambda_p)}.$$

(b) Using Exercises 5.2.4 and 5.2.5, prove the *hook length formula*

$$f^\lambda = \frac{n!}{\prod_{(i,j)\in D} h_{i,j}}.$$

5.3 Seminormal representations

We return to the Iwahori–Hecke algebras $H_n^R(q)$ of Section 4.2.2 and construct an $H_n^R(q)$-module V_λ^R for each partition λ of n. We begin with some notation.

5.3.1 q-integers and q-factorials

Fix a commutative ring R and an invertible element $q \in R$. For each integer $n \geq 1$, set

$$[n]_q = 1 + q + \cdots + q^{n-1} \in R \tag{5.7}$$

and $[n]!_q = [1]_q [2]_q \cdots [n]_q \in R$. We also set $[0]_q = 0$ and

$$[n]_q = -q^n[-n]_q \tag{5.8}$$

for $n < 0$. Observe that $[1]_q = 1$, $[-1]_q = -q^{-1}$, and

$$[m + n]_q = [m]_q + q^m[n]_q = q^n[m]_q + [n]_q \tag{5.9}$$

for all integers m and n.

Given a positive integer n, we say that q is *n-regular* if $[n]!_q$ is invertible in R or, equivalently, if the elements $[1]_q, [2]_q, \ldots, [n]_q$ are invertible in R. If q is n-regular, then it is k-regular for $k = 1, \ldots, n$.

Recall the integers $d_T(1), \ldots, d_T(n-1)$ defined by (5.1), and set

$$a_T(i) = \frac{q^{d_T(i)}}{[d_T(i)]_q} \in R \quad \text{and} \quad b_T(i) = a_T(i) - q \in R. \tag{5.10}$$

Since $1 \leq |d_T(i)| \leq n - 1$ by Lemma 5.3, the elements $a_T(i)$ and $b_T(i)$ of R are well defined, provided q is $(n-1)$-regular.

We shall later use the obvious implication

$$d_T(i) = d_{T'}(j) \implies \left(a_T(i) = a_{T'}(j) \quad \text{and} \quad b_T(i) = b_{T'}(j) \right).$$

Lemma 5.10. *If q is $(n-1)$-regular, then*

$$a_T(i) = q \Leftrightarrow d_T(i) = 1 \quad \text{and} \quad a_T(i) = -1 \Leftrightarrow d_T(i) = -1. \qquad (5.11)$$

Proof. Set $d = d_T(i)$. Then

$$a_T(i) = q \iff [d]_q = q^{d-1} \iff [d-1]_q = 0.$$

Since q is $(n-1)$-regular and $d < n$, the number $[d-1]_q$ vanishes if and only if $d = 1$. The second equivalence is proved in a similar way.   □

5.3.2 The module V_λ

We now assume that the element $q \in R$ is $(n-1)$-regular and construct an $H_n^R(q)$-module V_λ^R for each partition $\lambda \dashv n$.

Consider the free R-module $V_\lambda = V_\lambda^R$ with basis $\{v_T\}_{T \in \mathcal{T}_\lambda}$, where $\mathcal{T}_\lambda$ is the set of standard tableaux of shape λ. Using the previously defined elements $a_T(i)$, $b_T(i)$ of R, we let the generators $T_1, \ldots, T_{n-1}$ of $H_n^R(q)$ act on the basis of V_λ by

$$T_i\, v_T = a_T(i)\, v_T + b_T(i)\, v_{s_i T}. \qquad (5.12)$$

Here $s_i T$ is the tableau obtained from T by switching the labels i and $i+1$. If $s_i T$ is not standard, then we set $v_{s_i T} = 0$. Observe that $a_T(i)$ is invertible in R.

Theorem 5.11. *Formula (5.12) defines the structure of a left $H_n^R(q)$-module on V_λ.*

A proof of Theorem 5.11 will be given in Section 5.4. The module V_λ is called a *seminormal representation* of $H_n^R(q)$. By definition, its rank over R is equal to the number f^λ of standard tableaux of shape λ or, equivalently, to the number of oriented paths from $\emptyset$ to λ in the Bratteli diagram $\mathcal{Y}_n$.

Examples 5.12. (a) Consider the partition $\lambda = (n)$ corresponding to a single row of n boxes. There is a unique standard tableau T of shape (n). Therefore the module $V_{(n)}$ has a unique basis vector v_T. By Lemma 5.3 and formulas (5.10), (5.12), the generators of $H_n^R(q)$ act on v_T by

$$T_i\, v_T = q\, v_T \qquad (5.13)$$

for all $i = 1, \ldots, n-1$.

(b) For the conjugate partition $(1, \ldots, 1)$, there is also a unique standard tableau T'. The module $V_{(1,\ldots,1)}$ has a unique basis vector $v_{T'}$. By Lemma 5.3 and formulas (5.10), (5.12), the generators of $H_n^R(q)$ act on $v_{T'}$ by

$$T_i\, v_{T'} = -v_{T'} \qquad (5.14)$$

for all $i = 1, \ldots, n-1$ (here $v_{s_i T'} = 0$ for all i).

Since q and -1 are the only roots of the polynomial $X^2 - (q-1)X - q$, the modules $V_{(n)}$ and $V_{(1,\ldots,1)}$ are the only $H_n^R(q)$-modules of rank one over R.

5.3.3 Restriction to $H_{n-1}^R(q)$

We now state an important property of the seminormal representations. We use the following notation: when an $H_n^R(q)$-module V is considered as an $H_{n-1}^R(q)$-module via the natural injection $\iota : H_{n-1}^R(q) \hookrightarrow H_n^R(q)$ (see Proposition 4.21), we denote it by $V|_{H_{n-1}^R(q)}$.

Proposition 5.13. *For any partition λ of n, there is a canonical isomorphism of $H_{n-1}^R(q)$-modules*

$$V_\lambda|_{H_{n-1}^R(q)} = \bigoplus_{\mu \hookrightarrow \lambda} V_\mu .$$

Proof. We observed in Section 5.1.5 that the label n in a standard tableau of shape λ sits necessarily in a corner of λ. Therefore we can partition the set of standard tableaux of shape λ according to the corner in which n sits. We thus obtain the partition

$$\mathcal{T}_\lambda = \coprod_{\mu \hookrightarrow \lambda} \mathcal{T}_\mu . \tag{5.15}$$

Since the basis $\{v_T\}$ of V_λ is indexed by the elements of $\mathcal{T}_\lambda$, we obtain an R-module decomposition

$$V_\lambda = \bigoplus_{\mu \hookrightarrow \lambda} V_\mu .$$

It follows from (5.12) that the generators $T_1, \ldots, T_{n-2}$ preserve this decomposition (but the generator T_{n-1} does not). $\qquad \square$

Remark 5.14. The seminormal representations behave well under a change of scalars. Let $f : R \to S$ be a homomorphism of commutative rings and q an $(n-1)$-regular invertible element of R. Then $\tilde{q} = f(q)$ is $(n-1)$-regular in S. In this situation we have the $H_n^R(q)$-module V_λ^R and the $H_n^S(\tilde{q})$-module V_λ^S. By Proposition 4.56,

$$S \otimes_R H_n^R(q) \cong H_n^S(\tilde{q}) .$$

Similarly, there is an isomorphism of $H_n^S(\tilde{q})$-modules

$$S \otimes_R V_\lambda^R \cong V_\lambda^S . \tag{5.16}$$

Let $R_0 = \mathbf{Q}[q_0, q_0^{-1}, ([n-1]!_{q_0})^{-1}]$ be the smallest subring of the field of rational functions $\mathbf{Q}(q_0)$ containing the ring of Laurent polynomials $\mathbf{Q}[q_0, q_0^{-1}]$ and the fraction $1/[n-1]!_{q_0}$. Clearly, q_0 is an $(n-1)$-regular invertible element of R_0. For any partition $\lambda \vdash n$, the construction above yields an $H_n^{R_0}(q_0)$-module $V_\lambda^{R_0}$. This module is universal in the following sense. For any commutative ring R and any $(n-1)$-regular invertible element $q \in R$, there is a unique ring homomorphism $f : R_0 \to R$ sending q_0 to q. By (5.16), we have an isomorphism of $H_n^R(q)$-modules

$$V_\lambda^R \cong R \otimes_{R_0} V_\lambda^{R_0} . \tag{5.17}$$

Remark 5.15. Applying the constructions above to $R = \mathbf{Q}$ and $q = 1$, we obtain for every partition $\lambda \vdash n$ a module $V_\lambda^{\mathbf{Q}}$ over $H_n^{\mathbf{Q}}(1) \cong \mathbf{Q}[\mathfrak{S}_n]$. In this way, the $H_n^R(q)$-module V_λ^R specializes to a representation of $\mathfrak{S}_n$.

Remark 5.16. Let $H_n^R(q)^\times$ be the group of invertible elements in $H_n^R(q)$. Recall the group homomorphism $\omega : B_n \to H_n^R(q)^\times$ sending the generator σ_i of the braid group B_n to T_i for $i = 1, \ldots, n-1$. For a partition λ of n, let

$$\pi_\lambda : H_n^R(q) \to \mathrm{End}_R(V_\lambda)$$

be the algebra homomorphism induced by the action of $H_n^R(q)$ on V_λ. Composing π_λ with $\omega : B_n \to H_n^R(q)^\times$, we obtain a group homomorphism $\rho_\lambda : B_n \to \mathrm{Aut}_R(V_\lambda)$. Since V_λ is a free R-module of rank f^λ, we can identify $\mathrm{Aut}_R(V_\lambda)$ with the group of invertible $f^\lambda \times f^\lambda$ matrices over R. We thus obtain a representation ρ_λ of B_n by matrices of size f^λ. By definition of $H_n^R(q)$, the matrix $\rho_\lambda(\sigma_i)$ with $i = 1, \ldots, n-1$ satisfies the quadratic relation

$$\rho_\lambda(\sigma_i)^2 - (q-1)\,\rho_\lambda(\sigma_i) - q\,I_{f^\lambda} = 0$$

(here I_{f^λ} stands for the unit $f^\lambda \times f^\lambda$ matrix).

5.4 Proof of Theorem 5.11

Set $P = \{\pm 1, \pm 2, \ldots, \pm(n-1)\} \subset \mathbf{Z}$. For any $d \in P$, set

$$f(d) = \frac{q^d}{[d]_q} \in R,$$

where $[d]_q$ was defined in Section 5.3.1. The element $f(d)$ of R is well defined and invertible, since q is invertible and $(n-1)$-regular in R.

Lemma 5.17. *Let $d, e \in P$ be such that $d + e \in P$. Then*
   (a) $f(d) + f(-d) = q - 1$,
   (b) $f(d)f(e) = f(d+e)\Big(f(d) - f(-e)\Big)$.

Proof. (a) By (5.8),

$$
\begin{aligned}
f(d) + f(-d) &= \frac{q^d}{[d]_q} + \frac{q^{-d}}{[-d]_q} \\
&= \frac{q^d}{[d]_q} - \frac{q^{-d}}{q^{-d}[d]_q} \\
&= \frac{q^d - 1}{[d]_q} \\
&= q - 1.
\end{aligned}
$$

(b) Using (5.8), (5.9), we obtain

$$\frac{f(d)f(e)}{f(d+e)} = \frac{[d+e]_q}{[d]_q[e]_q} = \frac{[d]_q + q^d[e]_q}{[d]_q[e]_q}$$

$$= \frac{1}{[e]_q} + \frac{q^d}{[d]_q}$$

$$= f(d) - \frac{q^{-e}}{[-e]_q}$$

$$= f(d) - f(-e). \qquad \square$$

To prove Theorem 5.11, it suffices to show that the operators $T_1, \ldots, T_{n-1}$ defined by (5.12) satisfy (4.16), (4.17), and (4.20).

5.4.1 Proof of (4.16)

If $|i - j| \geq 2$, then by (5.12),

$$T_j T_i v_T = a_T(i) a_T(j) v_T + a_T(i) b_T(j) v_{s_j T}$$
$$+ b_T(i) a_{s_i T}(j) v_{s_i T} + b_T(i) b_{s_i T}(j) v_{s_j s_i T}$$
$$= a_T(i) a_T(j) v_T + a_T(i) b_T(j) v_{s_j T}$$
$$+ b_T(i) a_T(j) v_{s_i T} + b_T(i) b_T(j) v_{s_j s_i T}.$$

The last equality holds since by Lemma 5.2 (a),

$$d_{s_i T}(j) = d_T(j).$$

The scalars $a_T(j)$ and $b_T(j)$ being functions of $d_T(j)$, we obtain

$$a_{s_i T}(j) = a_T(j) \quad \text{and} \quad b_{s_i T}(j) = b_T(j).$$

Moreover, $s_j s_i = s_i s_j$. Therefore, the expression $T_j T_i v_T$ is symmetric in i and j. Hence, $T_j T_i v_T = T_i T_j v_T$ for all T.

5.4.2 Proof of (4.17)

Let $i \in \{1, \ldots, n-2\}$. We have

$$T_i T_{i+1} T_i v_T = \big(a_T(i) a_T(i+1) a_T(i) + b_T(i) a_{s_i T}(i+1) b_{s_i T}(i)\big) v_T$$
$$+ \big(a_T(i) a_T(i+1) b_T(i) + b_T(i) a_{s_i T}(i+1) a_{s_i T}(i)\big) v_{s_i T}$$
$$+ a_T(i) b_T(i+1) a_{s_{i+1} T}(i) v_{s_{i+1} T}$$
$$+ a_T(i) b_T(i+1) b_{s_{i+1} T}(i) v_{s_i s_{i+1} T}$$
$$+ b_T(i) b_{s_i T}(i+1) a_{s_{i+1} s_i T}(i) v_{s_{i+1} s_i T}$$
$$+ b_T(i) b_{s_i T}(i+1) b_{s_{i+1} s_i T}(i) v_{s_i s_{i+1} s_i T}.$$

Similarly,

$$T_{i+1}T_iT_{i+1}\, v_T$$

$$= \Big(a_T(i+1)\,a_T(i)\,a_T(i+1) + b_T(i+1)\,a_{s_{i+1}T}(i)\,b_{s_{i+1}T}(i+1)\Big)\, v_T$$

$$+ \Big(a_T(i+1)\,a_T(i)\,b_T(i+1) + b_T(i+1)\,a_{s_{i+1}T}(i)\,a_{s_{i+1}T}(i+1)\Big)\, v_{s_{i+1}T}$$

$$+ a_T(i+1)\,b_T(i)\,a_{s_iT}(i+1)\, v_{s_iT}$$

$$+ a_T(i+1)\,b_T(i)\,b_{s_iT}(i+1)\, v_{s_{i+1}s_iT}$$

$$+ b_T(i+1)\,b_{s_{i+1}T}(i)\,a_{s_is_{i+1}T}(i+1)\, v_{s_is_{i+1}T}$$

$$+ b_T(i+1)\,b_{s_{i+1}T}(i)\,b_{s_is_{i+1}T}(i+1)\, v_{s_{i+1}s_is_{i+1}T}\,.$$

In order to prove the vanishing of the vector

$$w = T_iT_{i+1}T_i\, v_T - T_{i+1}T_iT_{i+1}\, v_T\,,$$

it suffices to check that the coefficient of each of the six vectors v_T, v_{s_iT}, $v_{s_{i+1}T}$, $v_{s_is_{i+1}T}$, $v_{s_{i+1}s_iT}$, $v_{s_is_{i+1}s_iT}$ in w vanishes.
(a) The coefficient A of v_T in w is given by

$$A = a_T(i)\,a_T(i+1)\,a_T(i) + b_T(i)\,a_{s_iT}(i+1)\,b_{s_iT}(i)$$
$$- a_T(i+1)\,a_T(i)\,a_T(i+1) - b_T(i+1)\,a_{s_{i+1}T}(i)\,b_{s_{i+1}T}(i+1)\,.$$

By Lemma 5.2 (b), $a_{s_iT}(i+1) = a_{s_{i+1}T}(i)$ and

$$A = a_T(i)a_T(i+1)\big(a_T(i) - a_T(i+1)\big)$$
$$+ a_{s_iT}(i+1)\big(b_T(i)b_{s_iT}(i) - b_T(i+1)b_{s_{i+1}T}(i+1)\big)\,.$$

Set $d = d_T(i)$ and $e = d_T(i+1)$. By (5.10),

$$a_T(i) = f(d_T(i)) = f(d)\,,\quad b_T(i) = f(d) - q\,,$$
$$a_T(i+1) = f(d_T(i+1)) = f(e)\,,\quad\text{and}\quad b_T(i+1) = f(e) - q\,.$$

By Lemma 5.2 (b),

$$d_{s_iT}(i) = -d\,,\quad d_{s_iT}(i+1) = d_{s_{i+1}T}(i) = d+e\,,\quad d_{s_{i+1}T}(i+1) = -e\,,$$

so that $a_{s_iT}(i+1) = a_{s_{i+1}T}(i) = f(d+e)$ and

$$b_{s_iT}(i) = f(-d) - q\,,\quad b_{s_{i+1}T}(i+1) = f(-e) - q\,.$$

Therefore,

$$A = f(d)f(e)\big(f(d) - f(e)\big)$$
$$+ f(d+e)\big((f(d) - q)(f(-d) - q) - (f(e) - q)(f(-e) - q)\big)\,.$$

Using Lemma 5.17 (b), we obtain $A = f(d + e) A_0$, where

$$A_0 = \big(f(d) - f(-e)\big)\big(f(d) - f(e)\big)$$
$$+ \big(f(d) - q\big)\big(f(-d) - q\big) - \big(f(e) - q\big)\big(f(-e) - q\big)$$
$$= \big(f(d) - q\big)\Big(\big(f(d) + f(-d)\big) - \big(f(e) + f(-e)\big)\Big).$$

Using Lemma 5.17 (a), we obtain $A_0 = 0$. Hence, $A = f(d + e) A_0 = 0$.

(b) The coefficient B of $v_{s_i T}$ in w is

$$a_T(i)\, a_T(i + 1)\, b_T(i) + b_T(i)\, a_{s_i T}(i + 1)\, a_{s_i T}(i) - a_T(i + 1)\, b_T(i)\, a_{s_i T}(i + 1)$$
$$= b_T(i)\Big(a_T(i)\, a_T(i + 1) - \big(a_T(i + 1) - a_{s_i T}(i)\big)\, a_{s_i T}(i + 1)\Big).$$

We set $d = d_T(i)$ and $e = d_T(i + 1)$ as above. Then by Lemma 5.2 (b),

$$B = b_T(i)\Big(f(d)f(e) - \big(f(e) - f(-d)\big)f(d + e)\Big).$$

The latter vanishes by Lemma 5.17 (b) (with d and e exchanged).

(c) The coefficient C of $v_{s_{i+1} T}$ in w is

$$a_T(i)b_T(i+1)a_{s_{i+1}T}(i) - a_T(i+1)a_T(i)b_T(i+1) - b_T(i+1)a_{s_{i+1}T}(i)a_{s_{i+1}T}(i+1)$$
$$= b_T(i + 1)\big(a_T(i)\, a_{s_{i+1}T}(i) - a_T(i)\, a_T(i + 1) - a_{s_{i+1}T}(i)\, a_{s_{i+1}T}(i + 1)\big).$$

Using the same notation as in (a) and (b) and using Lemma 5.2 (b), we obtain

$$C = b_T(i + 1)\big(f(d)f(d + e) - f(d)f(e) - f(d + e)f(-e)\big),$$

which vanishes by Lemma 5.17 (b).

(d) The coefficient of $v_{s_i s_{i+1} T}$ in w is

$$a_T(i)\, b_T(i + 1)\, b_{s_{i+1}T}(i) - b_T(i + 1)\, b_{s_{i+1}T}(i)\, a_{s_i s_{i+1}T}(i + 1),$$

which is equal to

$$\big(a_T(i) - a_{s_i s_{i+1}T}(i + 1)\big)b_T(i + 1)\, b_{s_{i+1}T}(i) = 0$$

because $d_T(i) = d_{s_i s_{i+1}T}(i + 1)$ (Lemma 5.2 (b)).

(e) The coefficient of $v_{s_{i+1} s_i T}$ in w is

$$b_T(i)\, b_{s_i T}(i + 1)\, a_{s_{i+1} s_i T}(i) - a_T(i + 1)\, b_T(i)\, b_{s_i T}(i + 1),$$

which is equal to

$$\big(a_{s_{i+1} s_i T}(i) - a_T(i + 1)\big)b_T(i)\, b_{s_i T}(i + 1) = 0$$

because $d_{s_{i+1} s_i T}(i) = d_T(i + 1)$ (Lemma 5.2 (b)).

(f) The coefficient of $v_{s_i s_{i+1} s_i T}$ in w is

$$b_T(i)\, b_{s_i T}(i + 1)\, b_{s_{i+1} s_i T}(i) - b_T(i + 1)\, b_{s_{i+1}T}(i)\, b_{s_i s_{i+1}T}(i + 1).$$

It vanishes because of the following equalities of Lemma 5.2 (b):

$$d_T(i) = d_{s_i s_{i+1}T}(i + 1), \quad d_{s_i T}(i + 1) = d_{s_{i+1}T}(i), \quad d_{s_{i+1} s_i T}(i) = d_T(i + 1).$$

5.4.3 Proof of (4.20)

If i and $i+1$ are in the same row of T, then it follows from Lemma 5.3 (b) and (5.11) that $a_T(i) = q$ and $b_T(i) = 0$. Therefore, T_i acts on v_T by

$$T_i\, v_T = q\, v_T\,.$$

Then

$$\left(T_i^2 - (q-1)T_i - q\right) v_T = \left(q^2 - (q-1)q - q\right) v_T = 0\,.$$

If i and $i+1$ are in the same column of T, then by Lemma 5.3 (c) and (5.11), $a_T(i) = -1$. Since $s_i T$ is not standard, $v_{s_i T} = 0$ and

$$T_i\, v_T = -v_T\,,$$

from which it also follows that $\left(T_i^2 - (q-1)T_i - q\right) v_T = 0$.

If i and $i+1$ are neither in the same row nor in the same column of T, then $\{v_T, v_{s_i T}\}$ spans a rank-two free R-submodule of V_λ. The generator T_i acts on this based submodule via the matrix

$$M = \begin{pmatrix} a_T(i) & b_{s_i T}(i) \\ b_T(i) & a_{s_i T}(i) \end{pmatrix}.$$

In order to check (4.20) on this submodule, it suffices to prove that the trace of M equals $q-1$ and its determinant equals $-q$.

Set $d = d_T(i)$. It follows from Lemmas 5.2 (a) and 5.17 (a) that

$$\operatorname{Tr} M = a_T(i) + a_{s_i T}(i) = f(d) + f(-d) = q - 1$$

and

$$
\begin{aligned}
\det M &= a_T(i)\, a_{s_i T}(i) - b_T(i)\, b_{s_i T}(i) \\
&= f(d)f(-d) - (f(d) - q)(f(-d) - q) \\
&= (f(d) + f(-d))q - q^2 \\
&= (q-1)q - q^2 = -q\,.
\end{aligned}
$$

This completes the proof of relations (4.16), (4.17), (4.20) and of Theorem 5.11.   □

Exercise 5.4.1. Let f, g be functions from the set $P = \{\pm1, \pm2, \ldots, \pm(n-1)\}$ to the set of invertible elements of a commutative ring R. For any standard tableau T with n boxes and any $i = 1, \ldots, n-1$, set $a_T(i) = f(d_T(i)) \in R$ and $b_T(i) = g(d_T(i)) \in R$.

(a) Show that formula (5.12) defines the structure of a left $H_n^R(q)$-module on V_λ, provided f and g satisfy the following three conditions:

(i) $f(1) = q$ or $f(1) = -1$,
(ii) for all $d \in P$,

$$f(d) + f(-d) = q - 1 \quad \text{and} \quad g(d)\, g(-d) = f(d)f(-d) + q\,,$$

(iii) for all $d, e \in P$ such that $d + e \in P$,

$$f(d + e)\big(f(d) - f(-e)\big) = f(d)f(e).$$

(b) Show that if the conditions in (a) are satisfied, then for all $d \in P$,

$$f(d) = \begin{cases} q^d/[d]_q & \text{if } f(1) = q, \\ -1/[d]_q & \text{if } f(1) = -1, \end{cases}$$

and

$$g(d)\, g(-d) = q\, \frac{[d-1]_q\,[d+1]_q}{([d]_q)^2}.$$

Exercise 5.4.2. Let K be an algebraically closed field of characteristic zero and λ a partition of n. Show that the formulas

$$s_i\, v_T = \frac{1}{d_T(i)}\, v_T + \frac{1 - d_T(i)}{d_T(i)}\, v_{s_i T}$$

and

$$s_i\, v_T = \frac{1}{d_T(i)}\, v_T + \frac{\sqrt{d_T(i)^2 - 1}}{d_T(i)}\, v_{s_i T},$$

where $i = 1, \ldots, n - 1$, define two $K[\mathfrak{S}_n]$-module structures on the K-vector space with a basis $\{v_T\}_T$ indexed by the standard tableaux T of shape λ.

5.5 Simplicity of the seminormal representations

In this section K is an algebraically closed field whose characteristic does not divide $n!$, where n is a fixed positive integer. Let $q \in K - \{0\}$ be such that q is $(n-1)$-regular and the Iwahori–Hecke algebras $H_2^K(q), \ldots, H_n^K(q)$ are semisimple. By the definition of $(n-1)$-regularity and by Theorem 4.57, this holds for all values of q except a finite number of algebraic elements of $K - \{0, 1\}$. We freely use the definitions of Section 4.5. To simplify notation, set $V_\lambda = V_\lambda^K$ for any partition λ.

Theorem 5.18. *The $H_n^K(q)$-module V_λ is simple for any partition λ of n. For any simple finite-dimensional $H_n^K(q)$-module V, there is a unique partition $\lambda \vdash n$ such that $V \cong V_\lambda$.*

Since $H_n^K(1) \cong K[\mathfrak{S}_n]$, the theorem in particular provides a classification of the irreducible representations of the symmetric groups over K.

Proof. We proceed by induction on n. When $n = 1$, we have $\lambda = (1)$. As observed in Example 5.12 (a), the module $V_{(1)}$ is one-dimensional, hence simple. Since $H_1^K(q) = K$, it is clear that any simple $H_1^K(q)$-module is isomorphic to $V_{(1)} = K$.

Suppose that V_μ is a simple $H_{n-1}^K(q)$-module for any partition μ of $n-1$ and that any simple $H_{n-1}^K(q)$-module is isomorphic to a unique module of the form V_μ. The uniqueness in the latter assumption means that $V_\mu \cong V_{\mu'}$ implies $\mu = \mu'$.

Let us first show that if $V_\lambda \cong V_{\lambda'}$ is an isomorphism of $H_n^K(q)$-modules, where λ and λ' are partitions of n, then $\lambda = \lambda'$. Indeed, by Proposition 5.13, we have an isomorphism of $H_{n-1}^K(q)$-modules

$$\bigoplus_{\mu \hookrightarrow \lambda} V_\mu \cong \bigoplus_{\mu' \hookrightarrow \lambda'} V_{\mu'}.$$

By assumption, $H_{n-1}^K(q)$ is semisimple, and by induction, the modules V_μ and $V_{\mu'}$ are simple. Therefore, by Proposition 4.32,

$$\{\mu \dashv (n-1) \,|\, \mu \hookrightarrow \lambda\} = \{\mu' \dashv (n-1) \,|\, \mu' \hookrightarrow \lambda'\}.$$

Since the set of corners of λ is the complement of $\bigcap_{\mu \hookrightarrow \lambda} D(\mu)$ in $D(\lambda)$, the partitions λ and λ' have the same corners. Therefore, $\lambda = \lambda'$, since every partition is determined by its corners.

We next show that the $H_n^K(q)$-module V_λ is simple for any partition λ of n. Let V be a nonzero $H_n^K(q)$-submodule of V_λ. Consider V and V_λ as $H_{n-1}^K(q)$-modules. By Proposition 5.13,

$$V_\lambda = \bigoplus_{\mu \hookrightarrow \lambda} V_\mu.$$

By the induction hypothesis, this is a direct sum decomposition into simple $H_{n-1}^K(q)$-modules, and the modules V_μ in this decomposition are pairwise nonisomorphic. Pick a nonzero simple $H_{n-1}^K(q)$-submodule V' of V. We claim that there is $\mu \hookrightarrow \lambda$ such that $V' = V_\mu$. Indeed, since

$$\bigoplus_{\mu \hookrightarrow \lambda} \mathrm{Hom}_{H_{n-1}^K(q)}(V', V_\mu) = \mathrm{Hom}_{H_{n-1}^K(q)}(V', V_\lambda) \supset \mathrm{Hom}_{H_{n-1}^K(q)}(V', V)$$

is nonzero, by Proposition 4.30 (a), there is $\mu \hookrightarrow \lambda$ such that $V' \cong V_\mu$. If $\mu' \hookrightarrow \lambda$ is different from μ, then, since the modules V_μ are pairwise nonisomorphic, $\mathrm{Hom}_{H_{n-1}^K(q)}(V', V_{\mu'}) = 0$. Since the projection of V' on each summand $V_{\mu'}$ is zero except for $\mu' = \mu$, we conclude that $V' = V_\mu$.

If $\mu \dashv (n-1)$ is the only partition such that $\mu \hookrightarrow \lambda$, then $V' = V_\mu = V_\lambda$. Hence, $V = V_\lambda$, which shows that V_λ is simple.

Suppose that there is $\mu' \hookrightarrow \lambda$ distinct from μ. Assume that $D(\mu)$ is obtained from $D(\lambda)$ by removing the corner (r,s), and $D(\mu')$ is obtained from $D(\lambda)$ by removing the corner (r',s'). Clearly, $(r,s) \neq (r',s')$. Consider a standard tableau T of shape λ whose corner (r,s) is labeled n and whose corner (r',s') is labeled $n-1$ (such T obviously exists). Observe that the tableau $s_{n-1}T$ obtained from T by switching the labels $n-1$ and n is standard and consider the vector

$$T_{n-1} v_T = a_T(n-1) v_T + b_T(n-1) v_{s_{n-1}T} \in V_\lambda. \qquad (5.18)$$

Decompose this vector according to Proposition 5.13. By the definition of the inclusion $V_\mu \hookrightarrow V_\lambda$ given in the proof of Proposition 5.13,

$$v_T \in V' = V_\mu,$$

since removing the corner (r,s) with label n from λ yields μ. Similarly, $v_{s_{n-1}T} \in V_{\mu'}$. Since $n-1$ and n sit in corners of T and therefore sit neither in the same column nor in the same row, $d_T(n-1) \neq 1$ by Lemma 5.3 (c). This together with the equivalence (5.11) implies that $a_T(n-1) \neq q$, hence $b_T(n-1) \neq 0$. It then follows from (5.18) that $v_{s_{n-1}T} \in V_{\mu'}$ is a linear combination of v_T and $T_{n-1} v_T$, both belonging to V. Therefore, V contains a nonzero element of $V_{\mu'}$. Since $V_{\mu'}$ is a simple $H_{n-1}^K(q)$-module and $V \cap V_{\mu'}$ is a nonzero $H_{n-1}^K(q)$-submodule of $V_{\mu'}$, we have $V \cap V_{\mu'} = V_{\mu'}$, that is, $V \supset V_{\mu'}$. Since this holds for all $\mu' \hookrightarrow \lambda$ distinct from μ and $V \supset V' = V_\mu$,

$$V \supset V_\mu \oplus \bigoplus_{\substack{\mu' \hookrightarrow \lambda \\ \mu' \neq \mu}} V_{\mu'} = V_\lambda \supset V.$$

Thus, $V = V_\lambda$ and V_λ is simple.

We finally show that any simple finite-dimensional $H_n^K(q)$-module is isomorphic to V_λ for some $\lambda \vdash n$. This follows from a simple counting argument. Since $H_n^K(q)$ is semisimple, it has a finite number of simple finite-dimensional modules (considered up to isomorphism). Such modules include the modules V_λ, which are pairwise nonisomorphic. If $H_n^K(q)$ had at least one nonzero simple finite-dimensional module not isomorphic to a module of the form V_λ, then by Corollary 4.55 and Theorem 5.1 we would have

$$\dim_K H_n^K(q) > \sum_{\lambda \vdash n} (\dim_K V_\lambda)^2 = \sum_{\lambda \vdash n} (f^\lambda)^2 = n!.$$

This contradicts Theorem 4.17, which yields $\dim_K H_n^K(q) = n!$. $\qquad\square$

For any partition λ of n, let $\pi_\lambda : H_n^K(q) \to \mathrm{End}_K(V_\lambda)$ be the algebra homomorphism induced by the action of $H_n^K(q)$ on V_λ. The next result follows immediately from Theorem 5.18 and Corollary 4.55.

Corollary 5.19. *The algebra homomorphisms π_λ induce an algebra isomorphism*

$$H_n^K(q) \xrightarrow{\ \cong\ } \prod_{\lambda \vdash n} \mathrm{End}_K(V_\lambda).$$

Exercise 5.5.1. Determine the dimensions of all simple modules of $H_n^K(q)$ for $n \leq 5$. (Hint: Use Exercise 5.2.1.)

Exercise 5.5.2. Let K be an algebraically closed field containing $\mathbf{Z}[q, q^{-1}]$. Show that there is an algebra isomorphism $H_n^K(q) \cong K[\mathfrak{S}_n]$. (Hint: Use Corollary 5.19 and Remark 5.15.)

Exercise 5.5.3. Show that the two $K[\mathfrak{S}_n]$-modules of Exercise 5.4.2 are simple and isomorphic. (Hint: Restrict to $K[\mathfrak{S}_{n-1}]$ and use induction on n.)

Exercise 5.5.4 (Path algebras). Let K be a field and n a positive integer.

(a) Let $\mathcal{P}_n$ be the K-vector space with basis $\{E_{S,T}\}_{S,T}$ indexed by all couples (S, T) of standard tableaux of the same shape λ, where $\lambda \dashv n$. We endow $\mathcal{P}_n$ with the structure of an algebra by

$$E_{S,T} E_{S',T'} = \begin{cases} E_{S,T'} & \text{if } T = S', \\ 0 & \text{if } T \neq S'. \end{cases}$$

The vector $\sum_T E_{T,T}$ is the unit of this algebra. (The algebra $\mathcal{P}_n$ is called a *path algebra*.) Show that $\mathcal{P}_n$ is isomorphic to a product of matrix algebras

$$\mathcal{P}_n \cong \prod_{\lambda \dashv n} M_{f^\lambda}(K),$$

where f^λ is the number of standard tableaux of shape λ. (Hint: Consider first the elements $E_{S,T}$, where S, T are standard tableaux of a given shape λ and show that they span a subalgebra of $\mathcal{P}_n$, which is isomorphic to $M_{f^\lambda}(K)$.)

(b) To any basis element $E_{S,T}$ of $\mathcal{P}_{n-1}$ associate the element

$$i(E_{S,T}) = \sum_{S',T'} E_{S',T'} \in \mathcal{P}_n,$$

where S' and T' run over all standard tableaux with n labels obtained from S and T respectively by adding a box with label n. Show that this defines an injective algebra homomorphism $i : \mathcal{P}_{n-1} \to \mathcal{P}_n$.

(c) For $\lambda \dashv n$, define a $\mathcal{P}_n$-module U_λ as the K-vector space with basis $\{u_T\}_T$ indexed by all standard tableaux T of shape λ, and with $\mathcal{P}_n$-action

$$E_{S,T} u_{T'} = \begin{cases} u_S & \text{if } T = T', \\ 0 & \text{if } T \neq T'. \end{cases}$$

Show that, considered as a $\mathcal{P}_{n-1}$-module via the embedding $i : \mathcal{P}_{n-1} \hookrightarrow \mathcal{P}_n$,

$$U_\lambda \cong \bigoplus_{\mu \hookrightarrow \lambda} U_\mu.$$

(d) Prove that U_λ is a simple $\mathcal{P}_n$-module for each $\lambda \dashv n$, and that any simple $\mathcal{P}_n$-module is isomorphic to a module of this form.

5.6 Simplicity of the reduced Burau representation

In this section we show that the reduced Burau representation of the braid group introduced in Section 3.3.1 is irreducible.

We start with a property of the matrices $V_1, \ldots, V_{n-1} \in \mathrm{GL}_{n-1}(\Lambda)$ exhibited in Theorem 3.9, where $\Lambda = \mathbf{Z}[t, t^{-1}]$. Let K be an algebraically closed field containing Λ (we may take $K = \mathbf{C}$). Consider the $(n-1)$-dimensional vector space $\mathcal{L}_{n-1}$ consisting of all columns over K of height $n-1$. The matrices $V_1, \ldots, V_{n-1}$ act on $\mathcal{L}_{n-1}$ by left matrix multiplication.

Lemma 5.20. *Let $n \geq 3$ and $\alpha \in K$. The only vector $v \in \mathcal{L}_{n-1}$ satisfying $V_i v = \alpha v$ for all $i = 1, \ldots, n-1$ is zero.*

Proof. It is obvious from the form of V_1 that its only eigenvalues are 1 and $-t$. Therefore, it suffices to establish the lemma for $\alpha = 1$ and $\alpha = -t$.

It is easy to check that the eigenspace of the action of V_i on $\mathcal{L}_{n-1}$ for the eigenvalue -1 is the hyperplane of $\mathcal{L}_{n-1}$ consisting of the columns whose ith entries vanish. The intersection of these hyperplanes is clearly zero.

Consequently, the eigenspace of V_i for the second eigenvalue, that is, for $-t$, is one-dimensional. It suffices to prove that the one-dimensional subspaces corresponding to $i = 1$ and $i = 2$ do not coincide. A quick check shows that for V_1 (resp. for V_2), this eigenspace is spanned by $(1+t)v_1 - v_2$ (resp. by $tv_1 - (1+t)v_2$), where $(v_1, \ldots, v_{n-1})$ is the canonical basis of $\mathcal{L}_{n-1}$. We conclude by noting that these two vectors are not collinear (here we use the fact that $t^2 + t + 1 \neq 0$ in K). $\qquad\square$

We next relate the matrices $V_1, \ldots, V_{n-1}$ to the Iwahori–Hecke algebra $H_n^K(t)$. (To be consistent with the notation of Chapter 3, we use the parameter t rather than the parameter q used in the previous sections of the present chapter.) By Theorem 4.57, since K has characteristic zero and $t \in K$ is nonalgebraic, each $H_n^K(t)$ is a semisimple algebra. Recall the generators $T_1, \ldots, T_{n-1}$ of the Iwahori–Hecke algebra.

Proposition 5.21. *There is a unique structure of an $H_n^K(t)$-module on $\mathcal{L}_{n-1}$ such that each generator T_i $(i = 1, \ldots, n-1)$ acts on $\mathcal{L}_{n-1}$ by multiplication by the matrix $-V_i$.*

Proof. We know from Section 3.3.1 that the matrices $V_1, \ldots, V_{n-1}$ satisfy relations (4.16) and (4.17). So do the matrices $-V_1, \ldots, -V_{n-1}$. It is easy to check that each V_i satisfies the equation

$$(V_i - I_{n-1})(V_i + tI_{n-1}) = V_i^2 + (t-1)V_i - tI_{n-1} = 0.$$

Hence,

$$(-V_i)^2 = (t-1)(-V_i) + tI_{n-1}$$

for all $i = 1, \ldots; n-1$. In other words, the matrices $-V_1, \ldots, -V_{n-1}$ satisfy relation (4.20) with q replaced by t. $\qquad\square$

By Theorem 5.18, the $H_n^K(t)$-module $\mathcal{L}_{n-1}$ is a direct sum of simple modules of the form V_λ, where λ is a partition of n (see Section 5.3 for a definition of V_λ). As a matter of fact, as we shall see now, the module $\mathcal{L}_{n-1}$ is simple.

Theorem 5.22. *There is an isomorphism of $H_n^K(t)$-modules*

$$\mathcal{L}_{n-1} \cong V_{\lambda[n]},$$

where $\lambda[n]$ is the partition $(2,1,1,\ldots,1)$ of n.

Proof. We prove the theorem by induction on n.

(a) If $n = 2$, then by Section 3.3.1, T_1 acts via the 1×1 matrix $[t]$. Setting $t = q$ in (5.13), we see that $\mathcal{L}_1$ is the simple module $V_{\lambda[2]}$, where $\lambda[2] = (2)$ is the partition whose diagram consists of a single row of two boxes.

(b) Assume that the theorem holds for all positive $k < n$, where n is a given integer ≥ 3. Consider the natural projection $\mathcal{L}_{n-1} \to \mathcal{L}_{n-2}$ obtained by deleting the bottom entry of a column in $\mathcal{L}_{n-1}$. Observe that the matrices $-V_1, \ldots, -V_{n-2}$ are all of the form

$$-V_i = \begin{pmatrix} -V_i^0 & 0 \\ b_i & -1 \end{pmatrix},$$

where

$$V_i^0 \in \mathrm{GL}_{n-2}(\Lambda)$$

is the matrix defining the reduced Burau representation of B_{n-1}, and where b_i is the row of length $n - 2$ equal to 0 if $i < n - 2$ and to $(0, \ldots, 0, -1)$ if $i = n - 2$. Thus, the projection $\mathcal{L}_{n-1} \to \mathcal{L}_{n-2}$ induces an exact sequence of $H_{n-1}^K(t)$-modules

$$0 \to \mathcal{V} \to \mathcal{L}_{n-1}|_{H_{n-1}^K(t)} \to \mathcal{L}_{n-2} \to 0,$$

where $\mathcal{L}_{n-1}|_{H_{n-1}^K(t)}$ is $\mathcal{L}_{n-1}$ considered as an $H_{n-1}^K(t)$-module via the natural inclusion $H_{n-1}^K(t) \hookrightarrow H_n^K(t)$, and $\mathcal{V}$ is the one-dimensional $H_{n-1}^K(t)$-module consisting of the columns whose first $n - 2$ entries vanish. Since $T_1, \ldots, T_{n-2}$ act on $\mathcal{V}$ by -1, the $H_{n-1}^K(t)$-module $\mathcal{V}$ is isomorphic to $V_{\mu[n-1]}$, where $\mu[n-1]$ is the partition $(1, \ldots, 1)$ of $n - 1$. Since $H_{n-1}^K(t)$ is semisimple, the module $\mathcal{L}_{n-1}|_{H_{n-1}^K(t)}$ is semisimple, hence completely reducible by Proposition 4.28. Thus there is an isomorphism of $H_{n-1}^K(t)$-modules

$$\mathcal{L}_{n-1}|_{H_{n-1}^K(t)} \cong \mathcal{L}_{n-2} \oplus \mathcal{V}.$$

Using the induction hypothesis, we obtain the following isomorphisms of $H_{n-1}^K(t)$-modules:

$$\mathcal{L}_{n-1}|_{H_{n-1}^K(t)} \cong \mathcal{L}_{n-2} \oplus \mathcal{V} \cong V_{\lambda[n-1]} \oplus V_{\mu[n-1]}. \tag{5.19}$$

By (5.19) and Proposition 5.13, if V_λ occurs in $\mathcal{L}_{n-1}$, then the diagram of λ is such that by removing *any* of its corners, we obtain the diagram of $\mu[n-1]$ or the diagram of $\lambda[n-1]$ (and only those). Now, by Lemma 5.20, $\mathcal{L}_{n-1}$ cannot contain a one-dimensional representation. This fact, together with Examples 5.12, shows that the diagram of λ has at least two rows and two columns. We are thus left with a very little choice for λ: such a partition λ is necessarily equal to $\lambda[n]$ or to the partition $(2,2)$ of $n = 4$.

Therefore, if $n \neq 4$, then $\mathcal{L}_{n-1} \cong \left(V_{\lambda[n]}\right)^a$ for some nonnegative integer a. Restricting to $H_{n-1}^K(t)$, we obtain an isomorphism

$$\mathcal{L}_{n-1}|_{H_{n-1}^K(t)} \cong \left(V_{\lambda[n-1]}\right)^a \oplus \left(V_{\mu[n-1]}\right)^a.$$

Comparing with (5.19) and using Proposition 4.32, we obtain $a = 1$, which proves the theorem for $n \neq 4$.

For $n = 4$,

$$\mathcal{L}_3 \cong \left(V_{\lambda[4]}\right)^a \oplus \left(V_{(2,2)}\right)^b$$

for some nonnegative integers a, b. Restricting to $H_2^K(t)$, we obtain

$$\mathcal{L}_3|_{H_2^K(t)} \cong \left(V_{\lambda[3]}\right)^{(a+b)} \oplus \left(V_{\mu[3]}\right)^a.$$

Comparing with (5.19), we obtain $a + b = 1$ and $a = 1$, hence $b = 0$, which concludes the proof in the case $n = 4$. □

Corollary 5.23. *The reduced Burau representation* $\psi_n^r : B_n \to \mathrm{GL}_{n-1}(K)$ *is irreducible.*

Proof. The irreducibility of ψ_n^r means that the only subspaces of K^{n-1} preserved by ψ_n^r are 0 and K^{n-1}. If W is such a subspace, then

$$(-V_i)W = V_i W \subset W$$

for all $i = 1, \ldots, n - 1$. By definition of the action of $H_n^K(t)$ on $\mathcal{L}_{n-1}$, the vector space W is an $H_n^K(t)$-submodule of $\mathcal{L}_{n-1}$. Since $\mathcal{L}_{n-1}$ is simple by Theorem 5.22, we must have $W = 0$ or $W = K^{n-1}$. □

Exercise 5.6.1. Check that

$$I_{n-1} - V_1 - V_2 + V_1 V_2 + V_2 V_1 - V_1 V_2 V_1 = 0.$$

Deduce that $\mathcal{L}_{n-1}$ is a module over the quotient of the algebra $H_n^K(t)$ by the two-sided ideal generated by

$$1 + T_1 + T_2 + T_1 T_2 + T_2 T_1 + T_1 T_2 T_1.$$

(This quotient will be further discussed in Section 5.7.2.)

5.7 The Temperley–Lieb algebras

We end this chapter by presenting a family of algebras closely related to the Iwahori–Hecke algebras.

5.7.1 Definition and reduced words

For simplicity we work over the field $\mathbf{C}$ of complex numbers. We fix an integer $n \geq 2$ and a nonzero complex number a.

Definition 5.24. *The Temperley–Lieb algebra $A_n(a)$ is the $\mathbf{C}$-algebra generated by $n-1$ elements $e_1, \ldots, e_{n-1}$ subject to the relations*

$$e_i e_j = e_j e_i \tag{5.20}$$

for $i, j = 1, 2, \ldots, n-1$ such that $|i - j| \geq 2$,

$$e_i e_j e_i = e_i \tag{5.21}$$

for $i, j = 1, 2, \ldots, n-1$ such that $|i - j| = 1$, and

$$e_i^2 = a e_i \tag{5.22}$$

for $i = 1, \ldots, n-1$.

Any word $e_{i_1} \cdots e_{i_r}$ in the alphabet $\{e_1, \ldots, e_{n-1}\}$ represents an element of $A_n(a)$. The empty word represents the unit 1 of $A_n(a)$.

We define the *index* of a nonempty word $w = e_{i_1} \cdots e_{i_r}$ to be the maximum of all indices $i_1, \ldots, i_r$ appearing in w. If the index of w is equal to p, then we say that e_p is the *maximal generator* of w. We agree that the index of the empty word is 0.

Lemma 5.25. *Any nonempty word $w = e_{i_1} \cdots e_{i_r}$ is equal in $A_n(a)$ to a scalar multiple of a word in which the maximal generator appears exactly once.*

Proof. We proceed by induction on the index p of w. If $p = 1$, then w is a positive power of e_1. From (5.22) we derive $e_1^i = a^{i-1} e_1$ for all $i > 1$. Therefore, Lemma 5.25 holds for $p = 1$.

Suppose that Lemma 5.25 holds for all words of index $< p$. Consider a nonempty word $w = e_{i_1} \cdots e_{i_r}$ of index p. Suppose that e_p appears in w at least twice. Then w is of the form $w = w_1 e_p w' e_p w_2$, where w_1 and w_2 are arbitrary words, and w' is a word of index $\ell < p$.

If $\ell < p - 1$, then by (5.20), w' commutes with e_p. Therefore, by (5.22),

$$w = w_1 e_p w' e_p w_2 = w_1 w' e_p^2 w_2 = a\, w_1 e_p w_2.$$

In this way we have diminished the number of occurrences of e_p in w by one.

If $\ell = p-1$, then by the induction hypothesis we may assume that $e_\ell = e_{p-1}$ appears only once in w', so that $w' = w_3 e_{p-1} w_4$, where w_3 and w_4 are words of index $\leq p - 2$. Therefore, w_3 and w_4 commute with e_p. Using (5.21), we obtain

$$w = w_1 e_p w' e_p w_2 = w_1 e_p w_3 e_{p-1} w_4 e_p w_2$$
$$= w_1 w_3 e_p e_{p-1} e_p w_4 w_2 = w_1 w_3 e_p w_4 w_2 \, .$$

We have again diminished the number of occurrences of e_p in w by one.

Proceeding recursively, we can transform w into a scalar multiple of a word in which the maximal generator appears exactly once. $\qquad\square$

For $1 \leq k \leq n-1$, let $E_{n,k}$ be the set of $2k$-tuples $(i_1, \ldots, i_k, j_1, \ldots, j_k)$ of integers such that

$$0 < i_1 < i_2 < \cdots < i_k < n, \quad 0 < j_1 < j_2 < \cdots < j_k < n \, ,$$

and

$$j_1 \leq i_1, \quad j_2 \leq i_2, \quad \ldots, \quad j_k \leq i_k \, .$$

For such a tuple $\underline{s} = (i_1, \ldots, i_k, j_1, \ldots, j_k)$, set

$$e_{\underline{s}} = (e_{i_1} e_{i_1 - 1} \cdots e_{j_1})(e_{i_2} e_{i_2 - 1} \cdots e_{j_2}) \cdots (e_{i_k} e_{i_k - 1} \cdots e_{j_k}) \, .$$

In the expression for $e_{\underline{s}}$ the indices are decreasing from left to right between each pair of parentheses. Observe that the index of $e_{\underline{s}}$ is i_k. We say that a word of the form $e_{\underline{s}}$, where

$$\underline{s} \in E_n = E_{n,1} \amalg E_{n,2} \amalg \cdots \amalg E_{n,n-1}$$

is a *reduced word* in $A_n(a)$.

Lemma 5.26. *The set $\{e_{\underline{s}}\}_{\underline{s} \in E_n}$ of reduced words spans $A_n(a)$.*

Proof. It is enough to prove that any word $w = e_{i_1} \cdots e_{i_r}$ is a scalar multiple of a reduced word. We proceed by induction on the index p of w.

If $p = 1$, then w is a scalar multiple of e_1, which is a reduced word.

Let $p > 1$ and assume that any word of index $< p$ is a scalar multiple of a reduced word of index $< p$. Let w be a word of index p. By Lemma 5.25, w is a scalar multiple of some word $w_0 = w_1 e_p w_2$, where w_1 and w_2 are words of index $< p$. By the induction hypothesis, we may assume that w_2 is reduced. Suppose that

$$w_2 = e_{\underline{s}} = (e_{i_1} e_{i_1 - 1} \cdots e_{j_1})(e_{i_2} e_{i_2 - 1} \cdots e_{j_2}) \cdots (e_{i_k} e_{i_k - 1} \cdots e_{j_k})$$

for some

$$\underline{s} = (i_1, \ldots, i_k, j_1, \ldots, j_k) \in E_{n,k}$$

with $i_k < p$.

If $i_k \leq p - 2$, then w_2 commutes with e_p and

$$e_p w_2 = (e_{i_1} e_{i_1-1} \cdots e_{j_1})(e_{i_2} e_{i_2-1} \cdots e_{j_2}) \cdots (e_{i_k} e_{i_k-1} \cdots e_{j_k})(e_p).$$

If $i_k = p - 1$, then

$$e_p w_2 = (e_{i_1} e_{i_1-1} \cdots e_{j_1})(e_{i_2} e_{i_2-1} \cdots e_{j_2}) \cdots (e_p e_{i_k} e_{i_k-1} \cdots e_{j_k}).$$

In both cases, w_0 is equal in $A_n(a)$ to a word of the form $w'(e_p e_{p-1} \cdots e_q)$, where w' is a word of index $p' < p$, and $q \leq p$. By the induction hypothesis, we may restrict ourselves to the case $w' = e_{\underline{s}'}$ for some $\underline{s}' = (i'_1, \ldots, i'_\ell, j'_1, \ldots, j'_\ell) \in E_{n,\ell}$. We have $i'_\ell = p' < p$; set $q' = j'_\ell$.

(i) If $q' < q$, then

$$w_0 = w'(e_p e_{p-1} \cdots e_q) = e_{\underline{s}'}(e_p e_{p-1} \cdots e_q)$$

is reduced.

(ii) If $q' \geq q$, then $w' = w''(e_{p'} e_{p'-1} \cdots e_{q'})$, where $q \leq q' \leq p' < p$ and w'' has index $< p'$. If $q' \leq p - 2$, then by (5.20) and (5.21),

$$e_{q'}(e_p e_{p-1} \cdots e_q) = e_p e_{p-1} \cdots e_{q'+2}(e_{q'} e_{q'+1} e_{q'}) e_{q'-1} \cdots e_q$$
$$= e_p e_{p-1} \cdots e_{q'+2} e_{q'} e_{q'-1} \cdots e_q$$
$$= (e_{q'} e_{q'-1} \cdots e_q)(e_p e_{p-1} \cdots e_{q'+2}).$$

Therefore, $w_0 = w''(e_{q'} e_{q'-1} \cdots e_q)(e_p e_{p-1} \cdots e_{q'+2})$. Since $w''(e_{q'} e_{q'-1} \cdots e_q)$ has index $< p$, the word w_0 is of the form considered in (i) and the result follows from (i).

If $q' = p - 1$, then $p' = q' = p - 1$, and by (5.21),

$$e_{q'}(e_p e_{p-1} \cdots e_q) = (e_{p-1} e_p e_{p-1}) \cdots e_q = e_{p-1} \cdots e_q.$$

Therefore, $w_0 = w''(e_{p-1} \cdots e_q)$, where w'' has index $< p' = p - 1$. Thus, w_0 has index $p - 1$, and the result follows from the induction assumption. $\square$

Lemma 5.27. *We have*

$$\operatorname{card} E_n = \frac{1}{n+1} \binom{2n}{n}.$$

The integer $\binom{2n}{n}/(n+1)$ is called the nth *Catalan number*.

Proof. To any element $(i_1, \ldots, i_k, j_1, \ldots, j_k) \in E_{n,k}$ we associate the path

$$(0,0) \rightarrow (i_1, 0) \rightarrow (i_1, j_1) \rightarrow (i_2, j_1) \rightarrow (i_2, j_2) \rightarrow \cdots$$
$$\cdots \rightarrow (i_k, j_{k-1}) \rightarrow (i_k, j_k) \rightarrow (n, j_k) \rightarrow (n, n)$$

in the set $(\mathbf{R} \times \mathbf{Z}) \cup (\mathbf{Z} \times \mathbf{R}) \subset \mathbf{R}^2$. This path is an oriented polygonal line, alternating horizontal and vertical edges, all horizontal edges being directed to the right and all vertical edges directed upward. Let us call such a path an *admissible path* from $(0,0)$ to (n,n). An admissible path arising from an element of E_n lies under the diagonal $\{(x,y) \in \mathbf{R}^2 \,|\, x = y\}$, that is, it lies in the octant $\{(x,y) \in \mathbf{R}^2 \,|\, 0 \le y \le x\}$. It is clear that any admissible path from $(0,0)$ to (n,n) lying under the diagonal can be obtained from a unique element of E_n in this way.

We now count the admissible paths from $(0,0)$ to (n,n) lying under the diagonal. Translating an admissible path from $(0,0)$ to (n,n) along the vector $(1,0)$, we obtain an admissible path from $(1,0)$ to $(n+1,n)$ not intersecting the diagonal. Conversely, any admissible path from $(1,0)$ to $(n+1,n)$ not intersecting the diagonal is the translation of a unique admissible path from $(0,0)$ to (n,n) lying under the diagonal.

To count the admissible paths from $(1,0)$ to $(n+1,n)$ not intersecting the diagonal, we subtract from the number of all admissible paths from $(1,0)$ to $(n+1,n)$ the number of all admissible paths intersecting the diagonal.

An admissible path from $(1,0)$ to $(n+1,n)$ has n unit horizontal edges and n unit vertical edges. Therefore the number of admissible paths from $(1,0)$ to $(n+1,n)$ is the binomial coefficient $\binom{2n}{n}$.

To any admissible path γ from $(1,0)$ to $(n+1,n)$ intersecting the diagonal, we associate an admissible path γ' from $(0,1)$ to $(n+1,n)$ as follows: let (i,i) be the diagonal point on γ with smallest i; replace the subpath of γ from $(1,0)$ to (i,i) by its reflection in the diagonal; the path γ' is the union of the reflected subpath and the subpath of γ from (i,i) to $(n+1,n)$. It is clear that γ' is admissible. Any admissible path from $(0,1)$ to $(n+1,n)$ necessarily intersects the diagonal and therefore is obtained in this way from a unique admissible path from $(1,0)$ to $(n+1,n)$. Now, any admissible path from $(0,1)$ to $(n+1,n)$ has $n+1$ unit horizontal edges and $n-1$ unit vertical edges. Therefore the number of admissible paths from $(0,1)$ to $(n+1,n)$ is equal to the binomial coefficient $\binom{2n}{n+1}$. This is also the number of all admissible paths from $(1,0)$ to $(n+1,n)$ intersecting the diagonal.

Summing up, we see that

$$
\begin{aligned}
\operatorname{card} E_n &= \binom{2n}{n} - \binom{2n}{n+1} = \frac{(2n)!}{n!\,n!} - \frac{(2n)!}{(n+1)!\,(n-1)!} \\
&= \left(1 - \frac{n}{n+1}\right)\frac{(2n)!}{n!\,n!} = \frac{1}{n+1}\binom{2n}{n}.
\end{aligned}
$$

$\square$

The following inequality follows from Lemmas 5.26 and 5.27.

Proposition 5.28. *We have*

$$
\dim_{\mathbf{C}} A_n(a) \le \frac{1}{n+1}\binom{2n}{n}.
$$

We will see later (Corollary 5.32 or Remark 5.35) that this inequality is in fact an equality.

5.7.2 Relation to the Iwahori–Hecke algebras

We now establish a connection between $A_n(a)$ and the one-parameter Iwahori–Hecke algebra $H_n(q) = H_n^{\mathbf{C}}(q)$. Recall the generators $T_1, \ldots, T_{n-1}$ of $H_n(q)$.

Theorem 5.29. *Let q, $a \in \mathbf{C} - \{0\}$ satisfy $a^2 = (q+1)^2/q$.*

(a) There is a surjective algebra homomorphism $\Psi : H_n(q) \to A_n(a)$ such that

$$\Psi(T_i) = \frac{q+1}{a} e_i - 1 \tag{5.23}$$

for $i = 1, \ldots, n-1$.

(b) If $n = 2$, then $\Psi : H_n(q) \to A_n(a)$ is an isomorphism.

(c) If $n \geq 3$, then the kernel of Ψ is the two-sided ideal of $H_n(q)$ generated by $1 + T_1 + T_2 + T_1T_2 + T_2T_1 + T_1T_2T_1$.

Note that the conditions on a and q in the theorem imply that $q \neq -1$.

Proof. (a) For $i = 1, \ldots, n-1$, set

$$t_i = \Psi(T_i) = \frac{q+1}{a} e_i - 1 \in A_n(a). \tag{5.24}$$

Formula (5.23) defines an algebra homomorphism $\Psi : H_n(q) \to A_n(a)$, provided $t_1, \ldots, t_{n-1}$ satisfy relations (4.16), (4.17), and (4.20), where T_i is replaced by t_i. Let us check these relations.

Relation (4.16): This is an obvious consequence of (5.20).

Relation (4.17): If $|i - j| = 1$, then by (5.21) and (5.22),

$$t_i t_j t_i = \left(\frac{q+1}{a} e_i - 1\right)\left(\frac{q+1}{a} e_j - 1\right)\left(\frac{q+1}{a} e_i - 1\right)$$

$$= \left(\frac{q+1}{a}\right)^3 e_i e_j e_i - \left(\frac{q+1}{a}\right)^2 \left(e_i e_j + e_j e_i + e_i^2\right)$$

$$+ \left(\frac{q+1}{a}\right)\left(2e_i + e_j\right) - 1$$

$$= \left(\frac{q+1}{a}\right)^3 e_i - \left(\frac{q+1}{a}\right)^2 \left(e_i e_j + e_j e_i + a e_i\right)$$

$$+ \left(\frac{q+1}{a}\right)\left(2e_i + e_j\right) - 1.$$

It follows from this and the equality $(q+1)^2/a^2 = q$ that

$$t_i t_j t_i - t_j t_i t_j = \left(\frac{q+1}{a}\right)^3 (e_i - e_j) - \left(\frac{q+1}{a}\right)^2 a(e_i - e_j)$$

$$+ \left(\frac{q+1}{a}\right)(e_i - e_j)$$

$$= \frac{q+1}{a}\left(\left(\frac{q+1}{a}\right)^2 - (q+1) + 1\right)(e_i - e_j)$$

$$= \frac{q+1}{a}\left(q - (q+1) + 1\right)(e_i - e_j) = 0.$$

Relation (4.20): By (5.22),

$$t_i^2 - (q-1)t_i - q = \left(\frac{q+1}{a}e_i - 1\right)^2 - (q-1)\left(\frac{q+1}{a}e_i - 1\right) - q$$

$$= \left(\frac{q+1}{a}\right)^2 e_i^2 - 2\frac{q+1}{a}e_i + 1 \cdot$$

$$- \frac{(q-1)(q+1)}{a}e_i + (q-1) - q$$

$$= \left(\frac{q+1}{a}\right)^2 ae_i - 2\frac{q+1}{a}e_i - \frac{(q-1)(q+1)}{a}e_i$$

$$= \frac{q+1}{a}\left((q+1) - 2 - (q-1)\right)e_i = 0.$$

Formula (5.23) implies

$$e_i = \Psi\left(\frac{a}{q+1}(T_i + 1)\right)$$

for all $i = 1, \ldots, n-1$. Therefore, the generators e_i of $A_n(a)$ belong to the image of $\Psi : H_n(q) \to A_n(a)$, which proves that Ψ is surjective.

(b) The algebra $A_2(a)$ is generated by a single element e subject to the relation $e^2 = ae$. It is easy to check that the formula $e \mapsto a(T_1 + 1)/(q+1)$ defines an algebra homomorphism $A_2(a) \to H_2(q)$ inverse to Ψ.

(c) From (5.24) we derive

$$e_i = \frac{a}{q+1}(t_i + 1). \tag{5.25}$$

Substituting these expansions of $e_1, \ldots, e_{n-1}$ in (5.20)–(5.22), we obtain relations for $t_1, \ldots, t_{n-1}$. It is easy to see that the relation obtained in this way from relation (5.20) (resp. from relation (5.22)) is equivalent to relation (4.16) (resp. to relation (4.20)), where T_i is replaced by t_i.

Relations (5.21) with $|i - j| = 1$ and (4.20) yield

$$
\begin{aligned}
e_i e_j e_i - e_i &= \left(\frac{a}{q+1}\right)^3 (t_i + 1)(t_j + 1)(t_i + 1) - \frac{a}{q+1}(t_i + 1) \\
&= \frac{a}{(q+1)q}\left((t_i + 1)(t_j + 1)(t_i + 1) - q(t_i + 1)\right) \\
&= \frac{a}{(q+1)q}\left(t_i t_j t_i + t_i t_j + t_j t_i + t_i^2 + 2t_i + t_j + 1 - qt_i - q\right) \\
&= \frac{a}{(q+1)q}\left(t_i t_j t_i + t_i t_j + t_j t_i \right. \\
&\qquad\qquad \left. + (q-1)t_i + q + 2t_i + t_j + 1 - qt_i - q\right) \\
&= \frac{a}{(q+1)q}\left(t_i t_j t_i + t_i t_j + t_j t_i + t_i + t_j + 1\right).
\end{aligned}
$$

This shows that

$$
1 + t_i + t_j + t_i t_j + t_j t_i + t_i t_j t_i = 0
$$

for all i, j such that $|i - j| = 1$. Therefore the kernel I_n of $\Psi : H_n(q) \to A_n(a)$ is the two-sided ideal of $H_n(q)$ generated by the elements

$$
1 + T_i + T_j + T_i T_j + T_j T_i + T_i T_j T_i,
$$

for all i, j such that $|i - j| = 1$. Since $T_i T_j T_i = T_j T_i T_j$, it is enough to consider the generators corresponding to the pairs (i, j) with $j = i + 1$. Therefore, I_n is the two-sided ideal of $H_n(q)$ generated by the elements

$$
1 + T_i + T_{i+1} + T_i T_{i+1} + T_{i+1} T_i + T_i T_{i+1} T_i
$$

with $i = 1, \ldots, n - 1$.

Now, as observed in Exercise 1.1.4, for $i = 2, \ldots, n - 1$, we have

$$
\sigma_i = (\sigma_1 \sigma_2 \cdots \sigma_{n-1})^{i-1} \sigma_1 (\sigma_1 \sigma_2 \cdots \sigma_{n-1})^{-(i-1)}
$$

in the braid group B_n. Let ω be the image of $\sigma_1 \sigma_2 \cdots \sigma_{n-1}$ in $H_n(q)$ under the multiplicative homomorphism $B_n \to H_n(q)$ sending $\sigma_i^{\pm 1}$ to $T_i^{\pm 1}$ for all i. Clearly, ω is invertible in $H_n(q)$ and $T_i = \omega^{i-1} T_1 \omega^{-(i-1)}$. It follows that

$$
\begin{aligned}
&1 + T_i + T_{i+1} + T_i T_{i+1} + T_{i+1} T_i + T_i T_{i+1} T_i \\
&\qquad = \omega^{i-1}\left(1 + T_1 + T_2 + T_1 T_2 + T_2 T_1 + T_1 T_2 T_1\right)\omega^{-(i-1)}
\end{aligned}
$$

for all $i = 1, \ldots, n - 1$. Therefore, as a two-sided ideal, I_n is generated by the single element $1 + T_1 + T_2 + T_1 T_2 + T_2 T_1 + T_1 T_2 T_1$. $\square$

5.7.3 The semisimple case

Throughout this section, we assume that q, a are nonzero complex numbers related by the condition $a^2 = (q+1)^2/q$ of Theorem 5.29 and that the assumptions of Section 5.5 hold for q and an integer $n \geq 3$. In particular, the algebra $H_n(q)$ is semisimple, and then, by Corollary 4.51, so is $A_n(a)$.

By Theorem 5.18, any simple $H_n(q)$-module is of the form $V_\lambda = V_\lambda^{\mathbb{C}}$ for some partition λ of n. We may ask which V_λ is induced from an $A_n(a)$-module via the surjection $\Psi : H_n(q) \to A_n(a)$. In other words, for which partitions λ of n do we have $I_n V_\lambda = 0$, where $I_n = \mathrm{Ker}\,(\Psi : H_n(q) \to A_n(a))$?

Lemma 5.30. *If $\lambda = (\lambda_1, \lambda_2, \ldots, \lambda_p)$ is a partition of an integer $n \geq 3$ such that $\lambda_i \in \{1, 2\}$ for all $i = 1, \ldots, p$, then $I_n V_\lambda = 0$.*

Observe that the partitions in Lemma 5.30 are exactly those whose diagrams have one or two columns.

Proof. By Theorem 5.29 (c), it suffices to show that

$$X = 1 + T_1 + T_2 + T_1 T_2 + T_2 T_1 + T_1 T_2 T_1 \in H_3(q) \subset H_n(q)$$

acts trivially on V_λ. We proceed by induction on $n \geq 3$.

Suppose that $n = 3$. Then there are two partitions of n whose diagrams have one or two columns, namely $\lambda = (1, 1, 1)$ and $\mu = (2, 1)$. As we know, the module V_λ is one-dimensional and all T_i act by -1 on V_λ. It follows that X acts trivially on V_λ. The module V_μ is two-dimensional with basis $\{v_T, v_{T'}\}$, where T and T' are the standard tableaux of shape μ shown in Figure 5.7. Observe that $T' = s_2 T$ and that neither $s_1 T$ nor $s_1 T'$ is a standard tableau. We have $d_T(1) = 1 = -d_{T'}(1)$ and $d_T(2) = -2 = -d_{T'}(2)$. By (5.10)–(5.12), the generators T_1 and T_2 of $H_n(q)$ act on the basis $\{v_T, v_{T'}\}$ of V_μ by the matrices

$$T_1 = \begin{pmatrix} q & 0 \\ 0 & -1 \end{pmatrix} \quad \text{and} \quad T_2 = -\frac{1}{q+1} \begin{pmatrix} 1 & q \\ 1 + q + q^2 & -q^2 \end{pmatrix}.$$

We obtain

$$T_1 T_2 = -\frac{1}{q+1} \begin{pmatrix} q & q^2 \\ -(1 + q + q^2) & q^2 \end{pmatrix},$$

$$T_2 T_1 = -\frac{1}{q+1} \begin{pmatrix} q & -q \\ q(1 + q + q^2) & q^2 \end{pmatrix},$$

and

$$T_1 T_2 T_1 = -\frac{1}{q+1} \begin{pmatrix} q^2 & -q^2 \\ -q(1 + q + q^2) & -q^2 \end{pmatrix}.$$

From these computations it follows that $X = 0$ on V_μ.

Let $n \geq 4$ and let λ be a partition of n whose diagram has one or two columns. Now, $X \in H_3(q) \subset H_{n-1}(q)$. By Theorem 5.13,

$$V_\lambda|_{H_{n-1}(q)} = \bigoplus_{\mu \hookrightarrow \lambda} V_\mu,$$

where μ runs over all partitions of $n-1$ obtained from λ by removing a corner. The diagram of such a partition μ has one or two columns. Therefore, by the induction hypothesis, X acts as zero on V_μ. Hence, X acts as zero on V_λ. $\square$

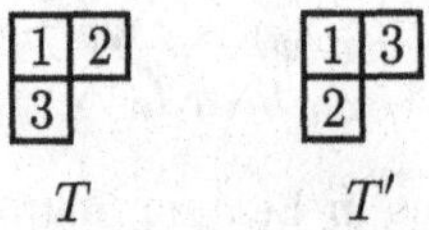

$$\begin{array}{|c|c|} \hline 1 & 2 \\ \hline 3 \\ \cline{1-1} \end{array} \qquad \begin{array}{|c|c|} \hline 1 & 3 \\ \hline 2 \\ \cline{1-1} \end{array}$$

$\qquad\qquad T \qquad\qquad\qquad T'$

Fig. 5.7. The two standard tableaux of shape μ

Proposition 5.31. *For any* $n \geq 3$,

$$\dim_{\mathbf{C}} I_n = n! - \frac{1}{n+1}\binom{2n}{n} > 0. \tag{5.26}$$

For a partition λ of n, we have $I_n V_\lambda = 0$ if and only if the diagram of λ has one or two columns.

Proof. Let $\pi_\lambda : H_n(q) \to \mathrm{End}_{\mathbf{C}}(V_\lambda)$ be the algebra homomorphism induced by the action of $H_n(q)$ on V_λ. Since $\pi_\lambda(I_n) = 0$ for all partitions λ whose diagrams have one or two columns, the isomorphism of Corollary 5.19 sends I_n injectively into the product algebra

$$\prod_{\lambda \in \Lambda_{\geq 3}(n)} \mathrm{End}_{\mathbf{C}}(V_\lambda),$$

where $\Lambda_{\geq 3}(n)$ is the set of partitions of n whose diagrams have at least three columns. Therefore, by Theorem 5.1,

$$\dim_{\mathbf{C}} I_n \leq \sum_{\lambda \in \Lambda_{\geq 3}(n)} (f^\lambda)^2 = n! - \sum_{\lambda \in \Lambda_{\leq 2}(n)} (f^\lambda)^2, \tag{5.27}$$

where $\Lambda_{\leq 2}(n)$ is the set of partitions of n whose diagrams have one or two columns. Recall that $f^\lambda = f^{\lambda^T}$, where λ^T is the conjugate partition (see Section 5.1.5). If $\lambda \in \Lambda_{\leq 2}(n)$, then λ^T has at most two parts, and we deduce from Exercise 5.2.3 (b) that

$$\sum_{\lambda \in \Lambda_{\leq 2}(n)} (f^\lambda)^2 = \sum_{\lambda^T} (f^{\lambda^T})^2 = \frac{1}{n+1}\binom{2n}{n}. \tag{5.28}$$

Therefore,

$$\dim_{\mathbf{C}} I_n \leq n! - \frac{1}{n+1}\binom{2n}{n}. \tag{5.29}$$

On the other hand, by definition of I_n and by Proposition 5.28,

$$\dim_{\mathbf{C}} I_n = \dim_{\mathbf{C}} H_n(q) - \dim_{\mathbf{C}} A_n(a) \geq n! - \frac{1}{n+1}\binom{2n}{n}. \tag{5.30}$$

Combining (5.29) and (5.30), we obtain the equality in (5.26).

It follows from (5.26), (5.27), and (5.28) that

$$\dim_{\mathbf{C}} I_n = \sum_{\lambda \in \Lambda_{\geq 3}(n)} (f^\lambda)^2. \tag{5.31}$$

Since $f^\lambda > 0$ for any $\lambda \neq \emptyset$ and the set $\Lambda_{\geq 3}(n)$ is nonempty, $\dim_{\mathbf{C}} I_n > 0$. Moreover, it follows from the computation of $\dim_{\mathbf{C}} I_n$ that the injection

$$I_n \rightarrow \prod_{\lambda \in \Lambda_{\geq 3}(n)} \mathrm{End}_{\mathbf{C}}(V_\lambda) \tag{5.32}$$

is an algebra isomorphism. Thus, $I_n V_\lambda = \pi_\lambda(I_n)V_\lambda = 0$ if and only if $\lambda \notin \Lambda_{\geq 3}(n)$ or, equivalently, $\lambda \in \Lambda_{\leq 2}(n)$. $\qquad\square$

Corollary 5.32. *Let* $n \geq 2$.

    (a) The dimension of $A_n(a)$ *as a complex vector space is given by*

$$\dim_{\mathbf{C}} A_n(a) = \frac{1}{n+1}\binom{2n}{n}.$$

    (b) The set $\{e_{\underline{s}}\}_{\underline{s} \in E_n}$ *of reduced words is a basis of* $A_n(a)$.

    (c) The algebra homomorphism $A_n(a) \rightarrow A_{n+1}(a)$ *defined by* $e_i \mapsto e_i$ *for* $i = 1, \ldots, n-1$ *is injective.*

    (d) The algebra $A_n(a)$ *is semisimple. Any simple* $A_n(a)$*-module is isomorphic to a unique module of the form* V_λ*, where* λ *is a partition of* n *whose diagram has one or two columns.*

Proof. (a) For $n \geq 3$, this follows from Proposition 5.31, since

$$\dim_{\mathbf{C}} A_n(a) = \dim_{\mathbf{C}} H_n(q) - \dim_{\mathbf{C}} I_n.$$

For $n = 2$, the claim (a) is straightforward.

    (b) By Lemmas 5.26 and 5.27, the set of reduced words spans $A_n(a)$ and consists of

$$\frac{1}{n+1}\binom{2n}{n} = \dim_{\mathbf{C}} A_n(a)$$

vectors. Therefore, it is a basis.

(c) This homomorphism sends a basis of $A_n(a)$ to a subset of a basis of $A_{n+1}(a)$; therefore it is injective.

(d) We have already observed that $A_n(a)$ is semisimple. Let λ be a partition of n whose diagram has one or two columns. By Lemma 5.30, the algebra $A_n(a) \cong H_n(q)/I_n$ acts on V_λ. If V_λ were not simple as as $A_n(a)$-module, then it would not be simple as a $H_n(q)$-module, which would contradict Theorem 5.18.

By Corollary 5.19 and (5.32) we have the algebra isomorphisms

$$A_n(a) \cong H_n(q)/I_n \cong \prod_{\lambda \in \Lambda_{\leq 2}(n)} \mathrm{End}_{\mathbf{C}}(V_\lambda).$$

Hence, any simple $A_n(a)$-module is isomorphic to V_λ for some $\lambda \in \Lambda_{\leq 2}(n)$. □

5.7.4 A graphical interpretation of the Temperley–Lieb algebras

We complete our survey of the Temperley–Lieb algebras by giving a graphical interpretation of their elements.

For $n \geq 1$, a *simple n-diagram* D is a disjoint union of n smoothly embedded arcs in $\mathbf{R} \times [0,1]$ such that the boundary ∂D of D consists of the points $(1,0),\ldots,(n,0)$ and $(1,1),\ldots,(n,1)$, and $D - \partial D \subset \mathbf{R} \times (0,1)$, and the tangent vector of D at each endpoint is parallel to $\{0\} \times \mathbf{R}$. Two simple n-diagrams are *isotopic* if they can be deformed into each other in the class of simple n-diagrams. Figures 5.8 and 5.9 show all simple n-diagrams up to isotopy for $n = 1, 2, 3$.

Fig. 5.8. The simple 1- and 2-diagrams

Fig. 5.9. The five simple 3-diagrams

Lemma 5.33. *The number of isotopy classes of simple n-diagrams is equal to the nth Catalan number*

$$\frac{1}{n+1}\binom{2n}{n}.$$

Proof. By a *semicircle* we shall mean a Euclidean semicircle in the upper half-plane $\mathbf{R} \times [0, +\infty)$ with endpoints (and center) on $\mathbf{R} \times \{0\}$. It is clear that pulling the upper endpoints of any simple n-diagram down as in Figure 5.10, we obtain a union of n disjoint embedded arcs in $\mathbf{R} \times [0, +\infty)$ with $2n$ endpoints on $\mathbf{R} \times \{0\}$. We can isotop such a union into a system of n disjoint semicircles. These transformations establish a bijective correspondence between the isotopy classes of simple n-diagrams and the isotopy classes of systems of n disjoint semicircles. Therefore it suffices to compute the number of (isotopy classes of) such systems.

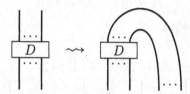

Fig. 5.10. Turning a simple n-diagram into a system of semicircles

We label an endpoint of a semicircle by L (resp. by R) if this point is the left (resp. right) endpoint of the semicircle. Reading the labels of the endpoints of a system of n disjoint semicircles from left to right along $\mathbf{R} \times \{0\}$, we obtain a word w of length $2n$ in the alphabet $\{L, R\}$. The word w is a *Dyck word*, i.e., w has as many occurrences of L as occurrences of R and no prefix of w has more occurrences of R than occurrences of L. It is easy to see that any Dyck word of length $2n$ comes from a system of n disjoint semicircles, which is unique up to isotopy.

Now to a Dyck word w of length $2n$ we associate a polygonal path Γ_w in $\mathbf{R}^2$ with consecutive vertices $(x_0, y_0), (x_1, y_1), \ldots, (x_{2n}, y_{2n})$. Here $x_0 = y_0 = 0$ and for $k \in \{1, \ldots, 2n\}$, the point (x_k, y_k) is defined inductively by $x_k = x_{k-1} + 1$ and $y_k = y_{k-1}$ if the kth letter in w is L, and $x_k = x_{k-1}$ and $y_k = y_{k-1} + 1$ if the kth letter in w is R. Since there are n occurrences of L and n occurrences of R in w, the path Γ_w leads from $(0,0)$ to (n,n). It is clear that Γ_w is admissible in the sense defined in the proof of Lemma 5.27. Moreover, Γ_w lies under the diagonal because of the condition on the prefixes of w. Conversely, any admissible path from $(0,0)$ to (n,n) lying under the diagonal is of the form Γ_w for a unique Dyck word w of length $2n$.

In conclusion, the number of isotopy classes of simple n-diagrams is equal to the number of admissible paths from $(0,0)$ to (n,n) lying under the diagonal. By Lemma 5.27, this number is equal to the nth Catalan number. $\quad\square$

Fix a nonzero complex number a. Let $A'_n(a)$ be the complex vector space spanned by the isotopy classes of simple n-diagrams. By Lemma 5.33, the dimension of $A'_n(a)$ is equal to the nth Catalan number. Every simple n-diagram D represents a vector in $A'_n(a)$, denoted by $[D]$.

Let us equip $A'_n(a)$ with the structure of an associative algebra. Given two simple n-diagrams D and D', define $D \natural D'$ to be the one-manifold in $\mathbf{R} \times [0,1]$ obtained by attaching D on top of D' and compressing the result into $\mathbf{R} \times [0,1]$. Then $D \natural D'$ is a disjoint union of n embedded arcs and a certain number $k(D, D') \geq 0$ of embedded circles. Removing the circles, we obtain a simple n-diagram, denoted by $D \circ D'$. Set

$$[D][D'] = a^{k(D,D')}[D \circ D'].$$

It is easy to check that this formula defines an associative product on $A'_n(a)$. The simple n-diagram

$$1_n = \{1, \dots, n\} \times [0,1]$$

represents the unit of $A'_n(a)$.

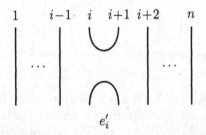

Fig. 5.11. The simple n-diagram e'_i

The following theorem gives a graphical interpretation for the Temperley–Lieb algebra $A_n(a)$.

Theorem 5.34. *For $i = 1, \dots, n-1$, let e'_i be the simple n-diagram in Figure 5.11. The assignment*

$$e_i \mapsto [e'_i] \quad (i = 1, \dots, n-1)$$

defines an algebra isomorphism $A_n(a) \to A'_n(a)$.

Proof. It is a pleasant exercise to verify that the elements $[e'_1], \dots, [e'_{n-1}]$ of $A'_n(a)$ satisfy the defining relations (5.20)–(5.22) of $A_n(a)$. Therefore there is an algebra homomorphism $f : A_n(a) \to A'_n(a)$ such that $f(e_i) = [e'_i]$ for all $i = 1, \dots, n-1$. We now verify that f is an isomorphism. It is enough to check that f is surjective, since

$$\dim_{\mathbf{C}} A_n(a) \leq \frac{1}{n+1}\binom{2n}{n} = \dim_{\mathbf{C}} A'_n(a)$$

by Proposition 5.28. It thus suffices to establish the following claim: if D is a simple n-diagram not isotopic to 1_n, then D is equal in $A'_n(a)$ to a product of elements of the form $[e'_1], [e'_2], \ldots, [e'_{n-1}]$.

We shall prove the claim by induction on n. If $n = 2$, then D is isotopic to e'_1 and the claim is true. Assume that the claim is true for simple diagrams with $n-1$ arcs and let us prove it for simple diagrams with n arcs. Let $P_1, \ldots, P_n$ be the bottom endpoints of D enumerated from left to right. Since $[D] \neq [1_n]$, there is an arc of D connecting two bottom endpoints of D. Since the arcs of D are disjoint, there is an arc of D connecting two consecutive bottom endpoints. Denote by $i = i(D)$ the minimal $i = 1, 2, \ldots, n-1$ such that there is an arc of D connecting P_i and P_{i+1}. Now we use induction on $i(D)$.

If $i(D) > 1$, then $[D] = [D'][e'_i]$, where D' is a simple n-diagram with $i(D') = i(D) - 1$. The diagram D' is obtained from D by the following transformation in a neighborhood of P_{i-1}, P_i, P_{i+1}: we slightly deform the arc of D issuing from P_{i-1} to produce a local maximum and a local minimum of the height function. We may assume that the local minimum lies strictly above the arc of D connecting P_i and P_{i+1}. Now we may strip e_i off and present $[D]$ as $[D] = [D'][e'_i]$ with $i(D') = i(D) - 1$.

It remains to consider the case $i(D) = 1$. We have $[D] = [D''][e'_1]$, where D'' is a simple n-diagram constructed as follows. Consider the arc of D descending from the leftmost top endpoint in $\mathbf{R} \times \{1\}$. We take a small subarc of this arc lying close to this top endpoint and push it down close to the arc of D connecting P_1 and P_2. This allows us to strip e_1 off and to present $[D]$ in the form $[D] = [D''][e'_1]$. It is clear that D'' contains a strand joining the leftmost bottom endpoint to the leftmost top endpoint. In other words, D'' is obtained by adding a vertical interval from the left to a simple $(n-1)$-diagram. The inductive assumption implies that $[D'']$ is a product of elements of the form $[e'_2], \ldots, [e'_{n-1}]$. This implies our claim and thus completes the proof of the theorem.   □

Remark 5.35. As a consequence of Theorem 5.34, the dimension of the Temperley–Lieb algebra $A_n(a)$ is equal to the nth Catalan number. This provides another proof of Corollary 5.32 (a). This proof is more general, since it holds for an arbitrary value of the complex parameter a.

Exercise 5.7.1. Let $K = \mathbf{C}$ and $\mathcal{L}_{n-1}$ be the $H_n^K(q)$-module introduced in Section 5.6. Use Theorem 5.22 and Lemma 5.30 to show that $\mathcal{L}_{n-1}$ is a module over the Temperley–Lieb algebra $A_n(a)$, where $a^2 = (q+1)^2/q$.

Exercise 5.7.2 (The Jones–Wenzl idempotents). Let u be a nonzero complex number such that $u^{2k} \neq 1$ for all $k = 1, \ldots, n$. Set $a = -(u + u^{-1})$. Define elements $f_1, \ldots, f_n \in A_n(a)$ inductively by $f_1 = 1$ and

$$f_k = f_{k-1} + \frac{u^{k-1} - u^{-(k-1)}}{u^k - u^{-k}} f_{k-1} e_{k-1} f_{k-1}.$$

for all $k = 2, \ldots, n$. Show that $f_k^2 = f_k$ for all $k = 1, \ldots, n$ and that $(f_1, \ldots, f_n)$ is the unique sequence of elements of $A_n(a)$ such that $f_k - 1$ is a linear combination of nonempty words in $\{e_1, \ldots, e_{k-1}\}$ for $k = 1, \ldots, n$ and for all $\ell < k$,

$$e_\ell f_k = f_k e_\ell = 0.$$

Exercise 5.7.3. Let $C_n = \binom{2n}{n}/(n+1)$ be the nth Catalan number.
  (a) Show that

$$C_{n+1} = \sum_{i=0}^{n} C_i C_{n-i}$$

for all $n \geq 0$. (Hint: Every Dyck word w of length ≥ 2 can be written uniquely in the form $w = L w_1 R w_2$ with (possibly empty) Dyck words w_1, w_2.)
  (b) Deduce the following generating function for the Catalan numbers:

$$\sum_{n=0}^{\infty} C_n x^n = \frac{1 - \sqrt{1 - 4x}}{2x}.$$

Notes

Before 1983, essentially the only interesting known linear representation of the braid group B_n was the Burau representation. The situation changed radically when Vaughan Jones introduced the Temperley–Lieb algebras and used them to construct new representations of the braid groups; see [Jon83], [Jon84], [Jon86], [Jon87], [Jon89]. Soon thereafter, inspired by Jones's work, Reshetikhin and Turaev showed how to obtain finite-dimensional representations of the braid groups from representations of quantum groups; see [Tur88], [RT90]. For comprehensive introductions to quantum groups and their connections to braids and links, see [Tur94], [Kas95], [KRT97]. In this chapter we followed a "dual" approach to the representations of B_n, based on the theory of Iwahori–Hecke algebras.

The content of Section 5.1 is standard; see, e.g., [Jam78], [FH91], [Ful97], [Sag01]. Our proof of Theorem 5.1 follows Stanley [Sta88]; see also [Sag01, Sect. 5.1]. This theorem can also be proved with the help of the Robinson–Schensted correspondence, which provides a bijection

$$\mathfrak{S}_n \simeq \coprod_{\lambda \vdash n} \mathcal{T}_\lambda \times \mathcal{T}_\lambda;$$

see [Knu73, Sect. 5.1.4], [Ful97, Chap. 4], [Sag01, Chap. 3].

The hook length formula in Exercise 5.2.6 is due to Frame, Robinson, and Thrall [FRT54]; for a proof, see, e.g., [Knu73, Sect. 5.1.4], [Gol93, Sect. 12], [Sag01, Chap. 3]. In the formulation of this exercise, we followed [Mat99, Chap. 3, Exer. 25]. Exercise 5.5.4 is taken from [GHJ89, Chap. 2] (see also [Ram97]).

The modules V_λ of Section 5.3 were constructed by Hoefsmit [Hoe74] as a generalization of Young's seminormal representations of the symmetric groups. A general theory of seminormal representations is given in [Ram97]. In Sections 5.3–5.5 we followed [Hoe74], [Wen88], [Ram97]. Lusztig [Lus81] gave an explicit construction of the isomorphism of Exercise 5.5.2.

For a study of the Iwahori–Hecke algebras and their representations without assuming semisimplicity, see, e.g., [DJ86], [DJ87], [Gec98], [Mat99].

The fact that the reduced Burau representation of B_n appears as the simple module associated to the partition $(2, 1, \ldots, 1)$ was pointed out by Jones [Jon84], [Jon86], [Jon87].

Linear representations of the Temperley–Lieb algebras first came up in physics in the work by Temperley and Lieb [TL71]. The Temperley–Lieb algebras themselves were introduced by Jones [Jon83] in his study of subfactors. Jones [Jon84], [Jon86] also related these algebras to the Iwahori–Hecke algebras and to the braid groups. The Jones–Wenzl idempotents of Exercise 5.7.2 were introduced by Jones in [Jon83]. The inductive formula defining them is due to Wenzl [Wen87]. These idempotents play an important role in the theory of invariants of three-dimensional manifolds (see [Tur94, Sect. XII.4]). The reader interested in Catalan numbers is encouraged to take a close look at [Sta99, Exercise 6.19], which lists 66 sets each of whose cardinal is equal to the nth Catalan number.

In Section 5.7 we essentially followed [GHJ89, Sects. 2.8–2.11]. For more on the graphical interpretation of the Temperley–Lieb algebras, see [Kau87], [Kau90], [Kau91], [Tur94, Sect. XII.3].

It should also be noted that Formanek et al. classified all complex irreducible representations of B_n of dimension $\leq n$; see [For96], [FLSV03]. For more on representations of B_3, see [TW01], [Tub01]. Quotients of the braid group algebras by cubic relations were investigated by Funar et al.; see [Fun95], [BF04].

6

Garside Monoids and Braid Monoids

Braid groups may be viewed as groups of fractions of certain monoids called braid monoids. The latter belong to a wider class of so-called Garside monoids. In this chapter we investigate properties of monoids and specifically of Garside monoids. As an application, we give a solution of the conjugacy problem in the braid groups. We also discuss generalized braid groups associated with Coxeter matrices.

6.1 Monoids

6.1.1 Definitions and examples

A *monoid* is a set M equipped with a binary operation (multiplication) $M \times M \rightarrow M$ that is associative and has a neutral element. For $a, b \in M$, the image of $(a, b) \in M \times M$ under the multiplication is denoted by ab and called the *product* of a and b. The associativity means that $(ab)c = a(bc)$ for all $a, b, c \in M$. The *neutral element* $1 \in M$ satisfies $a1 = 1a = a$ for all $a \in M$. Such an element is always unique.

A monoid M is *left* (resp. *right*) *cancellative* if for all $a, b, c \in M$,

$$ab = ac \implies b = c \quad (\text{resp.} \quad ba = ca \implies b = c).$$

An element a of a monoid M is *invertible* if there is $b \in M$ such that $ab = ba = 1$. A *group* is a monoid in which all elements are invertible.

A map f from a monoid M to a monoid M' is a *monoid homomorphism* if $f(ab) = f(a)f(b)$ for all $a, b \in M$ and f sends the neutral element of M to the neutral element of M'.

Examples 6.1. (a) The set of nonnegative integers with addition as a binary operation is a monoid denoted by $\mathbf{N}$.

(b) The set of positive integers with multiplication as a binary operation is a monoid denoted by $\mathbf{N}^\times$.

C. Kassel, V. Turaev, *Braid Groups*, DOI: 10.1007/978-0-387-68548-9_6,
© Springer Science+Business Media, LLC 2008

(c) A *free monoid* on a set X is a monoid X^* containing X as a subset and such that any set-theoretic map from X to a monoid M extends uniquely to a monoid homomorphism $X^* \to M$. This property defines X^* up to monoid isomorphism. It is easy to show that every element w of X^* can be expanded in a unique way as a *word* on the alphabet X, i.e., as a product of several elements of $X \subset X^*$. The number of elements of X in this expansion (counted with multiplicities) is called the *length* of w and is denoted by $l(w)$. Clearly, $l(1) = 0$, $l(x) = 1$ for $x \in X$, and $l(ww') = l(w) + l(w')$ for any $w, w' \in X^*$. For $X = \emptyset$, the monoid X^* consists only of the neutral element; this is the *trivial monoid*.

The monoids in examples (a), (b), (c) are left and right cancellative, and their neutral elements are the only invertible elements.

6.1.2 Divisibility in monoids

If $a = bc$, where a, b, c are elements of a monoid M, then we say that b is a *left divisor* of a and c is a *right divisor* of a. We also say that a is a *right multiple* of b and a *left multiple* of c. We write $b \preceq a$ and $a \succeq c$. For example, $1 \preceq a$ and $a \succeq 1$ for all $a \in M$, since $a = 1a = a1$.

Lemma 6.2. *The relations $\preceq$ and $\succeq$ in a monoid are reflexive and transitive.*

Proof. The reflexivity of $\preceq$ follows from the identity $a = a1$; and the transitivity, from the associativity of multiplication. The proofs for $\succeq$ are similar.   □

6.1.3 Atomic monoids

For any element $a \neq 1$ of a monoid M, set

$$\|a\| = \sup \left\{ r \geq 1 \mid a = a_1 \cdots a_r \text{ with } a_1, \ldots, a_r \in M - \{1\} \right\} \in \{1, 2, \ldots, \infty\}.$$

Also set $\|1\| = 0$. It is easy to check that for all $a, b \in M$,

$$\|ab\| \geq \|a\| + \|b\|.$$

Note that $\|a\| = 0$ if and only if $a = 1$.

An element $a \in M$ is called an *atom* if $\|a\| = 1$. In other words, $a \in M$ is an atom if $a \neq 1$ and $a = a_1 \cdots a_r$ implies that $a_i = 1$ for all i but one. Any $a \in M$ with finite $\|a\|$ expands as $a = a_1 \cdots a_r$, where $r = \|a\|$ and $a_1, \ldots, a_r$ are atoms. This justifies the following definition: a monoid M is *atomic* if $\|a\|$ is finite for all $a \in M$.

As an exercise, the reader may verify that the monoids $\mathbf{N}$, $\mathbf{N}^\times$ introduced above and all free monoids are atomic. The monoid $\{1, x\}$ with multiplication

$$x^2 = x1 = 1x = x, \quad 11 = 1,$$

is not atomic. Groups have no atoms and are not atomic (except the trivial group).

Lemma 6.3. *If elements a, b of an atomic monoid M satisfy $a \preceq b$ and $b \preceq a$, then $a = b$. Similarly, if $a \succeq b$ and $b \succeq a$, then $a = b$.*

Proof. Since $a \preceq b \preceq a$, there are u, $v \in M$ such that $b = au$ and $a = bv$. Then $a = auv$ and

$$\|a\| = \|auv\| \geq \|a\| + \|u\| + \|v\|.$$

This implies that $\|u\| = \|v\| = 0$. Hence $u = v = 1$ and $a = b$. The relation $\succeq$ is treated similarly. □

Lemmas 6.2 and 6.3 imply that the relations $\preceq$ and $\succeq$ on an atomic monoid are partial orders.

Given a subset E of an atomic monoid M, we say that an element $a \in E$ is *maximal* (resp. *minimal*) with respect to $\preceq$ if $b \preceq a$ (resp. $a \preceq b$) for all $b \in E$. A maximal (resp. minimal) element of E may not exist, but if it exists, it is unique by Lemma 6.3. Similar definitions apply to the relation $\succeq$.

The equation $ab = 1$ in an atomic module M has only one solution: $a = 1$, $b = 1$. Indeed, if $ab = 1$ for $a, b \in M$, then $1 \preceq a \preceq 1$, so that $a = 1$ and $b = 1$. In particular, the neutral element is the unique invertible element of M.

6.1.4 Presentations of a monoid

Consider a set X and a subset R of $X^* \times X^*$. Let $\sim$ be the smallest equivalence relation on X^* containing all pairs $(w_1 r w_2, w_1 r' w_2)$, where $(r, r') \in R$ and $w_1, w_2 \in X^*$. In other words, $\sim$ is the smallest equivalence relation on X^* such that $w_1 r w_2 \sim w_1 r' w_2$ for all $(r, r') \in R$ and $w_1, w_2 \in X^*$. We define M to be the set of equivalence classes for $\sim$. It is clear that M has a unique structure of a monoid such that the projection $P : X^* \to M$ is a monoid homomorphism. We say that $\langle X \,|\, R \rangle$ is a *monoid presentation* of M and call the elements of X *generators* and the elements of R *relations*.

It is clear that the set $P(X) \subset M$ generates M in the sense that every element of M is a product of elements of this set. For any relation $(r, r') \in R$, the element $P(r) = P(r')$ of M is called a *relator* associated with the presentation $\langle X \,|\, R \rangle$. In the sequel we shall often use the notation $r = r'$ for a relation $(r, r') \in R$ and make no distinction between a generator $x \in X$ and its projection $P(x)$ to M.

Note that a set-theoretic map f from the set X to a monoid M' induces a monoid homomorphism $M \to M'$ if and only if the monoid extension $f^* : X^* \to M'$ of f satisfies $f^*(r) = f^*(r')$ for all $(r, r') \in R$.

We introduce several useful classes of monoid presentations. A monoid presentation $\langle X \,|\, R \rangle$ is *finite* if both sets X and R are finite. A presentation $\langle X \,|\, R \rangle$ of a monoid M is *weighted* if there is a monoid homomorphism $\ell : M \to \mathbf{N}$ such that $\ell(x) \geq 1$ for all $x \in X$. The homomorphism ℓ is called the *weight*.

A presentation $\langle X \mid R \rangle$ of a monoid M is *length-balanced* if $l(r) = l(r')$ for all $(r, r') \in R$, where l is the length function on X^* introduced in Section 6.1.1. The formula $\ell(x) = 1$ for all $x \in X$ defines then a *canonical weight* $\ell : M \to \mathbf{N}$. Thus, all length-balanced presentations are weighted. The converse is not true; for instance, the presentation $\langle x, y \mid x^3 = y^2 \rangle$ is weighted but not length-balanced.

Lemma 6.4. *If a monoid M has a weighted presentation $\langle X \mid R \rangle$, then M is atomic and all its atoms are contained in the set X of generators. If M has a length-balanced presentation $\langle X \mid R \rangle$, then the set of atoms of M coincides with X and $\|a\| = \ell(a)$ for all $a \in M$, where ℓ is the canonical weight on M.*

Proof. Let $\ell : M \to \mathbf{N}$ be a monoid homomorphism such that $\ell(x) \geq 1$ for all generators $x \in X$. Then $\ell(a) \geq 1$ for all $a \in M - \{1\}$. If $a \in M$ expands as a product $a_1 \cdots a_r$ with $a_1, \ldots, a_r \in M - \{1\}$, then $\ell(a) = \ell(a_1) + \cdots + \ell(a_r) \geq r$. Hence $\ell(a) \geq \|a\|$, so that M is atomic. That all atoms of M belong to X follows from the fact that any generating subset of a monoid must contain all the atoms. The second claim of the lemma is a direct consequence of the definitions. $\qquad\square$

6.1.5 The word problem and the divisibility problem

The *word problem* for a presentation $\langle X \mid R \rangle$ of a monoid M is the following: given two words $w, w' \in X^*$ representing certain $a, a' \in M$, determine whether $a = a'$. The closely related *left* (resp. *right*) *divisibility problem* is the following: given two words $w, w' \in X^*$ representing $a, a' \in M$, determine whether $a \preceq a'$ (resp. $a' \succeq a$).

Both the word problem and the divisibility problem can easily be solved for a finite weighted presentation $\langle X \mid R \rangle$ of M. Let $\ell : M \to \mathbf{N}$ be a weight, so that $\ell(x) \geq 1$ for all $x \in X$. Observe that the value of ℓ on any $a \in M$ represented by a nonempty word $w \in X^*$ is greater than or equal to the length of w. Let $W(a) \subset X^*$ be the set of words representing a. All these words have length $\leq \ell(a)$. Since X is finite, the number of words of length $\leq \ell(a)$ is finite and the set $W(a)$ is finite. To list all elements of $W(a)$, one starts with the given word w representing a and consecutively applies all possible substitutions of the form

$$w_1 r w_2 \leftrightarrow w_1 r' w_2$$

(for $(r, r') \in R$) to any element of $W(a)$ already found. Since R is finite, this procedure is also finite. It gives a solution to the word problem: two elements $a, a' \in M$ are equal if and only if $W(a) = W(a')$.

We also obtain a solution of the left and right divisibility problems. Namely, $a \preceq a'$ if and only if some prefix (initial segment) of a word in $W(a')$ belongs to $W(a)$. Similarly, $a' \succeq a$ if and only if some suffix (final segment) of a word in $W(a')$ belongs to $W(a)$.

6.2 Normal forms and the conjugacy problem

We introduce and study a certain monoid M_Σ derived from a subset Σ of a given monoid M. Under favorable assumptions, we obtain a normal form for the elements of M_Σ and solve the conjugacy problem in M_Σ.

6.2.1 The monoid M_Σ

Let M be a monoid and let $\Sigma \subset M$ be a subset of M containing the neutral element 1. Let M_Σ be the monoid generated by the symbols $[a]$, where a runs over Σ, modulo the defining relations $[1] = 1$ and $[a][b] = [ab]$ whenever $a, b, ab \in \Sigma$. There is a monoid homomorphism $p : M_\Sigma \to M$ defined by $p([a]) = a$ for all $a \in \Sigma$.

The definition of M_Σ can be rephrased by identifying a product $[a_1] \cdots [a_r]$ in M_Σ (where $a_1, \ldots, a_r \in \Sigma$) with the sequence $(a_1, \ldots, a_r)$. Then M_Σ is the set of equivalence classes of finite sequences $(a_1, \ldots, a_r)$ of elements of Σ under the equivalence generated by the relations

$$(a_1, \ldots, a_{i-1}, a_i' a_i'', a_{i+1}, \ldots, a_r) \sim (a_1, \ldots, a_{i-1}, a_i', a_i'', a_{i+1}, \ldots, a_r)$$

whenever $a_i' a_i'' \in \Sigma$ and by the relation saying that the empty sequence is equivalent to the one-element sequence (1), where $1 \in \Sigma$. The product in M_Σ is induced by concatenation of sequences.

We formulate the main theorem on the structure of M_Σ.

Theorem 6.5. *Let M be an atomic monoid and Σ a subset of M such that $1 \in \Sigma$ and the following three conditions hold:*

$(*_1)$ *All left divisors and all right divisors of elements of Σ belong to Σ.*
$(*_2)$ *For any $a, b, c \in \Sigma$, if $ab = ac$ or $ba = ca$, then $b = c$.*
$(*_3)$ *For any $a, b \in \Sigma$, the set $\{x \in \Sigma \mid x \preceq b \text{ and } ax \in \Sigma\}$ has a maximal element (with respect to $\preceq$).*

Then for any $\xi \in M_\Sigma$, there is a unique $\alpha(\xi) \in \Sigma$ such that $[\alpha(\xi)]$ is a left divisor of ξ that is maximal among all left divisors of ξ lying in the set $\{[a]\}_{a \in \Sigma} \subset M_\Sigma$. Moreover, there is a unique $\omega(\xi) \in M_\Sigma$ such that $\xi = [\alpha(\xi)] \, \omega(\xi)$.

Proof. The proof goes in five steps.

Step 1. By $(*_3)$, for any $a, b \in \Sigma$, the set $\{x \in \Sigma \mid x \preceq b \text{ and } ax \in \Sigma\}$ has a maximal element $c \in \Sigma$. Then $b = cd$ for some $d \in M$. By $(*_1)$, $d \in \Sigma$, and by $(*_2)$, d is unique. Set $\alpha_2(a, b) = ac \in \Sigma$ and $\omega_2(a, b) = d \in \Sigma$. Clearly,

$$\alpha_2(a, b) \, \omega_2(a, b) = ab. \tag{6.1}$$

For instance, for all $a \in \Sigma$,

$$\alpha_2(a, 1) = \alpha_2(1, a) = a \quad \text{and} \quad \omega_2(a, 1) = \omega_2(1, a) = 1.$$

We claim that for any $a, b, c \in \Sigma$ such that $ab \in \Sigma$,

$$\alpha_2(ab, c) = \alpha_2(a, \alpha_2(b, c)) \tag{6.2}$$

and

$$\omega_2(ab, c) = \omega_2(a, \alpha_2(b, c))\, \omega_2(b, c). \tag{6.3}$$

The rest of Step 1 is devoted to the proof of this claim. We shall use the following observation: If $a, b, c \in M$ satisfy $ac \in \Sigma$ and $ab \preceq ac$, then $b \preceq c$. Indeed, if $ac = abd$ with $d \in M$, then the assumption $ac \in \Sigma$ and $(*_1)$ imply that $a, c, bd \in \Sigma$. By $(*_2)$, we have $c = bd$, so that $b \preceq c$.

By definition, $\alpha_2(b, c) = bd$, where d is maximal such that $d \preceq c$ and $bd \in \Sigma$. Similarly,

$$\alpha_2(a, \alpha_2(b, c)) = \alpha_2(a, bd) = ad',$$

where d' is maximal such that $d' \preceq bd$ and $ad' \in \Sigma$. Since $b \preceq bd$ and $ab \in \Sigma$ (the latter by hypothesis), $b \preceq d'$. Writing $d' = be$ with $e \in \Sigma$, we obtain $\alpha_2(a, bd) = abe$, with $be \preceq bd \in \Sigma$ and $abe \in \Sigma$. By the observation above, $e \preceq d \preceq c$, so that $e \preceq c$. Now $\alpha_2(ab, c) = abf$, where f is maximal such that $f \preceq c$ and $abf \in \Sigma$. Therefore, $e \preceq f$. On the other hand, $f \preceq c$ and $bf \in \Sigma$ imply $f \preceq d$ and $bf \preceq bd$. This and the inclusion $abf \in \Sigma$ imply that $bf \preceq d' = be$. Therefore, $f \preceq e$. By Lemma 6.3, $e = f$ and

$$\alpha_2(ab, c) = abf = abe = ad' = \alpha_2(a, \alpha_2(b, c)).$$

This proves (6.2). To prove (6.3), note that by (6.1) and (6.2),

$$\begin{aligned}
\alpha_2(ab, c)\, \omega_2(a, \alpha_2(b, c))\, \omega_2(b, c) &= \alpha_2(a, \alpha_2(b, c))\, \omega_2(a, \alpha_2(b, c))\, \omega_2(b, c) \\
&= a\, \alpha_2(b, c)\, \omega_2(b, c) \\
&= abc \\
&= \alpha_2(ab, c)\, \omega_2(ab, c).
\end{aligned}$$

By Condition $(*_2)$, to deduce (6.3) it is enough to prove that the product $\omega_2(a, \alpha_2(b, c))\, \omega_2(b, c)$ belongs to Σ. By definition, there is $d \in \Sigma$ such that $\alpha_2(b, c) = bd \in \Sigma$ and $c = d\omega_2(b, c)$. Let f be the maximal element of Σ such that $f \preceq bd$ and $af \in \Sigma$. Since $b \preceq bd$ and $ab \in \Sigma$, there is $e \in \Sigma$ such that $f = be$. Then

$$bd = f\omega_2(a, bd) = b\, e\, \omega_2(a, bd).$$

By $(*_2)$, $d = e\,\omega_2(a, bd)$ and

$$e\, \omega_2(a, \alpha_2(b, c))\, \omega_2(b, c) = e\, \omega_2(a, bd)\, \omega_2(b, c) = d\, \omega_2(b, c) = c.$$

This shows that $\omega_2(a, \alpha_2(b, c))\, \omega_2(b, c)$ is a right divisor of $c \in \Sigma$, hence an element of Σ.

Step 2. At this step we prove the following claim: there is a unique map $\alpha : M_\Sigma \to \Sigma$ such that

(i) $\alpha(1) = 1$ and
(ii) $\alpha([a]\eta) = \alpha_2(a, \alpha(\eta))$ for all $a \in \Sigma$ and $\eta \in M_\Sigma$.

Recall the monoid homomorphism $p : M_\Sigma \to M$. For any $\xi \in M_\Sigma$, we set $H(\xi) = \|p(\xi)\| \geq 0$. We call $H(\xi)$ the *height* of ξ. It is clear that

$$H(\xi\xi') \geq H(\xi) + H(\xi')$$

for any $\xi, \xi' \in M_\Sigma$. Note that $H(\xi) = 0$ if and only if $\xi = 1$, and $H(\xi) = 1$ if and only if $\xi = [a]$, where a is an atom of M belonging to Σ. To see this, pick an expansion $\xi = [a_1] \cdots [a_r]$ with $a_1, \ldots, a_r \in \Sigma$. Then

$$H(\xi) = \|p(\xi)\| \geq \|a_1\| + \cdots + \|a_r\|.$$

If $H(\xi) = 0$, then $a_1 = \cdots = a_r = 1$ and $\xi = 1$. If $H(\xi) = 1$, then all the elements $a_1, \ldots, a_r \in \Sigma$ are equal to 1 except one element, which is an atom.

For $\xi \in M_\Sigma$, we define $\alpha(\xi)$ by induction on the height of ξ. For $\xi = 1$, set $\alpha(\xi) = 1 \in \Sigma$. If $H(\xi) = 1$, then $\xi = [a] = [a]1$ for some atom $a \in \Sigma$, and to satisfy (ii) we have to set $\alpha(\xi) = \alpha_2(a, 1) = a$.

Pick an integer $k \geq 1$ and suppose that $\alpha(\xi)$ is defined for all ξ of height $\leq k$ so that conditions (i), (ii) are satisfied whenever $H([a]\eta) \leq k$. Let ξ be an element of M_Σ of height $k + 1$. We can expand $\xi = [a]\eta$, where $a \in \Sigma$, $a \neq 1$, and $\eta \in M_\Sigma$. Then $H([a]) \geq 1$ and $H(\eta) < H([a]\eta)$, so that $\alpha(\eta)$ is already defined. To satisfy (ii), we have to set $\alpha(\xi) = \alpha_2(a, \alpha(\eta))$. We must check that $\alpha_2(a, \alpha(\eta))$ does not depend on the choice of the expansion $\xi = [a]\eta$. By definition of M_Σ and the induction hypothesis, it is enough to check that

$$\alpha_2(a, \alpha(\eta)) = \alpha_2(a', \alpha([a'']\eta))$$

when $a = a'a''$ with $a', a'' \in \Sigma - \{1\}$. Since

$$H([a'']\eta) \leq H([a]\eta) - H([a']) < H([a]\eta),$$

the induction hypothesis yields that $\alpha([a'']\eta) = \alpha_2(a'', \alpha(\eta))$. By (6.2),

$$\begin{aligned}
\alpha_2(a', \alpha([a'']\eta)) &= \alpha_2(a', \alpha_2(a'', \alpha(\eta))) \\
&= \alpha_2(a'a'', \alpha(\eta)) \\
&= \alpha_2(a, \alpha(\eta)).
\end{aligned}$$

Therefore α is well defined on elements of M_Σ of height $\leq k + 1$ and satisfies conditions (i) and (ii). This completes the induction and proves our claim.

Step 3. We now check that for any $\xi \in M_\Sigma$, the element $[\alpha(\xi)]$ of M_Σ is a left divisor of ξ that is maximal among all left divisors of ξ lying in the set

$$\{[a]\}_{a \in \Sigma} \subset M_\Sigma.$$

Using the projection $p : M_\Sigma \to M$, it is easy to show that all divisors of 1 in M_Σ are equal to 1. Therefore if $H(\xi) = 0$, then $[\alpha(\xi)] = \xi = 1$ is the only left divisor of $1 \in M_\Sigma$. If $H(\xi) \geq 1$, write $\xi = [a]\eta$ for some $a \in \Sigma - \{1\}$ and $\eta \in M_\Sigma$. Then $H(\eta) < H(\xi)$ and $\alpha(\xi) = \alpha_2(a, \alpha(\eta))$. Therefore, $\alpha(\xi) = ab$ for some $b \in \Sigma$ such that $b \preceq \alpha(\eta)$. By the induction assumption, $[\alpha(\eta)]$ is a left divisor of η. Hence,

$$[\alpha(\xi)] = [ab] = [a][b] \preceq [a][\alpha(\eta)] \preceq [a]\eta = \xi.$$

This shows that $[\alpha(\xi)]$ is a left divisor of ξ belonging to the set $\{[a]\}_{a \in \Sigma}$. We now show that $[\alpha(\xi)]$ is maximal with these properties. Suppose that $\xi = [a']\eta'$ for some $a' \in \Sigma$, $\eta' \in M_\Sigma$. Then $\alpha(\xi) = \alpha_2(a', \alpha(\eta')) = a'b'$ for some $b' \in \Sigma$. Hence, $a' \preceq \alpha(\xi)$ and $[a'] \preceq [\alpha(\xi)]$.

Step 4. We claim that there is a unique map $\omega : M_\Sigma \hookrightarrow M_\Sigma$ such that

(i) $\omega(1) = 1$ and
(ii) $\omega([a]\eta) = [\omega_2(a, \alpha(\eta))]\,\omega(\eta)$ for all $a \in \Sigma$ and $\eta \in M_\Sigma$.

The value of ω on any $\xi \in M_\Sigma$ is defined by induction on the height of ξ. Set $\omega(1) = 1$. Pick an integer $k \geq 1$ and suppose that $\omega(\xi)$ is defined for all ξ of height $\leq k$, so that conditions (i) and (ii) are satisfied whenever $H([a]\eta) \leq k$. Let $\xi = [a]\eta$ be an element of M_Σ of height $k+1$ with $a \in \Sigma - \{1\}$ and $\eta \in M_\Sigma$. Then $H(\eta) < H(\xi)$ and $\omega(\eta)$ is defined. To satisfy (ii), we have to set

$$\omega(\xi) = [\omega_2(a, \alpha(\eta))]\,\omega(\eta).$$

We must check that $\omega(\xi)$ is independent of the choice of the expansion $\xi = [a]\eta$. By definition of M_Σ and the induction hypothesis, it is enough to check that

$$\omega_2(a, \alpha(\eta))\,\omega(\eta) = \omega_2(a', \alpha([a'']\eta))\,\omega([a'']\eta)$$

when $a = a'a''$ with $a', a'' \in \Sigma - \{1\}$. As we know, $\alpha([a'']\eta) = \alpha_2(a'', \alpha(\eta))$. Since $H([a'']\eta) < H([a]\eta)$, the induction hypothesis yields the equality $\omega([a'']\eta) = [\omega_2(a'', \alpha(\eta))]\,\omega(\eta)$. By (6.3),

$$
\begin{aligned}
\omega_2(a', \alpha([a'']\eta))\,\omega([a'']\eta) &= \omega_2(a', \alpha_2(a'', \alpha(\eta)))\,\omega([a'']\eta) \\
&= \omega_2(a', \alpha_2(a'', \alpha(\eta)))\,[\omega_2(a'', \alpha(\eta))]\,\omega(\eta) \\
&= \omega_2(a'a'', \alpha(\eta))\,\omega(\eta) \\
&= \omega_2(a, \alpha(\eta))\,\omega(\eta).
\end{aligned}
$$

Therefore ω is well defined on elements of M_Σ of height $\leq k+1$ and satisfies conditions (i) and (ii). This completes the induction and proves our claim.

Step 5. To complete the proof, it remains to show that for any $\xi \in M_\Sigma$, the element $\eta = \omega(\xi)$ is the unique element of M_Σ such that $\xi = [\alpha(\xi)]\,\eta$. We proceed by induction on $H(\xi)$. If $H(\xi) = 0$, then $\xi = 1$, $\alpha(\xi) = 1$, $\omega(\xi) = 1$, and the claim is obvious. Suppose that our claim holds for all ξ of height $\leq k$ for some integer $k \geq 1$. Let ξ be an element of M_Σ of height $k+1$. By Step 3, $\xi = [\alpha(\xi)]\eta$ with $\eta \in M_\Sigma$. Clearly, $\alpha(\xi) \neq 1$ and therefore $H(\eta) < H(\xi)$.

Set $\theta = [\alpha(\xi)]\,[\alpha(\eta)]$. By definition of the map α,

$$\alpha(\theta) = \alpha_2(\alpha(\xi), \alpha(\eta)) = \alpha(\xi)\, b$$

for some $b \in \Sigma$ such that $[b] \preceq [\alpha(\eta)] \preceq \eta$. Hence $\alpha(\xi) \preceq \alpha(\theta)$ and

$$[\alpha(\theta)] = [\alpha(\xi)]\,[b] \preceq [\alpha(\xi)]\,\eta = \xi\,.$$

Since $[\alpha(\xi)]$ is the maximal left divisor of ξ in the set $\{[a]\}_{a \in \Sigma}$, we have $[\alpha(\xi)] = [\alpha(\theta)]$. Projecting to M, we conclude that

$$\alpha(\xi) = \alpha(\theta) = \alpha_2(\alpha(\xi), \alpha(\eta))\,.$$

Therefore, $\alpha(\xi)\,\alpha(\eta) = \alpha_2(\alpha(\xi), \alpha(\eta))\,\alpha(\eta)$. On the other hand, by (6.1),

$$\alpha(\xi)\,\alpha(\eta) = \alpha_2(\alpha(\xi), \alpha(\eta))\,\omega_2(\alpha(\xi), \alpha(\eta))\,.$$

Combining these equalities, we obtain

$$\alpha_2(\alpha(\xi), \alpha(\eta))\,\alpha(\eta) = \alpha_2(\alpha(\xi), \alpha(\eta))\,\omega_2(\alpha(\xi), \alpha(\eta))\,.$$

By $(*_2)$, we may cancel $\alpha_2(\alpha(\xi), \alpha(\eta))$. Thus $\alpha(\eta) = \omega_2(\alpha(\xi), \alpha(\eta))$ and

$$\omega(\xi) = \omega([\alpha(\xi)]\,\eta) = [\omega_2(\alpha(\xi), \alpha(\eta))]\,\omega(\eta) = [\alpha(\eta)]\,\omega(\eta) = \eta\,,$$

where the last equality follows from the induction hypothesis. This shows that $\xi = [\alpha(\xi)]\,\omega(\xi)$ and that any $\eta \in M_\Sigma$ satisfying $\xi = [\alpha(\xi)]\,\eta$ is equal to $\omega(\xi)$. $\qquad\qquad\square$

6.2.2 The normal form in M_Σ

Under the assumptions of Theorem 6.5, any $\xi \in M_\Sigma$ may be inductively expanded as follows:

$$\begin{aligned}\xi &= [\alpha(\xi)]\,\omega(\xi) = [\alpha(\xi)]\,[\alpha(\omega(\xi))]\,\omega^2(\xi) \\ &= [\alpha(\xi)]\,[\alpha(\omega(\xi))]\,[\alpha(\omega^2(\xi))]\,\omega^3(\xi) = \cdots\,.\end{aligned}$$

This expansion process may be stopped at the rth step, where r is the minimal integer such that $\omega^{r+1}(\xi) = 1$. Such r exists and does not exceed $\|p(\xi)\|$, since in an expansion of ξ as a product of generators $[a]$ with $a \in \Sigma$ at most $\|p(\xi)\|$ of the generators may be distinct from 1. (Here it is useful to note that $\alpha(\eta) \neq 1$ for any $\eta \in M_\Sigma - \{1\}$.)

These observations lead us to a normal form for each element of M_Σ. A *normal form* for $\xi \in M_\Sigma$ is a sequence $(a_1, a_2, \ldots, a_r)$ of elements of Σ, all different from 1, such that $\xi = [a_1]\,[a_2] \cdots [a_r]$ and

$$a_i = \alpha([a_i][a_{i+1}] \cdots [a_r])$$

for all $i = 1, 2, \ldots, r$. The remarks above show that each $\xi \in M_\Sigma$ has a normal form. (The normal form of $1 \in M_\Sigma$ is the empty sequence.) The uniqueness in the last claim of Theorem 6.5 implies the uniqueness of the normal form.

6.2.3 The cancellativity of M_Σ

We use Theorem 6.5 and the map $\alpha_2 : \Sigma \times \Sigma \to \Sigma$ introduced in its proof to establish the left cancellativity of M_Σ.

Lemma 6.6. *Under the assumptions of Theorem 6.5, the monoid M_Σ is left cancellative.*

Proof. We need to show that $\xi\eta = \xi\theta \Rightarrow \eta = \theta$ for $\xi, \eta, \theta \in M_\Sigma$. Suppose first that $\xi = [a]$ for some $a \in \Sigma$. Then $\alpha(\xi\eta) = \alpha_2(a, \alpha(\eta)) = ab \in \Sigma$ for some $b \in \Sigma$ such that $b \preceq \alpha(\eta)$. The equalities

$$ab = \alpha(\xi\eta) = \alpha(\xi\theta) = \alpha([a]\theta) = \alpha_2(a, \alpha(\theta)) = ac$$

for some $c \preceq \alpha(\theta)$ imply that $b = c \preceq \alpha(\theta)$. Then there are $\eta', \theta' \in M_\Sigma$ such that $[b]\,\eta' = \eta$ and $[b]\,\theta' = \theta$. As we know, $\omega(\xi\eta)$ is the unique element $x \in M_\Sigma$ such that $\xi\eta = [\alpha(\xi\eta)]\,x = [ab]\,x$. Since

$$\xi\eta = [a][b]\,\eta' = [ab]\,\eta',$$

we have $\omega(\xi\eta) = \eta'$. Similarly, $\omega(\xi\theta) = \theta'$. Hence,

$$\eta' = \omega(\xi\eta) = \omega(\xi\theta) = \theta' \quad \text{and} \quad \eta = [b]\,\eta' = [b]\,\theta' = \theta.$$

In general, $\xi = [a_1][a_2]\cdots[a_r]$ with $a_1, a_2, \ldots, a_r \in \Sigma$. As we know, $[a_1][a_2]\cdots[a_r]\eta = [a_1][a_2]\cdots[a_r]\theta$ implies that $[a_2]\cdots[a_r]\eta = [a_2]\cdots[a_r]\theta$. Continuing inductively, we obtain $\eta = \theta$. $\qquad\square$

6.2.4 The word problem in M_Σ

We say that the set $\Sigma \subset M$ is *weighted* if there is a map $\ell : \Sigma \to \mathbf{N}$ such that $\ell(1) = 0$, $\ell(a) \geq 1$ for $a \neq 1$, and $\ell(a) + \ell(b) = \ell(ab)$ whenever $a, b, ab \in \Sigma$. The map ℓ extends then to a monoid homomorphism $M_\Sigma \to \mathbf{N}$ that turns the presentation of M_Σ above into a weighted presentation. If, in addition, Σ is finite, then Section 6.1.5 yields a solution of the word problem and of the divisibility problem in M_Σ.

6.2.5 The conjugacy problem in M_Σ

The *conjugacy problem in a group G* consists in finding a procedure that allows, given $\alpha, \beta \in G$, to decide whether there is $\gamma \in G$ such that $\alpha = \gamma\beta\gamma^{-1}$ or, equivalently, $\alpha\gamma = \gamma\beta$. By extension, the *conjugacy problem in a monoid M* consists in finding a procedure that allows, given $a, b \in M$, to decide whether there is $c \in M$ such that $ac = cb$. The following lemma yields the key to the conjugacy problem in M_Σ.

Lemma 6.7. *Let M, $\Sigma \subset M$ satisfy the assumptions of Theorem 6.5. Given a, $b \in M_\Sigma$, there is $c \in M_\Sigma$ such that $ac = cb$ if and only if there exist a sequence $a_0 = a$, $a_1, \ldots, a_r = b$ of elements of M_Σ and a sequence $c_1, \ldots, c_r$ of elements of Σ such that*

$$a_{i-1}[c_i] = [c_i]a_i$$

for all $i = 1, \ldots, r$.

Proof. If we have such sequences, then $ac = cb$ for $c = [c_1][c_2]\cdots[c_r]$. Conversely, let $c \in M_\Sigma$ be such that $ac = cb$. We prove the assertion by induction on the length r of the normal form $(c_1, \ldots, c_r)$ of c. If $r = 1$, then $c = [c_1]$ and we are done. Suppose that $r \geq 2$. Since

$$[c_1] = [\alpha(c)] \preceq c \preceq cb = ac,$$

we have

$$c_1 \preceq \alpha(ac) = \alpha_2(a, \alpha(c)) = \alpha_2(a, c_1).$$

Therefore,

$$[c_1] \preceq [\alpha_2(a, c_1)] \preceq ac_1.$$

Hence, there is $a_1 \in M_\Sigma$ such that

$$[c_1]a_1 = a[c_1].$$

We have

$$[c_1]a_1[c_2]\cdots[c_r] = a[c_1][c_2]\cdots[c_r] = ac = cb = [c_1][c_2]\cdots[c_r]b.$$

By Lemma 6.6, we may divide on the left by $[c_1]$. This gives $a_1c' = c'b$, where $c' = [c_2]\cdots[c_r]$ has a normal form of length $r - 1$. We conclude using the induction hypothesis. □

Lemma 6.7 provides a solution of the conjugacy problem in M_Σ. Suppose that M, Σ satisfy the conditions of Theorem 6.5 and Σ is finite. Suppose also that M_Σ admits a finite weighted presentation so that the word problem in M_Σ is solvable (for instance, it is enough to suppose that Σ is weighted in the sense of Section 6.2.4). To determine whether two elements a, $b \in M_\Sigma$ are conjugate (in the sense that there is $c \in M_\Sigma$ such that $ac = cb$), first observe that conjugate elements of M_Σ have the same weight. Since there are only finitely many elements of M_Σ of a given weight, a has only finitely many conjugates. Lemma 6.7 shows that in order to find them all, it is enough to apply all possible conjugacies by elements of Σ to all known conjugates of a until no new elements are found. We thus obtain a finite list $a_1, \ldots, a_s$ of conjugates of a in M_Σ. If $b = a_i$ for some i, then b is conjugate to a; otherwise, b is not conjugate to a.

6.2.6 Comprehensive sets

A set $\Sigma \subset M$ is *comprehensive* if $1 \in \Sigma$ and M has a presentation by generators and relations such that all generators and relators belong to Σ.

Lemma 6.8. *If Σ is a comprehensive subset of a monoid M such that all left divisors of elements of Σ belong to Σ, then the monoid homomorphism $p : M_\Sigma \to M$ is an isomorphism.*

Proof. Since Σ contains a set of generators of M, the homomorphism p is surjective. We need only to prove its injectivity. Observe first that if $a_1, a_2, \ldots, a_n$ are elements of Σ with $n \geq 2$ such that $a = a_1 a_2 \cdots a_n \in \Sigma$, then $[a] = [a_1][a_2] \cdots [a_n]$. Indeed, $a_1 a_2$ is a left divisor of a and therefore $a_1 a_2 \in \Sigma$. By definition of M_Σ, we have $[a_1][a_2] = [a_1 a_2]$. Continuing by induction, we obtain $[a_1][a_2] \cdots [a_n] = [a]$.

Consider now a presentation $\langle X \mid R \rangle$ of M by generators and relations and let $P : X^* \to M$ be the natural projection. We assume that $P(x) \in \Sigma$ for all $x \in X$ and $P(r) = P(r') \in \Sigma$ for all $(r, r') \in R$. Define a monoid homomorphism $Q : X^* \to M_\Sigma$ by $Q(x) = [P(x)]$ for $x \in X$. The observation above implies that for any $r \in X^*$ with $P(r) \in \Sigma$, we have $Q(r) = [P(r)]$. Therefore for any relation $(r, r') \in R$, we have

$$Q(r) = [P(r)] = [P(r')] = Q(r').$$

This implies that there is a monoid homomorphism $q : M \to M_\Sigma$ such that $Q = qP$. Then

$$qp([P(x)]) = q(P(x)) = Q(x) = [P(x)]$$

for all $x \in X$. Since the set $P(X)$ generates M, we have $qp = \mathrm{id}$. Therefore p is injective. $\qquad\square$

Lemma 6.8 shows that under appropriate assumptions on Σ we have $M_\Sigma \cong M$, so that all the properties of M_Σ obtained above hold for M.

Exercise 6.2.1. Show that $(a_1, a_2, \ldots, a_r)$ is a normal form of an element of M_Σ if and only if (a_i, a_{i+1}) is a normal form of the product $[a_i][a_{i+1}]$ for $i = 1, \ldots, r-1$.

Exercise 6.2.2. Observe that for any $\Sigma \subset M$ with $1 \in \Sigma$, the subset $\{[a]\}_{a \in \Sigma}$ of M_Σ is comprehensive. Deduce the following partial converse to Lemma 6.8: if the monoid homomorphism $p : M_\Sigma \to M$ is an isomorphism, then Σ is comprehensive.

Exercise 6.2.3. Verify that for any set X, the subset $\Sigma = X \cup \{1\}$ of the free monoid $M = X^*$ is weighted, comprehensive, and satisfies all the conditions of Theorem 6.5. For these M, Σ, verify directly all the claims established in this section.

6.3 Groups of fractions and pre-Garside monoids

6.3.1 Groups of fractions

A monoid homomorphism $i : M \to G$ is said to be *universal* if G is a group and for any monoid homomorphism f from M to an arbitrary group G', there is a unique group homomorphism $g : G \to G'$ such that $f = gi$. Every monoid M admits a universal homomorphism to a group. To see this, take an arbitrary presentation $\langle X \mid R \rangle$ of M by generators and relations and consider the group G defined by $\langle X \mid R \rangle$ viewed as a group presentation. The identity map $\mathrm{id}_X : X \to X$ extends to a monoid homomorphism $M \to G$, which is easily seen to be universal. The definition of a universal homomorphism $M \to G$ implies that it is unique up to composition with a group isomorphism. In particular, the group G is well defined up to isomorphism. This group is called the *group of fractions* of M and denoted by G_M. A presentation of G_M by generators and relations can be obtained by taking an arbitrary monoid presentation of M by generators and relations and viewing it as a group presentation.

A monoid M is *embeddable* if there is an injective monoid homomorphism from M into a group. It is clear that M is embeddable if and only if the universal homomorphism $M \to G_M$ is injective. For example, the inclusion $\mathbf{N} \hookrightarrow \mathbf{Z}$ shows that $\mathbf{N}$ is embeddable. It is easy to see that this inclusion is a universal homomorphism, so that $G_{\mathbf{N}} = \mathbf{Z}$.

It is clear that embeddable monoids are left cancellative and right cancellative. For example, the monoid $\{1, x\}$ with $xx = x$ is not left cancellative (since $x \neq 1$); therefore it is not embeddable. The group of fractions of this monoid is trivial.

6.3.2 Pre-Garside monoids

Definition 6.9. *A pre-Garside monoid is a pair consisting of a monoid M and an element Δ of M such that the set $\Sigma = \Sigma_\Delta$ of left divisors of Δ satisfies the following conditions:*

(a) Σ is finite, generates M, and coincides with the set of right divisors of Δ.
(b) If $a, b \in \Sigma$ are such that $\Delta a = \Delta b$ or $a\Delta = b\Delta$, then $a = b$.

The element $\Delta \in M$ is called the *Garside element* of M. Note that the set Σ of the divisors of Δ is closed under left and right divisibility, i.e., all left divisors and all right divisors of elements of Σ belong to Σ. Clearly, $1 \in \Sigma$ and $\Delta \in \Sigma$.

Examples 6.10. Any positive integer is a Garside element of the monoid $\mathbf{N}$. The monoid $\mathbf{N}^\times$ has no Garside elements. All elements of a finite group are Garside. More interesting examples will be given in subsequent sections.

Lemma 6.11. *Let (M, Δ) be a pre-Garside monoid and let $\Sigma \subset M$ be the set of divisors of Δ.*

(i) For all a, b, $c \in \Sigma$, if $ac = bc$ or $ca = cb$, then $a = b$.

(ii) There is a bijection $\delta : \Sigma \to \Sigma$ such that $\Delta a = \delta(a)\Delta$ for all $a \in \Sigma$.

(iii) If N is the order of δ (i.e., the minimal positive integer such that $\delta^N = \mathrm{id}$), then $\Delta^N a = a\Delta^N$ for all $a \in M$.

(iv) For any $a \in M$, there is an integer $r \geq 1$ such that $a \preceq \Delta^r$ and $\Delta^r \succeq a$.

Proof. (i) Since $c \in \Sigma$, there is $d \in M$ such that $cd = \Delta$. Then $ac = bc$ implies $a\Delta = acd = bcd = b\Delta$. Hence $a = b$ by condition (b) of Definition 6.9. The implication $ca = cb \Rightarrow a = b$ has a similar proof, using an element $e \in M$ such that $ec = \Delta$.

(ii) Since any left divisor of Δ is also a right divisor and vice versa, for any $a \in \Sigma$, there are $a', \delta(a) \in \Sigma$ such that $\Delta = a'a$ and $\Delta = \delta(a)a'$. By claim (i), a' and $\delta(a)$ are uniquely defined. We have

$$\Delta a = \delta(a)a'a = \delta(a)\Delta. \tag{6.4}$$

Since Σ is finite, in order to prove that the map $\delta : \Sigma \to \Sigma$ is bijective, it suffices to check that it is injective. The equality $\delta(a) = \delta(b)$ implies

$$\Delta a = \delta(a)\Delta = \delta(b)\Delta = \Delta b.$$

This implies $a = b$ by condition (b) of Definition 6.9.

(iii) By induction on n we derive from (6.4) that $\Delta^n a = \delta^n(a)\Delta^n$ for all $a \in \Sigma$. (Here δ^n is the composition of n copies of δ.) Since $\delta^N = \mathrm{id}$, we have $\Delta^N a = \delta^N(a)\Delta^N = a\Delta^N$ for all $a \in \Sigma$. In other words, Δ^N commutes with every element of the set Σ. Since this set generates the monoid M, we can conclude that Δ^N commutes with all elements of M.

(iv) Write a as a product $a = a_1 \cdots a_r$ of $r \geq 1$ elements of Σ. For each a_i let $b_i \in \Sigma$ be defined by $b_i a_i = \Delta$. Set $b = \delta^{r-1}(b_r) \cdots \delta(b_2)b_1$. We claim that $ba = \Delta^r$, which proves that $\Delta^r \succeq a$. Indeed,

$$
\begin{aligned}
ba &= \delta^{r-1}(b_r) \cdots \delta(b_2)b_1 a_1 \cdots a_r \\
&= \delta^{r-1}(b_r) \cdots \delta(b_2)\Delta a_2 \cdots a_r \\
&= \delta^{r-1}(b_r) \cdots \Delta b_2 a_2 \cdots a_r \\
&= \delta^{r-1}(b_r) \cdots \delta^2(b_3)\Delta^2 a_3 \cdots a_r \\
&= \delta^{r-1}(b_r) \cdots \Delta^2 b_3 a_3 \cdots a_r \\
&= \cdots = \Delta^{r-1}b_r a_r = \Delta^r.
\end{aligned}
$$

A similar proof shows that if $a_i c_i = \Delta$ for $i = 1, \dots, r$ and

$$c = c_r \delta^{-1}(c_{r-1}) \cdots \delta^{-(r-1)}(c_1),$$

then $ac = \Delta^r$. Thus $a \preceq \Delta^r$. □

6.3.3 Embeddability of pre-Garside monoids

Let (M, Δ) be a pre-Garside monoid. Under the assumption that M is left cancellative, we give an explicit construction of the group of fractions of M. The construction will imply that M is embeddable.

Let $N \geq 1$ be the order of $\delta : \Sigma \to \Sigma$. By Lemma 6.11, Δ^N is central in M. Consider the product $H = M \times \mathbf{N}$ of the monoids M and $\mathbf{N}$ with the coordinatewise multiplication

$$(a, p)(b, q) = (ab, p + q)$$

for all $a, b \in M$ and $p, q \in \mathbf{N}$. The neutral element of H is $(1, 0)$.

We define a relation $\sim$ on H by $(a, p) \sim (b, q)$ if $\Delta^{qN} a = \Delta^{pN} b$. For instance, $(\Delta^N, 1) \sim (1, 0)$. Let us show that $\sim$ is an equivalence relation. Reflexivity and symmetry are obvious. We check the transitivity. Suppose that $(a, p) \sim (b, q) \sim (c, r)$. Then $\Delta^{qN} a = \Delta^{pN} b$ and $\Delta^{rN} b = \Delta^{qN} c$. Therefore,

$$\begin{aligned} \Delta^{qN} \Delta^{rN} a = \Delta^{rN} \Delta^{qN} a &= \Delta^{rN} \Delta^{pN} b \\ &= \Delta^{pN} \Delta^{rN} b = \Delta^{pN} \Delta^{qN} c \\ &= \Delta^{qN} \Delta^{pN} c. \end{aligned}$$

Since M is left cancellative, we may divide both sides by Δ^{qN}, thus obtaining $\Delta^{rN} a = \Delta^{pN} c$. This gives $(a, p) \sim (c, r)$.

Let $G = H/\sim$ be the set of equivalence classes and let $\pi : H \to G$ be the projection. Since Δ^N is central in M, the set G has a unique monoid structure such that π is a monoid homomorphism. Define a monoid homomorphism $i : M \to G$ by $i(a) = \pi(a, 0)$ for $a \in M$.

Theorem 6.12. *Let* (M, Δ) *be a pre-Garside monoid such that* M *is left cancellative.*

(i) *The monoid* G, *constructed above, is a group, and the homomorphism* $i : M \to G$ *is an injection.*

(ii) *Any element of* G *can be written in the form* $i(\Delta)^s i(a)$, *where* $s \in \mathbf{Z}$ *and* $a \in M$.

(iii) *The monoid homomorphism* $i : M \to G$ *is universal, so that* G *is the group of fractions* G_M *of* M.

Proof. (i) If $i(a) = i(b)$ for $a, b \in M$, then $(a, 0) \sim (b, 0)$ in H. It follows that $a = \Delta^0 a = \Delta^0 b = b$. This proves the injectivity of i.

Any element $g \in G$ has the form $\pi(a, p)$ for certain $a \in M$ and $p \in \mathbf{N}$. Let us check that $g = \pi(a, p)$ is invertible. By Lemma 6.11 (iv) there are $b \in M$ and an integer $r \geq 1$ such that $ab = \Delta^r$. Multiplying b on the right by a power of Δ, we may assume that $r = qN$ for an integer $q \geq p$. Then

$$(a, p)(b, q - p) = (ab, q). \tag{6.5}$$

Since $\Delta^0 ab = \Delta^r = \Delta^{qN} 1$, we have $(ab, q) \sim (1, 0)$ and $\pi(ab, q) = \pi(1, 0) = 1$. This shows that $g = \pi(a, p)$ has a right inverse, say g'. In turn, g' has a right inverse g'' and

$$g = g(g'g'') = (gg')g'' = g''.$$

In other words, g' is also a left inverse to g. This shows that G is a group.

(ii) Let ξ be the central element $(1, +1) \in H = M \times \mathbf{N}$. Setting $a = 1$, $b = \Delta^N$, and $p = q = +1$ in (6.5), we obtain $\pi(\xi) i(\Delta)^N = 1$. Therefore, $\pi(\xi) = i(\Delta)^{-N}$. Any element of H is of the form $(a, p) = \xi^p(a, 0)$ for some $a \in M$ and $p \in \mathbf{N}$. Therefore, any element of G can be written in the form $\pi(\xi)^p i(a) = i(\Delta)^{-pN} i(a)$, where $a \in M$ and $p \in \mathbf{N}$.

(iii) Given a monoid homomorphism f from M to a group G', consider the map $H = M \times \mathbf{N} \to G'$ sending any pair $(a, p) \in M \times \mathbf{N}$ to $f(\Delta)^{-pN} f(a)$. This map is constant on the $\sim$-equivalence classes in H and induces a group homomorphism $H/\sim\, = G \to G'$. The composition of the latter with $i : M \to G$ is equal to f. The uniqueness of a group homomorphism $G \to G'$ whose composition with i is equal to f follows from the fact that the set $i(M)$ generates G as a group. $\qquad\square$

Corollary 6.13. *Left cancellative pre-Garside monoids are embeddable.*

This corollary shows in particular that for pre-Garside monoids, the left cancellativity implies the right cancellativity.

In the sequel we identify elements of a left cancellative pre-Garside monoid M with their images in G_M, so that M becomes a subset of G_M.

6.3.4 The conjugacy problem in the group of fractions

Let (M, Δ) be a left cancellative pre-Garside monoid. The conjugacy problem in its group of fractions $G = G_M$ can be reduced to the conjugacy problem in M as follows. As we know, for any $\alpha, \beta \in G$, there are $a, b \in M \subset G$ and $s, t \in \mathbf{Z}$ such that $\alpha = \Delta^s a$ and $\beta = \Delta^t b$. Pick an integer u such that $u \leq \min(s, t)$ and u is divisible by the number N from Lemma 6.11 (iii). Set $a' = \Delta^{s-u} a \in M$ and $b' = \Delta^{t-u} b \in M$. Clearly, $\alpha = \Delta^u a'$ and $\beta = \Delta^u b'$. We claim that α is conjugate to β in G if and only if a' is conjugate to b' in M. Indeed, suppose that $a'c = cb'$ for some $c \in M$. Since Δ^u is a power of Δ^N and is therefore central in M,

$$\alpha c = \Delta^u a'c = \Delta^u cb' = c\Delta^u b' = c\beta.$$

Conversely, if $\alpha\gamma = \gamma\beta$ with $\gamma \in G$, then $\gamma = \Delta^v c$ for some $c \in M$ and some integer v divisible by N. Replacing α, β, γ in the formula $\alpha\gamma = \gamma\beta$ by their expansions in a', b', c, and using the centrality of Δ^u, Δ^v, we obtain

$$\Delta^{u+v} a'c = \Delta^{u+v} cb'.$$

We divide by Δ^{u+v} and conclude that $a'c = cb'$.

6.3.5 The case of atomic M

For atomic M, the claim (ii) of Theorem 6.12 admits the following refinement.

Theorem 6.14. *Let (M, Δ) be a pre-Garside monoid such that M is nontrivial, left cancellative, and atomic. Then any element of $G = G_M \supset M$ can be written uniquely in the form $\Delta^s\, b$, where $s \in \mathbf{Z}$ and b is an element of M that is not a right multiple of Δ.*

Proof. Note first that $\|\Delta\| > 0$. Indeed, if $\|\Delta\| = 0$, then $\Delta = 1$. Since M is atomic, the remarks at the end of Section 6.1.3 imply that $\Sigma_\Delta = \{1\}$. Since Σ_Δ generates M, we have $M = \{1\}$. This contradicts the nontriviality of M.

By Theorem 6.12, any element of G has the form $\Delta^s\, a$ with $s \in \mathbf{Z}$ and $a \in M$. Let t be the greatest nonnegative integer such that $\Delta^t \preceq a$ in M; such t exists because the relation $\Delta^t \preceq a$ implies that

$$ t\,\|\Delta\| \le \|\Delta^t\| \le \|a\| < \infty. $$

Then $a = \Delta^t b$ for some $b \in M$ such that $\Delta \npreceq b$ and $\Delta^s\, a = \Delta^{s+t}\, b$. This proves the existence of the stated form.

Suppose that $\Delta^s\, b = \Delta^{s'}\, b'$ for some $s, s' \in \mathbf{Z}$ and some $b, b' \in M$ such that $\Delta \npreceq b$ and $\Delta \npreceq b'$. We may assume that $s \ge s'$. Dividing by $\Delta^{s'}$, we obtain $\Delta^{s-s'} b = b'$. Since b' is not a right multiple of Δ in M, we have $s - s' = 0$. Hence, $s = s'$ and $b = b'$, which proves the uniqueness. ☐

Exercise 6.3.1. Let (M, Δ) be a pre-Garside monoid such that M is nontrivial, left cancellative, and atomic. Prove that any element of the group of fractions G_M can be written uniquely in the form $a\,\Delta^s$, where $s \in \mathbf{Z}$ and $a \in M$ is not a left multiple of Δ.

Exercise 6.3.2. Generalize the construction of the group G in Section 6.3.3 to an arbitrary pre-Garside monoid (M, Δ). (Hint: Define the relation $\sim$ on H by $(a, p) \sim (b, q)$ if $\Delta^{s+qN} a = \Delta^{s+pN} b$ for some $s \ge 0$. Note that the resulting homomorphism $M \to G$ is injective if and only if M is left cancellative.)

6.4 Garside monoids

6.4.1 Definition and lemmas

Let (M, Δ) be a pre-Garside monoid and let Σ be the set of left (and right) divisors of Δ. Note that since Σ generates M, all atoms of M belong to Σ. In other words, all atoms of M are necessarily left divisors of Δ.

Definition 6.15. *The pair (M, Δ) is a Garside monoid if M is atomic and for any two atoms s, t of M, the set*

$$ \{a \in \Sigma \mid s \preceq a \text{ and } t \preceq a\} $$

has a minimal element $\Delta_{s,t}$ (with respect to $\preceq$).

By Lemma 6.3, the minimal element $\Delta_{s,t}$ is unique. Note for the record that $\Delta_{s,t} = \Delta_{t,s} \in \Sigma$, $s \preceq \Delta_{s,t}$, $t \preceq \Delta_{s,t}$, and

$$\{a \in \Sigma \mid s \preceq a \text{ and } t \preceq a\} = \{a \in \Sigma \mid \Delta_{s,t} \preceq a\}.$$

Any atom $s \in M$ is a minimal element of the set $\{a \in \Sigma \mid s \preceq a\}$, so that $\Delta_{s,s} = s$.

Lemma 6.16. *If (M, Δ) is a Garside monoid, then the set Σ satisfies all conditions of Theorem 6.5.*

This key lemma allows us to apply the results of Section 6.2 to Garside monoids. The rest of this subsection is devoted to the proof of Lemma 6.16. We need to verify that Σ satisfies conditions $(*_1)$–$(*_3)$ of Theorem 6.5. Condition $(*_1)$ directly follows from the definition of a pre-Garside monoid. Condition $(*_2)$ was verified in Lemma 6.11. The hard part is the verification of $(*_3)$. We begin with two lemmas. In both lemmas, we assume that (M, Δ) is a Garside monoid, Σ is the set of left (and right) divisors of Δ, and $S \subset \Sigma$ is the set of atoms of M.

Lemma 6.17. *Let E be a nonempty finite subset of M satisfying the following two conditions:*

(i) if $a \in M$ and $b \in E$ with $a \preceq b$, then $a \in E$;
(ii) if $a \in E$, $s, t \in S$ are such that $as, at \in E$, then $a\Delta_{s,t} \in E$.

Then E has a maximal element (with respect to $\preceq$).

Proof. Let c be an element of E such that $\|c\|$ is maximal (we say that c is of maximal height in E). We wish to show that $E = \{a \in M \mid a \preceq c\}$. By condition (i), $\{a \in M \mid a \preceq c\} \subset E$. Let us prove the opposite inclusion. Suppose it does not hold; then there is $b \in E$ such that $b \npreceq c$. Expand b as a product of atoms $b = s_1 \cdots s_n$ for some n and $s_1, \ldots, s_n \in S$. Set $a = s_1 \cdots s_k$, where $k < n$ is the maximal integer such that $a \preceq c$ (possibly $k = 0$, in which case $a = 1$). It is clear that $a \in E$ and there is an atom $s \in S$ (in fact $s = s_{k+1}$) such that $as \in E$ and $as \npreceq c$. We consider such $a \in E$ of maximal height. Since c is of maximal height in E, we have $\|a\| < \|as\| \leq \|c\|$. This and the relation $a \preceq c$ imply that there is $t \in S$ such that $at \preceq c$. Then, necessarily, $t \neq s$. We now have a, as, and at in E. By condition (ii), $a\Delta_{s,t} \in E$. The relations $as \preceq a\Delta_{s,t}$ and $as \npreceq c$ imply $a\Delta_{s,t} \npreceq c$. We can expand

$$\Delta_{s,t} = t s_1 s_2 \cdots s_m,$$

where $s_1, \ldots, s_m \in S$. There is $i = 1, \ldots, m$ such that $at s_1 s_2 \cdots s_{i-1} \preceq c$ and $at s_1 s_2 \cdots s_i \npreceq c$. Set $a' = at s_1 s_2 \cdots s_{i-1}$. The inclusion $a\Delta_{s,t} \in E$ implies that $a', a's_i \in E$. By the choice of i, we have $a's_i \npreceq c$. Thus, a' satisfies the same conditions as a, but $\|a'\| \geq \|at\| > \|a\|$. This yields a contradiction with the choice of a. $\square$

Lemma 6.18. *For any $a, b \in \Sigma$, the set*

$$E = \{x \in M \mid x \preceq a \text{ and } x \preceq b\} \subset \Sigma$$

has a maximal element (with respect to $\preceq$).

Proof. Since Σ is finite, so is E. Clearly, $1 \in E$. The set E is nonempty since $1 \in E$, and obviously satisfies condition (i) of Lemma 6.17. Let us check condition (ii). We have to show that if xs and xt are left divisors of both a and b for some $s, t \in S$, then so is $x\Delta_{s,t}$. Let $y \in \Sigma$ be such that $xy = a$. By hypothesis, $xs \preceq a = xy$ and $xt \preceq a = xy$. By Lemma 6.11 (i), this implies $s \preceq y$ and $t \preceq y$. By Definition 6.15, $\Delta_{s,t} \preceq y$, hence $x\Delta_{s,t} \preceq xy = a$. Similarly, $x\Delta_{s,t} \preceq b$. Now Lemma 6.17 implies that E has a maximal element. $\square$

We can now verify condition $(*_3)$ of Theorem 6.5. Pick any $a, b \in \Sigma$. Since $a \preceq \Delta$, we have $\Delta = aa'$ for some $a' \in \Sigma$. By Lemma 6.18, the set

$$\{x \in M \mid x \preceq a' \text{ and } x \preceq b\} \subset \Sigma$$

has an element c maximal with respect to $\preceq$. We claim that c is maximal in

$$\{x \in \Sigma \mid x \preceq b \text{ and } ax \in \Sigma\}.$$

Indeed, by definition, $c \preceq a'$, whence $ac \preceq aa' = \Delta$, so that $ac \in \Sigma$. Let $d \in \Sigma$ such that $d \preceq b$ and $ad \in \Sigma$. Then $ad \preceq \Delta = aa'$, which by Lemma 6.11 (i) (left cancellation in Σ) implies $d \preceq a'$. Therefore, $d \preceq c$. $\square$

6.4.2 Comprehensive Garside monoids

A Garside monoid (M, Δ) is *comprehensive* if the set $\Sigma \subset M$ of the divisors of Δ is comprehensive in the sense of Section 6.2.6. The results above yield the following properties of a comprehensive Garside monoid (M, Δ).

(1) We have $M_\Sigma \cong M$ (Lemma 6.8). In other words, M has a presentation with generators $[a]$, where a runs over Σ, and relations $[1] = 1$ and $[a][b] = [ab]$ whenever $a, b \in \Sigma$ satisfy $ab \in \Sigma$.

(2) For any $a \in M$, there is a unique left divisor $\alpha(a) \in \Sigma$ of a that is maximal among all left divisors of a lying in Σ (Theorem 6.5).

(3) Any $a \in M$ expands uniquely as a product $a = a_1 a_2 \cdots a_r$ of certain $a_1, a_2, \ldots, a_r \in \Sigma - \{1\}$ with $r \geq 0$ such that $a_i = \alpha(a_i a_{i+1} \cdots a_r)$ for all $i = 1, 2, \ldots, r$ (Section 6.2.2).

(4) The natural monoid homomorphism from M into its group of fractions G_M is injective (Lemma 6.6 and Corollary 6.13). In particular, M is left cancellative and right cancellative.

(5) If $M \neq \{1\}$, then any element of $G_M \supset M$ can be written uniquely in the form $\Delta^s b$, where $s \in \mathbf{Z}$ and $b \in M$ is not a right multiple of Δ (Theorem 6.14).

(6) The conjugacy problem in G_M is equivalent to the conjugacy problem in M (Section 6.3.4). The latter is solvable provided M admits a finite weighted presentation (Section 6.2.5).

6.4.3 Common divisors and multiples in Garside monoids

Given $k \geq 2$ elements $a_1, \ldots, a_k$ of a monoid M, we say that $d \in M$ is a *left greatest common divisor* (gcd) of $a_1, \ldots, a_k$ if $d \preceq a_i$ for all $i = 1, \ldots, k$, and $d' \preceq d$ for any $d' \in M$ such that $d' \preceq a_i$ for all $i = 1, \ldots, k$. Replacing $\preceq$ by $\succeq$, we obtain an analogous notion of a right gcd.

We say that $m \in M$ is a *right least common multiple* (lcm) of $a_1, \ldots, a_k$ if $a_i \preceq m$ for all $i = 1, \ldots, k$, and $m \preceq m'$ for any $m' \in M$ such that $a_i \preceq m'$ for all $i = 1, \ldots, k$. There is an analogous notion of a left lcm. If M is atomic, then the gcds and lcms are unique whenever they exist.

The condition in Definition 6.15 may be reformulated by saying that any two atoms have a right lcm. Property (2) in Section 6.4.2 may be reformulated by saying that Δ and any $a \in M$ have a left gcd. These properties of Garside monoids can be generalized as follows.

Theorem 6.19. *Let (M, Δ) be a comprehensive Garside monoid. Then any finite family of elements of M has a unique left gcd and a unique right lcm.*

Proof. Let $b, c \in M$. Consider the set

$$E = \{a \in M \mid a \preceq b \text{ and } a \preceq c\}.$$

In order to prove that b and c have a left gcd in M, it suffices to check that E satisfies the conditions of Lemma 6.17. The set E obviously satisfies condition (i). The set E is finite because $\|a\| \leq \|b\|$ for any $a \in E$, so that a is the product of at most $\|b\|$ atoms of M, and the set of atoms of M, being a subset of Σ, is finite.

Let us check condition (ii). Suppose we have $a \in E$ and atoms $s, t \in M$ such that $as, at \in E$. Write $b = ab_1$ with $b_1 \in M$. Since M is left cancellative, $as \preceq b = ab_1$ implies $s \preceq b_1$ and $at \preceq b = ab_1$ implies $t \preceq b_1$. Consider the maximal left divisor $\alpha(b_1)$ of b_1 in Σ. We have $s \preceq \alpha(b_1)$ and $t \preceq \alpha(b_1)$. Therefore $\Delta_{s,t} \preceq \alpha(b_1) \preceq b_1$. Hence, $\Delta_{s,t} \preceq b_1$ and $a\Delta_{s,t} \preceq ab_1 = b$. Similarly, $a\Delta_{s,t} \preceq c$. This proves that $a\Delta_{s,t} \in E$.

That any finite family of elements of M has a left gcd now easily follows by induction on the cardinal of the family.

Let us prove the existence of right lcms. Let $a_1, \ldots, a_k \in M$. In view of Lemma 6.11 (iv), there is $r \geq 1$ such that $a_i \preceq \Delta^r$ for all $i = 1, \ldots, k$. Consider the set

$$X = \{x \in M \mid a_i \preceq x \preceq \Delta^r \text{ for all } i = 1, \ldots, k\}.$$

Since the set of atoms of M is finite and all left divisors of Δ^r expand as products of $\leq r\|\Delta\|$ atoms, the set of left divisors of Δ^r is finite. Since it contains X, the latter is finite as well. Let m be a left gcd of the elements of X. We claim that m is a right lcm of $a_1, \ldots, a_k$. Indeed, $a_1, \ldots, a_k$ are left divisors of all elements of X; therefore, they are left divisors of m. This shows that m is a common right multiple of $a_1, \ldots, a_k$.

Let m' be another common right multiple of $a_1, \ldots, a_k$. Denote by m'' a left gcd of m' and Δ^r. Let us check that $m'' \in X$. First, $m'' \preceq \Delta^r$. Since a_i is a left divisor of m' and of Δ^r, it is a left divisor of m''. This proves that $m'' \in X$. By definition of m, we have $m \preceq m''$. Since $m'' \preceq m'$, we obtain $m \preceq m'$. This proves our claim. $\qquad\square$

Exercise 6.4.1. For $a \in \Sigma$, let $a' \in \Sigma$ be uniquely defined by $aa' = \Delta$ and let c be the left gcd of a' and $b \in \Sigma$. Prove that c is the maximal element of the set $\{x \in \Sigma \,|\, x \preceq b$ and $ax \in \Sigma\}$. (Hint: The proof is contained in the proof of Lemma 6.16.) Deduce that $\alpha_2(a, b) = ac$.

Exercise 6.4.2. Let (M, Δ) be a Garside monoid and let Σ be the set of divisors of Δ. Prove that $(M_\Sigma, [\Delta])$ is a comprehensive Garside monoid.

Exercise 6.4.3. Let M be the monoid with generators x, y and the defining relation $xyx = y^2$. Prove that the pair $(M, \Delta = y^3)$ is a comprehensive Garside monoid with atoms x, y. (Hint: To distinguish elements of M, use monoid homomorphisms to $\mathbf{N}$ and to the group $\langle a, b \,|\, a^2 = b^3 = 1\rangle$.)

6.5 The braid monoid

6.5.1 A presentation by generators and relations

For $n \geq 1$, denote by B_n^+ the monoid generated by $n - 1$ generators $\sigma_1, \sigma_2, \ldots, \sigma_{n-1}$ and the relations

$$\sigma_i \sigma_j = \sigma_j \sigma_i \quad \text{if } |i - j| \geq 2,$$
$$\sigma_i \sigma_j \sigma_i = \sigma_j \sigma_i \sigma_j \quad \text{if } |i - j| = 1,$$

where $i, j = 1, 2, \ldots, n - 1$. The monoid B_n^+ is called the *braid monoid* on n strings. The elements of B_n^+ are called *positive braids* on n strings. By definition, B_1^+ is the trivial monoid. The monoid B_2^+ is generated by a single generator σ_1 and an empty set of relations; it is isomorphic to the monoid $\mathbf{N}$ of nonnegative integers.

The presentation of B_n^+ given above is finite and length-balanced in the sense of Section 6.1.4. Section 6.1.5 yields a solution of the word problem for B_n^+. Moreover, Lemma 6.4 implies that the monoid B_n^+ is atomic with atoms $\sigma_1, \ldots, \sigma_{n-1}$ and $\|a\| = \ell(a)$ for all $a \in B_n^+$, where $\ell : B_n^+ \to \mathbf{N}$ is the monoid homomorphism defined by $\ell(\sigma_i) = 1$ for $i = 1, \ldots, n - 1$.

Set

$$\Delta_n = (\sigma_1 \cdots \sigma_{n-2}\sigma_{n-1})(\sigma_1 \cdots \sigma_{n-2}) \cdots (\sigma_1\sigma_2)\sigma_1 \in B_n^+.$$

The following theorem puts B_n^+ in the framework of the theory of Garside monoids and provides a fundamental example of Garside monoids.

Theorem 6.20. *For all $n \geq 1$, the pair (B_n^+, Δ_n) is a comprehensive Garside monoid.*

The proof of this theorem will be given in Section 6.5.3 using preliminary results from Section 6.5.2. In the proof we will use the terminology and results from Section 4.1. Applications of Theorem 6.20 will be discussed in Section 6.5.4.

6.5.2 Reduced braids

As in Section 4.1, consider the symmetric group $\mathfrak{S}_n$ consisting of all permutations of the set $\{1, \ldots, n\}$. We define a set-theoretic mapping $\rho : \mathfrak{S}_n \to B_n^+$ as follows. Consider the simple transpositions $s_1, \ldots, s_{n-1} \in \mathfrak{S}_n$, where s_i permutes i and $i+1$ and leaves the other elements of $\{1, \ldots, n\}$ fixed. The simple transpositions generate $\mathfrak{S}_n$, so that every element $w \in \mathfrak{S}_n$ can be expressed as a word $w = s_{i_1} s_{i_2} \cdots s_{i_r}$ with $i_1, i_2, \ldots, i_r \in \{1, 2, \ldots, n-1\}$. If r is minimal, then this is a reduced expression and we set $\rho(w) = \sigma_{i_1} \sigma_{i_2} \cdots \sigma_{i_r}$. By Theorem 4.12, $\rho(w)$ is a well-defined element of B_n^+. Let

$$\pi : B_n^+ \to \mathfrak{S}_n$$

be the monoid homomorphism defined by $\pi(\sigma_i) = s_i$ for all $i = 1, \ldots, n-1$. It is clear that $\pi \circ \rho = \mathrm{id}$, which implies that ρ is injective.

Set $B_n^{\mathrm{red}} = \rho(\mathfrak{S}_n) \subset B_n^+$. This is a finite set of cardinal $n!$ and the homomorphism $\pi : B_n^+ \to \mathfrak{S}_n$ is a bijection when restricted to B_n^{red}. We say that an element of B_n^+ is *reduced* if it lies in B_n^{red}. The atoms $\sigma_1, \ldots, \sigma_{n-1}$ of B_n^+ are reduced, since $\sigma_i = \rho(s_i)$ for $i = 1, \ldots, n-1$.

From Section 4.1.3 recall the length $\lambda(w)$ of $w \in \mathfrak{S}_n$: it is the length r of any reduced expression $s_{i_1} s_{i_2} \cdots s_{i_r}$ for w. It is clear from the definitions that

$$\lambda(\pi(a)) \leq \ell(a)$$

for all $a \in B_n^+$. The following is a useful algebraic characterization of B_n^{red}.

Lemma 6.21. *An element a of B_n^+ is reduced if and only if $\lambda(\pi(a)) = \ell(a)$.*

Proof. If $a = \rho(w)$ for some $w \in \mathfrak{S}_n$, then $\ell(a) = \lambda(w) = \lambda(\pi(a))$. Conversely, let $a = \sigma_{i_1} \cdots \sigma_{i_r} \in B_n^+$ with $r = \ell(a) = \lambda(\pi(a))$. Then $\pi(a) = s_{i_1} \cdots s_{i_r}$ is a reduced expression in $\mathfrak{S}_n$ and $a = \rho(\pi(a)) \in B_n^{\mathrm{red}}$. $\qquad\square$

Lemma 6.22. *A left or right divisor of a reduced element of B_n^+ is reduced.*

Proof. If $a, b \in B_n$ and $ab \in B_n^{\mathrm{red}}$, then

$$\ell(a) + \ell(b) = \ell(ab) = \lambda(\pi(ab)) = \lambda(\pi(a)\,\pi(b)) \leq \lambda(\pi(a)) + \lambda(\pi(b)).$$

Since $\ell(a) \geq \lambda(\pi(a))$ and $\ell(b) \geq \lambda(\pi(b))$, these inequalities are actually equalities. By Lemma 6.21, it follows that $a, b \in B_n^{\mathrm{red}}$. $\qquad\square$

Lemma 6.23. *For $u, v \in \mathfrak{S}_n$, we have $\rho(u)\rho(v) = \rho(uv)$ if and only if $\lambda(u) + \lambda(v) = \lambda(uv)$.*

Proof. Set $a = \rho(u)\rho(v) \in B_n^+$. We have $\pi(a) = uv$ and

$$\lambda(uv) = \lambda(\pi(a)) \le \ell(a) = \ell(\rho(u)) + \ell(\rho(v)) = \lambda(u) + \lambda(v).$$

Therefore $a \in B_n^{\text{red}}$ if and only if $\lambda(uv) = \lambda(u) + \lambda(v)$. On the other hand, $a \in B_n^{\text{red}}$ if and only if $a = \rho(\pi(a)) = \rho(uv)$. □

Recall the permutation $w_0 = (n, n-1, \ldots, 2, 1)$ from Section 4.1.6; it is the unique element $w_0 \in \mathfrak{S}_n$ of maximal length. It is easy to check that

$$w_0 = (s_1 \cdots s_{n-2}s_{n-1})(s_1 \cdots s_{n-2}) \cdots (s_1 s_2)s_1.$$

Since the word on the right-hand side has length $\lambda(w_0) = n(n-1)/2$, it is reduced. Therefore,

$$\rho(w_0) = (\sigma_1 \cdots \sigma_{n-2}\sigma_{n-1})(\sigma_1 \cdots \sigma_{n-2}) \cdots (\sigma_1\sigma_2)\sigma_1 = \Delta_n.$$

This shows that Δ_n is reduced.

Lemma 6.24. *An element of B_n^+ is reduced if and only if it is a left (or a right) divisor of Δ_n.*

Proof. Since Δ_n is reduced, all its left divisors and right divisors are reduced by Lemma 6.22.

Conversely, let $a = \rho(\pi(a)) \in B_n^{\text{red}}$. Set $b = \rho(\pi(a)^{-1}w_0) \in B_n^{\text{red}}$, $u = \pi(a)$, and $v = \pi(b) = \pi(a)^{-1}w_0$. We have $uv = w_0$. Hence, by Lemma 4.14,

$$\lambda(u) + \lambda(v) = \lambda(w_0).$$

This equality and Lemma 6.23 imply that $ab = \Delta_n$. Hence, a is a left divisor of Δ_n. A similar argument proves that a is a right divisor of Δ_n. □

6.5.3 Proof of Theorem 6.20

We observed in Section 6.5.1 that B_n^+ is atomic with atoms $\sigma_1, \ldots, \sigma_{n-1}$. Let us prove that (B_n^+, Δ_n) is a pre-Garside monoid by checking conditions (a), (b) of Definition 6.9.

By Lemma 6.24, the set of left divisors of Δ_n coincides with the set of right divisors of Δ_n and coincides with the set B_n^{red}. The latter is finite and contains the generators $\sigma_1, \ldots, \sigma_{n-1}$ of B_n^+. This verifies condition (a).

Condition (b): Let us prove that $\Delta_n a = \Delta_n b \Rightarrow a = b$ for $a, b \in B_n^{\text{red}}$. Applying the monoid homomorphism $\pi: B_n^+ \to \mathfrak{S}_n$, we obtain

$$\pi(\Delta_n)\pi(a) = \pi(\Delta_n a) = \pi(\Delta_n b) = \pi(\Delta_n)\pi(b) \in \mathfrak{S}_n.$$

Since $\mathfrak{S}_n$ is a group, $\pi(a) = \pi(b)$. This implies that

$$a = \rho(\pi(a)) = \rho(\pi(b)) = b.$$

The implication $a\Delta_n = b\Delta_n \Rightarrow a = b$ is proven similarly.

For any $i, j \in \{1, \ldots, n-1\}$, set

$$\sigma_{i,j} = \begin{cases} \sigma_i & \text{if } i = j, \\ \sigma_i \sigma_j \sigma_i = \sigma_j \sigma_i \sigma_j & \text{if } |i - j| = 1, \\ \sigma_i \sigma_j = \sigma_j \sigma_i & \text{if } |i - j| \geq 2. \end{cases}$$

Set $s_{i,j} = \pi(\sigma_{i,j}) \in \mathfrak{S}_n$. It is easy to check that $s_{i,j} = s_i s_j$ is a reduced expression when $|i - j| \geq 2$, and $s_{i,j} = s_i s_j s_i \in \mathfrak{S}_n$ is a reduced expression when $|i - j| = 1$. Then $\sigma_{i,j} = \rho(s_{i,j}) \in B_n^{\mathrm{red}}$ for all i, j. Therefore the set B_n^{red} is comprehensive.

To complete the proof of Theorem 6.20, it remains to check the condition in Definition 6.15. Observe that $\sigma_i \preceq \sigma_{i,j}$ and $\sigma_j \preceq \sigma_{i,j}$ for all i, j. We claim that $\sigma_{i,j}$ is the minimal element in the set

$$\{a \in B_n^{\mathrm{red}} \mid \sigma_i \preceq a \text{ and } \sigma_j \preceq a\}.$$

We must show that for any $a \in B_n^{\mathrm{red}}$ such that $\sigma_i \preceq a$ and $\sigma_j \preceq a$, we have $\sigma_{i,j} \preceq a$. The case $i = j$ being trivial, we consider the case $i \neq j$. Since the elements of B_n^{red} are in bijection with the elements of $\mathfrak{S}_n$ under the map $\pi : B_n^+ \to \mathfrak{S}_n$, it is enough to establish that if

$$w = \pi(a) = s_i u = s_j v$$

for some $u, v \in \mathfrak{S}_n$ with $\lambda(u) = \lambda(v) = \lambda(w) - 1$, then there is $w' \in \mathfrak{S}_n$ such that $w = s_{i,j} w'$ and $\lambda(w') = \lambda(w) - \lambda(s_{i,j})$.

We prove the latter assertion. First observe that $u \neq v$, since $s_i \neq s_j$. Let $s_{i_1} \cdots s_{i_r}$ be a reduced expression for v, where $r = \lambda(w) - 1$. We have

$$u = s_i w = s_i s_j v = s_i s_j s_{i_1} \cdots s_{i_r}.$$

Since $\lambda(u) < \lambda(w)$, it follows from Theorem 4.8 that u is obtained from $s_j s_{i_1} \cdots s_{i_r}$ by deleting one of the generators. If the deleted generator is the leftmost s_j, then $u = s_{i_1} \cdots s_{i_r} = v$, which is impossible. Therefore, $u = s_j w'$ with

$$w' = s_{i_1} \cdots \widehat{s_{i_p}} \cdots s_{i_r},$$

where some s_{i_p} is deleted. Therefore, $\lambda(w') \leq r - 1 = \lambda(w) - 2$. Since

$$w = s_i u = s_i s_j w',$$

we must have $\lambda(w') = \lambda(w) - 2$. This proves the desired assertion when $|i - j| \geq 2$, i.e., when $s_{i,j} = s_i s_j$.

Consider the case $|i - j| = 1$. By the previous computations,

$$v = s_j w = s_j s_i u = s_j s_i s_j w' = s_j s_i s_j s_{i_1} \cdots \widehat{s_{i_p}} \cdots s_{i_r} .$$

Since $\lambda(v) < \lambda(w) = r + 1$, it follows again from Theorem 4.8 that v is obtained from $s_i s_j s_{i_1} \cdots \widehat{s_{i_p}} \cdots s_{i_r}$ by deleting one of the generators. If the deleted generator is the leftmost s_i, then

$$v = s_j s_{i_1} \cdots \widehat{s_{i_p}} \cdots s_{i_r} = s_j w' = u ,$$

which is impossible. If the deleted generator is the generator s_j in the second position, then

$$v = s_i s_{i_1} \cdots \widehat{s_{i_p}} \cdots s_{i_r} = s_i w' .$$

We obtain $s_i s_j w' = w = s_j v = s_j s_i w'$, which implies $s_i s_j = s_j s_i$. This is impossible, since $|i - j| = 1$. Therefore $v = s_i s_j w''$, where w'' is obtained by deleting a generator from $w' = s_{i_1} \cdots \widehat{s_{i_p}} \cdots s_{i_r}$. Thus, $\lambda(w'') \leq r - 2 = \lambda(w) - 3$ and

$$w = s_j v = s_j s_i s_j w'' = s_{i,j} w'' .$$

Then of course $\lambda(w'') = \lambda(w) - 3$. □

6.5.4 Applications of Theorem 6.20

By Theorem 6.20, the pair (B_n^+, Δ_n) shares all properties of comprehensive Garside monoids; see Sections 6.4.2 and 6.4.3. We give here a summary of these properties.

(1) The monoid B_n^+ has a presentation with generators $[a]$, where a runs over B_n^{red}, and relations $[1] = 1$ and $[a][b] = [ab]$ whenever a, $b \in B_n^{\mathrm{red}}$ satisfy $ab \in B_n^{\mathrm{red}}$. Using the bijection $\rho : \mathfrak{S}_n \to B_n^{\mathrm{red}}$ and Lemma 6.23, we conclude that B_n^+ has a presentation with generators $[u]$, where u runs over $\mathfrak{S}_n$, and relations $[1] = 1$ and $[u][v] = [uv]$ whenever u, $v \in \mathfrak{S}_n$ satisfy $\lambda(u) + \lambda(v) = \lambda(uv)$.

(2) Any finite family of elements of B_n^+ has a unique left gcd and a unique right lcm.

(3) Any $a \in B_n^+$ has a *normal form* $(a_1, a_2, \ldots, a_r)$ with $r \geq 0$, where $a_1, a_2, \ldots, a_r$ are unique elements of $B_n^{\mathrm{red}} - \{1\}$ such that $a = a_1 a_2 \cdots a_r$ and a_i is the left gcd of $a_i a_{i+1} \cdots a_r$ and Δ_n for all $i = 1, 2, \ldots, r$.

(4) The monoid B_n^+ embeds into its group of fractions. By definition, the group of fractions of B_n^+ has the same presentation as B_n^+ and is nothing but the braid group B_n. Thus, the monoid homomorphism $B_n^+ \to B_n$ sending $\sigma_i \in B_n^+$ to $\sigma_i \in B_n$ for $i = 1, \ldots, n-1$ is injective. In the sequel we will identify the monoid B_n^+ with its image in B_n.

(5) For $n \geq 2$, any $\beta \in B_n$ can be written uniquely in the form $\beta = \Delta_n^s b$, where $s \in \mathbf{Z}$ and $b \in B_n^+ \subset B_n$ is not a right multiple of Δ_n.

(6) The conjugacy problem in B_n is equivalent to the conjugacy problem in B_n^+ and can be solved as in Section 6.2.5.

We complement this list with the following theorem.

Theorem 6.25. *Any finite family of elements of B_n^+ has a unique right gcd and a unique left lcm.*

Proof. Consider the map rev : $B_n^+ \to B_n^+$ obtained by reading the words in the generators $\sigma_1, \ldots, \sigma_{n-1}$ from right to left:

$$\mathrm{rev}(\sigma_{i_1}\sigma_{i_2}\cdots\sigma_{i_{r-1}}\sigma_{i_r}) = \sigma_{i_r}\sigma_{i_{r-1}}\cdots\sigma_{i_2}\sigma_{i_1}.$$

This map is well defined, since the defining relations of B_n^+, being read from right to left, give the same relations. The map rev is an involutive antiautomorphism of B_n^+ in the sense that $\mathrm{rev}^2 = \mathrm{id}$, $\mathrm{rev}(1) = 1$, and $\mathrm{rev}(ab) = \mathrm{rev}(b)\,\mathrm{rev}(a)$ for all $a, b \in B_n^+$. It is clear that $a \preceq b$ if and only if $\mathrm{rev}(a) \succeq \mathrm{rev}(b)$ for $a, b \in B_n^+$. Using these facts, it is easy to deduce the existence of right gcds and left lcms from the existence of left gcds and right lcms. The uniqueness follows from Lemma 6.3. $\qquad\square$

Note that the Garside element $\Delta_n \in B_n^+ \subset B_n$ was introduced as a braid in Section 1.3.3 (see Figure 1.11 for $n = 5$).

6.5.5 Computations

The expansion $\beta = \Delta^s\, b$ of a braid $\beta \in B_n$ provided by the item (5) of the previous subsection can be explicitly computed. Represent β by a word in the generators $\sigma_1, \ldots, \sigma_{n-1}$ and their inverses. Define $\nu_i \in B_n^+$ by $\nu_i\sigma_i = \Delta_n$. Then $\sigma_i^{-1} = \Delta_n^{-1}\nu_i$. In the word representing β replace all occurrences of σ_i^{-1} by $\Delta_n^{-1}\nu_i$ and expand all ν_i in terms of $\sigma_1, \ldots, \sigma_{n-1}$. In the resulting word we have only the generators σ_i (not their inverses) and negative powers of Δ_n. Using the identities

$$\sigma_i\Delta_n = \Delta_n\sigma_{n-i}, \tag{6.6}$$

where $i = 1, \ldots, n-1$ (cf. formula (1.8) in Section 1.3), we can move all powers of Δ_n to the left. In this way we obtain an expansion $\beta = \Delta_n^s b$ with $s \in \mathbf{Z}$ and $b \in B_n^+$. If b is not a right multiple of Δ_n, then we have the desired expansion of β. If $\Delta_n \preceq b$, then $b = \Delta_n b'$ with $b' \in B_n^+$ and $\beta = \Delta_n^{s+1} b'$. Note that to check whether $\Delta_n \preceq b$, it is enough to compute $\alpha(b) \in B_n^{\mathrm{red}}$ and to see whether $\alpha(b) = \Delta_n$. We then check whether b' is a right multiple of Δ_n, and so on. The process stops after at most $2\ell(b)/(n(n-1))$ steps.

To give an example, we apply this procedure to

$$\beta = \sigma_1^{-1}\sigma_2\sigma_1\sigma_2^{-2} \in B_4.$$

As in Section 6.1.5, denote by $W(a)$ the set of words in $\sigma_1, \ldots, \sigma_{n-1}$ representing an element $a \in B_n^+$. From

$$\Delta_4 = \sigma_1\sigma_2\sigma_3\sigma_1\sigma_2\sigma_1 = \sigma_1\sigma_2\sigma_3\sigma_2\sigma_1\sigma_2$$

we derive $\sigma_1^{-1} = \Delta_4^{-1}\sigma_1\sigma_2\sigma_3\sigma_1\sigma_2$ and $\sigma_2^{-1} = \Delta_4^{-1}\sigma_1\sigma_2\sigma_3\sigma_2\sigma_1$. Consequently,

$$\beta = (\Delta_4^{-1}\sigma_1\sigma_2\sigma_3\sigma_1\sigma_2)(\sigma_2\sigma_1)(\Delta_4^{-1}\sigma_1\sigma_2\sigma_3\sigma_2\sigma_1)(\Delta_4^{-1}\sigma_1\sigma_2\sigma_3\sigma_2\sigma_1)$$
$$= (\Delta_4^{-1}\sigma_1\sigma_2\sigma_3\sigma_1\sigma_2)(\sigma_2\sigma_1)(\Delta_4^{-2}\sigma_3\sigma_2\sigma_1\sigma_2\sigma_3)(\sigma_1\sigma_2\sigma_3\sigma_2\sigma_1)$$
$$= \Delta_4^{-3}abc^2,$$

where

$$a = \sigma_1\sigma_2\sigma_3\sigma_1\sigma_2, \quad b = \sigma_2\sigma_1, \quad \text{and} \quad c = \sigma_3\sigma_2\sigma_1\sigma_2\sigma_3 = \sigma_1\sigma_2\sigma_3\sigma_2\sigma_1.$$

Let us compute $\alpha(abc^2)$ in order to find out whether abc^2 is a right multiple of Δ_4. Observe that a, b, and c are reduced braids. By Exercise 6.4.1, since $c\sigma_2 = \Delta_4$, we have $\alpha(c^2) = \alpha_2(c,c) = cc'$, where c' is the left gcd of σ_2 and c. Now $W(c)$ consists of the six words

$$\sigma_3\sigma_2\sigma_1\sigma_2\sigma_3, \quad \sigma_1\sigma_2\sigma_3\sigma_2\sigma_1, \quad \sigma_1\sigma_3\sigma_2\sigma_3\sigma_1,$$
$$\sigma_1\sigma_3\sigma_2\sigma_1\sigma_3, \quad \sigma_3\sigma_1\sigma_2\sigma_3\sigma_1, \quad \sigma_3\sigma_1\sigma_2\sigma_1\sigma_3.$$

Therefore, $c' = 1$ and $\alpha(c^2) = c$. From $b(\sigma_2\sigma_3\sigma_2\sigma_1) = \Delta_4$, we obtain

$$\alpha(bc^2) = \alpha_2(b, \alpha(c^2)) = \alpha_2(b, c) = bb',$$

where b' is the left gcd of $\sigma_2\sigma_3\sigma_2\sigma_1$ and c. Now

$$W(\sigma_2\sigma_3\sigma_2\sigma_1) = \{\sigma_2\sigma_3\sigma_2\sigma_1, \sigma_3\sigma_2\sigma_3\sigma_1, \sigma_3\sigma_2\sigma_1\sigma_3\}.$$

Comparing with $W(c)$, we obtain $b' = \sigma_3\sigma_2\sigma_1$. Hence, $\alpha(bc^2) = d$, where $d = \sigma_2\sigma_1\sigma_3\sigma_2\sigma_1$. Finally, $a\sigma_1 = \Delta_4$ implies

$$\alpha(abc^2) = \alpha_2(a, \alpha(bc^2)) = \alpha_2(a, d) = aa',$$

where a' is the left gcd of σ_1 and d. The list

$$W(d) = \{\sigma_2\sigma_1\sigma_3\sigma_2\sigma_1, \sigma_2\sigma_3\sigma_1\sigma_2\sigma_1, \sigma_2\sigma_3\sigma_2\sigma_1\sigma_2, \sigma_3\sigma_2\sigma_3\sigma_1\sigma_2, \sigma_3\sigma_2\sigma_1\sigma_3\sigma_2\}$$

shows that $a' = 1$. Hence, $\alpha(abc^2) = a \neq \Delta_4$. Therefore, abc^2 is not a right multiple of Δ_4 and

$$\beta = \Delta_4^{-3}abc^2$$

is the required expansion of β.

The reader is invited to check that the normal form of abc^2 is (a, d, e, b), where a, b, d are as above and $e = \sigma_1\sigma_2\sigma_1\sigma_3\sigma_2$.

Exercise 6.5.1. (a) Give an algebraic proof of the identities (6.6) in B_n^+.

(b) Prove that Δ_n is the left (and the right) lcm of $\sigma_1, \ldots, \sigma_{n-1}$ in B_n^+.

(c) Deduce that the center of B_n^+ is generated by Δ_n^2.

(d) Show that $\text{rev}(\Delta_n) = \Delta_n$.

(e) Show that $\text{rev}(B_n^{\text{red}}) = B_n^{\text{red}}$. (Hint: $\rho(w^{-1}) = \text{rev}(\rho(w))$ for $w \in \mathfrak{S}_n$.)

6.6 Generalized braid groups

We introduce generalized braid groups and generalized braid monoids. Their definition is directly inspired by the theory of Coxeter groups. We begin with a short introduction to Coxeter groups.

6.6.1 Coxeter groups

A *Coxeter matrix* is a symmetric matrix $A = (a_{s,t})_{s,t \in S}$, where S is a finite set, $a_{s,s} = 1$ for all $s \in S$, and $a_{s,t} \in \{2, 3, \dots, \infty\}$ for all distinct $s, t \in S$. To such a matrix A we associate a graph Γ_A as follows: its vertices are the elements of S, and there is a (unique) edge between $s \in S$ and $t \in S$ whenever $a_{s,t} \geq 3$. We label the edge between s and t by $a_{s,t}$ whenever $a_{s,t} \geq 4$. The resulting labeled graph is the *labeled graph of* A. Every Coxeter matrix can be uniquely reconstructed from its labeled graph.

To a Coxeter matrix $A = (a_{s,t})_{s,t \in S}$ we associate the group W_A defined by the following presentation: the generators are the elements of S and the relations are

$$(st)^{a_{s,t}} = 1, \tag{6.7}$$

where s, t run over pairs of elements of S such that $a_{s,t} \neq \infty$.

Since $a_{s,s} = 1$, relation (6.7) for $s = t$ becomes $s^2 = 1$, which is equivalent to $s^{-1} = s$ for all $s \in S$. For $s \neq t$, relation (6.7) can be rewritten as

$$\underbrace{sts\cdots}_{a_{s,t} \text{ factors}} = \underbrace{tst\cdots}_{a_{s,t} \text{ factors}}, \tag{6.8}$$

where s, t run over S and both sides of (6.8) are defined when $2 \leq a_{st} < \infty$ and have $a_{s,t}$ factors. In other words, W_A is generated by the elements of S subject to the relations $s^2 = 1$ ($s \in S$) and relations (6.8). The group W_A is called the *Coxeter group* associated with A (or with the labeled graph Γ_A).

If $S = \{1, 2, \dots, n-1\}$ with $n \geq 1$ and $A = (a_{i,j})_{i,j \in S}$ is given by

$$a_{i,j} = \begin{cases} 1 & \text{if } i = j, \\ 3 & \text{if } |i - j| = 1, \\ 2 & \text{if } |i - j| \geq 2, \end{cases} \tag{6.9}$$

then W_A has a presentation that coincides with the presentation (4.1) of the symmetric group $\mathfrak{S}_n$. Thus, Coxeter groups generalize the symmetric groups.

Consider again an arbitrary Coxeter matrix $A = (a_{s,t})_{s,t \in S}$. Because of the relations $s^2 = 1$ for $s \in S$, any element $w \in W_A$ can be expanded as a product $w = s_1 \cdots s_r$ of elements $s_1, \dots, s_r$ of S. The minimal number r in such an expansion of w is called the *length* of w and denoted by $\lambda(w)$. An expansion $w = s_1 \cdots s_r$ with $r = \lambda(w)$ and $s_1, \dots, s_r \in S$ is called a *reduced expression* for w (in general it is nonunique).

The neutral element of W_A is the only element of length 0. The elements of length 1 in W_A are precisely the generators $s \in S$.

Many properties of the symmetric groups extend to Coxeter groups. Note the following generalization of Theorem 4.12 (for a proof, see [Mat64], [Bou68, Chap. IV, Sect. 1, Prop. 5], [GP00, Sect. 1.2]).

Theorem 6.26. *For any monoid M and any set of elements $x_s \in M$ indexed by $s \in S$ and satisfying the relations*

$$\underbrace{x_s x_t x_s \cdots}_{a_{s,t} \text{ factors}} = \underbrace{x_t x_s x_t \cdots}_{a_{s,t} \text{ factors}}$$

for all s, $t \in S$ such that $2 \le a_{s,t} < \infty$, there is a unique set-theoretic map $\rho : W_A \to M$ such that $\rho(w) = x_{s_1} \cdots x_{s_r}$ for an arbitrary reduced expression $w = s_1 \cdots s_r$ of any $w \in W_A$.

In Table 6.1 we give a list of labeled graphs consisting of four infinite families of graphs A_n $(n \ge 1)$, BC_n $(n \ge 2)$, D_n $(n \ge 4)$, $I_2(m)$ $(m = 5$ and $m \ge 7)$ and seven exceptional graphs. The subscripts in Table 6.1 indicate the number of vertices. It can be proved that the Coxeter groups associated to all these labeled graphs are finite. Moreover, any finite Coxeter group is a direct product of a finite family of Coxeter groups associated to graphs in Table 6.1, see [Bou68, Chap. VI, Sect. 4.1] or [Hum90, Sect. 2.7]. We record also the following lemma; see [GP00, Prop. 1.5.1].

Lemma 6.27. *A Coxeter group W_A is finite if and only if there is an element $w_0 \in W_A$ such that $\lambda(w_0 s) < \lambda(w_0)$ for all $s \in S$. Such w_0 (if it exists) is unique and satisfies $\lambda(w) < \lambda(w_0)$ for all $w \in W_A$, $w \ne w_0$.*

The element $w_0 \in W_A$ is called the *longest element* of W_A.

6.6.2 Generalized braid monoids and groups

Given a Coxeter matrix $A = (a_{s,t})_{s,t \in S}$, we define B_A^+ as the monoid (resp. B_A as the group) generated by the elements of S and relations (6.8). (The difference with W_A is that we now drop the relations $s^2 = 1$, $s \in S$.) The monoid B_A^+ is called the *generalized braid monoid*, and the group B_A is called the *generalized braid group* associated to A. It follows from Section 6.3.1 that B_A is the group of fractions of the monoid B_A^+.

By definition, the Coxeter group W_A is the quotient of B_A by the normal subgroup generated by s^2 for all $s \in S$. The composite map

$$\pi : B_A^+ \to B_A \to W_A$$

is clearly surjective. When A is the matrix (6.9),

$$B_A^+ \cong B_n^+ \quad \text{and} \quad B_A \cong B_n.$$

Table 6.1. Graphs of finite Coxeter groups

A_n

BC_n ⟨4⟩

D_n

E_6

E_7

E_8

F_4 ⟨4⟩

G_2 ⟨6⟩

H_3 ⟨5⟩

H_4 ⟨5⟩

$I_2(m)$ ⟨m⟩

The presentation of B_A^+ is finite and length-balanced in the sense of Section 6.1.4. Section 6.1.5 yields a solution of the word problem and of the divisibility problem for B_A^+. Lemma 6.4 implies that the monoid B_A^+ is atomic with $s \in S$ as atoms, and $\|a\| = \ell(a)$ for all $a \in B_A^+$, where $\ell : B_A^+ \to \mathbf{N}$ is the monoid homomorphism defined by $\ell(s) = 1$ for all $s \in S$. It is clear that the monoid B_A^+ is trivial if and only if $S = \emptyset$.

By Theorem 6.26, there is a unique set-theoretic map $\rho : W_A \to B_A^+$ such that $\rho(w) = s_1 \cdots s_r \in B_A^+$ for any reduced expression $w = s_1 \cdots s_r$. Clearly, $\pi \circ \rho = \mathrm{id}_{W_A}$, where $\pi : B_A^+ \to W_A$ is the projection. Hence ρ is injective.

Set $B_A^{red} = \rho(W_A) \subset B_A^+$. We say that an element of B_A^+ is *reduced* if it lies in B_A^{red}. For instance, the neutral element and the generators $s \in S$ of B_A^+ are reduced. Note also that $\lambda(\pi(a)) \leq \ell(a)$ for all $a \in B_A^+$.

The following lemma generalizes Lemmas 6.21–6.23 and is proven similarly.

Lemma 6.28. *(a) An element $a \in B_A^+$ is reduced if and only if $\lambda(\pi(a)) = \ell(a)$.*

(b) All left divisors and all right divisors of a reduced element of B_A^+ are reduced.

(c) For $u, v \in W_A$, we have $\rho(u)\,\rho(v) = \rho(uv) \iff \lambda(u) + \lambda(v) = \lambda(uv)$.

Now assume that the group W_A is finite. By Lemma 6.27, there is a unique element $w_0 \in W_A$ of maximal length. Set $\Delta = \rho(w_0) \in B_A^+$. The following lemma generalizes Lemma 6.24 and is proven similarly.

Lemma 6.29. *An element of B_A^+ is reduced if and only if it is a left (or a right) divisor of Δ.*

We can now state the main theorem of this section.

Theorem 6.30. *For any Coxeter matrix A such that the group W_A is finite, the pair (B_A^+, Δ) is a comprehensive Garside monoid.*

Proof. We have already observed that the monoid B_A^+ is atomic. Since both sides of (6.8) represent reduced expressions in W_A, the set B_A^{red} is comprehensive. The proof of conditions (a), (b) of Definition 6.9 reproduces the corresponding part of the proof of Theorem 6.20 with obvious changes. It remains to check the condition in Definition 6.15. For $s, t \in S$, set

$$\Delta_{s,t} = \begin{cases} s & \text{if } s = t, \\ \underbrace{sts\cdots}_{a_{st} \text{ factors}} = \underbrace{tst\cdots}_{a_{st} \text{ factors}} & \text{if } s \neq t. \end{cases}$$

The element $\Delta_{s,t}$ belongs to B_A^{red} and is a right common multiple of s and t. We claim that $\Delta_{s,t} \preceq a$ for any $a \in B_A^{\mathrm{red}}$ such that $s \preceq a$ and $t \preceq a$. Since the elements of B_A^{red} are in bijection with the elements of W_A under the map $\pi : B_A^+ \to W_A$, it is enough to establish that if $w = \pi(a) = su = tv$ for some $u, v \in W_A$ with $\lambda(u) = \lambda(v) = \lambda(w) - 1$, then there is $w' \in W_A$ such that $w = \pi(\Delta_{s,t})w'$ and $\lambda(w') = \lambda(w) - \lambda(\pi(\Delta_{s,t}))$. This reduces our claim to an assertion on Coxeter groups. For a proof of this assertion, see [GP00, Sect. 1.1.7 and Lemma 1.2.1]. □

Theorem 6.30 implies that the pair (B_A^+, Δ) with finite W_A shares all properties of comprehensive Garside monoids stated in Sections 6.4.2 and 6.4.3. We give here a summary of these properties.

(1) The monoid B_A^+ has a presentation with generators $[a]$, where a runs over B_A^{red}, and relations $[1] = 1$ and $[a][b] = [ab]$ whenever $a, b \in B_A^{\mathrm{red}}$ satisfy $ab \in B_A^{\mathrm{red}}$. Using the bijection $\rho : W_A \to B_A^{\mathrm{red}}$ and Lemma 6.28 (c), we conclude that B_A^+ has a presentation with generators $[u]$, where u runs over W_A, and relations $[1] = 1$ and $[u][v] = [uv]$ whenever $u, v \in W_A$ satisfy $\lambda(u) + \lambda(v) = \lambda(uv)$.

(2) Any finite family of elements of B_A^+ has a unique left gcd, right gcd, left lcm, right lcm. (The existence of the right gcd and left lcm is proven similarly to Theorem 6.25 using the involutive antiautomorphism of B_A^+ fixing S pointwise.)

(3) Any $a \in B_A^+$ has a *normal form* $(a_1, a_2, \ldots, a_r)$ with $r \geq 0$, where $a_1, a_2, \ldots, a_r$ are unique elements of $B_A^{\mathrm{red}} - \{1\}$ such that $a = a_1 a_2 \cdots a_r$ and a_i is the left gcd of $a_i a_{i+1} \cdots a_r$ and Δ for all $i = 1, 2, \ldots, r$.

(4) The natural monoid homomorphism $B_A^+ \to B_A$ is injective.

(5) If $S \neq \emptyset$, then any $\beta \in B_A$ can be written uniquely in the form $\beta = \Delta^s b$, where $s \in \mathbf{Z}$ and $b \in B_A^+ \subset B_A$ is not a right multiple of Δ.

(6) The conjugacy problem in B_A is equivalent to the conjugacy problem in B_A^+ and can be solved as in Section 6.2.5.

6.6.3 Brieskorn's theorem

In Section 1.4.3 we interpreted Artin's braid group B_n as the fundamental group of a configuration space. We give here a similar interpretation of the generalized braid group B_A associated to a Coxeter matrix $A = (a_{s,t})_{s,t \in S}$.

To begin with, we identify the Coxeter group W_A associated to A with a group of matrices. Let V be a real vector space with a basis $\{e_s\}_{s \in S}$ indexed by the set S. We define a symmetric bilinear form $\langle \, , \, \rangle$ on V by

$$\langle e_s, e_t \rangle = -\cos(\pi/a_{s,t}) = \cos(\pi - \pi/a_{s,t}),$$

where we use the convention that $\pi/a_{s,t} = 0$ if $a_{s,t} = +\infty$. In particular, we have $\langle e_s, e_s \rangle = \cos(0) = 1$ for all $s \in S$.

For each $s \in S$, define an endomorphism μ_s of V by

$$\mu_s(v) = v - 2\langle e_s, v \rangle e_s,$$

where $v \in V$. Since $\langle e_s, e_s \rangle \neq 0$, the subspace $H_s = \{v \in V \mid \langle e_s, v \rangle = 0\}$ orthogonal to e_s is a hyperplane, which does not contain e_s. We have an orthogonal decomposition

$$V = H_s \oplus \mathbf{R} e_s.$$

Since $\mu_s(e_s) = -e_s$ and $\mu_s(v) = v$ for all $v \in H$, the endomorphism μ_s is involutive and equal to the orthogonal reflection in the hyperplane H_s.

Lemma 6.31. *For all $s, t \in S$, the order of $\mu_s \mu_t$ is equal to $a_{s,t}$.*

Proof. (a) If $a_{s,t} = \infty$, then

$$(\mu_s \mu_t)(e_s) = \mu_s(e_s + 2e_t) = -e_s + 2(e_t + 2e_s) = 3e_s + 2e_t$$

and

$$(\mu_s \mu_t)(e_t) = -\mu_s(e_t) = -2e_s - e_t.$$

It follows that $\mu_s \mu_t$ fixes $e_s + e_t$. Using this fact and the equality

$$(\mu_s \mu_t)(e_s) = e_s + 2(e_s + e_t),$$

it is easy to check by induction that $(\mu_s \mu_t)^r(e_s) = e_s + 2r(e_s + e_t)$ for all $r \geq 0$. This shows that $\mu_s \mu_t$ is of infinite order.

(b) We noted above that μ_s is an involution. Therefore, for $s = t$, the order of $\mu_s \mu_t = \mu_s^2$ is $1 = a_{s,s}$.

(c) It remains to treat the case in which $s \neq t$ and $a_{s,t} < \infty$. Observe that $\mu_s \mu_t$ fixes $H_s \cap H_t$ pointwise and leaves the two-dimensional subspace $\Pi_{s,t}$ of V spanned by e_s and e_t invariant. We have $V = (H_s \cap H_t) \oplus \Pi_{s,t}$. Restricting the symmetric bilinear form $\langle \, , \, \rangle$ to $\Pi_{s,t}$, we obtain a symmetric bilinear form

$$\begin{pmatrix} \langle e_s, e_s \rangle & \langle e_s, e_t \rangle \\ \langle e_s, e_t \rangle & \langle e_t, e_t \rangle \end{pmatrix} = \begin{pmatrix} 1 & -\cos(\pi/a_{s,t}) \\ -\cos(\pi/a_{s,t}) & 1 \end{pmatrix}.$$

The inequalities $2 \leq a_{s,t} < \infty$ imply that $0 \leq \cos(\pi/a_{s,t}) < 1$, so that the latter bilinear form is positive definite. Therefore we can treat $\Pi_{s,t}$ as a Euclidean plane, where the vectors e_s and e_t are of norm one and the angle between them is $\pi - \pi/a_{s,t}$. It is well known that the composition of two planar reflections is a rotation by an angle equal to twice the angle between the vectors defining the reflections. Therefore, the restriction of $\mu_s \mu_t$ to $\Pi_{s,t}$ is a rotation by an angle of $2\pi - 2\pi/a_{s,t} = -2\pi/a_{s,t} \pmod{2\pi}$. Since $\mu_s \mu_t$ fixes $H_s \cap H_t$ pointwise, the order of $\mu_s \mu_t$ is equal to $a_{s,t}$.   □

By this lemma, the reflections μ_s satisfy (6.7). Therefore, there is a group homomorphism $\mu : W_A \to \mathrm{Aut}(V)$ defined by $\mu(s) = \mu_s$ for all $s \in S$. It can be shown that μ is an injective homomorphism onto a discrete subgroup of $\mathrm{Aut}(V)$. This realizes W_A as a group of matrices generated by reflections.

We assume until the end of the section that the Coxeter group $W = W_A$ is finite. Let $\{H_i\}_{i \in I}$ be the set of all hyperplanes of V obtained as the images of the hyperplanes H_s ($s \in S$) under the automorphisms of V lying in $\mu(W) \subset \mathrm{Aut}(V)$. Since W is finite, the set $\{H_i\}_{i \in I}$ is finite.

Let $V^{\mathbf{C}} = V \otimes_{\mathbf{R}} \mathbf{C}$ be the complexification of the real vector space V. The action μ of W on V extends to an action of W on $V^{\mathbf{C}}$ by $\mathbf{C}$-linear automorphisms. Consider the complex hyperplanes

$$H_i^{\mathbf{C}} = H_i \otimes_{\mathbf{R}} \mathbf{C} \subset V^{\mathbf{C}}$$

for $i \in I$. Since the action of W on V permutes the hyperplanes $\{H_i\}_{i \in I}$, the extended action on $V^{\mathbf{C}}$ permutes the complex hyperplanes $\{H_i^{\mathbf{C}}\}_{i \in I}$. Therefore, we obtain an action of W on the set

$$E = V^{\mathbf{C}} - \bigcup_{i \in I} H_i^{\mathbf{C}}.$$

The set E is an open subset of the complex vector space $V^{\mathbf{C}}$. Therefore, E is a complex manifold of complex dimension $\mathrm{card}(S)$.

The group W acts on E by fixed-point free homeomorphisms preserving the complex structure. The quotient space $W\backslash E$ naturally inherits the structure of a complex manifold of dimension $\text{card}(S)$. The projection $E \to W\backslash E$ is an unramified covering. Since the complex hyperplanes $H_i^{\mathbf{C}}$ are of real codimension two in $V^{\mathbf{C}}$, the manifolds E and $W\backslash E$ are connected.

Fix a point

$$p \in V \cap E = V - \bigcup_{i \in I} H_i.$$

For each $s \in S$, consider a broken line in $V^{\mathbf{C}}$ with consecutive vertices p, $p + \sqrt{-1}p$, $\mu_s(p) + \sqrt{-1}p$, and $\mu_s(p)$. This broken line lies in E and projects to a loop in $W\backslash E$ beginning and ending at the projection $\dot{p} \in W\backslash E$ of p. This loop represents an element of the fundamental group $\pi_1(W\backslash E, \dot{p})$ denoted by $\dot{s}$.

Theorem 6.32 (E. Brieskorn). *The map $S \to \pi_1(W\backslash E, \dot{p})$, $s \mapsto \dot{s}$ induces a group isomorphism $B_A \cong \pi_1(W\backslash E, \dot{p})$.*

For a proof, see Brieskorn [Bri71] or Deligne [Del72]. These authors also proved that the manifold $W\backslash E$ is aspherical, i.e., its higher homotopy groups vanish, see [Del72], [Bri73]. Since $E \to W\backslash E$ is a covering, the manifold E is also aspherical. Its fundamental group is isomorphic to the kernel of the projection $B_A \to W = W_A$. This kernel generalizes Artin's pure braid groups.

Notes

The word problem in the braid groups was first solved by Artin [Art25]. Garside [Gar69] introduced the braid monoids and studied their properties. This led him to a new solution of the word problem and a solution of the conjugacy problem in the braid groups. Garside [Gar69] also extended these results to some generalized braid monoids. Dehornoy and Paris [DP99] abstracted the ideas contained in [Gar69] and introduced the concept of a Garside monoid. We used the following sources while writing this chapter: [DP99], [Mic99], [GP00], [Deh02], [BDM02]. The definition of a (pre-)Garside monoid given in this chapter is slightly different from the definitions in these papers.

A systematic study of generalized braid monoids and groups associated with finite Coxeter groups was undertaken by Brieskorn [Bri71], [Bri73], Brieskorn and Saito [BS72], and Deligne [Del72]. Generalized braid groups are also called *Artin groups* or *Artin–Tits groups*. In the literature one also finds the expression "braid groups of spherical type," which designates generalized braid groups associated with finite Coxeter groups. Theorem 6.26 is due to Matsumoto [Mat64]. Theorem 6.32 was conjectured by J. Tits and first proven by Brieskorn [Bri71]. A description of generalized braid groups in terms of braid pictures can be found in [All02].

7

An Order on the Braid Groups

The principal aim of this chapter is to show that the braid groups have a natural total order.

7.1 Orderable groups

In this section we present generalities on orderable groups. All groups will be written multiplicatively and their neutral elements will be denoted by 1.

7.1.1 Orders

An *order* on a set X is a relation $\leq$ among elements of X satisfying the following properties for all x, y, $z \in X$:

(i) *(Reflexivity)* $x \leq x$,
(ii) *(Antisymmetry)* $(x \leq y$ and $y \leq x) \implies x = y$,
(iii) *(Transitivity)* $(x \leq y$ and $y \leq z) \implies x \leq z$.

We shall also write $y \geq x$ instead of $x \leq y$. We write $x < y$ or, equivalently, $y > x$ if $x \leq y$ and $x \neq y$. It is clear that there are no elements $x, y \in X$ such that simultaneously $x < y$ and $y < x$.

An order is said to be *total* (or *linear*) if for any $x, y \in X$, either $x = y$ or $x < y$ or $x > y$. An *order-preserving map* from an ordered set $(X, \leq)$ to an ordered set $(X', \leq')$ is a map $f : X \to X'$ such that $f(x) \leq' f(y)$ for all $x, y \in X$ such that $x \leq y$.

7.1.2 Basics on orderable groups

An order $\leq$ on a group G is *left-invariant* (resp. *right-invariant*) if

$$x \leq y \implies zx \leq zy \qquad (\text{resp.} \quad x \leq y \implies xz \leq yz)$$

for all x, y, $z \in G$. An order on a group that is both left- and right-invariant is said to be *bi-invariant*.

C. Kassel, V. Turaev, *Braid Groups*, DOI: 10.1007/978-0-387-68548-9_7,
© Springer Science+Business Media, LLC 2008

A group is *orderable* if it has a left-invariant total order. Note that if G is a group with left-invariant total order $\leq$, then G also admits a right-invariant total order $\leq'$ defined by $x \leq' y$ if $x^{-1} \leq y^{-1}$ for $x, y \in G$.

For example, the set of real numbers $\mathbf{R}$ is orderable, since the standard total order on $\mathbf{R}$ is left-invariant. Clearly, all subgroups of an orderable group are orderable.

From orderable groups we can construct new orderable groups as follows.

Lemma 7.1. *(a) If $G_1, \ldots, G_r$ are orderable groups, then their direct product $G_1 \times \cdots \times G_r$ is orderable.*

(b) Let G be a group and H a normal subgroup. If H and G/H have left-invariant total orders, then G has a unique left-invariant total order such that the inclusion $H \hookrightarrow G$ and the projection $p : G \to G/H$ are order-preserving. If furthermore the left-invariant total orders on H and G/H are bi-invariant and $zxz^{-1} > 1$ for all $z \in G$ and $x \in H$ with $x > 1$, then the associated left-invariant order on G is bi-invariant.

Proof. (a) Let $\leq_i$ be a left-invariant total order on G_i. We define a relation $\leq$ on $G = G_1 \times \cdots \times G_r$ by $(x_1, \ldots, x_r) \leq (y_1, \ldots, y_r)$ if either $x_i = y_i$ for all $i \in \{1, \ldots, r\}$ or there is $i \in \{1, \ldots, r\}$ such that $x_j = y_j$ for all $j < i$ and $x_i <_i y_i$. It is easy to check that this relation is an order on G. It is called the *lexicographic order*. Since the orders on $G_1, \ldots, G_r$ are total, so is the lexicographic order on G. Let us prove the left invariance. Let

$$x = (x_1, \ldots, x_r), \quad y = (y_1, \ldots, y_r), \quad z = (z_1, \ldots, z_r)$$

be three elements of G. If $x < y$, then there is $i \in \{1, \ldots, r\}$ such that $x_j = y_j$ for all $j < i$ and $x_i <_i y_i$. Consequently, $z_j x_j = z_j y_j$ for all $j < i$ and $z_i x_i <_i z_i y_i$ by the left invariance of $\leq_i$. Therefore, $zx < zy$.

(b) We define a relation $\leq$ on G by $x \leq y$ if either $p(x) < p(y)$ for the given order on G/H, or $p(x) = p(y)$ and $x^{-1}y \geq 1$ for the given order on H (observe that $x^{-1}y \in H$). It is an easy exercise to check that this is a left-invariant total order on G. Moreover, any left-invariant order on G such that $H \hookrightarrow G$ and $p : G \to G/H$ are order-preserving is necessarily of this form.

Assume that the given total orders on H and G/H are bi-invariant and $zxz^{-1} > 1$ for all $z \in G$ and $x \in H$ with $x > 1$. Let us check that the order $\leq$ on G is right-invariant. Let $x \leq y$ in G. Since the order on G/H is right-invariant, $p(x) < p(y)$ implies that

$$p(xz) = p(x)p(z) < p(y)p(z) = p(yz)$$

for all $z \in G$. Hence $xz \leq yz$. If $p(x) = p(y)$ and $x^{-1}y \geq 1$, then

$$p(xz) = p(x)p(z) = p(y)p(z) = p(yz)$$

and

$$(xz)^{-1}(yz) = z^{-1}(x^{-1}y)z \geq 1,$$

since conjugation preserves positive elements of H by hypothesis. Thus in this case again, $xz \leq yz$.   □

Lemma 7.1 (a) and the orderability of $\mathbf{R}$ imply that all finite-dimensional real vector spaces and their additive subgroups are orderable.

Not all groups are orderable. For instance, a finite group is orderable if and only if it is trivial (see Proposition 7.5 below).

7.1.3 The positive cone

For a subset $\mathcal{P}$ of a group G, set $\mathcal{P}^{-1} = \{x \in G \mid x^{-1} \in \mathcal{P}\}$ and

$$\mathcal{P}^2 = \{z \in G \mid \text{there are } x, y \in \mathcal{P} \text{ such that } z = xy\}.$$

Lemma 7.2. *For any subset $\mathcal{P}$ of a group G,*

$$\mathcal{P} \cap \{1\} = \emptyset \iff \mathcal{P}^{-1} \cap \{1\} = \emptyset \impliedby \mathcal{P} \cap \mathcal{P}^{-1} = \emptyset.$$

If $\mathcal{P}^2 \subset \mathcal{P}$, then $\mathcal{P} \cap \{1\} = \emptyset \implies \mathcal{P} \cap \mathcal{P}^{-1} = \emptyset$.

Proof. If $1 \in \mathcal{P}$, then $1 = 1^{-1} \in \mathcal{P}^{-1}$. This shows that $\mathcal{P}^{-1} \cap \{1\} = \emptyset \implies \mathcal{P} \cap \{1\} = \emptyset$. Replacing here $\mathcal{P}$ by $\mathcal{P}^{-1}$, we obtain the converse implication.

To prove the implication $\mathcal{P} \cap \mathcal{P}^{-1} = \emptyset \implies \mathcal{P} \cap \{1\} = \emptyset$, we check that $\mathcal{P} \cap \{1\} \neq \emptyset \implies \mathcal{P} \cap \mathcal{P}^{-1} \neq \emptyset$. If $\mathcal{P} \cap \{1\} \neq \emptyset$, then $1 \in \mathcal{P}$ and $1 = 1^{-1} \in \mathcal{P}^{-1}$. Hence, $1 \in \mathcal{P} \cap \mathcal{P}^{-1}$.

To prove the last claim of the lemma, we check that $\mathcal{P} \cap \mathcal{P}^{-1} \neq \emptyset \implies \mathcal{P} \cap \{1\} \neq \emptyset$. If $x \in \mathcal{P} \cap \mathcal{P}^{-1}$, then $x^{-1} \in \mathcal{P} \cap \mathcal{P}^{-1}$. Consequently,

$$1 = xx^{-1} \in \mathcal{P}^2 \subset \mathcal{P}.$$

Hence, $\mathcal{P} \cap \{1\} \neq \emptyset$.   □

Lemma 7.3. *Let $\leq$ be a left-invariant order on a group G. Set*

$$\mathcal{P} = \{x \in G \mid x > 1\}.$$

Then $\mathcal{P}^{-1} = \{x \in G \mid x < 1\}$, $\mathcal{P}^2 \subset \mathcal{P}$, and

$$\mathcal{P} \cap \{1\} = \mathcal{P}^{-1} \cap \{1\} = \mathcal{P} \cap \mathcal{P}^{-1} = \emptyset.$$

If the order $\leq$ is total, then $\mathcal{P} \cup \{1\} \cup \mathcal{P}^{-1} = G$.

Proof. If $x \in \mathcal{P}^{-1}$, then $x^{-1} \in \mathcal{P}$, so that $1 < x^{-1}$. Multiplying by x on the left, we obtain $x < 1$. Similarly, $x < 1$ implies $1 = x^{-1}x < x^{-1}1 = x^{-1}$, so that $x^{-1} \in \mathcal{P}$ and $x \in \mathcal{P}^{-1}$. This proves that $\mathcal{P}^{-1} = \{x \in G \mid x < 1\}$.

The antisymmetry axiom implies that $\mathcal{P}$ and $\mathcal{P}^{-1} = \{x \in G \mid x < 1\}$ are disjoint. That they are disjoint from $\{1\}$ follows from the definition of the relation $<$.

If $x, y \in \mathcal{P}$, then $xy > x1 = x > 1$, so that $xy \in \mathcal{P}$. Thus, $\mathcal{P}^2 \subset \mathcal{P}$.

If the order $\leq$ is total, then for any $x \in G$, necessarily $x > 1$ or $x = 1$ or $x < 1$. Therefore, $\mathcal{P} \cup \{1\} \cup \mathcal{P}^{-1} = G$.   □

The set $\mathcal{P} = \{x \in G \mid x > 1\}$ as in the previous lemma is called the *positive cone* associated to the order $\leq$. The elements of $\mathcal{P}$ are said to be *positive* with respect to $\leq$. The following theorem shows that a left-invariant total order on a group can be reconstructed from its positive cone.

Theorem 7.4. *Let $\mathcal{P}$ be a subset of a group G such that*

$$\mathcal{P}^2 \subset \mathcal{P} \quad and \quad 1 \notin \mathcal{P}.$$

Then G has a unique left-invariant order $\leq$ such that $\mathcal{P} = \{x \in G \mid x > 1\}$. If $z\mathcal{P}z^{-1} \subset \mathcal{P}$ for all $z \in G$, then the order $\leq$ is bi-invariant. If

$$\mathcal{P} \cup \{1\} \cup \mathcal{P}^{-1} = G,$$

then the order $\leq$ is total.

Proof. Let us first prove the uniqueness of the order. By the left invariance, the inequality $x < y$ is equivalent to the inequality $1 = x^{-1}x < x^{-1}y$. The latter is equivalent to the inclusion $x^{-1}y \in \mathcal{P}$. This shows that a left-invariant order on G with positive cone $\mathcal{P}$ is necessarily defined by

$$x \leq y \iff (x = y \text{ or } x^{-1}y \in \mathcal{P}). \qquad (7.1)$$

We next prove the existence. By Lemma 7.2, the assumptions $\mathcal{P}^2 \subset \mathcal{P}$ and $1 \notin \mathcal{P}$ imply that

$$\mathcal{P} \cap \{1\} = \mathcal{P}^{-1} \cap \{1\} = \mathcal{P} \cap \mathcal{P}^{-1} = \emptyset.$$

We define a binary relation $\leq$ on G by (7.1). Let us check that it satisfies the axioms of an order. The reflexivity follows from the definition.

Antisymmetry: If $x \leq y$ and $y \leq x$, then either $x = y$ or $x^{-1}y \in \mathcal{P}$, $y^{-1}x \in \mathcal{P}$. Since $y^{-1}x = (x^{-1}y)^{-1}$, we obtain $x^{-1}y \in \mathcal{P} \cap \mathcal{P}^{-1} = \emptyset$ in the second case, a contradiction. Therefore, $x = y$.

Transitivity: If $x^{-1}y, y^{-1}z \in \mathcal{P}$, then $x^{-1}z = (x^{-1}y)(y^{-1}z) \in \mathcal{P}^2 \subset \mathcal{P}$.

Let us show that the order $\leq$ is left-invariant. Pick $x, y \in G$ such that $x \leq y$. Then $x = y$ or $x^{-1}y \in \mathcal{P}$. If $x = y$, then $zx = zy$ for all $z \in G$. If $x^{-1}y \in \mathcal{P}$, then $(zx)^{-1}(zy) = x^{-1}y \in \mathcal{P}$. In both cases, $zx \leq zy$.

Assume that $z\mathcal{P}z^{-1} \subset \mathcal{P}$ for all $z \in G$. Let $x, y \in G$ be such that $x \leq y$. Then $x = y$ or $x^{-1}y \in \mathcal{P}$. If $x = y$, then $xz = yz$ for all $z \in G$. If $x^{-1}y \in \mathcal{P}$ and $z \in G$, then $(xz)^{-1}(yz) = z^{-1}(x^{-1}y)z$ belongs to $z^{-1}\mathcal{P}z$, hence to $\mathcal{P}$. This proves that $\leq$ is right-invariant.

If $\mathcal{P} \cup \{1\} \cup \mathcal{P}^{-1} = G$, then for all $x, y \in G$, we have $x^{-1}y \in \mathcal{P}$ or $x^{-1}y \in \mathcal{P}^{-1}$ or $x^{-1}y = 1$. In the first case, $x < y$; in the second case, $y^{-1}x = (x^{-1}y)^{-1} \in \mathcal{P}$, so that $y < x$; in the last case, $x = y$. This proves that the order $\leq$ is total. □

7.1.4 Properties of orderable groups

We state two properties of orderable groups.

Proposition 7.5. *Any orderable group G is torsion free.*

Proof. We have to show that $x^n \neq 1$ for any integer $n \geq 1$ and any $x \in G$ such that $x \neq 1$. Suppose that $x > 1$. Then by the left invariance,

$$x^n = (x^{n-1})x > x^{n-1}1 = x^{n-1}$$

for any $n \geq 1$. By induction, $x^n > x > 1$; hence $x^n \neq 1$. If $x < 1$, then $x^{-1} > 1$ and $x^{-n} = (x^{-1})^n \neq 1$. Hence, $x^n \neq 1$. □

Proposition 7.6. *If G is an orderable group and R is a ring without zero-divisors, then the group algebra $R[G]$ has no zero-divisors.*

Proof. Let $\omega = \sum_{i=1}^{p} r_i g_i$ and $\omega' = \sum_{j=1}^{q} s_j h_j$ be nonzero elements of $R[G]$, where $g_1, \ldots, g_p$ and $h_1, \ldots, h_q$ are elements of G, and $r_1, \ldots, r_p$ and $s_1, \ldots, s_q$ are elements of R. We may assume that $r_1, \ldots, r_p$ and $s_1, \ldots, s_q$ are all nonzero and the group elements h_i are numerated in such a way that $h_1 < h_2 < \cdots < h_q$. By the left invariance of the order, $g_i h_1 < g_i h_j$ for all $i = 1, \ldots, p$ and $j = 2, \ldots, q$. The order being total, there is a unique i_0 such that $g_{i_0} h_1 < g_i h_1$ for all $i \neq i_0$. We claim that $(i_0, 1)$ is the unique pair (i, j) such that $g_{i_0} h_1 = g_i h_j$ in G. Indeed, as observed above, $g_{i_0} h_1 < g_{i_0} h_j$ for all $j \neq 1$, and, if $i \neq i_0$, then $g_{i_0} h_1 < g_i h_1 < g_i h_j$. Therefore, the coefficient of $g_{i_0} h_1$ in $\omega \omega' \in R[G]$ is $r_{i_0} s_1$, which is nonzero, since R has no zero-divisors. Hence, $\omega \omega' \neq 0$. □

The *zero-divisor conjecture* (sometimes called Kaplansky's conjecture) states that if G is a torsion-free group and R is a ring without zero-divisors, then the group algebra $R[G]$ has no zero-divisors. Proposition 7.6 shows that this conjecture holds for orderable groups.

7.1.5 Biorderable groups

A group is *biorderable* if it has a bi-invariant total order. For example, any orderable abelian group is biorderable, since a left-invariant order on an abelian group is necessarily bi-invariant. All subgroups of a biorderable group are biorderable. We state one further property of biorderable groups.

Lemma 7.7. *Let G be a biorderable group. Then $x^n = y^n \Rightarrow x = y$ for any $x, y \in G$ and any positive integer n.*

Proof. We start with the following observation: in a biorderable group, $x < y$ together with $x' < y'$ implies $xx' < yy'$. Indeed, by the left and right invariance of the order, $xx' < xy' < x'y'$. From this an easy induction shows that $x < y \Rightarrow x^n < y^n$ for all positive integers n. Now let $x, y \in G$ be such that $x^n = y^n$. Since the order is total, we must have $x = y$ or $x < y$ or $y < x$. By the previous remark, the latter two cases cannot occur. Therefore, $x = y$. □

The group rings of biorderable groups have further interesting properties. For instance, Malcev [Mal48] and Neumann [Neu49] proved that the integral group ring of a biorderable group can be embedded into a division algebra.

The first two braid groups $B_1 = \{1\}$ and $B_2 = \mathbf{Z}$ are biorderable. The braid group B_n with $n \geq 3$ is not biorderable. Indeed, by Remark 1.30, $\sigma_1\sigma_2 \neq \sigma_2\sigma_1$, but $(\sigma_1\sigma_2)^3 = (\sigma_2\sigma_1)^3$. Lemma 7.7 implies that B_n is not biorderable.

7.2 Pure braid groups are biorderable

The main result of this section is the following theorem.

Theorem 7.8. *The pure braid group P_n is biorderable for all $n \geq 1$.*

To prove this theorem we first study Magnus expansions of free groups and then show that free groups are biorderable. Theorem 7.8 is proven in Section 7.2.3. Neither this theorem nor its proof will be used in the sequel.

7.2.1 The Magnus expansion

Fix a nonempty set X. We define a ring of (noncommutative) formal power series over X. Let X^* be the free monoid on X; see Example 6.1 (c). By a *formal power series* over X we mean an arbitrary formal sum $\sum_{W \in X^*} n_W W$, where W runs over X^* and $n_W \in \mathbf{Z}$. Such formal sums can be added in the obvious way and thus form an additive abelian group denoted by $\mathbf{Z}[[X]]$. The multiplication in X^* induces a multiplication in $\mathbf{Z}[[X]]$; this turns $\mathbf{Z}[[X]]$ into a ring whose unit is the neutral element $1 \in X^*$.

Recall the length function $\ell : X^* \to \mathbf{N}$, which is the unique morphism of monoids sending all elements of X to 1. We say that a formal power series $a = \sum_{W \in X^*} n_W W \in \mathbf{Z}[[X]]$ has *degree* $\geq r$, where r is a positive integer, if $n_W = 0$ for all $W \in X^*$ with $\ell(W) < r$. Clearly, the product of a formal power series of degree $\geq r$ with a formal power series of degree $\geq s$ is a formal power series of degree $\geq r + s$.

For a formal power series $a = \sum_{W \in X^*} n_W W$, let $\varepsilon(a) = n_1 \in \mathbf{Z}$ be the coefficient of the neutral element $1 \in X^*$. It is easy to show that a is invertible in $\mathbf{Z}[[X]]$ if and only if $\varepsilon(a) = \pm 1$. For instance, for any $x \in X$, the polynomial $1 + x \in \mathbf{Z}[[X]]$ is invertible and its inverse is the formal power series $\sum_{k \geq 0} (-1)^k x^k$.

The following lemma is left to the reader.

Lemma 7.9. *For any $x \in X$ and $k \in \mathbf{Z}$ there is a formal power series $h_k(x)$ in the variable x such that*

$$(1 + x)^k = 1 + kx + x^2 h_k(x).$$

Let $G(X) \subset \mathbf{Z}[[X]]$ be the set of all formal power series $a \in \mathbf{Z}[[X]]$ such that $\varepsilon(a) = 1$. This set is a group under multiplication.

Proposition 7.10. *Let F be a free group freely generated by a set X. The homomorphism of groups $\mu : F \to G(X)$ defined by $\mu(x) = 1 + x$ for all $x \in X$ is injective.*

The formal power series $\mu(w)$ is called the *Magnus expansion* of $w \in F$.

Proof. The existence and the uniqueness of μ follow from the definition of F. To check the injectivity of μ, pick a nontrivial element $w \in F$ and write it in the form

$$w = x_1^{k_1} x_2^{k_2} \cdots x_r^{k_r},$$

where $x_1, x_2, \ldots, x_r \in X$ satisfy $x_1 \neq x_2$, $x_2 \neq x_3$, $\ldots$, $x_{r-1} \neq x_r$, and all the integers $k_1, k_2, \ldots, k_r$ are nonzero. By Lemma 7.9,

$$
\begin{aligned}
\mu(w) &= (1 + x_1)^{k_1} (1 + x_2)^{k_2} \cdots (1 + x_r)^{k_r} \\
&= \left(1 + k_1 x_1 + x_1^2 h_{k_1}(x_1)\right)\left(1 + k_2 x_2 + x_2^2 h_{k_2}(x_2)\right) \\
&\quad \cdots \left(1 + k_r x_r + x_r^2 h_{k_r}(x_r)\right).
\end{aligned}
$$

Expanding the formal power series on the right-hand side, we see that it contains a unique monomial of the form $x_1 x_2 \cdots x_r$. The coefficient of this monomial is $k_1 k_2 \cdots k_r \neq 0$. Hence, $\mu(w) \neq 1$. $\square$

7.2.2 Free groups are biorderable

Proposition 7.11. *Let F be a free group freely generated by a set X. Any total order on X extends to a bi-invariant total order on F.*

Proof. A total order on X induces an order $\leq$ on X^* as follows:

(i) On $X \subset X^*$ the order $\leq$ is the given total order.
(ii) If $W_1, W_2 \in X^*$ satisfy $\ell(W_1) < \ell(W_2)$, then set $W_1 < W_2$.
(iii) If $W_1, W_2 \in X^*$ have the same length, then we order them lexicographically: if $W_1 = x_1 \cdots x_r$ and $W_2 = y_1 \cdots y_r$ with $x_i, y_i \in X$ for all i, then $W_1 < W_2$ provided there is $k \leq r$ such that $x_k < y_k$ and $x_i = y_i$ for all $i < k$.

The order $\leq$ on X^* is total and bi-invariant; the latter means that $W_1 < W_2$ implies $WW_1 < WW_2$ and $W_1 W < W_2 W$ for all $W \in X^*$.

By Proposition 7.10, if $w \in F$ is distinct from the neutral element 1, then $\mu(w) \neq 1 \in \mathbf{Z}[[X]]$. Write

$$\mu(w) - 1 = \sum_W n_W W,$$

where W runs over all nonempty words in X^* such that the integer n_W is nonzero.

One of the words W appearing in this expansion of $\mu(w) - 1$ has to be the smallest with respect to the above-defined total order on X^*. Denote this smallest word by $V(w)$ and set $n(w) = n_{V(w)} \neq 0$. Finally, set

$$\mathcal{P} = \{w \in F - \{1\} \mid n(w) > 0\}.$$

Clearly, an element w of $F - \{1\}$ lies in $\mathcal{P}$ if and only if $\mu(w)$ is of the form

$$1 + n(w) V + \sum_{W > V} n_W\, W,$$

where $V \neq 1$ and $n(w) > 0$.

We claim that $\mathcal{P}$ satisfies all conditions of Theorem 7.4 and therefore defines a bi-invariant total order on F. It follows from the definitions that $1 \notin \mathcal{P}$. To establish that $\mathcal{P}^2 \subset \mathcal{P}$, consider two elements $w, w' \in \mathcal{P}$ and their Magnus expansions

$$\mu(w) = 1 + n(w)V + \sum_{W > V} n_W W$$

and

$$\mu(w') = 1 + n(w')V' + \sum_{W > V'} n'_W W,$$

where $n(w) > 0$ and $n(w') > 0$. Expand $\mu(ww') = \mu(w)\mu(w')$ as a formal power series. Using the bi-invariance of the order on X^*, we easily obtain

$$n(ww') = \begin{cases} n(w) & \text{if } V < V', \\ n(w') & \text{if } V > V', \\ n(w) + n(w') & \text{if } V = V'. \end{cases}$$

In all cases, $n(ww') > 0$, which implies that $ww' \in \mathcal{P}$.

It is easy to deduce from the identity $\mu(w^{-1}) = (\mu(w))^{-1}$ that $\mathcal{P}^{-1}$ is the set of all $w \in F - \{1\}$ such that $n(w) < 0$. This together with the injectivity of μ implies that $\mathcal{P} \cup \{1\} \cup \mathcal{P}^{-1} = F$.

It remains to check that $w\mathcal{P}w^{-1} \subset \mathcal{P}$ for all $w \in F$. If $f \in \mathbf{Z}[[X]]$ is a formal power series without constant term and $W \in X^*$, then

$$(1 + f)W(1 + f)^{-1} = W + \sum_{W' > W} m_{W'}\, W'$$

for some integers $m_{W'}$. This implies that

$$n(ww'w^{-1}) = n(w')$$

for all w, w' in $F - \{1\}$. Therefore, $w\mathcal{P}w^{-1} \subset \mathcal{P}$ for all $w \in F$. $\square$

Corollary 7.12. *All free groups are biorderable.*

7.2.3 Proof of Theorem 7.8

We recall the notation and the results of Section 1.3. First of all, the pure braid group P_n is generated by the $n(n-1)/2$ braids $A_{i,j}$ $(1 \le i < j \le n)$ shown in Figure 1.10. Next, for each $n \ge 2$, we have an exact sequence

$$1 \to U_n \to P_n \to P_{n-1} \to 1,$$

where the map $P_n \to P_{n-1}$ is the homomorphism f_n that removes the rightmost string of a pure braid, and U_n is a free group on the $n-1$ generators $X_1 = A_{1,n}, \ldots, X_{n-1} \doteq A_{n-1,n}$. We give U_n the bi-invariant total order derived as above from the order $X_1 < X_2 < \cdots < X_{n-1}$ on the set of generators.

Since $P_1 = \{1\}$, we have $P_2 \cong U_2 \cong \mathbf{Z}$, which is biorderable. It follows from Lemma 7.1 (b) by induction on n that P_n has a unique left-invariant total order such that the homomorphisms $U_n \to P_n$ and $f_n : P_n \to P_{n-1}$ are order-preserving.

By Lemma 7.1 (b), this order on P_n is bi-invariant provided $\beta u \beta^{-1} > 1$ for any $\beta \in P_n$ and any $u \in U_n$ such that $u > 1$. To check this property, we observe from relations (1.7) that conjugating a generator $X_i = A_{i,n}$ of U_n by a generator $A_{r,s}$ of P_n with $s < n$ amounts to conjugating X_i by a product of generators of U_n. The same is true for $s = n$, since $A_{r,n} \in U_n$. Thus, in all cases, $A_{r,s}^{-1} X_i A_{r,s} = X_i$ modulo the commutator subgroup $[U_n, U_n]$ of U_n. It follows that $\beta X_i \beta^{-1} = X_i$ modulo $[U_n, U_n]$ for all $\beta \in P_n$ and $i \in \{1, \ldots, n-1\}$. In other words, $\beta X_i \beta^{-1} = X_i u_i$ for some $u_i \in [U_n, U_n]$. The Magnus expansion of $\beta X_i \beta^{-1}$ is computed by

$$\mu(\beta X_i \beta^{-1}) = \mu(X_i u_i) = (1 + X_i)\,\mu(u_i).$$

It follows from Exercise 7.2.1 below that $\mu(u_i) = 1 +$ (a formal power series of degree ≥ 2). Therefore, $\mu(\beta X_i \beta^{-1}) = 1 + X_i +$ (a formal power series of degree ≥ 2). The Magnus expansion of $\beta u \beta^{-1}$, where $u \in U_n$, is then obtained from $\mu(u)$ by replacing each X_i by the sum of X_i with a formal power series of degree ≥ 2. The Magnus expansions of u and $\beta u \beta^{-1}$ have therefore the same first nonconstant term. It follows that $\beta u \beta^{-1} > 1$ if and only if $u > 1$. $\square$

Exercise 7.2.1. Show that for $x, y \in X$,

$$\mu(x^{-1} y^{-1} x y) = 1 + (xy - yx) + \text{(a formal power series of degree } \ge 3).$$

Exercise 7.2.2. Find all biorderable groups G fitting in the exact sequence

$$0 \to \mathbf{Z}^r \to G \to \mathbf{Z} \to 0.$$

Exercise 7.2.3. Show that for any orderable group G and any ring R, the only invertible elements of the group algebra $R[G]$ are of the form rg, where r is an invertible element of R and $g \in G$.

Exercise 7.2.4. An element e of a ring is an idempotent if $e^2 = e$. Show that a ring having no zero-divisors has only two idempotents, 0 and 1.

7.3 The Dehornoy order

Fix an integer $n \geq 1$. The aim of this section is to construct the left-invariant total order on the braid group B_n due to P. Dehornoy.

7.3.1 Braid words

A *word of length* $m \geq 1$ on a set A is a mapping $w : \{1, 2, \ldots, m\} \rightarrow A$. Such a word is encoded by the expression $w(1)w(2) \cdots w(m)$. For example, for $a, b \in A$ the expression aba encodes the word $\{1, 2, 3\} \rightarrow A$ sending $1, 2, 3$ to a, b, a, respectively. By definition, there is a unique *empty word* $\emptyset$ of length 0.

For any $a \in A$ and $m \geq 1$, the word $aa \cdots a$ formed by m entries of a is denoted by a^m. Writing down consecutively the letters of two words v and w on A, we obtain their *concatenation* vw. For instance, for any $a \in A$ and $m, n \geq 1$, the concatenation of a^m and a^n is a^{m+n}.

We say that a word v is a *subword* of a word w if $w = w_1 v w_2$ for some (possibly empty) words w_1, w_2.

A *braid word* is a word on the set $\{\sigma_1, \ldots, \sigma_{n-1}, \sigma_1^{-1}, \ldots, \sigma_{n-1}^{-1}\}$. Every braid word w represents an element of the braid group B_n. Since w represents the same element of B_n as $w\sigma_1^k(\sigma_1^{-1})^k$ for all $k \geq 1$, any element of B_n can be represented by infinitely many braid words. The empty braid word represents the neutral element 1 of B_n.

The *inverse* of a nonempty braid word $w = \sigma_{i_1}^{\varepsilon_1} \cdots \sigma_{i_r}^{\varepsilon_r}$, where $\varepsilon_i = \pm 1$, is the braid word

$$w^{-1} = \sigma_{i_r}^{-\varepsilon_r} \cdots \sigma_{i_1}^{-\varepsilon_1}.$$

If w represents $\beta \in B_n$, then w^{-1} represents β^{-1}.

We define the *index* of a nonempty braid word w as the smallest integer $i \in \{1, \ldots, n-1\}$ such that σ_i or σ_i^{-1} appear in w. A nonempty braid word has the same index as its inverse. The empty braid word has no index.

7.3.2 σ-positive and σ-negative braids

We say that a braid word w is σ_i-*positive* if it is of index i and σ_i^{-1} does not appear in w. Neither the letter σ_i^{-1} nor the letters $\sigma_k^{\pm 1}$ with $k < i$ appear in a σ_i-positive braid word.

We say that a braid word w is σ_i-*negative* if its inverse is σ_i-positive. In other words, w is σ_i-negative if it is of index i and σ_i does not appear in w.

A braid word is said to be σ-*positive* (resp. σ-*negative*) if it is σ_i-positive (resp. σ_i-negative) for some $i \in \{1, \ldots, n-1\}$.

Definition 7.13. *An element of* B_n *is* σ_i-*positive (resp.* σ_i-*negative) if it is represented by a* σ_i-*positive (resp.* σ_i-*negative) braid word.*

An element of B_n *is* σ-*positive (resp.* σ-*negative) if it is* σ_i-*positive (resp.* σ_i-*negative) for some* $i \in \{1, \ldots, n-1\}$.

The generators $\sigma_1, \sigma_2, \ldots, \sigma_{n-1}$ of B_n are clearly σ-positive. More generally, any element of the submonoid B_n^+ of B_n introduced in Section 6.5 is σ-positive. There are σ-positive elements of B_n that do not belong to B_n^+ (for instance $\sigma_1 \sigma_2^{-1}$).

Warning: not all braid words representing σ-positive elements of B_n are σ-positive. For instance, take the braid word $w = \sigma_1 \sigma_2 (\sigma_1^{-1})^N$, where $N \geq 1$. The index of w is 1, but w is neither σ-positive nor σ-negative. Nevertheless, it represents a σ-positive braid $\beta \in B_n$. Indeed, a repeated application of the relation $\sigma_2 \sigma_1 \sigma_2 = \sigma_1 \sigma_2 \sigma_1$ yields

$$\sigma_2^N \sigma_1 \sigma_2 = \sigma_1 \sigma_2 \sigma_1^N$$

for all $N \geq 1$. Multiplying both sides by σ_2^{-N} on the left and by σ_1^{-N} on the right, we obtain

$$\beta = \sigma_1 \sigma_2 \sigma_1^{-N} = \sigma_2^{-N} \sigma_1 \sigma_2 .$$

The word $(\sigma_2^{-1})^N \sigma_1 \sigma_2$ representing β is σ-positive. Therefore, β is σ-positive.

Let $\mathcal{P}$ be the subset of B_n consisting of all σ-positive elements.

Lemma 7.14. *The subset of B_n consisting of all σ-negative elements is $\mathcal{P}^{-1}$, and $\mathcal{P}^2 \subset \mathcal{P}$.*

Proof. (a) Let β be a σ-negative element of B_n. Then it can be represented by a σ-negative braid word w. By definition, the inverse word w^{-1} is σ-positive. It represents $\beta^{-1} \in B_n$. Then $\beta^{-1} \in \mathcal{P}$ and hence $\beta \in \mathcal{P}^{-1}$. The converse inclusion is proved in a similar fashion.

(b) Let $\beta, \beta' \in \mathcal{P}$. Then β can be represented by a σ_i-positive braid word w and β' by a σ_j-positive braid word w' for some integers i, j. If $i \leq j$, then the word ww' is σ_i-positive. If $i > j$, then ww' is σ_j-positive. In all cases, $\beta\beta'$ is represented by a σ-positive braid word. $\qquad\square$

7.3.3 Definition of the Dehornoy order

We state the main result of this chapter.

Theorem 7.15. *For any $n \geq 1$, the braid group B_n has a left-invariant total order $\leq$ such that $1 < \beta$ if and only if β is σ-positive.*

The order $\leq$ on B_n is called the *Dehornoy order*. Note for the record that $\beta \leq \gamma$ if $\beta = \gamma$ or $\beta^{-1}\gamma \in \mathcal{P}$ for any $\beta, \gamma \in B_n$. Theorem 7.15 implies that B_n is orderable. Therefore, by Propositions 7.5 and 7.6, B_n is torsion free (this has been already proved in Chapter 1; see Corollary 1.29) and the group ring $\mathbf{Z}[B_n]$ has no zero-divisors.

When $n = 2$, any σ-positive braid word is necessarily of the form σ_1^k for some $k \geq 1$. Now, B_2 is isomorphic to $\mathbf{Z}$ via $\sigma_1^k \mapsto k$. The Dehornoy order on B_2 coincides with the standard total order on $\mathbf{Z}$ under this isomorphism.

Theorem 7.15 is an immediate consequence of Theorem 7.4, Lemma 7.14, and the following two lemmas.

Lemma 7.16. *We have* $1 \notin \mathcal{P}$.

Lemma 7.17. *Any element* $\beta \in B_n$, *distinct from 1, is σ-positive or σ-negative. In other words,* $\mathcal{P} \cup \{1\} \cup \mathcal{P}^{-1} = B_n$.

Lemma 7.16 will be proved in Section 7.4.2, and Lemma 7.17 at the end of Section 7.5.2.

7.3.4 Properties

We list a few properties of the Dehornoy order. First observe that σ_i^r and $\sigma_{i+1}^s \sigma_i^r$, with $r \geq 1$ and $s \in \mathbf{Z}$, are σ-positive elements of B_n. Therefore, for the Dehornoy order,

$$\cdots > \sigma_1^3 > \sigma_1^2 > \sigma_1 > \cdots > \sigma_2^3 > \sigma_2^2 > \sigma_2 > \cdots > \sigma_{n-1}^3 > \sigma_{n-1}^2 > \sigma_{n-1}.$$

Proposition 7.18. *(a)* σ_{n-1} *is the smallest σ-positive element of* B_n.
(b) B_n *has no maximal elements and no minimal elements.*

Proof. (a) Suppose that there is $\beta \in \mathcal{P}$ such that $\beta < \sigma_{n-1}$; this is equivalent to $\beta^{-1}\sigma_{n-1} \in \mathcal{P}$. Let w be a σ_i-positive word representing β. The braid $\beta^{-1}\sigma_{n-1}$ is represented by the word $w^{-1}\sigma_{n-1}$. If $i < n-1$, then $w^{-1}\sigma_{n-1}$ is σ_i-negative, which in view of Lemmas 7.2, 7.14, 7.16 contradicts the σ-positivity of $\beta^{-1}\sigma_{n-1}$. Therefore, $i = n-1$ and $w = (\sigma_{n-1})^r$ for some integer $r \geq 1$. Then $\beta^{-1}\sigma_{n-1}$ is 1 if $r = 1$ and belongs to $\mathcal{P}^{-1}$ if $r > 1$. This together with Lemmas 7.2, 7.14, 7.16 contradicts the σ-positivity of $\beta^{-1}\sigma_{n-1}$. So there is no $\beta \in \mathcal{P}$ such that $\beta < \sigma_{n-1}$.

(b) Since $\sigma_1 > 1 > \sigma_1^{-1}$, the left invariance of the order implies that $\beta\sigma_1 > \beta > \beta\sigma_1^{-1}$ for each $\beta \in B_n$. Thus, B_n has no maximal element and no minimal element. $\square$

The standard order on $\mathbf{Z}$ is Archimedian; translated to B_2, it means that given $1 < \alpha < \beta$ with $\alpha, \beta \in B_2$, there is an integer $r \geq 2$ such that $\beta < \alpha^r$. In other words, for any $\alpha \in B_2 \cap \mathcal{P}$ the disjoint intervals

$$\{\beta \in B_2 \mid \alpha^k \leq \beta < \alpha^{k+1}\}_{k \in \mathbf{Z}}$$

cover B_2. This property does not extend to B_n for $n \geq 3$, since $1 < \sigma_2 < \sigma_1$ and $\sigma_2^r < \sigma_1$ for all $r \geq 2$. Nevertheless, using the central element Δ_n^2 of B_n (see Theorem 1.24), we obtain the following result.

Proposition 7.19. *The intervals* $\{\beta \in B_n \mid \Delta_n^{2k} \leq \beta < \Delta_n^{2(k+1)}\}_{k \in \mathbf{Z}}$ *form a partition of* B_n.

Proof. Since Δ_n^2 belongs to B_n^+, we have $\Delta_n^2 > 1$. Hence,

$$\cdots < \Delta_n^{-6} < \Delta_n^{-4} < \Delta_n^{-2} < 1 < \Delta_n^2 < \Delta_n^4 < \Delta_n^6 < \cdots.$$

To prove the proposition it therefore suffices to prove that for any $\beta \in B_n$ there are positive integers r, s such that $\Delta_n^{-2r} \leq \beta$ and $\beta < \Delta_n^{2(s+1)}$. Indeed, suppose that these two inequalities hold. Then there is a largest integer k such that $\Delta_n^{2k} \leq \beta$. By definition of k, we do not have $\Delta_n^{2(k+1)} \leq \beta$. Since the order is total, $\Delta_n^{2(k+1)} > \beta$.

We now prove the existence of a positive integer s such that $\beta < \Delta_n^{2(s+1)}$. Consider a braid word w representing $\beta \in B_n$. Suppose that σ_1 occurs exactly s times in w (we may have $s = 0$). We can write $w = w_0 \sigma_1 w_1 \cdots \sigma_1 w_s$, where $w_0, \ldots, w_s$ are braid words in which σ_1 does not appear (but σ_1^{-1} may appear). In the braid monoid B_n^+ the generator σ_1 is a divisor of Δ_n, hence of Δ_n^2. Therefore, $\Delta_n^2 = \sigma_1 v$ for some $v \in B_n^+ \subset \mathcal{P}$. The braid $\beta^{-1} \Delta_n^{2(s+1)}$ is then represented by the word

$$w_s^{-1} \sigma_1^{-1} w_{s-1}^{-1} \cdots \sigma_1^{-1} w_0^{-1} \Delta_n^{2(s+1)}$$

and, since Δ_n^2 is central, by the words

$$w_s^{-1} \sigma_1^{-1} \Delta_n^2 w_{s-1}^{-1} \cdots \sigma_1^{-1} \Delta_n^2 w_0^{-1} \Delta_n^2 = w_s^{-1} v w_{s-1}^{-1} \cdots v w_0^{-1} \sigma_1 v.$$

In the latter word, σ_1 appears at least once, and σ_1^{-1} nowhere. Therefore, it is σ-positive, which implies that $1 < \beta^{-1} \Delta_n^{2(s+1)}$. Therefore, $\beta < \Delta_n^{2(s+1)}$.

We leave it to the reader to check in a similar fashion that if σ^{-1} occurs exactly r times in w, then $\Delta_n^{-2r} \leq \beta$. $\qquad\square$

Remark 7.20. Laver [Lav96] proved that $\sigma_i \beta > \beta$ for all $\beta \in B_n$ and $i \in \{1, \ldots, n-1\}$, from which it follows that the Dehornoy order has the so-called *subword property* (for other proofs, see [Bur97], [Wie99]). By a theorem of Higman's (see [Hig52]), this in turn implies that the restriction of the Dehornoy order to the braid monoid B_n^+ is a *well-ordering*, that is, any subset of B_n^+ has a minimal element. As a further consequence, the Dehornoy order $\leq$ extends the divisibility order of B_n^+ denoted by $\preceq$ in Chapter 6, that is, $a \preceq b \Rightarrow a \leq b$ for all $a, b \in B_n^+$.

7.3.5 The infinite braid group

Let $\leq_n$ be the Dehornoy order on B_n. Recall the inclusion $\iota : B_n \hookrightarrow B_{n+1}$ of Section 1.1.3. The following lemma is an immediate consequence of the definitions.

Lemma 7.21. *The inclusion $\iota : B_n \hookrightarrow B_{n+1}$ is order-preserving with respect to the Dehornoy order, that is,*

$$\beta \leq_n \beta' \implies \iota(\beta) \leq_{n+1} \iota(\beta')$$

for all $\beta, \beta' \in B_n$.

Let $B_\infty = \bigcup_{n \geq 1} B_n$ be the inductive limit of the groups B_n with respect to the inclusions ι. By definition, any element of B_∞ lies in some B_n. The group structures on the groups B_n naturally extend to a group structure on B_∞. The group B_∞ is called the *infinite braid group*.

Proposition 7.22. *There is a unique left-invariant total order on B_∞ such that the inclusions $B_n \hookrightarrow B_\infty$ are order-preserving. As an ordered set, B_∞ is isomorphic to the ordered set $\mathbf{Q}$ of rational numbers.*

Proof. (a) Let $\beta, \beta' \in B_\infty$. By definition, there is n such that $\beta, \beta' \in B_n$. We set $\beta \leq_\infty \beta'$ if $\beta \leq_n \beta'$. It follows from Lemma 7.21 that this is independent of the choice of n. We thus have a well-defined binary relation on B_∞. It is an easy exercise to check that $\leq_\infty$ is a left-invariant total order on B_∞, and that the inclusions $B_n \hookrightarrow B_\infty$ are order-preserving. It is also easy to check that $\leq_\infty$ is the unique order on B_∞ such that the inclusions $B_n \hookrightarrow B_\infty$ are order-preserving.

(b) It has been known since Cantor that a totally ordered set X is isomorphic to $\mathbf{Q}$ equipped with its standard order if and only if X is countable, has no maximal elements, has no minimal elements, and there is an element between any two elements. Let us check that B_∞ satisfies these conditions.

The group B_∞ is generated by the elements $\sigma_1, \sigma_2, \sigma_3, \ldots$. Since any group with a countable number of generators is countable (see Exercice 7.3.3), B_∞ is a countable set.

If B_∞ had a maximal (resp. minimal) element β, then β would be a maximal (resp. minimal) element in B_n, where n is the index of the braid group to which β belongs. This would contradict Proposition 7.18 (b). Hence, B_∞ has no maximal elements and no minimal elements.

To prove that there is an element between any two elements, it suffices by the left invariance to prove that for any $\beta \in B_\infty$ such that $1 < \beta$, there is α such that $1 < \alpha < \beta$. Let $\beta \in B_n$. We set $\alpha = \iota(\beta)\sigma_n^{-1} \in B_{n+1}$. Since the index of a σ-positive word representing β is $< n$, the braid α is σ-positive. Thus, $1 < \alpha$ in B_{n+1}, hence in B_∞. On the other hand, $\alpha^{-1}\beta = \sigma_n$ in B_∞, which shows that $\alpha^{-1}\beta$ is σ-positive. Therefore, $\alpha < \beta$. $\square$

Exercise 7.3.1. Show that B_∞ is isomorphic to the group generated by the countable set of generators $\{\sigma_1, \sigma_2, \sigma_3, \ldots\}$ subject to the braid relations of Definition 1.1.

Exercise 7.3.2. Let X be a countable totally ordered set without maximal or minimal elements such that there is an element between any two elements. Construct an order-preserving bijection $X \to \mathbf{Q}$, where $\mathbf{Q}$ is equipped with its natural order.

Exercise 7.3.3. Show that the free group on a countable number of generators is countable. Deduce that any group with a countable number of generators is countable.

7.4 Nontriviality of σ-positive braids

The aim of this section is to prove Lemma 7.16. To this end we introduce an action of B_n on a free group F_∞ with a countable basis.

7.4.1 An action of B_n on F_∞

In Section 1.5.1 we defined group automorphisms $\tilde{\sigma}_1, \ldots, \tilde{\sigma}_{n-1}$ of the free group F_n with free generators $x_1, \ldots, x_n$. We recall the formulas:

$$\tilde{\sigma}_i(x_k) = \begin{cases} x_{k+1} & \text{if } k = i, \\ x_k^{-1} x_{k-1} x_k & \text{if } k = i+1, \\ x_k & \text{otherwise.} \end{cases}$$

Their inverses $\tilde{\sigma}_i^{-1}$ are given by

$$\tilde{\sigma}_i^{-1}(x_k) = \begin{cases} x_k x_{k+1} x_k^{-1} & \text{if } k = i, \\ x_{k-1} & \text{if } k = i+1, \\ x_k & \text{otherwise.} \end{cases}$$

These formulas clearly extend to the free group F_∞ on the countable set of generators $\{x_1, x_2, x_3, \ldots\}$. This defines a group homomorphism $B_n \to \text{Aut}(F_\infty)$. We denote the image of $\beta \in B_n$ in $\text{Aut}(F_\infty)$ by $\tilde{\beta}$.

Let τ be the group endomorphism of F_∞ defined by $\tau(x_k) = x_{k+1}$ for all $k \geq 1$. The endomorphism τ is injective. Indeed, let τ_- be the group endomorphism of F_∞ defined by $\tau_-(x_k) = x_{k-1}$ for $k \geq 2$ and $\tau_-(x_1) = 1$; the injectivity of τ follows from the relation $\tau_- \circ \tau = \text{id}$.

Finally, for any group endomorphism φ of F_∞, we define another one, denoted by $T(\varphi)$, by

$$T(\varphi)(x_k) = \begin{cases} x_1 & \text{if } k = 1, \\ \tau(\varphi(x_{k-1})) & \text{if } k > 1. \end{cases}$$

Lemma 7.23. *(a)* $T(\tilde{\sigma}_i) = \tilde{\sigma}_{i+1}$ *for all* $i \in \{1, \ldots, n-2\}$.
(b) If $\varphi \neq \text{id}$, *then* $T(\varphi) \neq \text{id}$.
(c) If φ *is injective, then so is* $T(\varphi)$.

Proof. (a) This follows from the definitions.
(b) If $T(\varphi) = \text{id}$, then

$$\tau(\varphi(x_k)) = T(\varphi)(x_{k+1}) = x_{k+1} = \tau(x_k)$$

for all $k \geq 1$. Since the endomorphism τ of F_∞ is injective, $\varphi(x_k) = x_k$ for all $k \geq 1$. Hence, $\varphi = \text{id}$.

(c) We will show that $T(\varphi)(w) \neq 1$ for any $w \in F_\infty$ such that $w \neq 1$. Let us represent w by a word on the set $\{x_1, x_2, \ldots\} \cup \{x_1^{-1}, x_2^{-1}, \ldots\}$. We may assume that this word is nonempty and *reduced*, i.e., it contains no subword of the form $x_i x_i^{-1}$ or $x_i^{-1} x_i$ for some $i \geq 1$. (In the sequel we shall use the fact that a nonempty reduced word represents a nontrivial element in F_∞; for a proof, see [LS77, Sect. I.1], [Ser77, Sect. I.1].)

If the reduced word representing w does not contain any occurrences of x_1 or x_1^{-1}, then there is $w' \in F_\infty$ with $w' \neq 1$ such that $w = \tau(w')$. By definition of T, we have $T(\varphi)(w) = \tau(\varphi(w'))$. The injectivity of φ and τ then implies that $T(\varphi)(w) \neq 1$.

Suppose that the reduced word representing w contains occurrences of x_1^ε with $\varepsilon = \pm 1$. Then we can write it as

$$\tau(w_0)\, x_1^{k_1}\, \tau(w_1)\, x_1^{k_2} \cdots \tau(w_{r-1})\, x_1^{k_r}\, \tau(w_r)\,,$$

where $k_1, k_2, \ldots, k_r$ are nonzero integers and $w_0, w_1, \ldots, w_{r-1}, w_r$ are words in $x_1^{\pm 1}, x_2^{\pm 1}, x_3^{\pm 1}, \ldots$ such that $\tau(w_1), \ldots, \tau(w_{r-1})$ are nonempty and reduced. By definition of T,

$$T(\varphi)(w) = \tau(\varphi(w_0))\, x_1^{k_1}\, \tau(\varphi(w_1))\, x_1^{k_2} \cdots \tau(\varphi(w_{r-1}))\, x_1^{k_r}\, \tau(\varphi(w_r))\,.$$

Since the words $\tau(w_1), \ldots, \tau(w_{r-1})$ are nonempty and reduced, they represent nontrivial elements of F_∞. By the injectivity of τ and φ, the elements

$$\tau(\varphi(w_1)), \ldots, \tau(\varphi(w_{r-1}))$$

of F_∞ are nontrivial; hence they are represented by nonempty reduced words in $x_2^{\pm 1}, x_3^{\pm 1}, \ldots$. It follows that $T(\varphi)(w) \neq 1$.   □

Now let E be the set of elements of F_∞ that can be represented by a reduced word ending with x_1^{-1}.

Lemma 7.24. *We have*
   (a) $\widetilde{\sigma}_1^{-1}(E) \subset E$;
   (b) $T(\varphi)(E) \subset E$ for any injective endomorphism φ of F_∞.

Proof. (a) Let wx_1^{-1} be a reduced word representing an element of E. Then w is a reduced word not ending with x_1. Assume that

$$\widetilde{\sigma}_1^{-1}(wx_1^{-1}) = \widetilde{\sigma}_1^{-1}(w)\, x_1 x_2^{-1} x_1^{-1}$$

does not belong to E. Then $\widetilde{\sigma}_1^{-1}(w)$ must contain an occurrence of x_1 that cancels the final x_1^{-1}. It follows from the definition of $\widetilde{\sigma}_1^{-1}$ that w contains x_2 or x_1 or x_1^{-1}.

In the first case, write $w = w_1 x_2 w_2$ with x_2 such that $\widetilde{\sigma}_1^{-1}(x_2) = x_1$ cancels the final x_1^{-1} in $\widetilde{\sigma}_1^{-1}(wx_1^{-1})$. Since w is reduced, w_2 (which is also reduced) cannot begin with x_2^{-1}. Now,

$$\widetilde{\sigma}_1^{-1}(wx_1^{-1}) = \widetilde{\sigma}_1^{-1}(w)\,x_1x_2^{-1}x_1^{-1} = \widetilde{\sigma}_1^{-1}(w_1)\,x_1\,\widetilde{\sigma}_1^{-1}(w_2)\,x_1x_2^{-1}x_1^{-1}\,.$$

Since the leftmost x_1 on the right-hand side cancels the final x_1^{-1}, the word between these two letters must represent $1 \in F_\infty$, i.e., we must have

$$\widetilde{\sigma}_1^{-1}(w_2)\,x_1x_2^{-1} = 1$$

in F_∞. Hence,

$$\widetilde{\sigma}_1^{-1}(w_2) = x_2x_1^{-1} = x_1^{-1}x_1x_2x_1^{-1} = \widetilde{\sigma}_1^{-1}(x_2^{-1}x_1)\,.$$

Since $\widetilde{\sigma}_1^{-1}$ is bijective, $w_2 = x_2^{-1}x_1$, which is a reduced word beginning with x_2^{-1}, thus contradicting the hypothesis on w.

If w contains x_1^e with $e = \pm 1$, we similarly write $w = w_1x_1^e w_2$. Then

$$\widetilde{\sigma}_1^{-1}(wx_1^{-1}) = \widetilde{\sigma}_1^{-1}(w)\,x_1x_2^{-1}x_1^{-1} = \widetilde{\sigma}_1^{-1}(w_1)\,x_1x_2^e x_1^{-1}\,\widetilde{\sigma}_1^{-1}(w_2)\,x_1x_2^{-1}x_1^{-1}\,.$$

Since the leftmost x_1 on the right-hand side cancels the final x_1^{-1}, arguing as above, we obtain

$$x_2^e x_1^{-1}\widetilde{\sigma}_1^{-1}(w_2)x_1x_2^{-1} = 1$$

in F_∞. Hence,

$$\widetilde{\sigma}_1^{-1}(w_2) = x_1x_2^{1-e}x_1^{-1} = \widetilde{\sigma}_1^{-1}(x_1^{1-e})\,.$$

By the injectivity of $\widetilde{\sigma}_1^{-1}$, we obtain $w_2 = x_1^{1-e}$, hence $w = w_1x_1$, yielding a contradiction with the assumption on w. Thus, in all cases,

$$\widetilde{\sigma}_1^{-1}(wx_1^{-1}) \in E\,.$$

(b) As before, we represent an element of E by wx_1^{-1}, where w is a reduced word not ending with x_1. Suppose that $T(\varphi)(wx_1^{-1})$ does not belong to E. Since

$$T(\varphi)(wx_1^{-1}) = T(\varphi)(w)\,x_1^{-1}\,,$$

the final x_1^{-1} in $T(\varphi)(w)\,x_1^{-1}$ must be canceled by an x_1 appearing in $T(\varphi)(w)$.

We claim that w contains x_1. If not, then w contains only x_1^{-1} and $x_i^{\pm 1}$ with $i \geq 2$. By definition of $T(\varphi)$, this implies that $T(\varphi)(w)$ contains x_1^{-1} and $x_i^{\pm 1}$ with $i \geq 2$, but no x_1, a contradiction. We can thus write $w = w_1x_1w_2$ with x_1 such that its image $T(\varphi)(x_1) = x_1$ cancels the final x_1^{-1} in $T(\varphi)(wx_1^{-1})$. Therefore

$$T(\varphi)(wx_1^{-1}) = T(\varphi)(w)\,x_1^{-1} = T(\varphi)(w_1)\,x_1\,T(\varphi)(w_2)\,x_1^{-1}\,.$$

By assumption, the leftmost x_1 on the right-hand side cancels the final x_1^{-1}. Therefore, $T(\varphi)(w_2) = 1$. Since $T(\varphi)$ is injective by Lemma 7.23 (c), $w_2 = 1$. Therefore, $w = w_1x_1$ ends with x_1, which contradicts the hypothesis on w. $\square$

7.4.2 Proof of Lemma 7.16

We first prove that a σ_1-negative element $\beta \in B_n$ is nontrivial. It is enough to show that

$$\widetilde{\beta}(x_1) \neq x_1 \,,$$

where $\widetilde{\beta}$ is the image of $\beta \in B_n$ in $\mathrm{Aut}(F_\infty)$.

The σ_1-negative element β has an expansion of the form

$$\beta = \beta_0 \sigma_1^{-1} \beta_1 \sigma_1^{-1} \cdots \beta_{r-1} \sigma_1^{-1} \beta_r \,,$$

where $r \geq 1$ and $\beta_0, \beta_1, \ldots, \beta_{r-1}, \beta_r$ are words in the generators $\sigma_2, \ldots, \sigma_{n-1}$ and their inverses. By Lemma 7.23 (a), for each $k = 0, 1, \ldots, r$, there is an automorphism φ_k of F_∞ such that $\widetilde{\beta}_k = T(\varphi_k)$. Therefore,

$$\begin{aligned}
\widetilde{\beta} &= \widetilde{\beta}_0 \widetilde{\sigma}_1^{-1} \widetilde{\beta}_1 \widetilde{\sigma}_1^{-1} \cdots \widetilde{\beta}_{r-1} \widetilde{\sigma}_1^{-1} \widetilde{\beta}_r \\
&= T(\varphi_0) \, \widetilde{\sigma}_1^{-1} \, T(\varphi_1) \, \widetilde{\sigma}_1^{-1} \cdots T(\varphi_{r-1}) \, \widetilde{\sigma}_1^{-1} \, T(\varphi_r) \,.
\end{aligned}$$

Let us apply both sides of this equality to the generator x_1 of F_∞. Since

$$T(\varphi_r)(x_1) = x_1 \quad \text{and} \quad \widetilde{\sigma}_1^{-1}(x_1) = x_1 x_2 x_1^{-1} \,,$$

we have

$$\widetilde{\beta}(x_1) = \Big(T(\varphi_0) \, \widetilde{\sigma}_1^{-1} \, T(\varphi_1) \, \widetilde{\sigma}_1^{-1} \cdots T(\varphi_{r-1}) \Big) (x_1 x_2 x_1^{-1}) \,.$$

Since $x_1 x_2 x_1^{-1}$ belongs to the set E of reduced words in F_∞ ending with x_1^{-1}, Lemma 7.24 implies that $\widetilde{\beta}(x_1) \in E$ as well. Therefore, $\widetilde{\beta}(x_1) \neq x_1$.

To finish the proof, we use the group homomorphism $\mathrm{sh} : B_{n-1} \to B_n$ defined by $\mathrm{sh}(\sigma_i) = \sigma_{i+1}$ for all $i = 1, \ldots, n - 2$. In geometric language, the map sh shifts a geometric braid b to the right by adding on its left a vertical string completely unlinked with b. For this reason, we call sh the *shift* homomorphism. This homomorphism is injective: one can prove this using an argument similar to the one used in the proof of Corollary 1.14; one can also observe that sh is conjugate to the natural inclusion $\iota : B_{n-1} \to B_n$ (the conjugating element is $\sigma_1 \sigma_2 \cdots \sigma_{n-1}$; see Exercise 7.4.1).

We now prove that all elements of $\mathcal{P}$ are nontrivial. Let β be a σ_i-positive element of B_n with $i \geq 1$. By definition of a σ_i-positive element and of the shift sh, there is a σ_1-positive element $\alpha \in B_n$ such that $\beta = \mathrm{sh}^{i-1}(\alpha)$. Then α^{-1} is σ_1-negative, and by the argument above, $\alpha \neq 1$. Since sh is injective, $\beta \neq 1$. In conclusion, we have proved that $1 \notin \mathcal{P}$. $\qquad\square$

Exercise 7.4.1. Prove that for all $\beta \in B_{n-1}$,

$$\mathrm{sh}(\beta) = (\sigma_1 \sigma_2 \cdots \sigma_{n-1}) \, \iota(\beta) \, (\sigma_1 \sigma_2 \cdots \sigma_{n-1})^{-1} \,.$$

7.5 Handle reduction

The aim of this section is to prove Lemma 7.17, which states that any braid is σ-positive, σ-negative, or trivial. The proof requires some preliminary notions and auxiliary results.

Fix an integer $n \geq 1$. As in Section 7.3.1, by braid words we mean words in the letters

$$\sigma_1, \ldots, \sigma_{n-1}, \sigma_1^{-1}, \ldots, \sigma_{n-1}^{-1}.$$

We say that a braid word w *contains* a braid word v if v is a subword of w. A braid word w' is a *prefix* of w if there is a braid word w'' such that $w = w'w''$. Similarly, a braid word w'' is a *suffix* of w if there is a braid word w' such that $w = w'w''$.

7.5.1 Handles

Definition 7.25. *A σ_i-handle is a braid word of the form $\sigma_i u \sigma_i^{-1}$ or of the form $\sigma_i^{-1} u \sigma_i$, where $i \in \{1, \ldots, n-1\}$ and u is an empty word or a braid word of index $> i$. The sign of a σ_i-handle v is $+1$ if $v = \sigma_i u \sigma_i^{-1}$ and -1 if $v = \sigma_i^{-1} u \sigma_i$.*

By a *handle* we shall mean a σ_i-handle with $i \in \{1, \ldots, n-1\}$. Figure 7.1 represents two σ_i-handles, the left one of sign $+1$ and the right one of sign -1 (the empty boxes represent arbitrary braids on $n - i$ strings).

It is useful to note that a σ_{n-1}-handle is necessarily of the form $\sigma_{n-1}\sigma_{n-1}^{-1}$ or $\sigma_{n-1}^{-1}\sigma_{n-1}$.

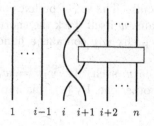 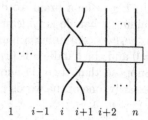

Fig. 7.1. σ_i-handles

The following lemma is an immediate consequence of the definitions.

Lemma 7.26. *A braid word of index $i \in \{1, \ldots, n-1\}$ that does not contain σ_i-handles is σ_i-positive or σ_i-negative.*

A concrete way to visualize the σ_i-handles contained in a braid word w is to delete from w all occurrences of $\sigma_j^{\pm 1}$ with $j > i$, thus obtaining a possibly shorter word $w[i]$. The braid word w contains a σ_i-handle each time $w[i]$ contains a subword of the form $\sigma_i \sigma_i^{-1}$ or $\sigma_i^{-1} \sigma_i$.

Consider, for instance, the braid word

$$w = \sigma_1\sigma_2\sigma_3\sigma_4\sigma_3^{-1}\sigma_1^{-1}\sigma_3^{-1}\sigma_2^{-1}\sigma_3\sigma_2\sigma_1\sigma_3\sigma_2^{-1}\sigma_1^{-1}.$$

Then

$$w[1] = \sigma_1\sigma_1^{-1}\sigma_1\sigma_1^{-1}, \quad w[2] = \sigma_1\sigma_2\sigma_1^{-1}\sigma_2^{-1}\sigma_2\sigma_1\sigma_2^{-1}\sigma_1^{-1},$$
$$w[3] = \sigma_1\sigma_2\sigma_3\sigma_3^{-1}\sigma_1^{-1}\sigma_3^{-1}\sigma_2^{-1}\sigma_3\sigma_2\sigma_1\sigma_3\sigma_2^{-1}\sigma_1^{-1}, \quad w[4] = w.$$

We see that w has three σ_1-handles, namely

$$\sigma_1\sigma_2\sigma_3\sigma_4\sigma_3^{-1}\sigma_1^{-1}, \quad \sigma_1^{-1}\sigma_3^{-1}\sigma_2^{-1}\sigma_3\sigma_2\sigma_1, \quad \sigma_1\sigma_3\sigma_2^{-1}\sigma_1^{-1},$$

one σ_2-handle $\sigma_2^{-1}\sigma_3\sigma_2$, one σ_3-handle $\sigma_3\sigma_4\sigma_3^{-1}$, and no σ_4-handles.

Definition 7.27. *A handle v contained in a braid word w is said to be prime if $w = w_1vw_2$, where w_1v is the shortest prefix of w containing a handle.*

Lemma 7.28. *(a) A prime handle contains no other handles.*

(b) Any braid word containing at least one handle contains a unique prime handle.

Proof. (a) Let $w = w_1vw_2$ be a braid word in which v is a prime handle. Suppose that $v = w'uw''$, where u is a handle. Then $w_1w'u$ is a prefix of w containing a handle. Since $w_1v = w_1w'uw''$ is the shortest prefix of w containing a handle, we must have $w_1w'u = w_1w'uw''$. This shows that w'' is empty. Since $v = w'u$ and u are handles, the first letter of v is the inverse of the last letter of u, which is the same as the first letter of u. Since v is a handle, $u = v$.

(b) Let w be a braid word containing a handle. The set of prefixes of w containing a handle is nonempty, since it contains w itself. Pick the shortest prefix w_1vw_2 containing a handle v. Since the prefix w_1v contains a handle, $w_2 = \emptyset$ and the handle v is prime.

Suppose that there is another prime handle v' such that $w_1v = w_1'v'$. Necessarily, one of the words v, v' contains the other one. By (a), this implies that $v' = v$. $\square$

In view of Lemma 7.28, we can speak of *the prime handle* of a braid word. We can paraphrase Definition 7.27 by saying that the prime handle of a braid word w is the first handle of w that appears entirely when one reads w from left to right. For instance, the prime handle of

$$w = \sigma_2\sigma_1\sigma_3^{-1}\sigma_4\sigma_3\sigma_1^{-1}\sigma_2\sigma_1$$

is $\sigma_3^{-1}\sigma_4\sigma_3$ (not $\sigma_1\sigma_3^{-1}\sigma_4\sigma_3\sigma_1^{-1}$).

7.5.2 Prime handle reduction

Our aim is to obtain σ-positive or σ-negative braid words by starting from arbitrary braid words and gradually getting rid of prime handles. We shall achieve this goal by an iterative process, which is repeated until no handles are left.

Definition 7.29. *Let v be a σ_i-handle of the form $v = \sigma_i^e u \sigma_i^{-e}$, where $i \in \{1, \ldots, n-1\}$, $e = \pm 1$, and u is the empty word or a word of index $> i$. The reduction of v is the braid word obtained from u by replacing each occurrence of $\sigma_{i+1}^{\pm 1}$ by $\sigma_{i+1}^{-e} \sigma_i^{\pm 1} \sigma_{i+1}^e$.*

Remarks 7.30. (i) If $v = \sigma_i^e u \sigma_i^{-e}$ is a σ_i-handle and u is a braid word of index $> i + 1$, then the reduction of v is u. In particular, the reduction of $\sigma_i^e \sigma_i^{-e}$ is the empty word.

(ii) The index of the reduction of a handle v is greater than or equal to the index of v.

Figure 7.2 shows the reduction of a σ_1-handle of sign $+1$ with no occurrences of $\sigma_2^{\pm 1}$, whereas Figure 7.3 shows the reduction of a σ_1-handle of sign $+1$ with two occurrences of σ_2 and no occurrences of σ_2^{-1}. The boxes u_0, u_1, u_2 in these figures represent braid words that are empty or have index ≥ 3.

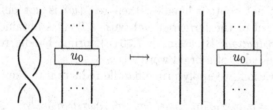

Fig. 7.2. Reduction of a σ_1-handle without occurrences of $\sigma_2^{\pm 1}$

The braids in Figure 7.2 are isotopic. The same holds for the braids in Figure 7.3. This is a special case of the following simple but fundamental property of reduction.

Lemma 7.31. *Any handle represents the same element of B_n as its reduction.*

Proof. This is a consequence of the relations

$$\sigma_i^e \sigma_j^{\pm 1} \sigma_i^{-e} = \begin{cases} \sigma_j^{\pm 1} & \text{if } j \geq i + 2, \\ \sigma_{i+1}^{-e} \sigma_i^{\pm 1} \sigma_{i+1}^e & \text{if } j = i + 1, \end{cases}$$

which follow from the braid relations of Definition 1.1 (here $e = \pm 1$). ☐

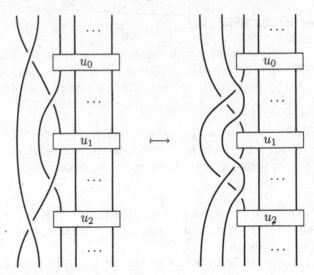

Fig. 7.3. Reduction of a σ_1-handle with occurrences of σ_2

Let w be a braid word containing at least one handle. We denote by $\mathrm{red}(w)$ the braid obtained from w by replacing the prime handle of w by its reduction. We define $\mathrm{red}^k(w)$ with $k \geq 0$ inductively as follows: $\mathrm{red}^0(w) = w$ and for $k \geq 1$, if $\mathrm{red}^{k-1}(w)$ contains a handle, then $\mathrm{red}^k(w) = \mathrm{red}(\mathrm{red}^{k-1}(w))$. If $\mathrm{red}^{k-1}(w)$ does not contain handles, then $\mathrm{red}^k(w)$ is not defined. We say that a braid word of the form $\mathrm{red}^k(w)$ with $k \geq 0$ is obtained from w by *prime handle reduction*. By Remark 7.30 (ii), prime handle reduction does not decrease the index of a braid word.

As an illustration, we apply prime handle reduction to the braid word

$$w = \sigma_1\sigma_2\sigma_3\sigma_4\sigma_3^{-1}\sigma_2\sigma_1^{-1}\sigma_3^{-1}\sigma_2^{-1}\sigma_3\sigma_2\sigma_1\sigma_3\sigma_2^{-1}\sigma_1^{-1}. \tag{7.2}$$

Indicating each prime handle with braces, we obtain

$$w = \sigma_1\sigma_2\underbrace{\sigma_3\sigma_4\sigma_3^{-1}}\sigma_2\sigma_1^{-1}\sigma_3^{-1}\sigma_2^{-1}\sigma_3\sigma_2\sigma_1\sigma_3\sigma_2^{-1}\sigma_1^{-1},$$

$$\mathrm{red}(w) = \sigma_1\sigma_2\underbrace{\sigma_4^{-1}\sigma_3\sigma_4\sigma_2\sigma_1^{-1}}\sigma_3^{-1}\sigma_2^{-1}\sigma_3\sigma_2\sigma_1\sigma_3\sigma_2^{-1}\sigma_1^{-1},$$

$$\mathrm{red}^2(w) = \sigma_2^{-1}\sigma_1\underbrace{\sigma_2\sigma_4^{-1}\sigma_3\sigma_4\sigma_2^{-1}}\sigma_1\sigma_2\sigma_3^{-1}\sigma_2^{-1}\sigma_3\sigma_2\sigma_1\sigma_3\sigma_2^{-1}\sigma_1^{-1},$$

$$\mathrm{red}^3(w) = \sigma_2^{-1}\sigma_1\sigma_4^{-1}\sigma_3^{-1}\sigma_2\sigma_3\sigma_4\sigma_1\underbrace{\sigma_2\sigma_3^{-1}\sigma_2^{-1}}\sigma_3\sigma_2\sigma_1\sigma_3\sigma_2^{-1}\sigma_1^{-1},$$

$$\mathrm{red}^4(w) = \sigma_2^{-1}\sigma_1\sigma_4^{-1}\sigma_3^{-1}\sigma_2\sigma_3\sigma_4\sigma_1\sigma_3^{-1}\underbrace{\sigma_2^{-1}\sigma_3\sigma_3\sigma_2}\sigma_1\sigma_3\sigma_2^{-1}\sigma_1^{-1},$$

$$\mathrm{red}^5(w) = \sigma_2^{-1}\sigma_1\sigma_4^{-1}\sigma_3^{-1}\sigma_2\sigma_3\sigma_4\sigma_1\underbrace{\sigma_3^{-1}\sigma_3}\sigma_2\sigma_3^{-1}\sigma_3\sigma_2\sigma_3^{-1}\sigma_1\sigma_3\sigma_2^{-1}\sigma_1^{-1},$$

$$\mathrm{red}^6(w) = \sigma_2^{-1}\sigma_1\sigma_4^{-1}\sigma_3^{-1}\sigma_2\sigma_3\sigma_4\sigma_1\sigma_2\underbrace{\sigma_3^{-1}\sigma_3}\sigma_2\sigma_3^{-1}\sigma_1\sigma_3\sigma_2^{-1}\sigma_1^{-1},$$

$$\mathrm{red}^7(w) = \sigma_2^{-1}\sigma_1\sigma_4^{-1}\sigma_3^{-1}\sigma_2\sigma_3\sigma_4\sigma_1\sigma_2\sigma_2\sigma_3^{-1}\underbrace{\sigma_1\sigma_3\sigma_2^{-1}\sigma_1^{-1}},$$

$$\mathrm{red}^8(w) = \sigma_2^{-1}\sigma_1\sigma_4^{-1}\sigma_3^{-1}\sigma_2\sigma_3\sigma_4\sigma_1\sigma_2\sigma_2\,\underbrace{\sigma_3^{-1}\sigma_3}\,\sigma_2^{-1}\sigma_1^{-1}\sigma_2,$$

$$\mathrm{red}^9(w) = \sigma_2^{-1}\sigma_1\sigma_4^{-1}\sigma_3^{-1}\sigma_2\sigma_3\sigma_4\sigma_1\sigma_2\,\underbrace{\sigma_2\sigma_2^{-1}}\,\sigma_1^{-1}\sigma_2,$$

$$\mathrm{red}^{10}(w) = \sigma_2^{-1}\sigma_1\sigma_4^{-1}\sigma_3^{-1}\sigma_2\sigma_3\sigma_4\,\underbrace{\sigma_1\sigma_2\sigma_1^{-1}}\,\sigma_2,$$

$$\mathrm{red}^{11}(w) = \sigma_2^{-1}\sigma_1\sigma_4^{-1}\sigma_3^{-1}\,\underbrace{\sigma_2\sigma_3\sigma_4\sigma_2^{-1}}\,\sigma_1\sigma_2\sigma_2,$$

$$\mathrm{red}^{12}(w) = \sigma_2^{-1}\sigma_1\sigma_4^{-1}\sigma_3^{-1}\sigma_3^{-1}\sigma_2\sigma_3\sigma_4\sigma_1\sigma_2\sigma_2.$$

The word $\mathrm{red}^{12}(w)$ has no handles; it is σ_1-positive.

Prime handle reduction has to stop, as stated in the following lemma.

Lemma 7.32. *For each braid word w, there is an integer $k \geq 0$ such that $\mathrm{red}^k(w)$ contains no handles.*

We are now able to prove Lemma 7.17, which is the last unproved ingredient in the proof of Theorem 7.15. Let w be a braid word representing $\beta \in B_n$. By Lemma 7.32, $\mathrm{red}^k(w)$ contains no handles for some k. Therefore, by Lemma 7.26, the braid word $\mathrm{red}^k(w)$ is empty, σ-positive, or σ-negative. But by Lemma 7.31, the word $\mathrm{red}^k(w)$ represents β. Hence, β is trivial, σ-positive, or σ-negative. This proves Lemma 7.17.

We are thus left with proving Lemma 7.32. The proof relies on four auxiliary results, namely Lemmas 7.35, 7.36, 7.37, and 7.39 below, and will be given in Section 7.5.8.

Remark 7.33. Lemma 7.32 provides an algorithm that turns any braid word w into a braid word that is empty, σ-positive, or σ-negative, and represents the same element of B_n as w. This algorithm gives an alternative solution to the word problem in B_n.

7.5.3 The Cayley graph

The four auxiliary results mentioned above make use of certain finite subgraphs of the Cayley graph of B_n.

Definition 7.34. *The Cayley graph of B_n is the graph Γ whose vertices are the elements of B_n and whose edges are defined as follows: for each $\beta \in B_n$ and $i = 1, \ldots, n-1$, there is a unique edge between the vertices β and $\beta\sigma_i$.*

An *oriented edge* in Γ is an edge for which one of its endpoints is distinguished and called *initial*, whereas the other one is called *terminal*. If we have an oriented edge a, then we denote the same edge with the reverse orientation by $\bar{a}$, i.e., the initial (resp. terminal) vertex of $\bar{a}$ is the terminal (resp. initial) vertex of a.

We label all oriented edges in Γ as follows. If the initial vertex of an oriented edge a is β and its terminal vertex is $\beta\sigma_i$ for some $i \in \{1, \ldots, n-1\}$, then its label is defined by $L(a) = \sigma_i$. If the terminal vertex of a is β and its initial vertex is $\beta\sigma_i$, then its label is defined by $L(a) = \sigma_i^{-1}$. In both cases, $L(a)$ is a one-letter braid word representing $\beta_0^{-1}\beta_1 \in B_n$, where β_0 is the initial vertex of a and β_1 its terminal vertex.

A *path* in the Cayley graph Γ is a finite sequence $a_1, a_2, \ldots, a_k$ of oriented edges of Γ such that for all $i = 1, \ldots, k-1$, the terminal vertex of a_i is the initial vertex of a_{i+1}. The *initial vertex* of the path is the initial vertex of a_1, and the *terminal vertex* of the path is the terminal vertex of a_k. The *reverse* of the path $a = (a_1, a_2, \ldots, a_k)$ is the path $\bar{a} = (\bar{a}_k, \ldots, \bar{a}_2, \bar{a}_1)$. By definition, an empty path in Γ is a vertex of Γ that is viewed as both the initial and the terminal vertex of the path. An empty path has no edges.

To a path $a = (a_1, a_2, \ldots, a_k)$ in Γ we associate its *label*, which is the braid word of length k

$$L(a) = L(a_1)L(a_2)\cdots L(a_k)$$

obtained as the concatenation of the labels of the oriented edges $a_1, a_2, \ldots, a_k$. If a is an empty path, then $L(a)$ is the empty word. It is clear that $L(a)$ represents the braid $\beta_0^{-1}\beta_1 \in B_n$, where β_0 is the initial vertex of a and β_1 its terminal vertex. Conversely, given a vertex β_0 in Γ and a braid word w, there is a unique path a in Γ with initial vertex β_0 such that $w = L(a)$. We thus have a bijection between the set of paths in Γ with initial vertex β_0 and terminal vertex β_1, and the set of braid words representing $\beta_0^{-1}\beta_1$. Observe that $L(\bar{a}) = (L(a))^{-1}$ for any path a.

7.5.4 The graph Γ_r

Consider the element Δ_n of the braid monoid B_n^+ introduced in Section 6.5.1. Recall from Lemma 6.11 (iv) that any element of B_n^+ is a left divisor of $\Delta_n^r = (\Delta_n)^r$ for some $r \geq 0$. For any integer $r \geq 0$, we define Γ_r to be the full subgraph of Γ whose vertices are the left divisors of Δ_n^r in the monoid B_n^+. Since the length of a left divisor of Δ_n^r cannot exceed the length of Δ_n^r and the set of elements of B_n^+ of a given length is finite, the set of vertices of Γ_r is finite. The number of edges ending in a given vertex being $\leq n-1$, the graph Γ_r is finite.

A *path* in Γ_r is a path in Γ whose vertices and edges belong to Γ_r.

Lemma 7.35. *Let N_r be the number of edges of Γ_r. For $i \in \{1, \ldots, n-1\}$, any σ_i-positive (resp. σ_i-negative) braid word that is the label of a path in Γ_r contains the letter σ_i (resp. σ_i^{-1}) at most N_r times.*

Proof. We give the proof for the σ_i-positive case. The σ_i-negative case can be treated in a similar way.

Let $a = (a_1, a_2, \ldots, a_k)$ be a path in Γ_r whose label is a σ_i-positive word w. The word w has no occurrences of σ_i^{-1}, and each occurrence of σ_i in w is the label of some oriented edge in the path a. To prove the lemma, it is enough to check that the edges with label σ_i in this path are all different. Suppose that it is not the case and that $a_s = a_t$ for some s and t such that $1 \leq s < t \leq k$ and $L(a_s) = L(a_t) = \sigma_i$. Consider the nonempty subpath

$$a' = (a_s, \ldots, a_{t-1}).$$

This subpath lies in Γ_r and its label is a subword u of w. Since u is a subword of a σ_i-positive word and contains at least one occurrence of σ_i, namely $L(a_s)$, it is σ_i-positive. On the other hand, the terminal vertex of a' is the initial vertex of the oriented edge $a_t = a_s$. In other words, a' is a loop. Therefore, the word u represents the trivial braid. But by Lemma 7.16, a σ-positive word cannot represent the trivial braid. Therefore, the edges with label σ_i in the path a are all different. $\square$

Lemma 7.36. *For any braid word w, there is an integer $r \geq 0$ and a path in Γ_r whose label is w.*

Proof. Let $a = (a_1, a_2, \ldots, a_k)$ be a path in Γ with label w. We denote the initial vertex of a_1 by β_0 and the terminal vertex of a_i by β_i $(1 \leq i \leq k)$. By definition of a path, the initial vertex of a_i is β_{i-1} for $i = 1, \ldots, k$. By Section 6.5.4, there is $s \geq 0$ such that $\Delta_n^s \beta_i \in B_n^+$ for all $i = 0, 1, \ldots, k$. Consider the "translated" path

$$\Delta_n^s(a) = (\Delta_n^s a_1, \ldots, \Delta_n^s a_k),$$

where $\Delta_n^s a_i$ is the oriented edge of Γ with initial vertex $\Delta_n^s \beta_{i-1}$ and terminal vertex $\Delta_n^s \beta_i$ for $i = 1, \ldots, k$. The path $\Delta_n^s(a)$ also has w as its label. The vertices of this path belong to B_n^+. By Lemma 6.11 (iv), there is an integer $r \geq 0$ such that

$$\Delta_n^s \beta_0, \Delta_n^s \beta_1, \ldots, \Delta_n^s \beta_k$$

are left divisors of Δ_n^r. It follows that the translated path $\Delta_n^s(a)$ is in Γ_r. $\square$

7.5.5 Performing prime handle reduction in Γ_r

Let a be a path in Γ with initial vertex β_0 and label w. By Section 7.5.3 there is a unique path in Γ with initial vertex β_0 and label $\text{red}(w)$. We denote this path by $\text{red}(a)$. By Lemma 7.31, $\text{red}(w)$ represents the same element of B_n as w. Therefore, the terminal vertices of a and $\text{red}(a)$ coincide.

Lemma 7.37. *If a is a path in Γ_r with $r \geq 0$, then so is $\text{red}(a)$.*

The proof of this lemma given in Section 7.5.6 is based on the following decomposition of prime handle reductions into elementary steps.

Let w, w' be braid words. We say that w' is obtained from w by an *elementary reduction* if w' is obtained from w by replacing some subword u of w by the word u', where $u \mapsto u'$ is one of the following substitutions:

$$\sigma_i^e \sigma_i^{-e} \mapsto \emptyset \quad \text{with } e = \pm 1, \tag{7.3}$$

$$\sigma_i^e \sigma_j^k \mapsto \sigma_j^k \sigma_i^e \quad \text{with } e = \pm 1, \ k = \pm 1, \text{ and } |i - j| \geq 2, \tag{7.4}$$

$$\sigma_i \sigma_{i+1}^{-1} \mapsto \sigma_{i+1}^{-1} \sigma_i^{-1} \sigma_{i+1} \sigma_i, \tag{7.5}$$

$$\sigma_i^{-1} \sigma_{i+1} \mapsto \sigma_{i+1} \sigma_i \sigma_{i+1}^{-1} \sigma_i^{-1}, \tag{7.6}$$

$$\sigma_{i+1}^{-1} \sigma_i \mapsto \sigma_i \sigma_{i+1} \sigma_i^{-1} \sigma_{i+1}^{-1}, \tag{7.7}$$

$$\sigma_{i+1} \sigma_i^{-1} \mapsto \sigma_i^{-1} \sigma_{i+1}^{-1} \sigma_i \sigma_{i+1}. \tag{7.8}$$

Lemma 7.38. *For any braid word w, one can pass from w to $\mathrm{red}(w)$ by a finite sequence of elementary reductions.*

Proof. It is enough to check that one can pass from a prime handle v to its reduction v' by a finite sequence of elementary reductions. Now, a prime σ_i-handle v is necessarily of one of the following two forms:

(i) If v does not contain any occurrences of $\sigma_{i+1}^{\pm 1}$, then

$$v = \sigma_i^e u_0 \sigma_i^{-e}, \tag{7.9}$$

where $e = \pm 1$ and u_0 is empty or of index $> i + 1$. In this case, $v' = u_0$. We use (7.4) to transform v into $\sigma_i^e \sigma_i^{-e} u_0$. We then use (7.3) to transform the latter into u_0.

(ii) If v contains an occurrence of σ_{i+1}^k with $k = \pm 1$, then it contain no occurrences of σ_{i+1}^{-k}; otherwise, it would contain σ_{i+1}-handles, which contradicts Lemma 7.28 (a). It follows that v is of the form

$$v = \sigma_i^e u_0 \sigma_{i+1}^k u_1 \sigma_{i+1}^k u_2 \cdots u_{r-1} \sigma_{i+1}^k u_r \sigma_i^{-e}, \tag{7.10}$$

where $e = \pm 1$, $k = \pm 1$, $r \geq 1$, and $u_0, \ldots, u_r$ are empty or of index $> i + 1$. In this case, the reduction of v is

$$v' = u_0 \sigma_{i+1}^{-e} \sigma_i^k \sigma_{i+1}^e u_1 \sigma_{i+1}^{-e} \sigma_i^k \sigma_{i+1}^e u_2 \cdots u_{r-1} \sigma_{i+1}^{-e} \sigma_i^k \sigma_{i+1}^e u_r. \tag{7.11}$$

We now distinguish four cases depending on the values of e and k.

(a) If $e = 1$ and $k = -1$, then

$$v = \underbrace{\sigma_i u_0}\, \sigma_{i+1}^{-1} u_1 \sigma_{i+1}^{-1} u_2 \cdots u_{r-1} \sigma_{i+1}^{-1} u_r \sigma_i^{-1}.$$

Since u_0 is of index $> i+1$, we can apply substitutions (7.4) to the underbraced subword of v; we thus transform v into

$$u_0 \underbrace{\sigma_i \sigma_{i+1}^{-1}}\, u_1 \sigma_{i+1}^{-1} u_2 \cdots u_{r-1} \sigma_{i+1}^{-1} u_r \sigma_i^{-1}.$$

Applying substitution (7.5) to the underbraced subword, we obtain

$$u_0 \left(\sigma_{i+1}^{-1}\sigma_i^{-1}\sigma_{i+1}\right)[\sigma_i u_1 \sigma_{i+1}^{-1} u_2 \cdots u_{r-1}\sigma_{i+1}^{-1} u_r \sigma_i^{-1}].$$

The subword in square brackets is of the same form as v, but shorter. Iterating substitutions (7.4), (7.5), we obtain the word

$$u_0 \left(\sigma_{i+1}^{-1}\sigma_i^{-1}\sigma_{i+1}\right) u_1 \left(\sigma_{i+1}^{-1}\sigma_i^{-1}\sigma_{i+1}\right) u_2 \cdots u_{r-1} \left(\sigma_{i+1}^{-1}\sigma_i^{-1}\sigma_{i+1}\right)[\sigma_i u_r \sigma_i^{-1}].$$

We finally apply substitutions (7.3), (7.4) to the subword $\sigma_i u_r \sigma_i^{-1}$, and obtain the word v' as in (7.11).

(b) If $e = -1$ and $k = 1$, then we proceed as in the previous case using (7.6) instead of (7.5).

(c) If $e = -1$ and $k = -1$, then

$$v = \sigma_i^{-1} u_0 \sigma_{i+1}^{-1} u_1 \sigma_{i+1}^{-1} u_2 \cdots u_{r-1}\sigma_{i+1}^{-1} u_r \sigma_i.$$

Here we start from the right: we use (7.4) to transform v into

$$\sigma_i^{-1} u_0 \sigma_{i+1}^{-1} u_1 \sigma_{i+1}^{-1} u_2 \cdots u_{r-1} \underbrace{\sigma_{i+1}^{-1}\sigma_i}\, u_r.$$

Next we use (7.7) to transform the latter into

$$[\sigma_i^{-1} u_0 \sigma_{i+1}^{-1} u_1 \sigma_{i+1}^{-1} u_2 \cdots u_{r-1}\sigma_i](\sigma_{i+1}\sigma_i^{-1}\sigma_{i+1}^{-1}) u_r.$$

Now the word in square brackets is of the same form as v, but shorter. We then iterate substitutions (7.4), (7.7), and obtain the word

$$[\sigma_i^{-1} u_0 \sigma_i](\sigma_{i+1}\sigma_i^{-1}\sigma_{i+1}^{-1}) u_1 \left(\sigma_{i+1}\sigma_i^{-1}\sigma_{i+1}^{-1}\right) u_2 \cdots u_{r-1} \left(\sigma_{i+1}\sigma_i^{-1}\sigma_{i+1}^{-1}\right) u_r.$$

We finally apply substitutions (7.3), (7.4) to the subword $\sigma_i^{-1} u_0 \sigma_i$, and obtain v' as in (7.11).

(d) If $e = 1$ and $k = 1$, then we proceed as in the previous case using (7.8) instead of (7.7). $\square$

7.5.6 Proof of Lemma 7.37

Let w' be obtained from the label w of a by an elementary reduction and let a' be the path in Γ with label w' and the same initial vertex as a. By Lemma 7.38, it suffices to prove that a' lies in Γ_r. We consider successively each of the substitutions (7.3)–(7.8).

(a) Substitution (7.3): If w' is obtained from w by (7.3), then a' is obtained from a by removing a loop. Since a lies in Γ_r, so does a'.

(b) Substitution (7.4): We may assume that the word $\sigma_i^e \sigma_j^k$ ($e = \pm 1$, $k = \pm 1$) is the label of a path in Γ_r with initial vertex β_0 and terminal vertex β_1. By assumption, β_0, $\beta_0 \sigma_i^e$, and $\beta_1 = \beta_0 \sigma_i^e \sigma_j^k$ are vertices of Γ_r.

Since (7.4) substitutes $\sigma_j^k \sigma_i^e$ for $\sigma_i^e \sigma_j^k$, we have to check that $\beta_0 \sigma_j^k$ is a vertex of Γ_r, i.e., that it is a left divisor of Δ_n^r in B_n^+.

If $e = k = 1$, then we have to check that $\beta_0 \sigma_j$ is a left divisor of Δ_n^r. But $\beta_0 \sigma_j$ is a left divisor of $\beta_0 \sigma_j \sigma_i = \beta_0 \sigma_i \sigma_j = \beta_1$, which by assumption is a left divisor of Δ_n^r.

Let $e = 1$ and $k = -1$. By definition of β_0 and β_1, we have $\beta_1 \sigma_j = \beta_0 \sigma_i$. In particular, σ_i and σ_j are right divisors of $\beta_0 \sigma_i$. We proved in Section 6.5 that $\sigma_i \sigma_j = \sigma_j \sigma_i$ is the left lcm of σ_i and σ_j. Hence there is $\beta \in B_n^+$ such that $\beta_0 \sigma_i = \beta \sigma_j \sigma_i$. Consequently, $\beta_0 = \beta \sigma_j$. The vertex $\beta_0 \sigma_j^{-1} = \beta$ lies in B_n^+ and is a left divisor of Δ_n^r, as desired.

The case $e = -1$ reduces to the previous ones by reversing the paths.

(c) Substitution (7.5): Assume that the braid word $\sigma_i \sigma_{i+1}^{-1}$ is the label of a path in Γ_r with initial vertex β_0 and terminal vertex β_1. This means that the braids β_0, $\beta_0 \sigma_i$, and $\beta_1 = \beta_0 \sigma_i \sigma_{i+1}^{-1}$ belong to B_n^+ and are left divisors of Δ_n^r. We have to show that the braids

$$\beta_0 \sigma_{i+1}^{-1}, \quad \beta_0 \sigma_{i+1}^{-1} \sigma_i^{-1}, \quad \beta_0 \sigma_{i+1}^{-1} \sigma_i^{-1} \sigma_{i+1} \tag{7.12}$$

also belong to B_n^+ and are left divisors of Δ_n^r.

The element $\beta_0 \sigma_i = \beta_1 \sigma_{i+1}$ of B_n^+ is a left multiple of σ_i and σ_{i+1}. It follows that $\beta_0 \sigma_i$ is a left multiple of the left lcm of σ_i and σ_{i+1}, which by Section 6.5 is $\sigma_i \sigma_{i+1} \sigma_i$. Therefore, there is $\beta \in B_n^+$ such that $\beta_0 \sigma_i = \beta \sigma_i \sigma_{i+1} \sigma_i$. We thus have $\beta_0 = \beta \sigma_i \sigma_{i+1}$. The braids (7.12) can be expressed in terms of β as follows:

$$\beta_0 \sigma_{i+1}^{-1} = \beta \sigma_i \sigma_{i+1} \sigma_{i+1}^{-1} = \beta \sigma_i ,$$
$$\beta_0 \sigma_{i+1}^{-1} \sigma_i^{-1} = \beta \sigma_i \sigma_i^{-1} = \beta ,$$
$$\beta_0 \sigma_{i+1}^{-1} \sigma_i^{-1} \sigma_{i+1} = \beta \sigma_{i+1} .$$

Clearly these braids belong to B_n^+. Since they are left divisors of

$$\beta \sigma_i \sigma_{i+1} \sigma_i = \beta \sigma_{i+1} \sigma_i \sigma_{i+1} = \beta_0 \sigma_i ,$$

they are left divisors of Δ_n^r.

(d) Substitution (7.6): Assume that $\sigma_i^{-1} \sigma_{i+1}$ is the label of a path in Γ_r with initial vertex β_0 and terminal vertex β_1. Then the braids

$$\beta_0 , \quad \beta = \beta_0 \sigma_i^{-1} , \quad \beta_1 = \beta_0 \sigma_i^{-1} \sigma_{i+1} = \beta \sigma_{i+1}$$

belong to B_n^+ and are left divisors of Δ_n^r. We have to show that the braids $\beta_0 \sigma_{i+1}, \beta_0 \sigma_{i+1} \sigma_i$, and $\beta_0 \sigma_{i+1} \sigma_i \sigma_{i+1}^{-1}$ belong to B_n^+ and are left divisors of Δ_n^r.

It is clear that $\beta_0 \sigma_{i+1}, \beta_0 \sigma_{i+1} \sigma_i$, and

$$\beta_0 \sigma_{i+1} \sigma_i \sigma_{i+1}^{-1} = \beta_0 \sigma_i^{-1} \sigma_{i+1} \sigma_i = \beta_1 \sigma_i$$

belong to B_n^+. We know that $\beta_0 = \beta \sigma_i$ and $\beta_1 = \beta \sigma_{i+1}$ are left divisors of Δ_n^r. This implies that the right lcm of $\beta \sigma_i$ and $\beta \sigma_{i+1}$ in B_n^+ is a left divisor of Δ_n^r.

We claim that the right lcm of $\beta\sigma_i$ and $\beta\sigma_{i+1}$ is $\beta\mu$, where $\mu = \sigma_i\sigma_{i+1}\sigma_i$ is the right lcm of σ_i and σ_{i+1}. Indeed, $\beta\mu$ is clearly a right multiple of $\beta\sigma_i$ and $\beta\sigma_{i+1}$. Let $\nu = \beta\sigma_i\beta' = \beta\sigma_{i+1}\beta''$ be a right multiple of $\beta\sigma_i$ and $\beta\sigma_{i+1}$, where $\beta', \beta'' \in B_n^+$. Since B_n^+ is left cancellative, $\sigma_i\beta' = \sigma_{i+1}\beta''$, which is a right multiple of σ_i and σ_{i+1} and hence a right multiple of μ. Therefore, ν is a right multiple of $\beta\mu$, which proves the claim. It follows from the previous arguments that $\beta\sigma_i\sigma_{i+1}\sigma_i = \beta\sigma_{i+1}\sigma_i\sigma_{i+1}$ is a left divisor of Δ_n^r. So are then

$$\beta_0\sigma_{i+1} = \beta\sigma_i\sigma_{i+1}, \quad \beta_0\sigma_{i+1}\sigma_i = \beta\sigma_i\sigma_{i+1}\sigma_i,$$

and

$$\beta_0\sigma_{i+1}\sigma_i\sigma_{i+1}^{-1} = \beta_0\sigma_i^{-1}\sigma_{i+1}\sigma_i = \beta\sigma_{i+1}\sigma_i.$$

(e) Substitutions (7.7) and (7.8): The two words in (7.7) and in (7.8) are inverses of the words in (7.6) and in (7.5), respectively. So we may argue as above after reversing the paths. $\qquad\square$

7.5.7 Critical prefixes and critical handles

Consider a braid word w of index $i \in \{1, \ldots, n-1\}$. Let $e(w) = \pm 1$ be the integer such that the leftmost occurrence of $\sigma_i^{\pm 1}$ in w is $\sigma_i^{e(w)}$. We define the *critical prefix* $P(w)$ of w as the longest prefix of w such that its last letter is $\sigma_i^{e(w)}$ and it contains no occurrences of $\sigma_i^{-e(w)}$. For example, if $i = 1$ and

$$w = \sigma_1\sigma_2\sigma_3\sigma_2^{-1}\sigma_1\sigma_3^{-1}\sigma_1\sigma_3\sigma_2^{-1}\sigma_1^{-1}\sigma_2^{-1}\sigma_3^{-1}\sigma_1\sigma_2,$$

then $e(w) = 1$ and $P(w) = \sigma_1\sigma_2\sigma_3\sigma_2^{-1}\sigma_1\sigma_3^{-1}\sigma_1$.

We denote by $h(w)$ the number of σ_i-handles contained in w. If $h(w) \geq 1$, then there is a unique σ_i-handle whose first letter, $\sigma_i^{e(w)}$, is the last letter of the critical prefix $P(w)$. We call this handle the *critical handle* of w. It is easy to see that the critical handle of w is the unique σ_i-handle v such that $w = w_1vw_2$, where w_1v is the shortest prefix of w containing a σ_i-handle. The essential difference between the critical handle and the prime handle (as introduced in Definition 7.27) of a braid word of index i is that the critical handle is always a σ_i-handle, whereas the prime handle may be a σ_j-handle with $j > i$. It follows from the definitions that the prime handle of w is contained in the critical handle, and if the prime handle is a σ_i-handle, then it coincides with the critical handle. Let us illustrate the difference between critical and prime handles on the following three words of index 1:

(i) If $w = \sigma_1\sigma_2\sigma_3\sigma_2^{-1}\sigma_1\sigma_3^{-1}\sigma_1\sigma_3\sigma_2^{-1}\sigma_1^{-1}\sigma_2^{-1}$, then its critical handle is $\sigma_1\sigma_3\sigma_2^{-1}\sigma_1^{-1}$; its prime handle is $\sigma_2\sigma_3\sigma_2^{-1}$.

(ii) If $w = \sigma_1\sigma_2\sigma_3\sigma_2^{-1}\sigma_1\sigma_3^{-1}\sigma_1\sigma_3\sigma_2^{-1}$, then it has no σ_1-handles, hence no critical handles; its prime handle is $\sigma_2\sigma_3\sigma_2^{-1}$.

(iii) If $w = \sigma_1\sigma_2\sigma_3\sigma_2\sigma_1\sigma_3^{-1}\sigma_1\sigma_3\sigma_2^{-1}\sigma_1^{-1}\sigma_2^{-1}$, then its prime handle is the σ_1-handle $\sigma_1\sigma_3\sigma_2^{-1}\sigma_1^{-1}$. This handle is also the critical handle of w.

Observe that if a word is the label of a path a in Γ, then all its subwords, in particular the critical prefix, the prime handle, the critical handle, are labels of subpaths of a.

Lemma 7.39. *Let w be a braid word of index i containing at least one handle. Assume that w is the label of a path a in Γ_r with initial vertex β_0. Then $h(\mathrm{red}(w)) \leq h(w)$. If $h(\mathrm{red}(w)) = h(w) \geq 1$, then $e(\mathrm{red}(w)) = e(w)$, and there is a path $a(w)$ in Γ_r such that*

(i) *the initial vertex of $a(w)$ is the terminal vertex of the path $p(w)$ with initial vertex β_0 and label $P(w)$, and the terminal vertex of $a(w)$ is the terminal vertex of the path $p(\mathrm{red}(w))$ with initial vertex β_0 and label $P(\mathrm{red}(w))$,*

(ii) *if the index of the prime handle of w is $> i$, then the path $a(w)$ is empty; if the index of the prime handle of w is i, then the label of $a(w)$ contains exactly one occurrence of $\sigma_i^{-e(w)}$ and no occurrences of $\sigma_i^{e(w)}$.*

Figure 7.4 shows the paths a, $\mathrm{red}(a)$, the subpath $p(w)$ of a, the subpath $p(\mathrm{red}(w))$ of $\mathrm{red}(a)$, and the path $a(w)$.

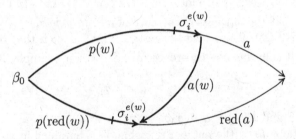

Fig. 7.4. The path $a(w)$

Proof. If $h(w) = 0$, then w contains no σ_i-handles, and one passes from w to $\mathrm{red}(w)$ by reducing some σ_j-handle with $j > i$. It is clear that $\mathrm{red}(w)$ contains no σ_i-handles. Hence, $h(\mathrm{red}(w)) = 0 = h(w)$.

Now assume that $h(w) \geq 1$. We can write

$$w = v_0 \sigma_i^e v_1 \sigma_i^e \cdots v_{p-1} \sigma_i^e v_p \underbrace{\sigma_i^e v_{p+1} \sigma_i^{-e}}\ v_{p+2} \sigma_i^f \cdots, \qquad (7.13)$$

where $p \geq 0$, $v_0, v_1, \ldots, v_{p-1}, v_p, v_{p+1}, v_{p+2}$ are braid words of index $> i$, $e = \pm 1$, and the subword indicated by braces is the critical handle of w, which exists since $h(w) \geq 1$. By σ_i^f in (7.13), we mean the first letter $\sigma_i^{\pm 1}$ appearing to the right of the critical handle of w if such a letter exists, and the empty word if no such letter exists, i.e., if the letters to the right of the critical handle are $\sigma_j^{\pm 1}$ with $j > i$. Clearly,

$$P(w) = v_0 \sigma_i^e v_1 \sigma_i^e \cdots v_{p-1} \sigma_i^e v_p \sigma_i^e .$$

Assume first that the index of the prime handle of w is $> i$. Then the prime handle of w must be a subword of v_r for some $r \in \{0, 1, \ldots, p+1\}$ (by definition of the prime handle, it cannot lie to the right of the critical handle). The word $\mathrm{red}(w)$ is then obtained from w by replacing v_r with $\mathrm{red}(v_r)$. This operation does not affect the σ_i-handles. Hence, $h(\mathrm{red}(w)) = h(w)$ and $e(\mathrm{red}(w)) = e(w)$. The critical prefix behaves under the reduction as follows: if $r = p + 1$, then $P(\mathrm{red}(w)) = P(w)$; if $r \leq p$, then $P(\mathrm{red}(w))$ is obtained from $P(w)$ by replacing v_r with $\mathrm{red}(v_r)$. In both cases, $P(\mathrm{red}(w))$ represents the same element of B_n as $P(w)$. Therefore, the paths $p(w)$ and $p(\mathrm{red}(w))$ have the same terminal vertex and we take $a(w)$ to be the empty path.

Now assume that the index of the prime handle of w is i. Then this handle has to be the critical handle $\sigma_i^e v_{p+1} \sigma_i^{-e}$. The word $\mathrm{red}(w)$ is obtained by reducing this handle. By Lemma 7.28, the prime handle does not contain any other handles. Therefore, the word v_{p+1} either contains no occurrences of $\sigma_{i+1}^{\pm 1}$, or contains occurrences of $\sigma_{i+1}^{\pm 1}$, but no occurrences of $\sigma_{i+1}^{\mp 1}$. Let us consider these cases separately.

(A) Suppose that v_{p+1} contains no occurrences of $\sigma_{i+1}^{\pm 1}$. The word $\mathrm{red}(w)$ is then obtained from w by replacing the prime handle

$$\sigma_i^e v_{p+1} \sigma_i^{-e}$$

by v_{p+1}. If $p = 0$, then

$$h(\mathrm{red}(w)) < h(w)$$

and we are done. Assume that $p \geq 1$. Then

$$\mathrm{red}(w) = v_0 \sigma_i^e v_1 \sigma_i^e \cdots v_{p-1} \underbrace{\sigma_i^e v_p v_{p+1} v_{p+2} \sigma_i^f} \cdots . \tag{7.14}$$

Comparing (7.13) and (7.14), we see that $h(\mathrm{red}(w)) < h(w)$ unless $\sigma_i^f = \sigma_i^{-e}$. In the latter case, the subword indicated by braces in (7.14) is the critical handle of $\mathrm{red}(w)$, and

$$h(\mathrm{red}(w)) = h(w) \geq 1 \quad \text{and} \quad e(\mathrm{red}(w)) = e(w).$$

Since

$$P(\mathrm{red}(w)) = v_0 \sigma_i^e v_1 \sigma_i^e \cdots v_{p-1} \sigma_i^e,$$

we have

$$P(w) = P(\mathrm{red}(w)) \, v_p \sigma_i^e.$$

Moreover, the path $p(\mathrm{red}(w))$ is a subpath of the path $p(w)$, hence a subpath of a. Let $a(w)$ be the path whose label is $\sigma_i^{-e} v_p^{-1}$ and whose initial vertex is the terminal vertex of $p(w)$. It is clear that $a(w)$ is a subpath of $\bar{a}$; hence $a(w)$ lies in Γ_r. The terminal vertex of $a(w)$ is the terminal vertex of $p(\mathrm{red}(w))$. Figure 7.5 shows parts of the paths a and $\mathrm{red}(a)$ in Γ_r. The path $a(w)$ appears in the gray zone of the figure (with reverse orientation). The label $\sigma_i^{-e} v_p^{-1}$ of $a(w)$ contains exactly one occurrence of σ_i^{-e} and no occurrences of σ_i^e.

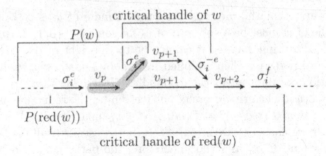

critical handle of red(w)

Fig. 7.5. Proof of Lemma 7.39: Case (A)

(B) Suppose that v_{p+1} contains occurrences of σ_{i+1}^{-e} and no occurrences of σ_{i+1}^{e}, i.e.,

$$v_{p+1} = u_0 \sigma_{i+1}^{-e} u_1 \cdots u_{q-1} \sigma_{i+1}^{-e} u_q,$$

where $q \geq 1$ and $u_0, u_1, \ldots, u_{q-1}, u_q$ are braid words of index $\geq i+2$. If $p = 0$, then

$$\mathrm{red}(w) = v_0 u_0 \sigma_{i+1}^{-e} \sigma_i^{-e} \sigma_{i+1}^{e} u_1 \cdots u_{q-1} \sigma_{i+1}^{-e} \sigma_i^{-e} \sigma_{i+1}^{e} u_q v_2 \sigma_i^{f} \cdots .$$

Clearly, $h(\mathrm{red}(w)) < h(w)$ and we are done. If $p \geq 1$, then $\mathrm{red}(w)$ is equal to

$$v_0 \sigma_i^{e} v_1 \sigma_i^{e} \cdots v_{p-1} \underbrace{\sigma_i^{e} v_p u_0 \sigma_{i+1}^{-e} \sigma_i^{-e}}_{} \sigma_{i+1}^{e} u_1 \cdots u_{q-1} \sigma_{i+1}^{-e} \sigma_i^{-e} \sigma_{i+1}^{e} u_q v_{p+2} \sigma_i^{f} \cdots .$$

The subword indicated by braces is the critical handle of $\mathrm{red}(w)$, and we have $h(\mathrm{red}(w)) = h(w) \geq 1$ and $e(\mathrm{red}(w)) = e(w)$. Moreover,

$$P(w) = P(\mathrm{red}(w)) \, v_p \sigma_i^{e}$$

as in (A) and we can conclude in the same way. Figure 7.6 shows parts of the paths a and $\mathrm{red}(a)$ in Γ_r. The path $a(w)$ appears in the gray zone of the figure (with reverse orientation).

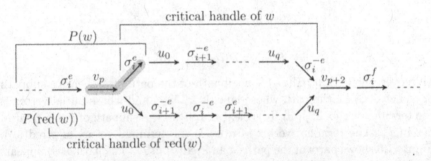

critical handle of red(w)

Fig. 7.6. Proof of Lemma 7.39: Case (B)

(C) We finally suppose that v_{p+1} contains occurrences of σ_{i+1}^e and no occurrences of σ_{i+1}^{-e}, i.e.,

$$v_{p+1} = u_0\sigma_{i+1}^e u_1 \cdots u_{q-1}\sigma_{i+1}^e u_q\,,$$

where $q \geq 1$ and $u_0, u_1, \ldots, u_{q-1}, u_q$ are braid words of index $\geq i+2$. Then red(w) is equal to

$$v_0\sigma_i^e v_1\sigma_i^e \cdots v_{p-1}\sigma_i^e v_p u_0\sigma_{i+1}^{-e}\sigma_i^e\sigma_{i+1}^e u_1 \cdots u_{q-1}\sigma_{i+1}^{-e}\underbrace{\sigma_i^e\sigma_{i+1}^e u_q v_{p+2}\sigma_i^f} \cdots.$$

If σ_i^f is the empty word or $f = e$, then $h(\text{red}(w)) < h(w)$ and we are done. If $f = -e$, then $h(\text{red}(w)) = h(w) \geq 1$ and the critical handle of red(w) is the one indicated by braces. We then have $e(\text{red}(w)) = e(w)$. Setting $v = v_0\sigma_i^e v_1\sigma_i^e \cdots v_{p-1}\sigma_i^e v_p$, we obtain

$$P(w) = v\sigma_i^e \quad \text{and} \quad P(\text{red}(w)) = vu_0\sigma_{i+1}^{-e}\sigma_i^e\sigma_{i+1}^e u_1 \cdots u_{q-1}\sigma_{i+1}^{-e}\sigma_i^e.$$

Let $a(w)$ be the path whose label is

$$L = u_0\sigma_{i+1}^e u_1 \cdots u_{q-1}\sigma_{i+1}^e u_q\sigma_i^{-e}u_q^{-1}\sigma_{i+1}^{-e}$$

and whose initial vertex is the terminal vertex of the subpath $p(w)$ of a. In Figure 7.7 the path $a(w)$ appears in the gray zone. We see that the terminal vertex of $a(w)$ is the terminal vertex of the subpath $p(\text{red}(w))$ of red(a), and the edges of $a(w)$ are edges either of a or of red(a). The path red(a) lies in Γ_r by Lemma 7.37; hence, so does $a(w)$. The word L contains exactly one occurrence of σ_i^{-e} and no occurrences of σ_i^e. □

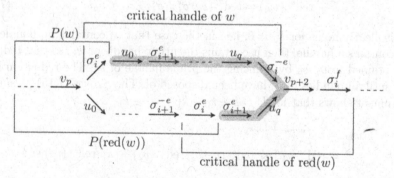

Fig. 7.7. Proof of Lemma 7.39: Case (C)

7.5.8 Proof of Lemma 7.32

We now use the previous lemmas to prove that prime handle reduction eventually stops. Let us proceed by descending induction on the index i of w.

If $i = n - 1$, then w is a word in the letters $\sigma_{n-1}^{\pm 1}$, and any handle is of the form $\sigma_{n-1}^{\pm 1} \sigma_{n-1}^{\mp 1}$. Reducing it means deleting it, hence shortening the length of the word by 2. It is obvious that $\text{red}^k(w)$ contains no handles for sufficiently large k.

Suppose that the lemma holds for all braid words of index $> i$ and let w be a braid word of index i. Assume that Lemma 7.32 does not hold for w. This means that $\text{red}^k(w)$ exists for all $k \geq 0$, that is, every braid word $w_k = \text{red}^k(w)$ has at least one handle. By Lemma 7.39, the nonnegative integers $h(w_k)$ form a nonincreasing sequence, which eventually must be constant. After discarding a finite number of w_k, we may assume that there is an integer h such that $h(w_k) = h$ for all $k \geq 0$. By definition, w_{k+1} is obtained from w_k by reducing the prime handle, which is either a σ_i-handle or a σ_j-handle for some $j > i$. Let K be the set of all integers k such that the prime handle of w_k is a σ_i-handle. In the sequel we shall prove first that K is infinite, then that K is finite. This will give a contradiction, so that Lemma 7.32 must hold for w.

We first prove that K is infinite and $h \geq 1$. For any $k \geq 0$, the braid word w_k is of the form

$$w_k = v_0 \sigma_i^e v_1 \sigma_i^e v_2 \cdots \sigma_i^e v_p w',$$

where $e = \pm 1$, the words $v_0, v_1, v_2, \ldots, v_p$ are of index $> i$, and the word w' either begins with the letter σ_i^{-e} (in which case $h = h(w_k) > 0$) or is empty (in which case $h = 0$). By the induction assumption, for each $r \in \{0, 1, \ldots, p\}$, there is $k_r \geq 0$ such that $\text{red}^{k_r}(v_r)$ contains no handles. We claim that

$$\text{red}^{k_0}(w_k) = \text{red}^{k_0}(v_0) \sigma_i^e v_1 \sigma_i^e v_2 \cdots \sigma_i^e v_p w'. \tag{7.15}$$

This clearly holds for $k_0 = 0$, i.e., in the case that v_0 contains no handles. If v_0 contains a handle, then it contains the prime handle of w_k, so that $\text{red}(w_k)$ is obtained from w_k by reducing the prime handle of v_0. The reduction goes on until all handles in v_0 have been disposed of. This proves (7.15). A similar argument shows that for $k' = k + k_0 + k_1 + \cdots + k_p$,

$$
\begin{aligned}
w_{k'} &= \text{red}^{k_0 + k_1 + \cdots + k_p}(w_k) \\
&= \text{red}^{k_0}(v_0) \sigma_i^e \, \text{red}^{k_1}(v_1) \sigma_i^e \, \text{red}^{k_2}(v_2) \cdots \sigma_i^e \, \text{red}^{k_p}(v_p) \, w'.
\end{aligned}
$$

If $w' = \emptyset$, then $w_{k'}$ contains no handles, contradicting our hypothesis that the sequence $(w_k)_k$ is infinite. Hence, w' must begin with σ_i^{-e}. One sees immediately that the σ_i-handle $\sigma_i^e \, \text{red}^{k_p}(v_p) \sigma_i^{-e}$ is the prime handle of $w_{k'}$. Hence, $k' \in K$. Thus for any $k \geq 0$ there is $k' \in K$ such that $k' \geq k$. This proves that K is infinite. The argument also shows that $h = h(w_k) \geq 1$.

We now claim that K is finite. By Lemma 7.36, the braid word w is the label of a path in Γ_r for some $r \geq 0$. It follows from Lemma 7.37 that for each $k \geq 0$ the word w_k is the label of a path in Γ_r. Let us apply Lemma 7.39 to w_k. We observed above that $h \geq 1$. Let e be the common value of $e(w_k)$ for all k. Consider the path $a(w_k)$ produced by Lemma 7.39 and its label $L_k = L(a(w_k))$. If $k \notin K$, then $L_k = \emptyset$; if $k \in K$, then L_k contains exactly one occurrence of σ_i^{-e} and no occurrences of σ_i^{e}. For any integer $\ell \geq 0$, the paths $a(w_0), a(w_1), \ldots, a(w_\ell)$ can be concatenated, since by Lemma 7.39 the initial vertex of each $a(w_s)$ is the terminal vertex of $a(w_{s-1})$. Each path $a(w_0), a(w_1), \ldots, a(w_\ell)$ being in Γ_r, so is the concatenated path $a(w_0)\, a(w_1) \cdots a(w_\ell)$. The label of the latter is the braid word $L_0 L_1 \cdots L_\ell$, which by Lemma 7.39 contains no occurrences of σ_i^{e} and as many occurrences of σ_i^{-e} as there are elements of K in $\{0, 1, \ldots, \ell\}$. By Lemma 7.35, the number of such occurrences of σ_i^{-e} is bounded from above by an integer N_r. It follows that

$$\operatorname{card}(K \cap \{0, 1, \ldots, \ell\}) \leq N_r$$

for each $\ell \geq 0$. Therefore K is a finite set. We have thus reached the desired contradiction.   □

Remark 7.40. Prime handle reduction allows us to get rid of all handles in a braid word. Actually, in order to prove Lemma 7.17, we need only to kill the σ_i-handles of braid words of index i. Killing σ_i-handles can be achieved by reducing only the critical handles. The latter can be reduced after the σ_{i+1}-handles that they contain have previously been disposed of. The reader is encouraged to make the reduction of critical handles work in a proper way. The appropriately defined critical handle reduction is faster than prime handle reduction since there are fewer handles to kill, as can be seen for instance when one applies both prime handle reduction and critical handle reduction to the braid word (7.2).

7.6 The Nielsen–Thurston approach

To end this chapter we outline a geometric method to order the braid groups. The method, based on Proposition 7.41 below, requires some familiarity with hyperbolic geometry and with Nielsen's classical work on homeomorphisms of surfaces [Nie27].

Proposition 7.41. *Let G be a group acting on a totally ordered set X by order-preserving bijections such that there is an element of X whose stabilizer is trivial. Then G is orderable.*

Recall that the *stabilizer* of $a \in X$ is the subgroup of G consisting of all elements fixing a.

Proof. For $f \in G$ and $b \in X$, let $f(b) \in X$ be the result of the action of f on b. By assumption, $b < b' \Rightarrow f(b) < f(b')$ in X for all $b, b' \in X$ and $f \in G$, and there is $a \in X$ such that $f(a) = a \Rightarrow f = 1$. For $f, g \in G$, set $f \leq_a g$ if $f(a) \leq g(a)$ for the given total order on X. It is clear that the relation $\leq_a$ on G is reflexive and transitive. Let us show that it is antisymmetric. Indeed, $f \leq_a g$ and $g \leq_a f$ imply $f(a) \leq g(a) \leq f(a)$. Hence, $f(a) = g(a)$, which is equivalent to $(g^{-1}f)(a) = a$. Therefore, $g^{-1}f = 1$; hence $f = g$. We have thus checked that $\leq_a$ is an order on G. Since the order on X is total, so is the order $\leq_a$ on G.

It remains to prove that $\leq_a$ is left-invariant. Let $f \leq_a g$ in G and $h \in G$. Since $f(a) \leq g(a)$ and h acts on X by an order-preserving bijection,

$$(hf)(a) = h(f(a)) \leq h(g(a)) = (hg)(a).$$

Thus, $hf \leq_a hg$. $\qquad\qquad\qquad\qquad\qquad\qquad\qquad\qquad\qquad\qquad\square$

Let S be a closed connected oriented surface of genus one with $n \geq 1$ marked points $P_1, \ldots, P_n$. Let C be a simple closed curve on S separating S into a genus-one surface S_1 and a disk S_2 containing all marked points (see Figure 7.8 for $n = 3$). By Theorem 1.33, the braid group B_n is isomorphic to the mapping class group $\mathcal{M}$ of the orientation-preserving self-homeomorphisms of S that are the identity on S_1 and permute the marked points.

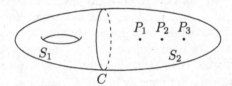

Fig. 7.8. The surface S

Equip $S - \{P_1, \ldots, P_n\}$ with a complete hyperbolic metric for which the curve C is a geodesic and the marked points are cusps. Fix a basepoint x_0 on C. The hyperbolic metric allows us to identify the universal covering of $S - \{P_1, \ldots, P_n\}$ with the interior $D^\circ = D - \partial D$ of the unit disk D in $\mathbf{C}$. Moreover, we can assume that the center 0 of D projects to x_0. Any orientation-preserving self-homeomorphism φ of S fixing x_0 and permuting the marked points can be lifted uniquely to a self-homeomorphism $\widetilde{\varphi}$ of D° fixing 0. Nielsen [Nie27, Sect. 10] showed that $\widetilde{\varphi}$ extends to an orientation-preserving self-homeomorphism Φ of D. He also proved that $\partial \varphi = \Phi|_{\partial D}$ depends only on the isotopy class of φ. Consequently, the mapping class group $\mathcal{M} \cong B_n$ acts by orientation-preserving homeomorphisms on the circle ∂D. This action fixes a point $z \in \partial D$. In the role of z we can take one of the endpoints of the component of the preimage of C in D passing through 0. We thus have an action of B_n on $\mathbf{R} = \partial D - \{z\}$ by orientation-preserving homeomorphisms, hence by order-preserving homeomorphisms.

We now apply Proposition 7.41 to $G = B_n$ acting on $X = \mathbf{R}$ via the above-defined action. In order to be able to conclude that B_n is orderable, we have to check that the subset Y of $\mathbf{R}$ consisting of the points with trivial stabilizer is nonempty. The complement Z of Y in X is the union of the sets of fixed points of $\partial\varphi$, where φ runs over all elements of $\mathcal{M} \cong B_n$ distinct from the identity. Since B_n is countable, Z is a countable union of such fixed-point sets. By [Nie27, Sect. 14], if $\varphi \neq 1$, then the set of fixed points of $\partial\varphi$ is a closed subset with empty interior. It follows from Baire's theorem (see [Kel55, Chap. 6] or [Rud66, Th. 5.6]) that Z has an empty interior. Therefore, its complement Y is dense in $\mathbf{R}$ and hence is nonempty.

Notes

For general references on orderable groups, see [MR77] or [Pas77]. A number of groups arising in topology are orderable; see [RW00], [SW00], [RW01], [Gon02], [BRW05]. For the biorderability of the pure braid groups, we followed [KR03], [DDRW02, Sect. 9.2].

The left-invariant total order presented in Section 7.3 was discovered by Dehornoy in 1991–1992; see [Deh94]. Until then it was not known whether braids groups were orderable. We followed [Deh00], [DDRW02, Chap. 1] for most of Section 7.3. Theorem 7.15 is due to Dehornoy. Proposition 7.19 is established in [MN03]. For the proof of Lemma 7.16 we followed [Lar94].

Handle reduction was introduced in [Deh97]; see also [Deh00, Chap. III], [DDRW02, Chap. 3]. In practice, the algorithm provided by Lemma 7.32 turns out to be very efficient, faster than other available algorithms.

The geometric approach in Section 7.6 is based on an observation of W. Thurston recorded by H. Short and B. Wiest. This approach leads to a family of left-invariant total orders of B_n including the Dehornoy order. A classification of these orders is given in [SW00]; see also [DDRW02, Chap. 7]. Proposition 7.41 has a nice converse when $X = \mathbf{R}$: any countable orderable group acts on $\mathbf{R}$ by order-preserving homeomorphisms such that there is a point on $\mathbf{R}$ whose stabilizer is trivial (see [Ghy01] or [DDRW02, Prop. 7.1.1]).

There are other proofs of the orderability of the braid groups, notably by Fenn, Greene, Rolfsen, Rourke, Wiest [FGRRW99], by Short and Wiest [SW00], by Funk [Fun01], and by I. Dynnikov (unpublished). See also the monographs [Deh00], [DDRW02], [DDRW08], and the survey [Kas02].

A

Presentations of $\mathrm{SL}_2(\mathbf{Z})$ and $\mathrm{PSL}_2(\mathbf{Z})$

Let $\mathrm{SL}_2(\mathbf{Z})$ be the group of 2×2 matrices with entries in $\mathbf{Z}$ and with determinant 1. The center of $\mathrm{SL}_2(\mathbf{Z})$ is the group of order 2 generated by the scalar matrix $-I_2$, where I_2 is the unit matrix. The quotient group

$$\mathrm{PSL}_2(\mathbf{Z}) = \mathrm{SL}_2(\mathbf{Z})/\langle -I_2 \rangle$$

is called the *modular group*; it can be identified with the group of rational functions on $\mathbf{C}$ of the form $(az+b)/(cz+d)$, where a, b, c, d are integers such that $ad - bc = 1$.

Consider the following three group presentations:

$$\langle a, b \mid aba = bab,\ (aba)^4 = 1 \rangle, \tag{A.1}$$

$$\langle s, t \mid s^3 = t^2,\ t^4 = 1 \rangle, \tag{A.2}$$

$$\langle s, t \mid s^3 = t^2 = 1 \rangle. \tag{A.3}$$

Lemma A.1. *(a) The presentations (A.1) and (A.2) define the same group G up to isomorphism. The group G is isomorphic to the quotient of the braid group B_3 by the central subgroup generated by $(\sigma_1\sigma_2\sigma_1)^4$.*

(b) The group H defined by (A.3) is isomorphic to the quotient of B_3 by its center.

Proof. (a) It is easy to check that the mutually inverse substitutions

$$s = ab,\ t = aba \quad \text{and} \quad a = s^{-1}t,\ b = t^{-1}s^2$$

transform (A.1) into (A.2). This proves that the presentations (A.1) and (A.2) define isomorphic groups.

Replacing a by σ_1 and b by σ_2 in (A.1), we see that G is isomorphic to the quotient of B_3 by the normal subgroup generated by $(\sigma_1\sigma_2\sigma_1)^4$. The latter is the square of $(\sigma_1\sigma_2\sigma_1)^2$, which by Theorem 1.24 generates the center $Z(B_3)$ of B_3.

C. Kassel, V. Turaev, *Braid Groups*, DOI: 10.1007/978-0-387-68548-9_8,
© Springer Science+Business Media, LLC 2008

(b) It is clear from the presentations (A.2) and (A.3) that H is the quotient of G by the normal subgroup generated by $s^3 = t^2 \in G$. Under the identifications

$$s = ab = \sigma_1\sigma_2, \quad t = aba = \sigma_1\sigma_2\sigma_1,$$

we have $H = B_3/Z(B_3)$.   □

Consider the matrices $A, B \in SL_2(\mathbf{Z})$ defined by

$$A = \begin{pmatrix} 1 & 1 \\ 0 & 1 \end{pmatrix} \quad \text{and} \quad B = \begin{pmatrix} 1 & 0 \\ -1 & 1 \end{pmatrix}.$$

It is easy to check the following relations in $SL_2(\mathbf{Z})$:

$$ABA = BAB \quad \text{and} \quad (ABA)^4 = 1.$$

Hence there is a group homomorphism $f : G \to SL_2(\mathbf{Z})$ such that $f(a) = A$ and $f(b) = B$. For $s = ab$ and $t = aba$, a quick computation gives

$$f(s) = AB = \begin{pmatrix} 0 & 1 \\ -1 & 1 \end{pmatrix}, \quad f(t) = ABA = \begin{pmatrix} 0 & 1 \\ -1 & 0 \end{pmatrix}, \tag{A.4}$$

$$f(t^2) = (f(t))^2 = (ABA)^2 = \begin{pmatrix} -1 & 0 \\ 0 & -1 \end{pmatrix} = -I_2. \tag{A.5}$$

By (A.5), f induces a group homomorphism $\overline{f} : H = G/\langle z \rangle \to PSL_2(\mathbf{Z})$.

Theorem A.2. *The group homomorphisms*

$$f : G \to SL_2(\mathbf{Z}) \quad and \quad \overline{f} : H = B_3/Z(B_3) \to PSL_2(\mathbf{Z})$$

are isomorphisms.

Proof. We claim that $f : G \to SL_2(\mathbf{Z})$ is injective (resp. surjective) if and only if $\overline{f} : H \to PSL_2(\mathbf{Z})$ is injective (resp. surjective). Indeed, f sends the subgroup $\langle t^2 \rangle \subset G$ onto the group of order 2 generated by $-I_2$. Since $t^4 = 1$, the subgroup $\langle t^2 \rangle$ is of order at most 2. Therefore, f induces an isomorphism from $\langle t^2 \rangle$ onto $\{\pm I_2\}$. The claim follows immediately.

To prove the theorem, it therefore suffices to show that $f : G \to SL_2(\mathbf{Z})$ is surjective and $\overline{f} : H \to PSL_2(\mathbf{Z})$ is injective.

We first check that the matrices $A = f(a)$ and $B = f(b)$ generate $SL_2(\mathbf{Z})$, which implies that $f : G \to SL_2(\mathbf{Z})$ is surjective. To this end we show that any $M \in SL_2(\mathbf{Z})$ can be expressed as a word in $A^{\pm 1}$ and $B^{\pm 1}$. In the argument below it will be convenient to denote the entries b and d of

$$M = \begin{pmatrix} a & b \\ c & d \end{pmatrix} \in SL_2(\mathbf{Z})$$

respectively by $b(M)$ and $d(M)$. Set $T = f(t) = ABA \in SL_2(\mathbf{Z})$.

If $b = 0$, then $a = d = \pm 1$ and either $M = B^{-c}$ or $M = -I_2 B^c = T^2 B^c$. Thus, M can be expressed as a word in $A^{\pm 1}$ and $B^{\pm 1}$.

If $d = 0$, then $bc = -1$. Either $b = -c = 1$ and then $M = A^{-a}T$, or $b = -c = -1$ and then $M = A^a T^3$. In both cases M can be expressed as a word in $A^{\pm 1}$ and $B^{\pm 1}$.

Assume now that neither $b = b(M)$ nor $d = d(M)$ is zero. Observe that

$$b(AM) = b(M) + d(M) \quad \text{and} \quad d(AM) = d(M), \tag{A.6}$$

and

$$b(TM) = d(M) \quad \text{and} \quad d(TM) = -b(M). \tag{A.7}$$

From (A.6) we deduce that by multiplying M on the left by a suitable positive or negative power of A, we obtain a matrix $A^n M$ such that

$$0 \le |b(A^n M)| < |d(A^n M)|.$$

Using (A.7), we may exchange the roles of $\pm b$ and $\pm d$ via left multiplication by T. In this way we can decrease the absolute values of b and d until one of them vanishes. Consequently, multiplying M on the left by powers of A or T, we can reduce the proof to the case $b = 0$ or $d = 0$ considered above.

We now prove that $\overline{f} : H \to PSL_2(\mathbf{Z})$ is injective. The group H presented by (A.3) is the free product of the cyclic group of order 3 generated by s and the cyclic group of order 2 generated by t. Any element of H distinct from the neutral element has a unique expression of one of the following forms:

$$w = s^{\varepsilon_1} t s^{\varepsilon_2} t \cdots t s^{\varepsilon_r}, \quad wt, \quad tw, \quad twt, \quad t,$$

where $\varepsilon_i = \pm 1$ $(i = 1, \ldots, r)$ (for a definition of free products and a description of normal forms for their elements, see for instance [LS77, Sect. I.11], [Ser77, Sect. I.1]). It is therefore enough to show that none of these elements is in the kernel of $\overline{f}$.

The element t is not in the kernel of $\overline{f}$ by (A.4). Since $twt = twt^{-1}$ is a conjugate of w and tw a conjugate of wt, it is enough to check that $\overline{f}(w) \neq 1$ and $\overline{f}(wt) \neq 1$.

Let us begin with $wt = (s^{\varepsilon_1} t)(s^{\varepsilon_2} t) \cdots (s^{\varepsilon_r} t)$. Since $s^{-1} t = a$ and

$$st = (t^{-1} s^2)^{-1} = b^{-1} \in H,$$

we have $\overline{f}(s^{-1} t) = \overline{A}$ and $\overline{f}(st) = \overline{B}^{-1}$, where $\overline{A}$ and $\overline{B}$ are the images of A and B in $PSL_2(\mathbf{Z})$, respectively. It follows that $\overline{f}(wt)$ is a nonempty product of the matrices $\overline{A}$ and $\overline{B}^{-1}$. It suffices then to check that no nonempty product of the matrices

$$A = \begin{pmatrix} 1 & 1 \\ 0 & 1 \end{pmatrix} \quad \text{and} \quad B^{-1} = \begin{pmatrix} 1 & 0 \\ 1 & 1 \end{pmatrix}$$

is equal to $\pm I_2$. Such a product has only nonnegative entries, and after each multiplication by A or B^{-1} the sum of the nondiagonal entries strictly increases. Therefore no such product may be equal to $\pm I_2$.

If $\overline{f}(w) = 1$, then

$$\overline{f}(wt) = \overline{f}(t) = \begin{pmatrix} 0 & 1 \\ -1 & 0 \end{pmatrix}.$$

This is impossible because by the previous argument, $\overline{f}(wt)$ is a product of the matrices A and B^{-1}, and has only nonnegative entries, whereas the matrix on the right has entries with opposite signs. This contradiction proves that $\overline{f}(w) \neq 1$. $\qquad\qquad\qquad\qquad\qquad\qquad\qquad\qquad\qquad\qquad\qquad\qquad\qquad\square$

Notes

The above proofs were inspired by [Rei32, 2.8–2.9]. There are alternative proofs using the action of the group $\mathrm{PSL}_2(\mathbf{Z})$ on the Poincaré upper half-plane (see [Ser70, Sect. VII.1]) or algebraic K-theory methods (see [Mil71, Chap. 10]).

B

Fibrations and Homotopy Sequences

We recall several basic notions from the theory of fibrations needed in the main text. For details, the reader is referred, for instance, to [FR84, Chap. 5].

A (continuous) map $p : E \to B$ is called a *locally trivial fibration with fiber* F if for every point of B there is a neighborhood $U \subset B$ of this point together with a homeomorphism $U \times F \to p^{-1}(U)$ whose composition with p is the projection to the first factor $U \times F \to U$. It is clear then that F is homeomorphic to $p^{-1}(b)$ for any $b \in B$. The spaces E and B are called respectively the *total space* and the *base* of p. A map $\widehat{f}$ from a topological space X to E is said to be a *lifting*, or *lift*, of a map $f : X \to B$ if

$$p \circ \widehat{f} = f.$$

Set $I = [0, 1]$. A map $p : E \to B$ has the *homotopy lifting property with respect to a topological space* X if for any maps $\widehat{f} : X \to E$ and $g : X \times I \to B$ such that $g(x, 0) = p(\widehat{f}(x))$ for all $x \in X$, there is a lift

$$\widehat{g} : X \times I \to E$$

of g such that $\widehat{g}(x, 0) = \widehat{f}(x)$ for all $x \in X$.

More generally, a map $p : E \to B$ has the *homotopy lifting property with respect to a topological pair* $(X, A \subset X)$ if for arbitrary maps $\widehat{f} : X \to E$, $g : X \times I \to B$ and any lift $h : A \times I \to E$ of $g|_{A \times I}$ such that $g(x, 0) = p(\widehat{f}(x))$ for all $x \in X$ and $h(x, 0) = \widehat{f}(x)$ for all $x \in A$, there is a lift

$$\widehat{g} : X \times I \to E$$

of g such that $\widehat{g}(x, 0) = \widehat{f}(x)$ for all $x \in X$ and $\widehat{g}|_{A \times I} = h$.

A map $p : E \to B$ is a *Serre fibration* if it has the homotopy lifting property with respect to all cubes I^n with $n = 0, 1, \ldots$. For example, all locally trivial fibrations are Serre fibrations. It is known that each Serre fibration has the homotopy lifting property with respect to any pair (a polyhedron, a subpolyhedron).

C. Kassel, V. Turaev, *Braid Groups*, DOI: 10.1007/978-0-387-68548-9_9,
© Springer Science+Business Media, LLC 2008

The key property of a Serre fibration $p : E \to B$ is the existence of an exact sequence involving the homotopy groups of the total space, the base, and the fiber of p. More precisely, pick a point $e \in E$, set $b = p(e) \in B$, and let $F = p^{-1}(b) \subset E$ be the fiber of p over b. Then we have an infinite (to the left) sequence

$$\cdots \xrightarrow{\partial} \pi_2(F, e) \xrightarrow{i_\#} \pi_2(E, e) \xrightarrow{p_\#} \pi_2(B, b) \xrightarrow{\partial} \pi_1(F, e)$$
$$\xrightarrow{i_\#} \pi_1(E, e) \xrightarrow{p_\#} \pi_1(B, b) \xrightarrow{\partial} \pi_0(F, e) \xrightarrow{i_\#} \pi_0(E, e) \xrightarrow{p_\#} \pi_0(B, b),$$

where the morphisms $i_\#$ and $p_\#$ are induced by the inclusion $i : F \hookrightarrow E$ and the projection $p : E \to B$, respectively. The terms of this sequence are groups except the last three terms, which are sets with a distinguished element represented by the base point. The morphisms in this sequence are group homomorphisms except the three rightmost arrows, which are set-theoretic mappings preserving the distinguished elements. The sequence above is called the *homotopy sequence* of p. It is exact in the sense that the image of each morphism is equal to the kernel of the next morphism (for the three rightmost arrows, by the kernel we mean the preimage of the distinguished element).

The boundary homomorphism $\partial : \pi_n(B, b) \to \pi_{n-1}(F, e)$ with $n \geq 1$ is defined as follows. Represent any $a \in \pi_n(B, b)$ by a map $\alpha : I^n \to B$ with $\alpha(\partial I^n) = b$. The homotopy lifting property of p with respect to the pair $(I^{n-1}, \partial I^{n-1})$ implies that α has a lift $\widehat{\alpha} : I^n = I^{n-1} \times I \to E$ such that

$$\widehat{\alpha}(I^{n-1} \times \{1\}) = \widehat{\alpha}(\partial I^{n-1} \times I) = e.$$

The restriction of $\widehat{\alpha}$ to $I^{n-1} \times \{0\} = I^{n-1}$ yields a map $I^{n-1} \to E$ sending I^{n-1} to $p^{-1}(b) = F$ and sending ∂I^{n-1} to e. This map represents $\partial(a) \in \pi_{n-1}(F, e)$.

C

The Birman–Murakami–Wenzl Algebras

We briefly discuss a family of finite-dimensional quotients of the braid group
algebras due to J. Murakami, J. Birman, and H. Wenzl. We also outline an interpretation of the Lawrence–Krammer–Bigelow representation of Section 3.5
in terms of representations of these algebras.

J. Murakami [Mur87] and independently J. Birman and H. Wenzl [BW89]
introduced a two-parameter family of finite-dimensional **C**-algebras

$$C_n(\alpha, \ell),$$

where α and ℓ are nonzero complex numbers such that $\alpha^4 \neq 1$ and $\ell^4 \neq 1$.
For $i = 1, \ldots, n-1$, set

$$e_i = \frac{\sigma_i + \sigma_i^{-1}}{\alpha + \alpha^{-1}} - 1 \in \mathbf{C}[B_n].$$

The algebra $C_n(\alpha, \ell)$ is the quotient of the group algebra $\mathbf{C}[B_n]$ by the relations

$$e_i \sigma_i = \ell^{-1} e_i, \quad e_i \sigma_{i-1} e_i = \ell\, e_i, \quad e_i \sigma_{i-1}^{-1} e_i = \ell^{-1} e_i,$$

where $i = 1, \ldots, n-1$ in the first relation and $i = 2, \ldots, n-1$ in the last two
relations. Note that the original definition in [BW89] involves more relations;
for the shorter list given above, see [Wen90]. The algebra $C_n(\alpha, \ell)$ is called
the *Birman–Murakami–Wenzl algebra* (BMW algebra for short). It admits a
geometric interpretation in terms of so-called Kauffman skein classes of tangles
in Euclidean 3-space. This family of algebras is a deformation of an algebra
introduced by R. Brauer [Bra37].

The algebraic structure and representations of $C_n(\alpha, \ell)$ were studied by
Wenzl [Wen90], who established the following three facts.

(i) For generic α and ℓ, the algebra $C_n(\alpha, \ell)$ is semisimple. Here "generic"
means that α is not a root of unity and $\sqrt{-1}\,\ell$ is not an integral power
of $-\sqrt{-1}\,\alpha$. (The latter two numbers correspond to r and q in Wenzl's
notation.) In the sequel we assume that α and ℓ are generic in this sense.

C. Kassel, V. Turaev, *Braid Groups*, DOI: 10.1007/978-0-387-68548-9_10,
© Springer Science+Business Media, LLC 2008

(ii) Simple finite-dimensional $C_n(\alpha, \ell)$-modules are indexed by partitions λ of nonnegative integers m such that $m \leq n$ and $m \equiv n$ (mod 2). The simple $C_n(\alpha, \ell)$-module corresponding to λ will be denoted by $V_{n,\lambda}$. Composing the natural homomorphism $\mathbf{C}[B_n] \to C_n(\alpha, \ell)$ with the action of $C_n(\alpha, \ell)$ on the module $V_{n,\lambda}$, we obtain an irreducible representation $B_n \to \mathrm{Aut}(V_{n,\lambda})$.

(iii) The natural inclusion $B_{n-1} \hookrightarrow B_n$ induces an inclusion

$$C_{n-1}(\alpha, \ell) \hookrightarrow C_n(\alpha, \ell)$$

for all $n \geq 2$. Moreover, the $C_n(\alpha, \ell)$-module $V_{n,\lambda}$, where $\lambda \vdash m$, decomposes as a $C_{n-1}(\alpha, \ell)$-module into a direct sum

$$\bigoplus_\mu V_{n-1, \mu},$$

where μ ranges over all partitions whose diagrams have been obtained from the diagram of λ by removing or (if $m < n$) adding one box. Each such μ appears in this decomposition with multiplicity 1.

The assertions (ii) and (iii) allow us to draw the Bratteli diagram for the sequence

$$C_1(\alpha, \ell) \subset C_2(\alpha, \ell) \subset \cdots.$$

On the level $n = 1, 2, \ldots$ of this diagram we place all partitions $\lambda \vdash m$ such that $m \leq n$ and $m \equiv n$ (mod 2). Then we connect each λ on the nth level by an edge to every partition on the $(n-1)$st level whose diagram has been obtained from the diagram of λ by removing or (if $m < n$) adding one box. For instance, the $n = 1$ level consists of the partition (1) corresponding to the tautological one-dimensional representation of $C_1(\alpha, \ell) = \mathbf{C}$. The $n = 2$ level contains the partitions (2), $(1,1)$, and the empty partition $\emptyset$ of zero. All three are connected to the unique partition on the level 1. Each partition $\lambda \vdash m$ with $m \geq 0$ appears on the levels $m, m+2, m+4, \ldots$.

As in the case of the Iwahori–Hecke algebras, the Bratteli diagram of the BMW algebras yields a useful method for computing the dimension of $V_{n,\lambda}$, where λ is a partition on the nth level. It is clear from (iii) that $\dim V_{n,\lambda}$ is the number of paths on the Bratteli diagram leading from the unique partition on the level 1 to λ. Here by a path we mean a path with vertices lying on consecutively increasing levels. We illustrate this computation with a few examples.

(a) Let $\mu[n] = (1, \ldots, 1)$ be the partition of n whose diagram is a single column of n boxes. Let $\mu'[n] = (n)$ be the conjugate partition of n whose diagram is a single row of n boxes. There is only one path from the unique partition on the level 1 to $\mu[n]$ placed on the level n. Hence,

$$\dim V_{n, \mu[n]} = 1$$

for all $n \geq 1$. Similarly, $\dim V_{n, \mu'[n]} = 1$.

For $n \geq 3$, the algebra $C_n(\alpha, \ell)$ has two one-dimensional representations. In both of them all e_i act as 0 and all σ_i act as multiplication by one and the same number equal either to α or to α^{-1}. We choose the correspondence between the irreducible $C_n(\alpha, \ell)$-modules and the partitions so that all σ_i act as multiplication by α on $V_{n, \mu[n]}$ and as multiplication by α^{-1} on $V_{n, \mu'[n]}$.

(b) For $n \geq 2$, let $\lambda[n] = (2, 1, \ldots, 1)$ be the partition of n whose diagram has two columns with $n-1$ boxes in the first column and one box in the second column. For $n \geq 3$, the partition $\lambda[n]$, placed on the level n, is connected to only two partitions on the level $n-1$, namely to $\lambda[n-1]$ and $\mu[n-1]$. Hence,

$$\dim V_{n, \lambda[n]} = \dim V_{n-1, \lambda[n-1]} + \dim V_{n-1, \mu[n-1]}$$
$$= \dim V_{n-1, \lambda[n-1]} + 1.$$

We have $\lambda[2] = \mu'[2]$, so that $\dim V_{2, \lambda[2]} = 1$. Hence $\dim V_{n, \lambda[n]} = n - 1$ for all $n \geq 2$.

(c) For $n \geq 3$, consider the partition $\mu[n-2]$ placed on the level n. It is connected to three partitions on the level $n-1$, namely to $\mu[n-1]$, $\mu[n-3]$, and $\lambda[n-1]$. Hence,

$$\dim V_{n, \mu[n-2]} = \dim V_{n-1, \mu[n-1]} + \dim V_{n-1, \mu[n-3]} + \dim V_{n-1, \lambda[n-1]}$$
$$= 1 + \dim V_{n-1, \mu[n-3]} + n - 2$$
$$= \dim V_{n-1, \mu[n-3]} + n - 1.$$

We set $\mu[0] = \emptyset$ and deduce from (iii) above that $\dim V_{2, \mu[0]} = \dim V_{1, \mu[1]} = 1$. Therefore for all $n \geq 2$,

$$\dim V_{n, \mu[n-2]} = \frac{n(n-1)}{2}.$$

We conclude that the dimension of $V_{n, \mu[n-2]}$ coincides with the rank of the Lawrence–Krammer–Bigelow representation of B_n over $\mathbf{Z}[q^{\pm 1}, t^{\pm 1}]$. This suggests that these two representations may be related. To describe their relationship, we rescale the representation $B_n \rightarrow \mathrm{Aut}(V_{n, \mu[n-2]})$ by dividing the action of each σ_i by α.

Theorem C.1 (M. Zinno [Zin01]). *The Lawrence–Krammer–Bigelow representation computed at $q = -\alpha^{-2}$ and $t = \alpha^3 \ell^{-1}$ is isomorphic to the rescaled representation $B_n \rightarrow \mathrm{Aut}(V_{n, \mu[n-2]})$.*

This theorem implies that the Lawrence–Krammer–Bigelow representation is irreducible and that after the substitution $q = -\alpha^{-2}$, $t = \alpha^3 \ell^{-1} \in \mathbf{C}$, this representation factors through the projection $B_n \rightarrow C_n(\alpha, \ell)$.

D

Left Self-Distributive Sets

We give here a brief introduction to so-called left self-distributive sets, which are closely related to braid groups.

D.1 LD sets, racks, and quandles

A *left self-distributive set* (LD set) is a pair $(X, *)$, where X is a set and $* : X \times X \to X$ is a binary operation satisfying

$$a * (b * c) = (a * b) * (a * c) \tag{D.1}$$

for all $a, b, c \in X$. A *morphism* $f : (X, *) \to (X', *)$ of LD sets is a set-theoretic map $f : X \to X'$ such that $f(a * b) = f(a) * f(b)$ for all $a, b \in X$.

The idea of an LD set is very natural: for any element a of a set X equipped with a binary operation $* : X \times X \to X$, consider the left multiplication $L_a : X \to X$ defined by $L_a(b) = a * b$ for all $b \in X$. The equation (D.1) can be reformulated as

$$L_a(b * c) = L_a(b) * L_a(c) .$$

Thus, an LD set is a set equipped with a binary operation that is preserved by all left multiplications. The terminology "left self-distributive" arises from the fact that a binary operation satisfying (D.1) is left distributive with respect to itself.

An LD set $(X, *)$ is a *rack* if the left multiplication $b \mapsto a * b$ is bijective for all $a \in X$. A *quandle* is a rack satisfying $a * a = a$ for all $a \in X$.

Examples D.1. (a) The formula $a * b = b$ defines a left self-distributive operation on any set. This is a quandle.

(b) Given a monoid M together with an element $e \in M$, set $a * b = be$ $(a, b \in M)$. Then $(M, *)$ is an LD set. It is a rack if and only if e has a left inverse in the monoid. It is a quandle if and only if $be = b$ for all $b \in M$.

C. Kassel, V. Turaev, *Braid Groups*, DOI: 10.1007/978-0-387-68548-9_11,
© Springer Science+Business Media, LLC 2008

(c) Given a group G, set $a * b = aba^{-1}$ for $a, b \in G$. The pair $(G, *)$ is a quandle.

(d) Let R be a ring and $t \in R$. For $a, b \in R$, set

$$a * b = (1 - t)a + tb. \tag{D.2}$$

This is an LD operation. The pair $(R, *)$ is a rack (actually a quandle) if and only if t is invertible in R.

D.2 An action of the braid monoid

We relate LD sets to the braid monoids B_n^+ introduced in Section 6.5. Given an LD set $(X, *)$ and an integer $n \geq 2$, consider the product $X^n = X \times X \times \cdots \times X$ of n copies of X. For $i = 1, \ldots, n - 1$, set

$$\sigma_i(a_1, \ldots, a_n) = (a_1, \ldots, a_{i-1}, a_i * a_{i+1}, a_i, a_{i+2}, \ldots, a_n), \tag{D.3}$$

where $\sigma_1, \ldots, \sigma_{n-1}$ are the standard generators of B_n^+ and $a_1, \ldots, a_n \in X$.

Lemma D.2. *Formula* (D.3) *equips the set X^n with a left action of B_n^+. This action extends to a left action of the braid group B_n if and only if $(X, *)$ is a rack.*

By a left action of B_n^+ on X^n we mean a map

$$B_n^+ \times X^n \to X^n, \quad (\beta, A) \mapsto \beta A$$

such that $1A = A$ and $\beta(\beta' A) = (\beta\beta')A$ for all $A \in X^n$ and $\beta, \beta' \in B_n^+$.

We can give a geometric description of this action. Represent $\beta \in B_n^+$ by a braid diagram $\mathcal{D}$ with n strands and only positive crossings. Color the n lower endpoints of $\mathcal{D}$ from left to right by $a_1, \ldots, a_n \in X$. Let the colors flow up along the strands of $\mathcal{D}$ subject to the following rule: the colors remain unchanged as long as they do not meet a crossing of $\mathcal{D}$. At a crossing the color a of the overgoing strand remains unchanged whereas the color b of the undergoing strand becomes $a * b$. The n-tuple $(b_1, \ldots, b_n) \in X^n$ of colors of the upper endpoints of $\mathcal{D}$ satisfies

$$(b_1, \ldots, b_n) = \beta(a_1, \ldots, a_n).$$

See Figure D.1 for $n = 3$ and $\beta = \sigma_1$.

Proof. (a) To prove that (D.3) equips X^n with a left action of B_n^+, it suffices to check that for all $A = (a_1, \ldots, a_n) \in X^n$,

$$\sigma_i(\sigma_j A) = \sigma_j(\sigma_i A)$$

for $i, j \in \{1, \ldots, n - 1\}$ with $|i - j| \geq 2$, and

$$\sigma_i(\sigma_{i+1}(\sigma_i A)) = \sigma_{i+1}(\sigma_i(\sigma_{i+1} A))$$

$$a_1 * a_2 \quad a_1 \quad a_3$$

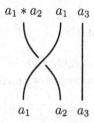

$$a_1 \qquad a_2 \quad a_3$$

Fig. D.1. The rule for braid coloring

for $i \in \{1, \ldots, n-2\}$. The first identity is a triviality. For the second one, we obtain

$$\sigma_i(\sigma_{i+1}(\sigma_i A))$$
$$= (a_1, \ldots, a_{i-1}, (a_i * a_{i+1}) * (a_i * a_{i+2}), a_i * a_{i+1}, a_i, a_{i+3}, \ldots, a_n),$$

whereas

$$\sigma_{i+1}(\sigma_i(\sigma_{i+1} A))$$
$$= (a_1, \ldots, a_{i-1}, a_i * (a_{i+1} * a_{i+2}), a_i * a_{i+1}, a_i, a_{i+3}, \ldots, a_n).$$

These expressions are equal by (D.1).

(b) The action of B_n^+ on X^n extends to a left action of B_n if and only if the maps $A \mapsto \sigma_i A$ are bijective for all $i = 1, \ldots, n-1$. It is clear from the definitions that this is equivalent to the bijectivity of all left multiplications $b \mapsto a * b$. □

D.3 Orderable LD sets

Given an LD set $(X, *)$ and elements $a, c \in X$, we write $a \prec c$ if $a * b = c$ for some $b \in X$. For example, if X is a rack, then $a \prec c$ for all $a, c \in X$.

We define a binary relation $\preceq$ on an LD set X by $a \preceq b$ if $a = b$ or there are $a_0, a_1, \ldots, a_r \in X$ such that $a = a_0 \prec a_1 \prec \cdots \prec a_r = b$. We say that a LD set X is *orderable* if the relation $\preceq$ is an order on X. In this case, the relation $\preceq$ is called the *canonical order* of X. For example, a rack $(X, *)$ is orderable if and only if the set X consists of only one element. This suggests that orderable LD sets are very different from racks.

We give three examples of orderable LD sets. In Section 7.4.1 we considered the free group F_∞ on the countable set $\{x_1, x_2, x_3, \ldots\}$ of generators. Resuming the notation of that section, we define a binary operation $*$ on the automorphism group $\operatorname{Aut}(F_\infty)$ by

$$\varphi * \psi = \varphi \circ T(\psi) \circ \widetilde{\sigma}_1 \circ T(\varphi^{-1}), \tag{D.4}$$

for any $\varphi, \psi \in \mathrm{Aut}(F_\infty)$. The reader may check that (D.4) is a left self-distributive operation and that the LD set $(\mathrm{Aut}(F_\infty), *)$ is orderable (see Exercise D.3.4).

A second example of an orderable LD set is given by the infinite braid group B_∞ (see Section 7.3.5), equipped with the binary operation

$$\beta * \beta' = \beta \operatorname{sh}(\beta') \sigma_1 \operatorname{sh}(\beta^{-1}), \qquad\qquad (D.5)$$

where β, $\beta' \in B_\infty$ and sh is the shift introduced in Section 7.4.2. The group homomorphism $B_\infty \to \mathrm{Aut}(F_\infty)$ that is the direct limit of the injective homomorphisms $B_n \to \mathrm{Aut}(F_n)$ defined in Section 1.5.1 is a morphism of LD sets. This observation can be used to check that (D.5) is a left self-distributive operation and that the LD set $(B_\infty, *)$ is orderable (see Exercise D.3.5).

The third example of an orderable LD set is provided by the *free LD set* on one generator. This LD set is characterized by the following universal property.

Proposition D.3. *There is an LD set $(D, *)$ with distinguished element $x \in D$ such that for any LD set $(X, *)$ and any $a \in X$, there is a unique morphism of LD sets $f : D \to X$ such that $f(x) = a$. The LD set $(D, *)$ is unique up to isomorphism.*

Proof. Following Bourbaki, define a *magma* to be a set equipped with a binary operation $*$. Consider the magma Mag that is free on one generator x (for details, see [Bou70, Chap. 1, Sect. 7]). An element of Mag can be viewed as a positive power of x equipped with a full set of parentheses, e.g., x, $x * x$, $(x * x) * x$, $x * (x * x)$, $((x * x) * x) * x$, $(x * (x * x)) * x$, $x * ((x * x) * x)$, $x * (x * (x * x))$, $(x * x) * (x * x), \ldots$. The binary operation $*$ on Mag is the concatenation of parenthesized words.

Let $\sim$ be the smallest equivalence relation on Mag such that

$$t_1 * (t_2 * t_3) \sim (t_1 * t_2) * (t_1 * t_3)$$

and $t_1 * t_2 \sim t_1' * t_2'$ whenever $t_1 \sim t_1'$ and $t_2 \sim t_2'$. We define D as the set of equivalence classes in Mag with respect to $\sim$. By definition of $\sim$, the binary operation $*$ of Mag induces a left self-distributive operation, still denoted by $*$, on D.

For any LD set $(X, *)$ and $a \in X$, we define a map $f' : \mathrm{Mag} \to X$ inductively by $f'(x) = a$ and by

$$f'(t_1 * t_2) = f'(t_1) * f'(t_2)$$

for all t_1, $t_2 \in \mathrm{Mag}$. Since X is an LD set, the map f' induces a morphism $f : D \to X$ of LD sets such that $f(x) = a$. It is easy to show that such a morphism f is unique.

The uniqueness of D up to isomorphism follows from the universal property of D. $\qquad\Box$

Theorem D.4. *The LD set* $(D, *)$ *is orderable and its canonical order is total.*

For a proof, see [Deh94] or [Deh00, Chap. V].

Exercise D.3.1. Let $E^\times$ be the set of nonzero vectors of a Euclidean vector space. For $a, b \in E^\times$ define $a * b$ to be the image of b under the orthogonal symmetry with respect to the hyperplane orthogonal to a. Show that $(E^\times, *)$ is a rack.

Exercise D.3.2. (a) Let F_S be the free group on a set of generators S. Equip $X_S = F_S \times S$ with the binary operation

$$(w_1, s_1) * (w_2, s_2) = (w_1 s_1 w_1^{-1}, s_2),$$

where $w_1, w_2 \in F_S$ and $s_1, s_2 \in S$. Show that $(X_S, *)$ is a rack.
(b) Show that any rack X is the quotient of the rack X_S, where S is a generating set of X.

Exercise D.3.3. Let $\Lambda = \mathbf{Z}[t, t^{-1}]$ be the ring of Laurent polynomials with integer coefficients. It is a rack under the binary operation (D.2). Show that the corresponding action of B_n on Λ^n is linear and is isomorphic to the Burau representation of Section 3.1.

Exercise D.3.4. Show that $(\mathrm{Aut}(F_\infty), *)$, where $*$ is defined by (D.4), is an orderable LD set. (Hint: Use the set E of Section 7.4.1.)

Exercise D.3.5. Show that $(\mathrm{Aut}(B_\infty), *)$, where $*$ is defined by (D.5), is an orderable LD set.

Exercise D.3.6. Show that there is a bijection between the free magma Mag, defined in the proof of Proposition D.3, and the set of planar rooted binary trees. Show that the number of elements of Mag containing n occurrences of x is equal to the Catalan number $\binom{2n}{n}/(n+1)$.

Notes

The idea of using racks to construct representations of the braid groups can be found, e.g., in Joyce [Joy82], Matveev [Mat82], Brieskorn [Bri88] (Brieskorn calls them "automorphic sets"; see [FR92] for a historical presentation of racks). Joyce and Matveev have associated to each knot a quandle that determines the knot up to isotopy and mirror reflection. Racks and quandles have therefore been familiar to topologists for quite a while.

On the other hand, orderable LD sets, especially the ones whose canonical orders are total, have been studied only recently, mainly by set theorists. The reason is that the first observed orderable LD set appeared in the theory of large cardinals, and its first construction relied on a large-cardinal axiom.

To avoid the use of this axiom, Dehornoy investigated the free LD set D of Section D.3. For details about the flow of ideas from set theory to braid groups, see [Lav92], [Deh00, Chap. XII]. Note that Laver [Lav92] proved that any orderable LD set generated by a single element is isomorphic to the free LD set D.

Theorem D.4 is due to Dehornoy [Deh94]. Exercise D.3.1 is from [Bri88], Exercises D.3.2–D.3.5 are from [Deh00].

References

[Ale23a]   J. W. Alexander, *A lemma on systems of knotted curves*, Proc. Nat. Acad. Sci. 9 (1923), 93–95.

[Ale23b]   J. W. Alexander, *Deformations of an n-cell*, Proc. Nat. Acad. Sci. 9 (1923), 406–407.

[Ale28]   J. W. Alexander, *Topological invariants of knots and links*, Trans. Amer. Math. Soc. 30 (1928), 275–306.

[All02]   D. Allcock, *Braid pictures for Artin groups*, Trans. Amer. Math. Soc. 354 (2002), 3455–3474.

[AAG99]   I. Anshel, M. Anshel, D. Goldfeld, *An algebraic method for public-key cryptography*, Math. Res. Lett. 6 (1999), 287–291.

[Arn70]   V. I. Arnold, *On some topological invariants of algebraic functions* (Russian), Trudy Moskov. Mat. Obshch. 21 (1970), 27–46. English translation: Trans. Moscow Math. Soc. 21 (1970), 30–52.

[Art25]   E. Artin, *Theorie der Zöpfe*, Abh. Math. Sem. Univ. Hamburg 4 (1925), 47–72.

[Art47a]   E. Artin, *Theory of braids*, Ann. of Math. (2) 48 (1947), 101–126.

[Art47b]   E. Artin, *Braids and permutations*, Ann. of Math. (2) 48 (1947), 643–649.

[Bau63]   G. Baumslag, *Automorphism groups of residually finite groups*, J. London Math. Soc. 38 (1963), 117–118.

[Bax72]   R. J. Baxter, *Partition function for the eight-vertex lattice model*, Ann. Physics 70 (1972), 193–228.

[Bax82]   R. J. Baxter, *Exactly solved models in statistical mechanics*, Academic press, London, 1982.

[BF04]   P. Bellingeri, L. Funar, *Polynomial invariants of links satisfying cubic skein relations*, Asian J. Math. 8 (2004), 475–509.

[Ben83]   D. Bennequin, *Entrelacements et équations de Pfaff*, Third Schnepfenried geometry conference, Vol. 1 (Schnepfenried, 1982), 87–161, Astérisque, vol. 107–108, Soc. Math. France, Paris, 1983.

[Ben98]   D. J. Benson, *Representations and cohomology. I: Basic representation theory of finite groups and associative algebras*, second edition, Cambridge Studies in Advanced Mathematics, 30, Cambridge University Press, Cambridge, 1998.

[BDM02]   D. Bessis, F. Digne, J. Michel, *Springer theory in braid groups and the Birman–Ko–Lee monoid*, Pacific J. Math. 205 (2002), 287–309.

[Big99]   S. Bigelow, *The Burau representation is not faithful for n = 5*, Geom. Topol. 3 (1999), 397–404 (electronic).

[Big01]   S. Bigelow, *Braid groups are linear*, J. Amer. Math. Soc. 14 (2001), 471–486.

[Big02]   S. Bigelow, *Representations of braid groups*, Proceedings of the International Congress of Mathematicians, Vol. II (Beijing, 2002), 37–45, Higher Ed. Press, Beijing, 2002.

[Big03]   S. Bigelow, *The Lawrence-Krammer representation*, Topology and geometry of manifolds (Athens, GA, 2001), 51–68, Proc. Sympos. Pure Math., 71, Amer. Math. Soc., Providence, RI, 2003.

[Bir69a]   J. S. Birman, *Mapping class groups and their relationship to braid groups*, Comm. Pure Appl. Math. 22 (1969), 213–238.

[Bir69b]   J. S. Birman, *Automorphisms of the fundamental group of a closed, orientable 2-manifold*, Proc. Amer. Math. Soc. 21 (1969), 351–354.

[Bir74]   J. Birman, *Braids, links and mapping class groups*, Ann. of Math. Studies, vol. 82, Princeton University Press, Princeton, NJ, 1974.

[BB05]   J. S. Birman, T. E. Brendle, *Braids: a survey*, in Handbook of knot theory, 19–103, Elsevier B. V., Amsterdam, 2005.

[BKL98]   J. S. Birman, K. H. Ko, S. J. Lee, *A new approach to the word and the conjugacy problem in the braid groups*, Adv. Math. 139 (1998), 322–353.

[BW89]   J. S. Birman, H. Wenzl, *Braids, link polynomials and a new algebra*, Trans. Amer. Math. Soc. 313 (1989), 249–273.

[Bou58]   N. Bourbaki, *Algèbre, chapitre 8*, Hermann, Paris, 1958.

[Bou68]   N. Bourbaki, *Groupes et algèbres de Lie*, Hermann, Paris, 1968.

[Bou70]   N. Bourbaki, *Algèbre, chapitres 1–3*, Hermann, Paris, 1970.

[BRW05]   S. Boyer, D. Rolfsen, B. Wiest, *Orderable 3-manifold groups*, Ann. Inst. Fourier (Grenoble) 55 (2005), 243–288.

[Bra37]   R. Brauer, *On algebras which are connected with the semisimple continuous groups*, Ann. of Math. (2) 38 (1937), 857–872.

[Bri71]   E. Brieskorn, *Die Fundamentalgruppe des Raumes des regulären Orbits einer endlichen komplexen Spiegelungsgruppe*, Invent. Math. 12 (1971), 57–61.

[Bri73]   E. Brieskorn, *Sur les groupes de tresses [d'après Arnold]*, Séminaire Bourbaki (1971/1972), Exp. No. 401, 21–44, Lecture Notes in Math., vol. 317, Springer-Verlag, Berlin, 1973.

[Bri88]   E. Brieskorn, *Automorphic sets and braids and singularities*, Braids (Joan S. Birman, Anatoly Libgober, eds.), Contemp. Math. 78, Amer. Math. Soc., Providence, RI, 1988, 45–115.

[BS72]   E. Brieskorn, K. Saito, *Artin-Gruppen und Coxeter-Gruppen*, Invent. Math. 17 (1972), 245–271.

[Bud05]   R. D. Budney, *On the image of the Lawrence-Krammer representation*, J. Knot Theory Ramifications 14 (2005), 773–789.

[Bur32]   W. Burau, *Über Zöpfinvarianten*, Abh. Math. Sem. Univ. Hamburg 9 (1932), 117–124.

[Bur36]   W. Burau, *Über Zopfgruppen und gleichsinnig verdrillte Verkettungen*, Abh. Math. Sem. Univ. Hamburg 11 (1936), 179–186.

[Bur97]   S. Burckel, *The wellordering on positive braids*, J. Pure Appl. Algebra 120 (1997), 1–17.

[BZ85]   G. Burde, H. Zieschang, *Knots*, de Gruyter Studies in Mathematics, 5, Walter de Gruyter, Berlin, 1985.

[CP94]   V. Chari, A. Pressley, *A guide to quantum groups*, Cambridge University Press, Cambridge, 1994.

[Cho48]   W.-L. Chow, *On the algebraical braid group*, Ann. of Math. (2) 49 (1948), 654–658.

[CS96]   C. de Concini, M. Salvetti, *Cohomology of Artin groups*, Math. Res. Lett. 3 (1996), 296–297.

[Con70]   J. H. Conway, *An enumeration of knots and links, and some of their algebraic properties*, Computational Problems in Abstract Algebra, Pergamon Press, Oxford, 1970, 329–358.

[CR62]   C. W. Curtis, I. Reiner, *Representation theory of finite groups and associative algebras*, Interscience Publishers, John Wiley & Sons, New York, London, 1962.

[Deh94]   P. Dehornoy, *Braid groups and left distributive operations*, Trans. Amer. Math. Soc. 345 (1994), 115–150.

[Deh97]   P. Dehornoy, *A fast method for comparing braids*, Adv. Math. 125 (1997), 200–235.

[Deh00]   P. Dehornoy, *Braids and self-distributivity*, Progress in Math., vol. 192, Birkhäuser, Basel, Boston, 2000.

[Deh02]   P. Dehornoy, *Groupes de Garside*, Ann. Scient. Éc. Norm. Sup. 4e série, 35 (2002), 267–306.

[DDRW02]   P. Dehornoy, I. Dynnikov, D. Rolfsen, B. Wiest, *Why are braid groups orderable?* Panoramas et Synthèses, 14, Soc. Math. France, Paris 2002.

[DDRW08]   P. Dehornoy, with I. Dynnikov, D. Rolfsen, B. Wiest, *Ordering braids*, Math. Surveys and Monographs, Amer. Math. Soc., 2008.

[DP99]   P. Dehornoy, L. Paris, *Gaussian groups and Garside groups, two generalisations of Artin groups*, Proc. London Math. Soc. (3) 79 (1999), 569–604.

[Del72]   P. Deligne, *Les immeubles des groupes de tresses généralisés*, Invent. Math. 17 (1972), 273–302.

[DJ86]   R. Dipper, G. James, *Representations of Hecke algebras of general linear groups*, Proc. London Math. Soc. (3) 52 (1986), 20–52.

[DJ87]   R. Dipper, G. James, *Blocks and idempotents of Hecke algebras of general linear groups*, Proc. London Math. Soc. (3) 54 (1987), 57–82.

[Dra97]   A. Drápal, *Finite left distributive algebras with one generator*, J. Pure Appl. Algebra 121 (1997), 233–251.

[DK94]   Yu. A. Drozd, V. V. Kirichenko, *Finite-dimensional algebras*, translated from the 1980 Russian original and with an appendix by Vlastimil Dlab, Springer-Verlag, Berlin, 1994.

[Eps66]   D. B. A. Epstein, *Curves on 2-manifolds and isotopies*, Acta Math. 115 (1966), 83–107.

[ECHLPT92]   D. B. A. Epstein, J. W. Cannon, D. F. Holt, S. V. F. Levy, M. S. Paterson, W. P. Thurston, *Word processing in groups*, Jones and Bartlett Publishers, Boston, MA, 1992.

[ES98]   P. Etingof, O. Schiffmann, *Lectures on quantum groups*, Lectures in Mathematical Physics, International Press, Boston, MA, 1998.

[FV62]   E. Fadell, J. Van Buskirk, *The braid groups of E^2 and S^2*, Duke Math. J. 29 (1962), 243–257.

[FaN62]   E. Fadell, L. Neuwirth, *Configuration spaces*, Math. Scand. 10 (1962), 111–118.

[FGRRW99]   R. Fenn, M. T. Greene, D. Rolfsen, C. Rourke, B. Wiest, *Ordering the braid groups*, Pacific J. Math. 191 (1999), 49–74.

[FR92]   R. Fenn, C. Rourke, *Racks and links in codimension two*, J. Knot Theory Ramifications 1 (1992), 343–406.

[For96]   E. Formanek, *Braid group representations of low degree*, Proc. London Math. Soc. (3) 73 (1996), 279–322.

[FLSV03]   E. Formanek, W. Lee, I. Sysoeva, M. Vazirani, *The irreducible complex representations of the braid group on n strings of degree $\leq n$*, J. Algebra Appl. 2 (2003), 317–333.

[FoN62]   R. Fox, L. Neuwirth, *The braid groups*, Math. Scand. 10 (1962), 119–126.

[FRT54]   J. S. Frame, G. de B. Robinson, R. M. Thrall, *The hook graphs of the symmetric groups*, Canadian J. Math. 6 (1954), 316–324.

[FYHLMO85] P. J. Freyd, D. N. Yetter, J. Hoste, W. B. R. Lickorish, K. C. Millett, A. Ocneanu, *A new polynomial invariant of knots and links*, Bull. Amer. Math. Soc. 12 (1985), 239–246.

[Frö36]   W. Fröhlich, *Über ein spezielles Transformationsproblem bei einer besonderen Klasse von Zöpfen*, Monatsh. Math. Phys. 44 (1936), 225–237.

[FR84]   D. B. Fuks, V. A. Rokhlin, *Beginner's course in topology. Geometric chapters*, translated from the Russian by A. Iacob, Universitext, Springer Series in Soviet Mathematics, Springer-Verlag, Berlin, 1984.

[Ful97]   W. Fulton, *Young tableaux. With applications to representation theory and geometry*, London Mathematical Society Student Texts, 35, Cambridge University Press, Cambridge, 1997.

[FH91]   W. Fulton, J. Harris, *Representation theory. A first course*, Graduate Texts in Mathematics, 129, Springer-Verlag, New York, 1991.

[Fun95]   L. Funar, *On the quotients of cubic Hecke algebras*, Comm. Math. Phys. 173 (1995), 513–558.

[Fun01]   J. Funk, *The Hurwitz action and braid group orderings*, Theory Appl. Categ. 9 (2001/02), 121–150 (electronic).

[Gar69]   F. A. Garside, *The braid group and other groups*, Quart. J. Math. Oxford Ser. (2) 20 (1969), 235–254.

[Gas62]   B. J. Gassner, *On braid groups*, Abh. Math. Sem. Univ. Hamburg 25 (1962), 10–22.

[Gec98]   M. Geck, *Representations of Hecke algebras at roots of unity*, Séminaire Bourbaki, Exposé n° 836 (1997/98), Astérisque, vol. 252, Soc. Math. France, Paris 1998, 33–55.

[GP00]   M. Geck, G. Pfeiffer, *Characters of finite Coxeter groups and Iwahori–Hecke algebras*, London Math. Soc. Monographs, New Series 21, Clarendon Press, Oxford, 2000.

[Ghy01]   E. Ghys, *Groups acting on the circle*, Enseign. Math. (2) 47 (2001), 329–407.

[Gol93]   D. M. Goldschmidt, *Group characters, symmetric functions, and the Hecke algebras*, Univ. Lect. Series, vol. 4, Amer. Math. Soc., Providence, RI, 1993.

[Gon02]   J. González-Meneses, *Ordering pure braid groups on compact, connected surfaces*, Pacific J. Math. 203 (2002), 369–378.

[Gon03]   J. González-Meneses, *The nth root of a braid is unique up to conjugacy*, Algebr. Geom. Topol. 3 (2003), 1103–1118 (electronic).

[GHJ89]   F. M. Goodman, P. de la Harpe, V. F. R. Jones, *Coxeter graphs and towers of algebras*, MSRI Publ., vol. 14, Springer-Verlag, New York, 1989.

[GL69]   E. A. Gorin, V. Ja. Lin, *Algebraic equations with continuous coefficients and some problems of the algebraic theory of braids* (Russian), Mat. Sb. 78 (1969) 579–610. English translation: Math. USSR Sbornik 7 (1969), 569–596.

[Han89]   V. L. Hansen, *Braids and coverings: selected topics*. With appendices by Lars Gæde and Hugh R. Morton. London Math. Soc. Student Texts, 18. Cambridge Univ. Press, Cambridge, 1989.

[HKW86]   P. de la Harpe, M. Kervaire, C. Weber, *On the Jones polynomial*, Enseign. Math. 32 (1986), 271–335.

[Hig52]   G. Higman, *Ordering by divisibility in abstract algebras*, Proc. London Math. Soc. (3) 2 (1952), 326–336.

[Hoe74]   P. N. Hoefsmit, *Representations of Hecke algebras of finite groups with BN-pairs of classical type*, Ph.D. thesis, University of British Columbia, 1974.

[Hum90]   J. E. Humphreys, *Reflection groups and Coxeter groups*, Cambridge Stud. Adv. Math., vol. 29, Cambridge University Press, 1990.

[Hur91]   A. Hurwitz, *Über Riemann'sche Flächen mit gegebenen Verzweigungspunkten*, Math. Ann. 39 (1891), 1–61.

[Iva88]   N. V. Ivanov, *Automorphisms of Teichmüller modular groups*, Topology and geometry—Rohlin Seminar, 199–270, Lecture Notes in Math., 1346, Springer, Berlin, 1988.

[Iva92]   N. V. Ivanov, *Subgroups of Teichmüller modular groups*, translated from the Russian by E. J. F. Primrose and revised by the author, Translations of Mathematical Monographs, 115, Amer. Math. Soc., Providence, RI, 1992.

[Iva02]   N. V. Ivanov, *Mapping class groups*. Handbook of geometric topology, 523–633, North-Holland, Amsterdam, 2002.

[Iwa64]   N. Iwahori, *On the structure of a Hecke ring of a Chevalley group over a finite field*, J. Fac. Sci. Univ. Tokyo Sect. I 10 (1964), 215–236.

[Jam78]   G. D. James, *The representation theory of the symmetric groups*, Lecture Notes in Mathematics, vol. 682, Springer-Verlag, Berlin, 1978.

[Jon83]   V. F. R. Jones, *Index for subfactors*, Invent. Math. 72 (1983), 1–25.

[Jon84]   V. F. R. Jones, *Groupes de tresses, algèbres de Hecke et facteurs de type II₁*, C. R. Acad. Sci. Paris Sér. I Math. 298 (1984), 505–508.

[Jon85]   V. F. R. Jones, *A polynomial invariant for links via von Neumann algebras*, Bull. Amer. Math. Soc. 12 (1985), 103–111.

[Jon86]   V. F. R. Jones, *Braid groups, Hecke algebras and type II₁ factors*, Geometric methods in operator algebras (Kyoto, 1983), 242–273, Pitman Res. Notes Math. Ser., 123, Longman Sci. Tech., Harlow, 1986.

[Jon87]   V. F. R. Jones, *Hecke algebra representations of braid groups and link polynomials*, Ann. of Math. (2) 126 (1987), 335–388.

[Jon89]   V. F. R. Jones, *On knot invariants related to some statistical mechanical models*, Pacific J. Math. 137 (1989), 311–334.

[Joy82]   D. Joyce, *A classifying invariant of knots, the knot quandle*, J. Pure Appl. Algebra 23 (1982), 37–65.

[Kas95]   C. Kassel, *Quantum groups*, Graduate Texts in Mathematics, 155, Springer-Verlag, New York, 1995.

[Kas02]   C. Kassel, *L'ordre de Dehornoy sur les tresses*, Séminaire Bourbaki, Exposé n° 865 (1999/2000), Astérisque, vol. 276, Soc. Math. France, Paris 2002, 7–28.

[KR07]   C. Kassel, C. Reutenauer, *Sturmian morphisms, the braid group B_4, Christoffel words and bases of F_2*, Ann. Mat. Pura Appl. 186 (2007), 317–339.

[KRT97]   C. Kassel, M. Rosso, V. Turaev, *Quantum groups and knot invariants*, Panoramas et Synthèses, 5, Soc. Math. France, Paris, 1997.

[Kau87]   L. H. Kauffman, *State models and the Jones polynomial*, Topology 26 (1987), 395–407.

[Kau90]   L. H. Kauffman, *An invariant of regular isotopy*, Trans. Amer. Math. Soc. 318 (1990), 417–471.

[Kau91]   L. H. Kauffman, *Knots and physics*, Series on Knots and Everything, vol. 1, World Scientific Publishing Co., Inc., River Edge, NJ, 1991.

[Kaw96]   A. Kawauchi, *A survey of knot theory*, Birkhäuser Verlag, Basel, 1996.

[Kel55]   J. L. Kelley, *General topology*, D. Van Nostrand Company, Inc., Toronto-New York-London, 1955.

[KR03]   D. J. Kim, D. Rolfsen, *An ordering for groups of pure braids and fibre-type hyperplane arrangements*, Canad. J. Math. 55 (2003), 822–838.

[Knu73]   D. E. Knuth, *The art of computer programming*, Vol. 3, Sorting and searching, Addison-Wesley Publishing Co., Reading, Mass.-London-Don Mills, Ont., 1973.

[KLCHKP00] K. H. Ko, S. J. Lee, J. H. Cheon, J. W. Han, J.-S. Kang, C. Park, *New public-key cryptosystem using braid groups*, Advances in cryptology—CRYPTO 2000 (Santa Barbara, CA), 166–183, Lecture Notes in Comput. Sci., 1880, Springer, Berlin, 2000.

[Kra00]   D. Krammer, *The braid group B_4 is linear*, Invent. Math. 142 (2000), 451–486.

[Kra02]   D. Krammer, *Braid groups are linear*, Ann. of Math. (2) 155 (2002), 131–156.

[Lan02]   S. Lang, *Algebra*, Revised third edition, Graduate Texts in Mathematics, 211, Springer-Verlag, New York, 2002.

[Lar94]   D. M. Larue, *On braid words and irreflexivity*, Algebra Universalis 31 (1994), 104–112.

[Lav92]   R. Laver, *The left distributive law and the freeness of an algebra of elementary embeddings*, Adv. Math. 91 (1992), 209–231.

[Lav95]   R. Laver, *On the algebra of elementary embeddings of a rank into itself*, Adv. Math. 110 (1995), 334–346.

[Lav96]   R. Laver, *Braid group actions on left distributive structures, and well orderings in the braid groups*, J. Pure Appl. Algebra 108 (1996), 81–98.

[Law90]   R. J. Lawrence, *Homological representations of the Hecke algebra*, Comm. Math. Phys. 135 (1990), 141–191.

[Lic97]   W. B. R. Lickorish, *An introduction to knot theory*, Graduate Texts in Mathematics, 175, Springer-Verlag, New York, 1997.

[Lin79]   V. Ja. Lin, *Artin braids and related groups and spaces* (Russian), Itogi Nauki i Techniki, Algebra. Topology. Geometry, Vol. 17 (1979), 159–227. English translation: J. of Soviet Mathematics 18 (1982), 736–788.

[Lin96]   V. Lin, *Braids, Permutation, Polynomials -I*, Preprint Max-Planck Institut 118 (1996), Bonn.

[LP93]   D. D. Long, M. Paton, *The Burau representation is not faithful for* $n \geq 6$, Topology 32 (1993), 439–447.

[Lus81]   G. Lusztig, *On a theorem of Benson and Curtis*, J. Algebra 71 (1981), 490–498.

[Lus93]   G. Lusztig, *Introduction to quantum groups*, Progress in Mathematics, 110, Birkhäuser Boston, Inc., Boston, MA, 1993.

[LS77]   R. C. Lyndon, P. E. Schupp, *Combinatorial group theory*, Springer-Verlag, Berlin, Heidelberg, New York, 1977.

[Mag72]   W. Magnus, *Braids and Riemann surfaces*, Comm. Pure Appl. Math. 25 (1972), 151–161.

[MKS66]   W. Magnus, A. Karrass, D. Solitar, *Combinatorial group theory: Presentation of groups by generators and relations*, Interscience Publishers, John Wiley and Sons, Inc., New York, London, Sydney, 1966.

[MP69]   W. Magnus, A. Peluso, *On a theorem of V. I. Arnold*, Commun. Pure Appl. Math. 22 (1969), 683–692.

[Maj95]   S. Majid, *Foundations of quantum group theory*, Cambridge University Press, Cambridge, 1995.

[Mal40]   A. Malcev, *On isomorphic matrix representations of infinite groups* (Russian), Rec. Math. [Mat. Sbornik] N.S. 8 (50) (1940), 405–422.

[Mal48]   A. I. Malcev, *On the embedding of group algebras in division algebras* (Russian), Doklady Akad. Nauk SSSR (N.S.) 60 (1948), 1499–1501.

[MN03]   A. V. Malyutin, N. Yu. Netsvetaev, *Dehornoy order in the braid group and transformations of closed braids* (Russian), Algebra i Analiz 15 (2003), no. 3, 170–187. English translation: St. Petersburg Math. J. 15 (2004), no. 3, 437–448.

[Mar36]   A. Markov, *Über die freie Äquivalenz der geschlossenen Zöpfe*, Recueil Mathématique Moscou, Mat. Sb. 1 (43) (1936), 73–78.

[Mar45]   A. Markov, *Foundations of the algebraic theory of braids* (Russian), Trudy Math. Inst. Steklov 16 (1945), 1–54.

[Mas98]   H. Maschke, *Über den arithmetischen Charakter der Coefficienten der Substitutionen endlicher linearer Substitutionsgruppen*, Math. Ann. 50 (1898), 482–498.

[Mat99]   A. Mathas, *Iwahori–Hecke algebras and Schur algebras of the symmetric group*, Univ. Lect. Series, vol. 15, Amer. Math. Soc., Providence, RI, 1999.

[Mat64]   H. Matsumoto, *Générateurs et relations des groupes de Weyl généralisés*, C. R. Acad. Sci. Paris 258 (1964), 3419–3422.

[Mat82]   S. V. Matveev, *Distributive groupoids in knot theory*, Mat. Sb. (N.S.) 119 (161) (1982), 78–88. English translation: Math. USSR Sbornik 47 (1984), 73–83.

[Mic99]   J. Michel, *A note on words in braid monoids*, J. of Algebra 215 (1999), 366–377.

[Mil71]   J. Milnor, *Introduction to algebraic K-theory* Annals of Math. Studies, No. 72, Princeton University Press, Princeton, N.J.; University of Tokyo Press, Tokyo, 1971.

[Moo91]   J. A. Moody, *The Burau representation of the braid group B_n is unfaithful for large n*, Bull. Amer. Math. Soc. (N.S.) 25 (1991), 379–384.

[Moo97]   E. H. Moore, *Concerning the abstract group of order k! and $\frac{1}{2}k!$ . . .*, Proc. London Math. Soc. (1) 28 (1897), 357–366. ·

[Mor78]   H. R. Morton, *Infinitely many fibred knots having the same Alexander polynomial*, Topology 17 (1978), 101–104.

[Mor86]   H. R. Morton, *Threading knot diagrams*, Math. Proc. Camb. Phil. Soc. 99 (1986), 247–260.

[Mos95]   L. Mosher, *Mapping class groups are automatic*, Ann. of Math. (2) 142 (1995), 303–384.

[MR77]   R. Botto Mura, A. Rhemtulla, *Orderable groups*, Lecture Notes in Pure and Applied Mathematics, vol. 27, Marcel Dekker, Inc., New York-Basel, 1977.

[Mur87]   J. Murakami, *The Kauffman polynomial of links and representation theory*, Osaka J. Math. 24 (1987), 745–758.

[Mur96]   K. Murasugi, *Knot theory and its applications*, translated from the 1993 Japanese original by Bohdan Kurpita, Birkhäuser Boston, Inc., Boston, MA, 1996.

[MK99]   K. Murasugi, B. I. Kurpita, *A study of braids*, Mathematics and its Applications, 484, Kluwer Academic Publishers, Dordrecht, 1999.

[Neu49]   B. H. Neumann, *On ordered division rings*, Trans. Amer. Math. Soc. 66 (1949), 202–252.

[Neu67]   H. Neumann, *Varieties of groups*, Springer-Verlag New York, Inc., New York, 1967.

[Nie27]   J. Nielsen, *Untersuchungen zur Topologie der geschlossener zweiseitigen Flächen*, Acta Math. 50 (1927), 189–358. (English translation by John Stillwell in Jakob Nielsen, *Collected mathematical papers*, Birkhäuser, Boston, Basel, Stuttgart, 1986).

[PP02]   L. Paoluzzi, L. Paris, *A note on the Lawrence–Krammer–Bigelow representation*, Algebr. Geom. Topol. 2 (2002), 499–518.

[PR00]   L. Paris, D. Rolfsen, *Geometric subgroups of mapping class groups*, J. reine angew. Math. 521 (2000), 47–83.

[Pas77]   D. S. Passman, *The algebraic structure of group rings*, Pure and Applied Mathematics, Wiley-Interscience, New York-London-Sydney, 1977.

[Per06]   B. Perron, *A homotopic intersection theory on surfaces: applications to mapping class group and braids*, Enseign. Math. (2) 52 (2006), 159–186.

[Pie88]   R. S. Pierce, *Associative algebras*, Graduate Texts in Mathematics, 88, Springer-Verlag, New York-Berlin, 1982.

[PT87]   J. H. Przytycki, P. Traczyk, *Invariants of links of Conway type*, Kobe J. Math. 4 (1987), 115–139.

[Ram97]   A. Ram, *Seminormal representations of Weyl groups and Iwahori–Hecke algebras*, Proc. London Math. Soc. (3) 75 (1997), 99–133.

[Rei32]   K. Reidemeister, *Einführung in die kombinatorische Topologie*, Friedr. Vieweg & Sohn, Braunschweig, 1932.

[Rei83]   K. Reidemeister, *Knot theory*, translated from the German by L. Boron, C. Christenson and B. Smith, BCS Associates, Moscow, Idaho, 1983.

[RT90]   N. Yu. Reshetikhin, V. G. Turaev, *Ribbon graphs and their invariants derived from quantum groups*, Comm. Math. Phys. 127 (1990), 1–26.

[Rol76]   D. Rolfsen, *Knots and links*, Mathematics Lecture Series, No. 7. Publish or Perish, Inc., Berkeley, Calif., 1976.

[RW00]   C. Rourke, B. Wiest, *Order automatic mapping class groups*, Pacific J. Math. 194 (2000), 209–227.

[RW01]   D. Rolfsen, B. Wiest, *Free group automorphisms, invariant orderings and topological applications*, Algebr. Geom. Topol. 1 (2001), 311–320 (electronic).

[Rud66]   W. Rudin, *Real and complex analysis*, McGraw-Hill Book Co., New York-Toronto, Ont.-London, 1966.

[Sag01]   B. E. Sagan, *The symmetric group. Representations, combinatorial algorithms, & symmetric functions*, Graduate Texts in Mathematics, 203, Springer-Verlag, New York, 2001 (first published by Wadsworth & Brooks/Cole Advanced Books & Software, Pacific Grove, CA, 1991).

[Sal94]   M. Salvetti, *The homotopy type of Artin groups*, Math. Res. Lett. 1 (1994), 565–577.

[Ser70]   J.-P. Serre, *Cours d'arithmétique*, Presses Univ. de France, Paris, 1970. English translation: *A course in arithmetic*, Graduate Texts in Mathematics, 7, Springer-Verlag, New York-Heidelberg, 1973.

[Ser77]   J.-P. Serre, *Arbres, amalgames, SL₂*, Astérisque, No. 46, Soc. Math. France, Paris, 1977. English translation: *Trees*, Springer-Verlag, Berlin-New York, 1980.

[Ser93]   V. Sergiescu, *Graphes planaires et présentations des groupes de tresses*, Math. Z. 214 (1993), 477–490.

[Shi59]   G. Shimura, *Sur les intégrales attachées aux formes automorphes*, J. Math. Soc. Japan 11 (1959), 291–311.

[SW00]   H. Short, B. Wiest, *Orderings of mapping class groups after Thurston*, Enseign. Math. (2) 46 (2000), 279–312.

[Shp01]   V. Shpilrain, *Representing braids by automorphisms*, Internat. J. Algebra Comput. 11 (2001), 773–777.

[SCY93]   V. M. Sidelnikov, M. A. Cherepnev, V. Y. Yashchenko, *Public key distribution systems based on noncommutative semigroups*, Dokl. Akad. Nauk 332 (1993), no. 5, 566–567. English translation: Russian Acad. Sci. Dokl. Math. 48 (1994), no. 2, 384–386.

[Smi63]   D. M. Smirnov, *On the theory of residually finite groups* (Russian), Ukrain. Math. Zh. 15 (1963), 453–457.

[Squ84]   C. C. Squier, *The Burau representation is unitary*, Proc. Amer. Math. Soc. 90 (1984), 199–202.

[Sta88]   R. P. Stanley, *Differential posets*, J. Amer. Math. Soc. 1 (1988), 919–961.

[Sta99]   R. P. Stanley, *Enumerative combinatorics*, vol. 2, Cambridge Studies in Advanced Mathematics, 62, Cambridge University Press, Cambridge, 1999.

[TL71]   H. N. V. Temperley, E. H. Lieb, *Relations between the "percolation" and "colouring" problem and other graph-theoretical problems associated with regular planar lattices: some exact results for the "percolation" problem*, Proc. Roy. Soc. London Ser. A 322 (1971), 251–280.

[Tie14]   H. Tietze, *Über stetige Abbildungen einer Quadratfläche auf sich selbst*, Rend. Circ. Math. Palermo 38 (1914), 1–58.

[Tra79]   *Travaux de Thurston sur les surfaces*, Séminaire Orsay, Astérisque, 66-67, Société Mathématique de France, Paris, 1979.

[Tra98]   P. Traczyk, *A new proof of Markov's braid theorem*, Knot theory (Warsaw, 1995), 409–419, Banach Center Publ., 42, Polish Acad. Sci., Warsaw, 1998.

[Tub01]   I. Tuba, *Low-dimensional unitary representations of B_3*, Proc. Amer. Math. Soc. 129 (2001), 2597–2606.

[TW01]   I. Tuba, H. Wenzl, *Representations of the braid group B_3 and of $SL(2, \mathbf{Z})$*, Pacific J. Math. 197 (2001), 491–510.

[Tur88]   V. G. Turaev, *The Yang–Baxter equation and invariants of links*, Invent. Math. 92 (1988), 527–553.

[Tur94]   V. Turaev, *Quantum invariants of knots and 3-manifolds*, W. de Gruyter, Berlin 1994.

[Tur02]   V. Turaev, *Faithful linear representations of the braid groups*, Séminaire Bourbaki, Exposé n° 865 (1999/2000), Astérisque, vol. 276, Soc. Math. France, Paris 2002, 389–409.

[Vai78]   F. V. Vainstein, *Cohomologies of braid groups*, Funktsional. Anal. i Prilozhen. 12 (1978), no. 2, 72–73. English translation: Functional Anal. Appl. 12 (1978), 135–137.

[Ver99]   V. V. Vershinin, *Braid groups and loop spaces*, Uspekhi Mat. Nauk 54 (1999), no. 2 (326), 3–84. English translation in Russian Math. Surveys 54 (1999), no. 2, 273–350.

[Vog90]   P. Vogel, *Representation of links by braids: a new algorithm*, Comment. Math. Helv. 65 (1990), 104–113.

[Wad92]   M. Wada, *Group invariants of links*, Topology 31 (1992), 399–406.

[Wen87]   H. Wenzl, *On a sequence of projections*, C. R. Math. Rep. Can. J. Math. 9 (1987), 5–9.

[Wen88]   H. Wenzl, *Hecke algebras of type A_n and subfactors*, Invent. Math. 92 (1988), 349–383.

[Wen90]   H. Wenzl, *Quantum groups and subfactors of type B, C, and D*, Comm. Math. Phys. 133 (1990), 383–432.

[Wie99]   B. Wiest, *Dehornoy's ordering of the braid groups extends the subword ordering*, Pacific J. Math. 191 (1999), 183–188.

[Yam87]   S. Yamada, *The minimal number of Seifert circles equals the braid index of a link*, Invent. Math. 89 (1987), 347–356.

[Yan67]   C. N. Yang, *Some exact results for the many-body problem in one dimension with repulsive delta-function interaction*, Phys. Rev. Lett. 19 (1967), 1312–1315.

[Zin01]   M. G. Zinno, *On Krammer's representation of the braid group*, Math. Ann. 321 (2001), 197–211.

Index

Graduate Texts in Mathematics

(*continued from page ii*)